IET MATERIALS, CIRCUITS AND DEVICES SERIES 54

Advances in High-Power Fiber and Diode Laser Engineering

Other volumes in this series:

Volume 2	**Analogue IC Design: The current-mode approach** C. Toumazou, F.J. Lidgey and D.G. Haigh (Editors)
Volume 3	**Analogue–Digital ASICs: Circuit techniques, design tools and applications** R.S. Soin, F. Maloberti and J. France (Editors)
Volume 4	**Algorithmic and Knowledge-Based CAD for VLSI** G.E. Taylor and G. Russell (Editors)
Volume 5	**Switched Currents: An analogue technique for digital technology** C. Toumazou, J.B.C. Hughes and N.C. Battersby (Editors)
Volume 6	**High-Frequency Circuit Engineering** F. Nibler *et al.*
Volume 8	**Low-Power High-Frequency Microelectronics: A unified approach** G. Machado (Editor)
Volume 9	**VLSI Testing: Digital and mixed analogue/digital techniques** S.L. Hurst
Volume 10	**Distributed Feedback Semiconductor Lasers** J.E. Carroll, J.E.A. Whiteaway and R.G.S. Plumb
Volume 11	**Selected Topics in Advanced Solid State and Fibre Optic Sensors** S.M. Vaezi-Nejad (Editor)
Volume 12	**Strained Silicon Heterostructures: Materials and devices** C.K. Maiti, N.B. Chakrabarti and S.K. Ray
Volume 13	**RFIC and MMIC Design and Technology** I.D. Robertson and S. Lucyzyn (Editors)
Volume 14	**Design of High Frequency Integrated Analogue Filters** Y. Sun (Editor)
Volume 15	**Foundations of Digital Signal Processing: Theory, algorithms and hardware design** P. Gaydecki
Volume 16	**Wireless Communications Circuits and Systems** Y. Sun (Editor)
Volume 17	**The Switching Function: Analysis of power electronic circuits** C. Marouchos
Volume 18	**System on Chip: Next generation electronics** B. Al-Hashimi (Editor)
Volume 19	**Test and Diagnosis of Analogue, Mixed-Signal and RF Integrated Circuits: The system on chip approach** Y. Sun (Editor)
Volume 20	**Low Power and Low Voltage Circuit Design with the FGMOS Transistor** E. Rodriguez-Villegas
Volume 21	**Technology Computer Aided Design for Si, SiGe and GaAs Integrated Circuits** C.K. Maiti and G.A. Armstrong
Volume 22	**Nanotechnologies** M. Wautelet *et al.*
Volume 23	**Understandable Electric Circuits** M. Wang
Volume 24	**Fundamentals of Electromagnetic Levitation: Engineering sustainability through efficiency** A.J. Sangster
Volume 25	**Optical MEMS for Chemical Analysis and Biomedicine** H. Jiang (Editor)
Volume 26	**High Speed Data Converters** A.M.A. Ali
Volume 27	**Nano-Scaled Semiconductor Devices** E.A. Gutiérrez-D (Editor)
Volume 28	**Security and Privacy for Big Data, Cloud Computing and Applications** L. Wang, W. Ren, K.R. Choo and F. Xhafa (Editors)
Volume 29	**Nano-CMOS and Post-CMOS Electronics: Devices and modelling** Saraju P. Mohanty and Ashok Srivastava
Volume 30	**Nano-CMOS and Post-CMOS Electronics: Circuits and design** Saraju P. Mohanty and Ashok Srivastava
Volume 32	**Oscillator Circuits: Frontiers in design, analysis and applications** Y. Nishio (Editor)
Volume 33	**High Frequency MOSFET Gate Drivers** Z. Zhang and Y. Liu
Volume 34	**RF and Microwave Module Level Design and Integration** M. Almalkawi
Volume 35	**Design of Terahertz CMOS Integrated Circuits for High-Speed Wireless Communication** M. Fujishima and S. Amakawa
Volume 38	**System Design with Memristor Technologies** L. Guckert and E.E. Swartzlander Jr.
Volume 39	**Functionality-Enhanced Devices: An alternative to Moore's law** P.-E. Gaillardon (Editor)
Volume 40	**Digitally Enhanced Mixed Signal Systems** C. Jabbour, P. Desgreys and D. Dallett (Editors)
Volume 43	**Negative Group Delay Devices: From concepts to applications** B. Ravelo (Editor)
Volume 45	**Characterisation and Control of Defects in Semiconductors** F. Tuomisto (Editor)
Volume 47	**Understandable Electric Circuits: Key concepts, 2nd Edition** M. Wang
Volume 53	**VLSI Architectures for Future Video Coding** M. Martina (Editor)
Volume 58	**Magnetorheological Materials and their Applications** S. Choi and W. Li (Editors)
Volume 60	**IP Core Protection and Hardware-Assisted Security for Consumer Electronics** A. Sengupta and S. Mohanty
Volume 68	**High Quality Liquid Crystal Displays and Smart Devices: Vol. 1 and Vol. 2** S. Ishihara, S. Kobayashi and Y. Ukai (Editors)
Volume 69	**Fibre Bragg Gratings in Harsh and Space Environments: Principles and applications** B. Aïssa, E I. Haddad, R.V. Kruzelecky, W.R. Jamroz
Volume 70	**Self-Healing Materials: From fundamental concepts to advanced space and electronics applications, 2nd Edition** B. Aïssa, E I. Haddad, R.V. Kruzelecky, W.R. Jamroz
Volume 71	**Radio Frequency and Microwave Power Amplifiers: Vol. 1 and Vol. 2** A. Grebennikov (Editor)
Volume 73	**VLSI and Post-CMOS Electronics Volume 1: VLSI and post-CMOS electronics and Volume 2: Materials, devices and interconnects** R. Dhiman and R. Chandel (Editors)

Advances in High-Power Fiber and Diode Laser Engineering

Edited by
Ivan Divliansky

The Institution of Engineering and Technology

Published by The Institution of Engineering and Technology, London, United Kingdom

The Institution of Engineering and Technology is registered as a Charity in England & Wales (no. 211014) and Scotland (no. SC038698).

First published 2019

The Institution of Engineering and Technology
Michael Faraday House
Six Hills Way, Stevenage
Herts, SG1 2AY, United Kingdom

www.theiet.org

British Library Cataloguing in Publication Data
A catalogue record for this product is available from the British Library

ISBN 978-1-78561-751-5 (hardback)
ISBN 978-1-78561-752-2 (PDF)

Typeset in India by MPS Limited
Printed in the UK by CPI Group (UK) Ltd, Croydon

Contents

Chapter 1

Diode laser: fundamentals and improving the brightness

Manoj Kanskar[1]

Diode lasers started out with a humble beginning. They were perceived as impractical devices suitable only for scientific curiosity at cryogenic temperatures. Over the past half a century, remarkable improvements in diode laser performance took place throughout the world. Realization of quantum-well gain medium was an early seminal achievement. This ushered in an era where it became possible to operate laser diodes at room temperature, but they were used in limited niche applications due to exorbitantly high cost. Confluence of numerous technological advancements such as defect-free crystal growth, as well as device fabrication, facet passivation, and stress-free die-bonding led to unprecedented device reliability. Long lifetime is one of the trademarks of semiconductor lasers. These devices became ubiquitous in data storage for mass media applications and 980 nm single mode pump diode lasers became the workhorse in telecom industry. As a result, volume production of diode lasers took place. Once it became clear to the Defense community that further improving the electrical-to-optical power conversion efficiency (PCE) of diode lasers was the key to realizing compact high energy lasers (HEL), race for achieving super high efficiency began and record high efficiency of ~73 percent was realized. This was a second seminal technological advancement in diode lasers leading to unprecedented power-scaling which drove down the cost of laser diodes. As the diode laser output power improved, fiber laser power increased commensurately; very low dollars-per-watt, both direct diode and fiber lasers became affordable. This led to displacement of legacy CO_2 and flash-lamp pumped solid-state lasers and flood gates opened for new applications. Diode lasers and fiber lasers enable low-cost manufacturing of hand-held mobile devices, HD TV screens, virtual reality glasses, highly reliable and light-weight and efficient cars, ships and airplanes, 3D-printed metal parts, HEL for Defense, and many types of lasers for scientific and industrial applications. There is much room for improvement in diode laser brightness which will certainly lower the cost thus enabling further new applications. This chapter is dedicated to briefly looking back at the evolution of rise of a disruptive technology and surveying the current

[1]nLIGHT, Inc., Vancouver, WA, USA

challenges and roadblocks to improving brightness as well as understanding the current state-of-the-art of diode lasers followed by a summary of outlook of its brightness improvement.

1.1 A brief history of high-power semiconductor laser: the rise of a disruptive technology

1.1.1 The beginning of a semiconductor laser

The initial observations of the conversion of charge carriers into photons through a semiconductor device date back over 100 years. Henry J. Round working for the Marconi Company in New York in 1907 experimented with cat's whisker detectors using carborundum when he applied a forward bias that caused electrons and holes to recombine at the junction to produce a "glow." A systematic investigation of semiconductor materials was launched in the early 1930s at the Physico-Technical Institute (FTI) under the direction of A.F. Ioffe [1]. Conductivities of semiconductors, current rectification, and excitonic phenomena were developed during this period laying the groundwork for experimental realization of excitons by E.F. Gross [2] reported in 1952. This observation of limited emission and unknown applications attracted limited attention until the 1950s when Rubin Braunstein observed infrared emission from gallium arsenide (GaAs) semiconductor device at RCA laboratories in Princeton, New Jersey.

This early work did not attract much attention until 1962 when Robert Keyes presented results from work at MIT Lincoln Laboratory on diodes developed by diffusing zinc impurities into GaAs. This chapter showed the promise for higher power and higher efficiency light emitting diodes. Robert Hall from General Electric Research labs in Schenectady, NY saw this presentation and began developing concepts to create a laser cavity by polishing the edges of his GaAs device. Two years earlier the first ruby laser had been demonstrated using a five-meter square amplification apparatus and a significant amount of energy input. By comparison, Hall began developing a concept for a heavily doped diode with a resonant cavity along the junction plane that eventually fit on a chip smaller than a finger nail and efficient enough to run with a small battery. On the morning of Sunday, 16 September 1962, Hall demonstrated a bright horizontal line of infrared emission that clearly wasn't spontaneous emission—the first semiconductor laser was born [3]. Later Nick Holonyak also at General Electric developed the first visible wavelength laser by increasing the bandgap with gallium arsenide-phosphide.

These early demonstrations of a semiconductor laser involved simple p–n homojunction devices. These devices had very poor carrier and optical confinement. The concepts for low-threshold double-heterostructure lasers were suggested by Herbert Kroemer [4] at the Varian Central Research Laboratory in Palo Alto and R.F. Kazarinov and Zhores Alferov [5] of the A.F. Ioffe Physico-Technical Institute in St. Petersberg in 1963. This development led to the first continuous-wave room-temperature diode lasers and for this work the two shared the Nobel Prize in 2000.

1.1.2 Era of high-temperature operation—birth of quantum-well gain medium

Initial crystals for semiconductor lasers were produced using liquid phase epitaxy (LPE). These devices suffered tremendously from carrier leakage. These early double-heterostructures did not provide sufficient carrier confinement. To mitigate carrier leakage, double heterostructure laser diodes had to be operated at cryogenic temperature. Furthermore, threshold current densities were many orders of magnitude higher by today's standards. These factors resulted in rapid device degradation. They were not practical for commercial products. It took several more years to refine double-heterostructure lasers to overcome cryogenic operation. Both a Bell Laboratory research team and researchers at the Ioffe Institute demonstrated the first room temperature double-heterostructure laser in 1970 [6,7]. Although this technology still was not ready for commercial use, it was a critical turning point in the development of semiconductor lasers. This proof-of-concept stimulated vigorous competition among companies to develop high power diodes operating at room temperature. The era of high-temperature operating diodes began. US electronic, telecommunication and computing enterprises such as GE, RCA, Bell Laboratory and IBM as well as Japanese firms such as Hitachi, Toshiba, Mitsubishi Electric, Nippon Electric Company (NEC), Fujitsu, and Nippon Telegraph and Telephone (NTT) participated in the race for high power and room temperature operating laser diodes. Many of these companies ultimately developed lasers either for communications or data-storage. The media hailed this invention as one that would fundamentally transform Optical Engineering akin to the transistor changing the face of Electronics Engineering. But this harbinger of good news had to wait for another critical technology development—the invention of quantum-well gain medium.

At about this same time, in 1966, another revolution was brewing. Charles K. Kao working at Standard Telecommunication Laboratories in Harlow, UK, makes a discovery that leads to a breakthrough in fiber optics. His calculations demonstrated that light could travel over 100 km distance in a glass fiber for which he received the 2009 Novel Prize in physics. Neither of these inventions could have foreseen an eventual conjugation of these two technologies giving rise to high-power fiber lasers after decades of continuous improvements. This union had to wait for the advent of high brightness laser diodes and rare-earth doped fibers to mature beyond what was sufficient for telecom and data storage applications.

Researchers had been exploring superlattices before the dawn of quantum-well gain medium. But in 1972 Charles H. Henry at AT&T Bell Laboratory realized that "a heterostructure is a waveguide for electrons" with a complete analogy between the confinement of photons in a slab waveguide and the confinement of electrons by the quantum-well potential that is formed from the difference in bandgaps in a heterostructure. From this perspective, a semiconductor laser is an electron-hole (bipolar) waveguide superimposed inside a photon waveguide. He proposed quantum-wells as gain medium for semiconductor laser to overcome the carrier leakage problem that limited laser diodes to cryogenic temperatures in its early days. This is perhaps the first key breakthrough on the road to a practical semiconductor laser. Due to quantum-confinement of the bipolar carriers, these lasers required many orders of magnitude lower threshold current density. They could now be operated at room temperature with reasonable efficiency. Going from Henry's theoretical prediction to actualization of quantum-well gain medium took some nontrivial material development efforts. However, quantum-well technology allowed the crystal growers to control the range, depth, and arrangement of quantum mechanical potential wells for the first time giving rise to a very broadband gain medium.

Early crystal growth using LPE was capable of growing materials with thicknesses down to about 2000 Å which was, at least, an order of magnitude thicker than needed to form useful quantum-wells in GaAs or InP semiconductor lasers. Fortunately, another innovation in more precise crystal growth technology was emerging. A method that originated in J.R. Arthur's laboratory in 1968 was turned into a practical crystal growth technique called Molecular Beam Epitaxy (MBE) by A.Y. Cho. By 1971 Cho had demonstrated MBE growth of hetero-structure with optical absorption edge features that were as good as LPE. There were several MBE efforts at Bell Labs in the early 1970s. Finally, quantum-well samples grown by W. Wiegmann and tested by R. Dingle definitively showed Henry's predicted spectra due to quantum confinement of bipolar carriers [8]. However, first operational demonstration of quantum-well physics in hetero-structure was achieved three months prior to Henry's effort by Chang, Esaki, and Tsu [9] at IBM Watson Center in Yorktown Heights by demonstrating resonant tunneling through levels of a quantum well. Dingle and Henry filed a patent on use of quantum-well as gain medium semiconductor laser in 1975 and the first optically pumped quantum-well laser was demonstrated that same year by J.P. Van der Ziel's group [10]. There were early concerns that strain in the quantum-wells would lead to defects and poor reliability. However, contrary to this belief, strain in the quantum-well unexpectedly pinned line defects that double-heterostructure lasers were replete with. Quantum-well lasers turned out to be much more reliable compared to double heterostructure lasers.

A new epitaxial growth technique using Metal Organic Chemical Vapor Deposition (MOCVD) first used to make AlGaAs heterostructure lasers by R.D. Dupuis and P.D. Dapkus of Rockwell International came onto the scene in 1977 [11]. Dupuis and Dapkus in collaboration with N. Holonyak and coworkers at University of Illinois at Urbana-Champaign demonstrated the first electrically injected quantum-well laser (grown by MOCVD) at the end of 1977 [12]. MBE and MOCVD technologies went through numerous improvements over the next few decades enabling high volume manufacturing. Both technologies are prevalent in laser diode production throughout the world.

Little mentioned, but well understood amongst crystal growers, is a need for very low dislocation-density substrates to grow high-quality semiconductor crystals especially for reliable operation. Typically, less than several hundred per square centimeter of etch-pit-density (EPD) are acceptable for high power and high-reliability laser diodes. Improvement in these areas by GaAs substrate vendors paved the path for a rapid rise in power and reliability of semiconductor lasers in the following decades.

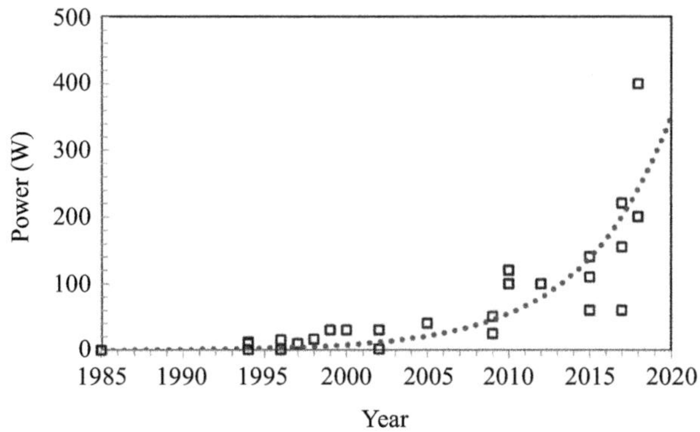

Figure 1.1 Plot showing nonlinear increase in fiber-coupled power of multimode diode lasers from 105 μm and 0.15 NA beam over three decades

In the early 1980s, 8xx nm wavelength was popular for pumping Nd-doped solid-state gain media. In the mid-1990s 9xx nm laser diodes emerged as a higher brightness source driven by Yb-doped fiber application. Eventually, the available spectral range spanned from 63x nm to 9xx nm on GaAs substrate to enable data storage and telecom pump diodes and from 13xx to 15xx nm on InP substrate to enable telecom transmitters. This breadth of spectral range helped to serve other applications such as optical storage, solid-state laser pumping, medical therapeutics, image recording, barcode scanning, metrology tools, materials processing, and basic scientific research. The highest power semiconductor lasers remained in the 8xx nm to 9xx nm spectral range due to their superior electrical, thermal and optical material properties. A plot of fiber-coupled output power increase over time from a 105 μm diameter fiber is shown in Figure 1.1. It is evident from this plot that until about 2,000, a 100 μm emitter broad area laser (BAL) could produce only several watts of ex-facet output power at about 50 percent electrical-to-optical PCE. Facet fluence and current density were low enough that reliability was not an issue even for demanding industrial applications. There was a period of lull in power and efficiency improvement since laser diodes seemed to be "good enough" for most applications. But visionaries for HEL for directed energy weapons (DEW) recognized that practical multihundred kilowatt-class DEW could not be realized without further improving diode efficiency.

1.1.3 Era of high-reliability operation

One of the most appealing aspects of laser diodes is their unprecedented reliability under very high-power operation when designed and fabricated appropriately. Current high-power diode lasers are over thousand times brighter than our sun. Irradiance of 15×10^6 W per cm^2 is not uncommon at the output facet of a laser diode. At this high irradiance, there are several mechanisms that lead to facet damage from catastrophic optical mirror damage (COMD). Near the cleaved surfaces, semiconductor crystal relaxes which reduces the bandgap near the surface thus making it absorptive to exiting photons. Furthermore, if the surface of the cleaved surface were not passivated, it leaves dangling bonds with surface states that absorb photons exiting the facet. In addition, photolysis leads to oxidation on the surface of the facet which over time and activated by temperature leads to solid–solid reaction of III–V compounds. Final precipitate is typically disordered elements such as arsenic clusters which are very absorptive. These mechanisms have an exponential deleterious impact on the facet-power-density-induced degradation. This type of problem had been encountered and mitigated by 980 nm pump developers targeting Er-doped Fiber Amplifier (EDFA) market in the 1980s and 1990s. IBM (Zurich) developed the so-called "E2" process which used vacuum-cleaving and Si-passivation layer to suppress facet damage. Others used quantum well intermixing (disordering) techniques such as vacancy-induced disordering (VID) or impurity-induced disordering (IID) that increased the bandgap near the facet making it transparent. Techniques such as Zn diffusion near the facet or regrowth of high bandgap epitaxial layers were also employed to achieve transparency at the facet. Some of these techniques were cumbersome and painstakingly slow. To increase throughput, some companies also developed "cleaving-in-air" techniques followed by oxide-cleaning and deposition of a passivating dielectric or semiconductor film. These techniques provided varying degrees of protection at the facet and allowed harnessing commensurately more useful reliable power.

The laser design must provide ample guard band to withstand high current density and high facet power density. The substrates need to be of high quality with low EPD. Crystal growth and growth processes must be refined enough to achieve high material quality, good interfaces and precise control of thicknesses and doping profiles. Wafer fabrication must not introduce any additional defects. Facet-passivation must withstand high power operation. And last, but not least, bonding and packaging of laser chip onto submount and heatsink must not introduce any residual strain or stress. Over the past decades, all these technologies have come to fruition with a culmination of a very useful technology that we take for granted now. However, at its inception, laser diodes based on homojunction or heterojunctions performed only at cryogenic temperatures and at many kA per cm^2 current density and lasted for a few seconds. In the 1970s and 1980s use of quantum-well gain medium and proper heatsinking-enabled room temperature operation with under one watt of reliable power. Improvement in crystal growth and substrate quality further pushed this power to several watts. Improvement in efficiency, large optical cavity (LOC) and facet passivation raised this power by another order of magnitude. Laser diodes are used for their reliability in some of the most demanding applications. Single mode 980 nm pump diodes are the workhorse of telecom's long-haul communication backbone operating around the clock for many decades. Semiconductor laser chips are also used for many data storage applications for mass media. Hard disk drive companies are considering laser diodes to be implemented in heat-assisted magnetic recording (HAMR) technology to increase storage capacity. Laser diodes are used as tools for cutting and welding in industrial environments. In the future, unattended autonomous vehicles (UAVs) may rely on diode lasers for navigation.

There was one additional key technology, LOC, that contributed to the improved performance in output power while maintaining high reliability especially for industrial applications requiring multimode pump diodes. New designs emerged that reduced facet power density by using LOC that produced large effective spot size with reduced optical confinement factor at the quantum well. Several successful designs led to use of LOC in the fast-axis direction [13,14] which dramatically improved COMD power threshold. Today reliability is one of the hallmarks of laser diodes.

1.1.4 Race for super high efficiency for defense application

The advantage to the Defense community for a truly low size, weight, and power-consumption (SWaP) HEL was very clear. In the early 2000s, diode-pumped solid-state lasers (DPSSLs) were the only viable contenders for 100-kW class HEL since fiber laser was a nascent technology and diode brightness was not sufficient for fiber laser power to be of interest to HEL community. With diode ex-facet efficiency at 50 percent, a 100 kW DPSSL with 33 percent optical-to-optical PCE would have required 300 kW of diode power and 600 kW of input electrical power and a total waste heat generation of 500 kW that would have to be managed. The SWaP of power supplies and thermal management systems were not practical for military applications. In 2003, Defense Advanced Research Program Agency (DARPA), launched Super High Efficiency Diode Sources Program (SHEDS Program) to improve diode efficiency. By the end of the program, a record NIST-verified 73 percent [15] efficiency was achieved near 976 nm wavelength. Post SHEDS Program, efficiency of other wavelengths improved similarly limited only by their inherent material properties. This ceiling in diode efficiency is currently mostly restricted by the residual free-carrier absorption in the quantum-well gain medium and waveguide structure. However, the jump from 50 percent to 73 percent in efficiency allowed diode developers to explore new designs to scale up in power and brightness. Post SHEDS Program, reliable output power increased from just a few watts to over 15 watts from a typical 100-μm stripe-width laser diode.

1.1.5 Era of high brightness—the rise of a disruptive technology

Once the ex-facet efficiency reached 73 percent and facet power-handling techniques were under control using passivation techniques, quantum-well lasers continued to produce progressively higher reliable power. However, around 2010, power that could be coupled into a given beam-parameter-product (BPP) started to saturate. For example, although, a single-emitter laser diode produced >15 watt of reliable ex-facet output power, maximum useful fiber-coupled power in these fiber-coupled packages with an industry standard 105 μm/0.15 NA beam was restricted to about 12 watts of ex-facet power per emitter. Beyond this power level, a rapid increase in slow-axis divergence prevented fiber-coupling higher power because the slow-axis BPP exceeded the required value of 4.5 mm-mrad that was needed to fit in a 105 μm/0.15 NA fiber. A 75-μm-wide stripe device on a submount with 2.5 C/W thermal resistance produced a maximum power of 12 watts at a BPP of 4.5 mm-mrad. Smaller stripe-width devices were brighter but produced lower output power; whereas, the larger stripe-width devices produced higher power but their BPP exceeded the value required to couple into a 105 μm/0.15 NA fiber. Therefore, a maximum slow-axis brightness (power divided by slow-axis BPP) of 2.8 W/mm-mrad was useable to couple into 105 μm/0.15 NA fiber with a 75 μm stripe-width BAL at 12 watt of output power.

This chapter is devoted to understanding the key areas of diode laser physics and methods of overcoming the current obstacles to achieve the ultimate potential brightness of these lasers. Broad area diode laser brightness scaling has far-reaching implication especially for power-scaling of cladding-pumped fiber lasers.

1.2 High power and high brightness broad area diode lasers

1.2.1 Fundamentals of diode lasers

Basic structure of an edge-emitting Fabry Perot quantum-well diode laser is shown in Figure 1.2. A planar optical slab waveguide and the quantum-well active gain region embedded in the waveguide are formed by epitaxial crystal growth using either MOCVD or MBE by lattice matching to substrates such as GaAs or InP. The epitaxial growth direction which is orthogonal to the substrate surface crystalline plane is referred to as the transverse or vertical direction. The lateral direction is in the plane normal to the axial propagation direction. The volume of the active region is formed in the lateral direction by emitter width *W*; a vertical height *d* the

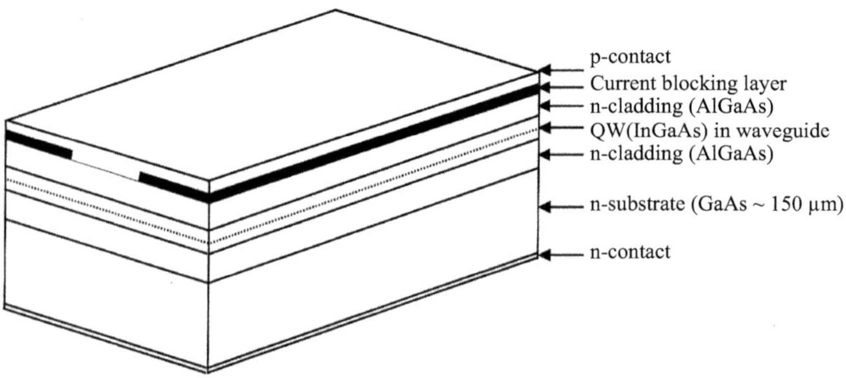

Figure 1.2 Basic structure of an edge-emitting Fabry Perot quantum-well diode laser showing quantum-well gain medium buried in a waveguide with p–n junction residing on GaAs substrate

quantum-well thickness and an axial length L which is the cavity length determined by the length between the cleaved facets. These facets form an atomically smooth surface for depositing high reflection (HR) and partial reflection (PR) dielectric mirrors to form a Fabry Perot resonator. This figure illustrates a BAL with stripe width W defined by the top ohmic electrical-contact on the p-side. Under forward bias conditions, due to the requirement for electrical neutrality, the minority and majority carriers are injected into the quantum-well through the p–n junction. For stimulated emission, very high density ($>>10^{18}$ cm^{-3}) of injected carriers is required for population inversion. The coherent light generated from stimulated emission process in the quantum-well layer is emitted from the partially reflecting facet.

1.2.2 Optical gain medium—quantum well

The optical gain medium in semiconductor lasers comprises an undoped layer of direct bandgap semiconductor material inserted between high bandgap p- and n-doped regions forming a quantum well. When this p–i–n junction is forward biased, the bipolar carriers of electrons and holes are injected into the quantum well. The double hetero-barriers confine the bipolar carriers in the quantum well; hence, this structure forms a "waveguide" for the electron and hole wavefunction in the conduction and valence bands, respectively. Typically, the quantum well is a ternary In_xGa_{1-x} As alloy with thickness d which is less than 10 nm thick for generating 9xx nm emission wavelength. The quantum well is embedded in a dielectric waveguide comprised of high index semiconductor material flanked by lower index cladding semiconductor materials. Optical gain is produced by overlapping portion of this mode interacting with the quantum well to generate stimulated emission. The emission wavelength is determined by the choice of a combination of bandgap of the quantum well and barrier materials which depend on the alloy composition and the thicknesses of the layers.

1.2.3 Optical waveguide

The light generated in the quantum well from stimulated emission is guided by the transverse waveguide and propagates in the Fabry Perot waveguide resonator with an intensity profile $I(x)$ for the fundamental mode. For 9xx nm diode lasers, the waveguide and cladding layers are made of $In_xGa_{1-x}As_yP_{1-y}$ or $Al_xGa_{1-x}As$ materials or a combination thereof. Typically, the transverse waveguide is designed to operate only in a single mode by employing a thin waveguide with a small index contrast between the guide and the cladding layers. However, most modern high power BALs typically resort to LOC design in the transverse direction to minimize optical confinement factor, Γ, in the quantum well by increasing the effective spot size d/Γ. As a result, the waveguide is large and supports numerous transverse modes. To force single transverse mode operation, the location of the quantum well is judiciously chosen and loss for the higher order modes are precisely designed to preferentially suppress higher order modes. One of the most critical parameters for designing a high-power semiconductor laser is the thickness of the waveguide and the choice of the optical confinement factor in the quantum well. The optical confinement factor, Γ, is typically less than 1.5 percent. This leads to low modal gain; however, these structures have very low internal free-carrier absorption and scattering loss due to substantial confinement of the fundamental mode in the undoped waveguide. As a result, the cladding layers can be heavily doped without incurring high loss. Very long cavity (>4 mm) devices can be fabricated providing a path to high power, high efficiency, and lower lateral divergence angle:

$$\Gamma = \frac{\int_{-\frac{d}{2}}^{+\frac{d}{2}} I(x)dx}{\int_{-\infty}^{+\infty} I(x)dx} \tag{1.1}$$

In the lateral direction, the simplest BALs use gain guiding effect to confine the photons. In some designs, lateral index-guiding is introduced by either ion-implantation, mesa etch, or regrowth of lower index material as lateral cladding.

1.2.4 Optical feedback

Most high-power diode lasers use Fabry Perot cavity as the resonator. Figure 1.2 shows a schematic of a Fabry Perot resonator comprised of two cleaved facets (mirrors) separated by a cavity length L in an active gain medium with an optical waveguide with effective modal index n_{eff}. This resonator provides feedback when a standing wave is formed between the mirrors. Typical cavity lengths are >3 mm for high power diode lasers. Near 976 nm emission wavelength and for effective index of 3.355, longitudinal mode order number m is large and the mode spacing is very dense (<<0.1 nm). Many of these longitudinal modes near the peak of the quantum-well gain spectrum lase producing a broad emission width:

$$L = m\frac{\lambda_0}{2n_{\text{eff}}} \quad m = 1,\ 2,\ 3\ldots \tag{1.2}$$

The envelope of the emission bandwidth is typically 2.5 nm at the FWHM. The center wavelength is determined by the location of the maximum gain. Typically, the center wavelength shifts at 0.32 nm K^{-1}. In many applications one needs to pump at the upper-state of Yb-doped solid-state gain medium such as Yb-doped large mode area fibers near 976 nm to keep the quantum defect to a minimum, but the absorption band is less than 6 nm. Hence, it is increasingly more important to both narrow the spectral width to less than 0.5 nm and reduce the temperature-shift coefficient. External cavity grating coupling can be used to accomplish these features but with cost of reduced efficiency, higher cost, and some reliability risk. These external cavity lasers do not lock at the desired emission wavelength over all operating power either. In the future, broad area diode lasers with integrated internal gratings such as distributed feedback (DFB) or distributed Bragg reflection (DBR) will become indispensable to achieve these functionalities. Such devices with on-chip wavelength-stabilization techniques will have low temperature sensitivity and narrow spectral width [16] since these lasers typically emit less than 0.5 nm bandwidth and wavelengths shift at 0.065 nm K^{-1} near $\lambda_0 \sim 976$ nm. Furthermore, the center wavelength is tightly controlled by the pitch Λ of the grating which can be controlled to less than 0.1 nm accuracy and the effective index of the transverse fundamental mode, n_{eff}. The relationship between the emission wavelength and the pitch of the grating is shown below:

$$\Lambda = k\frac{\lambda_0}{2n_{\text{eff}}} \text{ where } k \text{ is the grating order} \tag{1.3}$$

1.2.5 Electrical-to-optical power conversion efficiency

Improvements to diode laser PCE can have far-reaching implications, including lower cost, improved reliability, lower SWaP, and higher-power laser systems. Power losses in diode lasers stem from several factors such as scattering and free-carrier absorption losses, band misalignment, Joule heating, carrier leakage, and below-threshold losses. These loss mechanisms are described below.

Below-threshold losses: A certain amount of the drive current is consumed simply to attain population inversion in the laser. Furthermore, there are non-radiative recombination mechanisms that consume carriers especially near the quantum well and barrier interfaces.

Band misalignment: Diode laser must overcome voltage deficit resulting from misalignment of the band structure of various double-heterostructure interfaces. Until this deficit is overcome, useful optical power cannot be generated.

Carrier leakage: This refers to bipolar carriers that do not make it to the quantum well to effectively produce photons due to thermionic emission out of the quantum well and nonradiative recombination in the barrier and cladding layers.

Scattering and absorption: This accounts for generated photons that do not stay in the waveguide due to interface alloy scattering or roughness scattering. There is also residual free-carrier absorption by the carriers that is required to reach transparency and overcome all the other loss mechanisms. Additionally, photons are absorbed by the free-carriers from the *p*- and *n*-dopants in the waveguide and cladding layers.

Joule heating: This results from the effective series resistance of the diode, which includes contact resistances as well as bulk resistance in the heterostructure itself leading to I^2R loss.

Power dissipation in diode laser is described more precisely by 1.4 and 1.5 I^2R_s is the Joule heating term, where I is the operating current and R_s is the series resistance, and V_0 is the built-in voltage which has a value given by extrapolating back to zero current from the linear portion of the V–I curve. The IV_0 captures the stimulated emission, scattering and absorption losses, hetero-junction band misalignment, carrier leakage, and below-threshold losses as described in those equations. V_F is the voltage associated with the quasi-Fermi energy difference corresponding to the lasing transition, V_{hj} is the voltage penalty required to overcome hetero-junction barrier discontinuities, η_i is the internal injection efficiency, and I_{th} is the threshold current.

$$\begin{aligned}
P_{\text{Total}} &= I^2 R_s + I V_0 \\
&= I^2 R_s + I(V_{\text{hj}} + V_F) \\
&= P_{\text{Joule}} + P_{\text{hj}} + P_{\text{carrier leakage}} + P_{\text{belowthreshold}} + P_{\text{stimulated}} \\
&= P_{\text{Joule}} + P_{\text{hj}} + P_{\text{carrier leakage}} + P_{\text{belowthreshold}} + P_{\text{laser output}} + P_{\text{absorption+scattering}}
\end{aligned} \tag{1.4}$$

$$\begin{aligned}
P_{\text{Joule}} &= I^2 R_s \\
P_{hj} &= I(V_0 - V_F) \\
P_{\text{carrier leakage}} &= I V_F (1 - \eta_i) \\
P_{\text{belowthreshold}} &= I_{th} V_F \eta_i \\
P_{\text{stimulated}} &= V_F (I - I_{th}) \eta_i \\
P_{\text{absorption+scattering}} &= P_{\text{stimulated}} - P_{\text{laser}}
\end{aligned} \tag{1.5}$$

The expression for the maximum PCE, $\eta_{p,\text{max}}$, is given by [17]

$$\eta_{p,\text{max}} = \eta_{\text{ext}} \frac{V_F}{V_o} \left(1 - 2\sqrt{R_s I_{\text{th}} / V_0}\right) \tag{1.6}$$

It is apparent that the four key means for improving $\eta_{p,\text{max}}$ are (1) maximizing η_{ext}, (2) minimizing V_0, (3) minimizing R_s, and (4) minimizing I_{th}. In a CW operation, η_{ext} is a function of temperature and is given by

$$\eta_{\text{ext}}(T) = \eta_{\text{ext}}(T_h) \exp\left(-\frac{\Delta T_j}{T_1}\right) \tag{1.7}$$

where T_h is the heatsink temperature, ΔT_j is the junction-temperature rise (i.e., $T = T_h + \Delta T_j$), and T_1 is the characteristic temperature coefficient for η_{ext}. Since V_0 is weakly dependent on temperature, the temperature dependence of $\eta_{p,\text{max}}$ is primarily determined by the temperature dependence of η_{ext}. Therefore, it is of paramount importance to improve the thermal properties of diode laser by maximizing T_1 and minimizing ΔT_j.

It is clear from (1.6) that maximum PCE is directly proportional to external differential quantum efficiency which, in turn, is proportional to η_i, as shown in (1.8) where α_m is mirror loss and α_i is the total internal loss:

$$\eta_{\text{ext}} = \eta_i \frac{\alpha_m}{\alpha_m + \alpha_i} \tag{1.8}$$

As shown in (1.6) reduction in V_0 has a profound impact on the peak PCE. Though it is a weaker dependence relative to η_{ext} and V_0, R_s has a significant contribution to overall loss mechanism especially at the high operating current since the loss increases with the square of the drive current with R_s as the prefactor. One cannot simply reduce R_s to arbitrarily small values without incurring higher loss due to free-carrier loss associated with higher doping that would be required to do so. The choice of doping profile requires judicious choice of a profile which balances R_s of the device, which favors high doping levels, with α_i, which favors low doping levels. The ideal doping profile minimizes the sum of the resistive loss and the bulk optical absorption, thus maximizing the PCE of the diode laser.

The resistive power loss of the device can be expressed as:

$$P_R = I^2 R = I^2 \frac{\rho \cdot \Delta y}{WL} = \int_0^h \frac{I^2}{q\mu(y) D(y) W(y) L} dy \tag{1.9}$$

where I (Amperes) is the operating current of the device, ρ (Ω·cm) is the resistivity of a layer of the device, L (cm) is the length of the device, W (cm) is the width of the pumped region of the device, q(C) is the charge on an electron, y (cm) is the transverse coordinate of the device extending from 0 to h, $\mu(y)$(cm^2/V·s) is the mobility of the charge carriers, and $D(x)$ (cm^{-3}) is the doping density of the charge carriers—either p or n type. Solving for the doping distribution $D(y)$ [18] leads to:

$$D(y) = \sqrt{\frac{\eta(y)}{q\mu(y) W(y) W_{\text{opt}} L^2 k(y)} \frac{I}{E(y)}} \tag{1.10}$$

which is the doping distribution which minimizes the total resistive plus optical power losses. It is interesting to note that the appearance of the operating current in 1.10 implies that the expected operating point must be known for specifying the doping levels and the profile. Since the mobility for electrons and holes are quite different, this naturally leads to asymmetric doping profiles on either side of the p–n junction. Figure 1.3 shows a typical LOC cavity diode laser design with

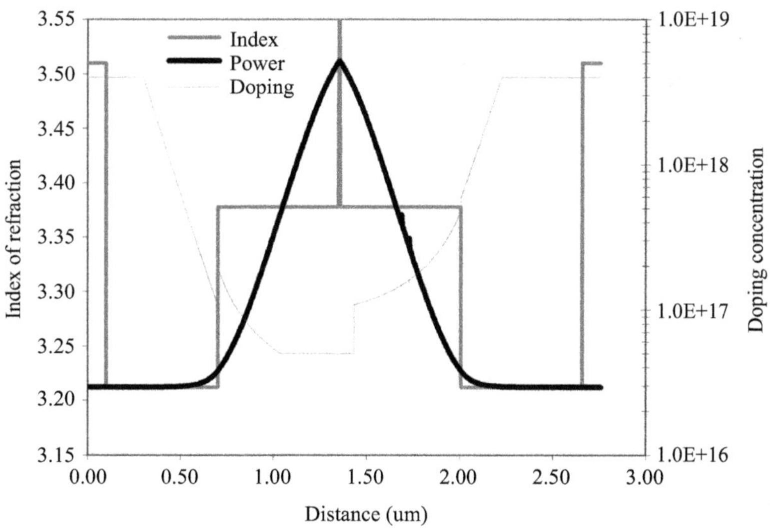

Figure 1.3 Schematic of a typical diode laser structure showing refractive index profile along with fundamental mode shape and optimized doping profile

optimized doping profile. Using these techniques, diode lasers with record 73 percent efficiency has been demonstrated [19].

1.3 Mitigation of slow-axis divergence blooming in broad area diode lasers

Most broad-stripe diode lasers are designed to operate with a diffraction-limited beam in the fast-axis direction. Over the past several decades, an emergence of a variety of designs producing large effective spot size in the fast-axis direction has ensued in reliable operation of BALs with >15 MW per cm^2 of facet power density. There were two key technological advancements that contributed to this improved performance in output power. First, several successful designs led to use of large effective spot size in the fast-axis direction which dramatically improved the COMD power threshold [20–23]. Second, the electrical-to-optical PCE increased from approximately 50 percent to 73 percent [24,25] due to overall optimization of a quantum-well diode laser structure. These advancements led to the advent of very low-loss laser diode structures that are capable of greater than 15 watts of reliable output power near 9xx nm emission wavelength [26,27] that is suitable for industrial applications. As the power and brightness of single-emitter broad area diodes continue to improve, there has been a shift away from using bars to using multiple single-emitter-based fiber-coupled packages for better cost-per-bright-watt. The fiber laser technology is predominantly using high output power packages with 105 μm/0.15 NA fiber-coupled beam as the building block for power-scaling. In the past decade, the amount of power coupled into a single 105 μm/0.15 NA fiber has increased by over a factor of ten through improved diode laser brightness and development of techniques for efficiently coupling multiple emitters into a single fiber [28,29]. Single-emitter BAL is the building block for these fiber-coupled pump modules due to their high brightness, excellent reliability, low production cost, and high electrical-to-optical efficiency. Although a single-emitter laser diode can produce >15 watt of reliable output power, maximum useful fiber-coupled power in these packages with an industry standard 105 μm/0.15 NA beam has been restricted to about 12 watts per emitter. Beyond this power level, a rapid increase in slow-axis divergence limits the useful fiber-coupled power because the slow-axis beam-parameter-product (BPP) exceeds the required value of approximately 4.5 mm-mrad to couple multiple-emitters in a stacked configuration. The data shown in Figure 1.4 for 9xx nm diodes illustrate this point. A 75 μm stripe-wide device on a submount (2.5 C/W thermal resistance) produces a maximum power of 12 watts at a BPP of 4.5 mm-mrad. Smaller stripe-width devices are brighter but produce lower output power; whereas, the larger stripe-width devices produce higher power but their BPP exceeds the value required to couple into a 105 μm/0.15 NA fiber. Therefore, a maximum brightness of 2.8 W/mm-mrad is useable to couple into this fiber with a 75-μm stripe-width BAL at 12 watts of power. Compared to this geometry, larger stripe-width BALs produce more output power but exhibit progressively worse slow-axis brightness under all operating conditions. Similarly, a

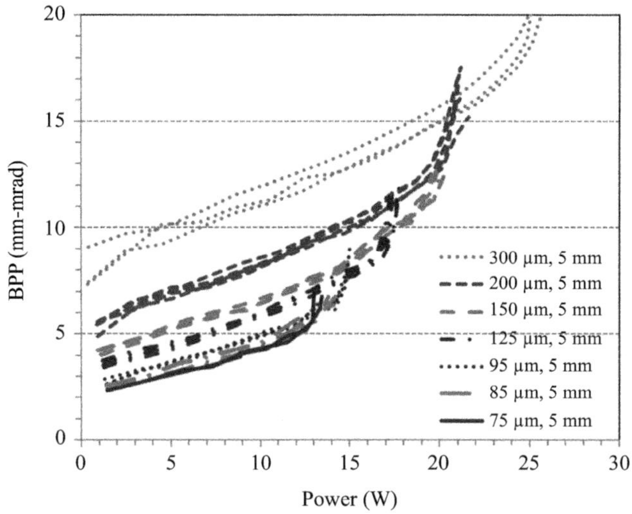

Figure 1.4 *Plots of slow-axis beam-parameter-product (BPP) as a function of output power for broad area lasers with different stripe widths and 5 mm long cavity. For each geometry, there is a corresponding power at which the BPP sharply increases*

smaller stripe-width BAL such as a 50 μm stripe-width diode produces 10 watt of output power at its brightest operation. This shows that the rate of slow-axis divergence increase versus operating power has two undesirable consequences. First, higher output power does not necessarily lead to higher brightness operation once the BPP turning-point is reached. Second, when wider emitter width devices are used to scale power, the maximum achievable brightness progressively degrades. Neither of these effects is conducive for producing high-brightness fiber-coupled modules. This so-called "blooming" of the slow-axis divergence as a function of output power has been a formidable challenge for power-scaling high brightness diode laser pumps.

Most broad-stripe diode lasers are designed to operate with a diffraction-limited beam in the fast-axis direction. Over the past several decades, an emergence of a variety of designs producing large effective spot size in the fast-axis direction has ensued in reliable operation of BALs with >15 MW per cm^2 of facet power density. In addition, over the past decade, a steady improvement of electrical-to-optical PCE as well as semiconductor crystal quality have produced single emitter BALs with >15 watts of power from a ~100 μm emitter width in the 780 nm to 980 nm wavelength emission range. Although, these advances continue to push the maximum ex-facet output power that can be extracted from a single diode laser, the useful power that can be fiber-coupled into a given beam-parameter product has become prohibitively challenging. The key reason for this obstacle is the fact that slow-axis divergence increases nonlinearly at higher output power. As a result, the slow-axis brightness degrades. Hence, higher output power from BALs does not translate to increase in brightness (See Figure 1.5). Several different techniques have been reported to overcome the slow-axis brightness bottleneck. This section focuses on work that addresses those techniques as well as accounts of the most recent advances.

The key performance metric for high-power diode lasers is a combination of power and beam quality also known as brightness. Typically, broad-stripe diode lasers are designed so that power-scaling is achieved by increasing the emission stripe-width for an optimized cavity length. Therefore, we focus on linear slow-axis brightness, B_{SA} (W/mm-mrad), and it is defined as:

$$\boldsymbol{B}_{\mathbf{SA}} = \frac{\boldsymbol{P}}{\mathbf{BPP}_{\mathbf{SA}}} = \frac{\boldsymbol{P}}{(\boldsymbol{w} \times \boldsymbol{\theta}_{\mathbf{SA}})} \tag{1.11}$$

where P is the total optical output power, the slow-axis beam-parameter-product (BPP_{SA}) is the product of the emitter stripe half-width w, and the 95 percent power enclosed slow-axis divergence half-angle in radians θ_{SA}. For example, a purely single mode laser diode commercial product is available with reliable power of 0.75 watt and 1-watt ex-fiber and ex-facet power respectively which corresponds to chip B_{SA} of 3.3 W/mm-mrad. An industry standard 100-μm stripe-width diode with 12 watt of output power at BPP_{SA} = 4.5 mm-mrad corresponds to B_{SA} of 2.7 W/mm-mrad. A higher brightness pump source has multiple advantages. Not only do they open new applications due to affordability, but brightness is

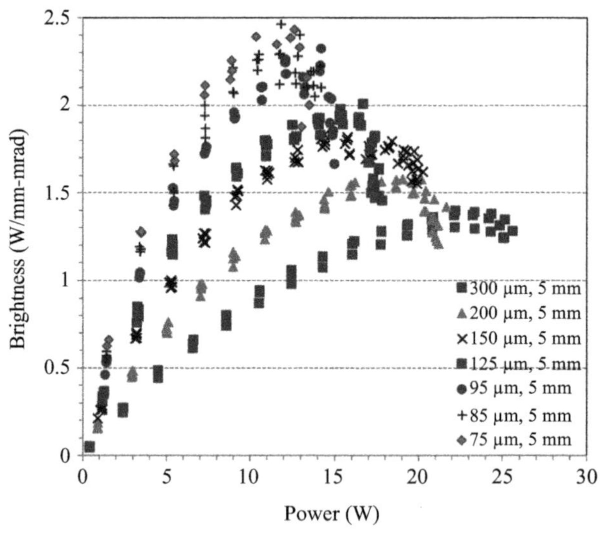

Figure 1.5 Plots showing slow-axis brightness as a function of output power for different size broad-stripe lasers with 5 mm long cavity. Brightness degrades as the stripe-width is increased at any given output power

indispensable in terms of mitigating deleterious nonlinear effects such as SBS, SRS, and SPM in high average power fiber lasers [30]. Especially for the multikilowatt-class CW fiber laser pumping application, slow-axis diode brightness, $B_{SA} > 3$W/mm-mrad is highly desirable. The demand for even higher brightness will continue to rise.

1.3.1 Why brightness degrades as BPP increases

Output power and slow-axis brightness of a typical 976 nm 95 μm stripe-width and 3.8 mm long cavity laser diode is plotted in Figure 1.6. The thermal rollover power of this device is not shown but estimated to be over 16 watts and occurs at greater than 18 A. However, the slow-axis brightness begins to roll over sooner at around 14 A. More importantly, the brightness rolls much faster compared to output power. From (1.11) it is clear that, at least, if power were to scale proportionately with the emitter stripe-width, the brightness would remain constant and not degrade. A plot of output power at the brightest operation value versus emitter stripe-width is shown in Figure 1.7. This output power scales with stripe-width up to approximately 50 μm emitter width. Beyond this dimension, output power scales sub-linearly up to 200 μm and then begins to saturate with diminishing return at larger emitter widths. This suboptimal power-scaling with stripe-width is an even larger contributor to the slow-axis brightness degradation. In large part, the origin of this lower performance stems from higher average junction temperature for wider stripe devices at the same current density. The focus in this section is not to dwell on how

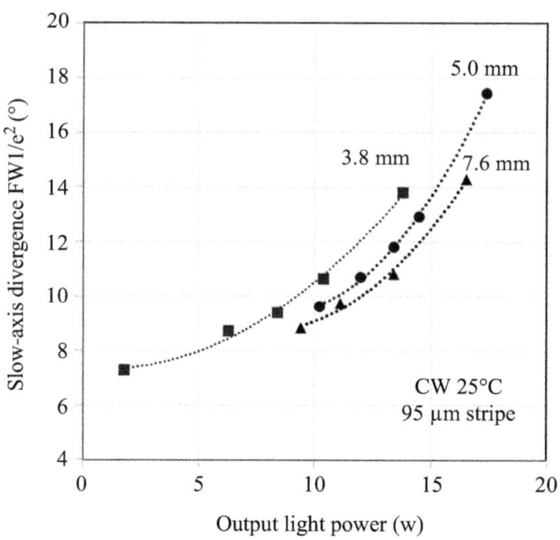

Figure 1.6 Plots showing power and brightness as a function of current for a 3.8 mm and 95 μm stripe width broad area laser

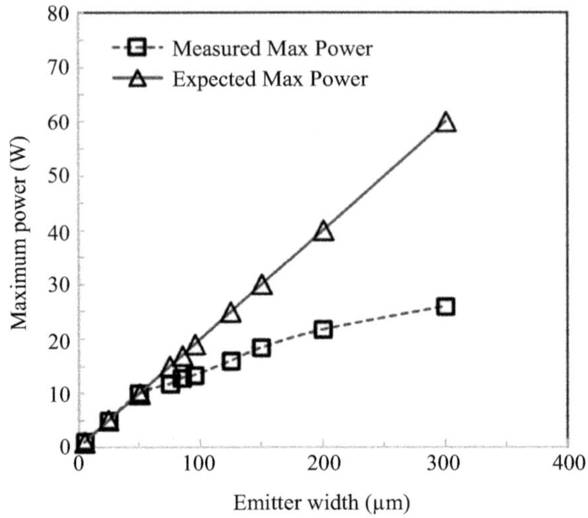

Figure 1.7 Plots showing expected maximum power as a function of broad area laser emitter width compared to experimentally measure values

to improve the thermal conductivity of submounts to increase power-scaling but to address how to reduce slow-axis divergence instead as the emitter stripe-width gets larger. Improving the junction temperature is a separate topic that pertains to primarily improving the submount thermal conductivity and it will certainly improve divergence angle. We will discuss improving thermal performance by engineering the chip dimensions and microscopic thermal transport path from the chip onto the underlying heatsink to flatten the lateral index profile.

1.3.2 The origin of slow-axis divergence blooming in high-power diode lasers

Identification of the exact wave-guiding mechanism in BALs is nontrivial and all its complexities have been studied and reported over the past numerous decades [31–34]. There is a wide variety of lateral index-guiding techniques used in practice. In its simplest form, a gain-guided structure is used which relies on thermally induced index contrast for the lateral wave-guiding. More commonly, etched ridges or trenches are fabricated to create lateral index contrast. In some cases, buried regrown structures are implemented for lateral index step to form a lateral waveguide. Below 0.5 kA per cm^2 of current density, the effective index profile is intricate, and it requires very detailed knowledge of cold-cavity index profile which is modified by both gain-guiding and thermally induced index. Up to these low current densities, the slow-axis divergence increases by less than a few degrees and therefore poses no serious constraint on brightness scaling. In the past, attempts have been made to characterize the lateral guided modes in a broad-stripe laser diode with deep mesa structures with index contrast $\delta n > 1$. In such a case, an ensemble of modes will be supported, and each guided mode will have a wavelength shift relative to the one transverse fundamental mode of $\Delta\lambda$. This relative wavelength shift can be approximated as follows:

$$\Delta\lambda(p,q) = -\frac{\lambda_0}{2n_g}\left(\frac{p}{L}\lambda_0 + \frac{(q^2-1)\lambda_0^2}{4nw^2}\right) \tag{1.12}$$

The wavelength of the fundamental mode is λ_0, the longitudinal and lateral mode orders are given by $p = 0, 1, 2, \ldots$ and $q = 1, 2, 3 \ldots$ respectively, n is the effective index, n_g the group index, L is the cavity length, and w is the stripe-width. As can be seen from (1.12), each guided mode has a unique wavelength. From a 94-μm stripe-width device producing up to 4 watt of CW output power, Stelmakh[34] showed that by performing a spectrally resolved near field and far field measurements comparison could be made between this model and spatially periodic profiles representing lateral modes measured in the experiment. He also showed that higher order modes have larger divergence angles as expected. Current state-of-the-art laser diodes operate at even higher (>12 watt) CW power and at greater than 3 kA/cm^2. Measurements of dramatic increase in slow-axis divergence as a function of current is shown in Figure 1.8. A 95 μm × 3.8 mm device at 25°C heatsink temperature shows a rapid increase in slow-axis divergence from about 4° to a fitted-value of 16° at 20 A and 12° at the 4 kA/cm^2 current density. The exact rate of increase of slow-axis divergence depends on the lateral thermal profile ensuing from the cold-cavity index profile, thermal resistance of the device under test,

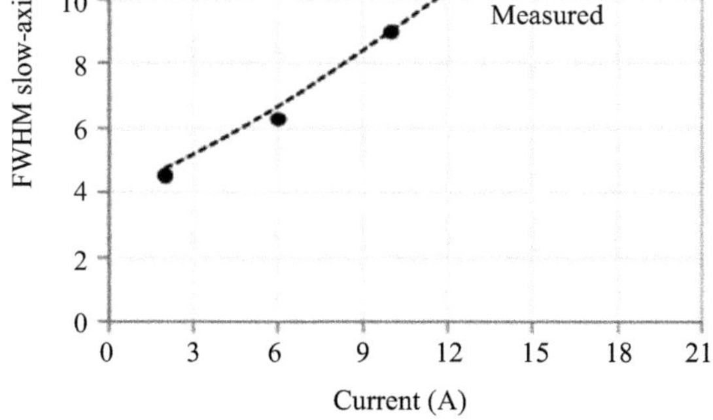

Figure 1.8 Plot showing measured and modelled slow-axis divergence angle increase as a function of operating current for 95 μm broad-stripe lasers with 3.8 mm long cavity

stripe-width, and cavity length. An estimate of the maximum angle which can be supported in the lateral thermal waveguide is described by:

$$\begin{aligned} 2\theta_{\text{Crit},FF} &= 2 \cdot \arcsin\left(\sqrt{n_1^2 - n_2^2}\right) \\ &\approx 2 \cdot \arcsin\left(n\sqrt{2\Delta n}\right) \\ &\approx 2 \cdot \arcsin\left(n\sqrt{2\left(\frac{\partial n}{\partial T}\Delta T + \Delta n_{\text{etch}}\right)}\right) \end{aligned} \tag{1.13}$$

where, $\theta_{\text{Crit},FF}$ is the maximum external angle based on the total internal reflection inside the waveguide, n_1 and n_2 are effective index of the thermal guide and the cladding respectively, $\partial n/\partial T = 2.5 \times 10^{-4}\ \text{K}^{-1} - 3.0 \times 10^{-4}\ \text{K}^{-1}$ is the thermo-optic coefficient, ΔT is the rise in junction temperature, and Δn_{etch} is the index contrast introduced by the etched trench in the cold cavity. Finite element analysis (FEA) can be performed to predict both the rise in junction temperature as well as the lateral thermal profile as shown in Figure 1.13. Using (1.13) one can directly predict the observed behavior in divergence angle versus current, as shown in Figure 1.8. This is further supported by comparison of slow-axis divergence of two diodes that differ only in cavity length as shown in Figure 1.9. There is a strong correlation between the rate at which the divergence angle and the junction temperature rises.

To emphasize further, primary interest is in devices operating at much higher current densities with emitter stripe-widths that are ~100 μm or wider that are essential for achieving the power per emitter required for the demanding applications previously discussed. When broad-stripe laser diodes are operated at a

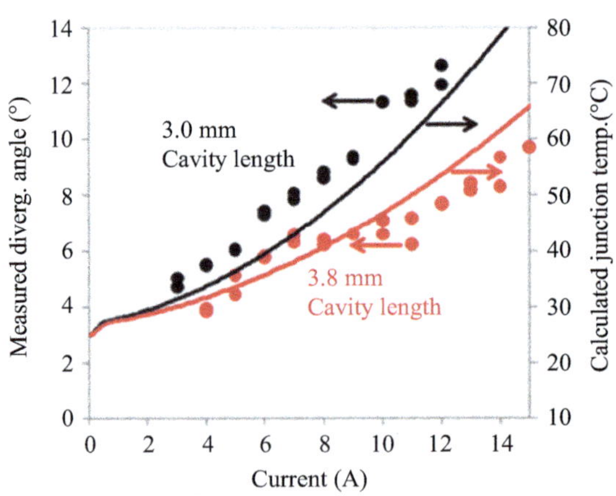

Figure 1.9 Plots showing measured slow-axis divergence angle for 3.0 mm and 3.8 mm cavity length devices along with calculated junction temperature as a function of drive current

1–5 kA/cm^2 of current density and greater than 50 percent electrical-to-optical PCE, thermal flux of a few kW per cm^2 is typically generated. Consequently, at these high thermal flux, self-heating leads to a pronounced thermal profile along the lateral direction of broad-stripe lasers ensuing in a thermal lens. Since the junction temperature can rise by greater than 50 K relative to the ambient, $\delta n \geq 1.5 \times 10^{-2}$ in lateral index-contrast from the center to the edge of the emitter width is not uncommon. This thermally induced strong index contrast overwhelms other effects at current density greater than 0.5 kA/cm^2. Consequently, the cold-cavity index-profile which is modified by the internal lateral thermal profile captures the dominant physical process governing the slow-axis blooming phenomenon.

1.3.3 Mitigating slow-axis divergence blooming

Since the slow-axis blooming is a result of the rise in junction temperature as well as formation of lateral thermal guide, mechanisms for reducing the junction temperature as well as flattening the lateral thermal profile has been explored with some success in the past. Besides thermal management, there are alternative methods to reduce the number of modes inside the waveguide to reduce the divergence at high output powers. In fact, the optimum is to implement a combination of these techniques without incurring additional penalties.

1.3.3.1 Reducing junction temperature via thermal management

In principle, reducing the thermal resistance of the diode submount alone can mitigate slow-axis blooming if sufficient cooling is provided. However, it requires very aggressive thermal management techniques. Since the state-of-the-art junction temperature rise is more than 50 K at high power operation, thermal resistance of less than 0.5 K W^{-1} would be required to maintain a low divergence angle. In the past, diode lasers, predominantly an array of them in a bar-form on micro-channel coolers aided by chillers able to handle high pressure and high flow rates were employed to achieve such low thermal resistance. However, many applications cannot afford such aggressive cooling architecture or complexity in the diode-pumped laser systems. Furthermore, due to thermal cross-talk between the emitters in a bar, brightness-scaling has been limited and its performance has been subpar to single-emitters. Single emitters are thermally isolated from each other in multi-emitter packaging configuration and produce over 50 percent more power per emitter. Reducing the overall thermal resistance of the laser diode package is certainly a worthwhile method; however, due to lack of high thermal conductivity materials with matched coefficient of thermal expansion (CTE) to GaAs, passively-cooled architectures face some challenges. An alternative strategy is to improve the thermal resistance by using a longer cavity length device. In other words, given the same stripe width of a diode laser, a longer cavity makes more area over which to spread heat, thereby reducing thermal resistance and thus the thermal lensing effect. Figure 1.10 illustrates the dependence of thermal resistance on the cavity length. The modeled divergence angle as a function of operating current for

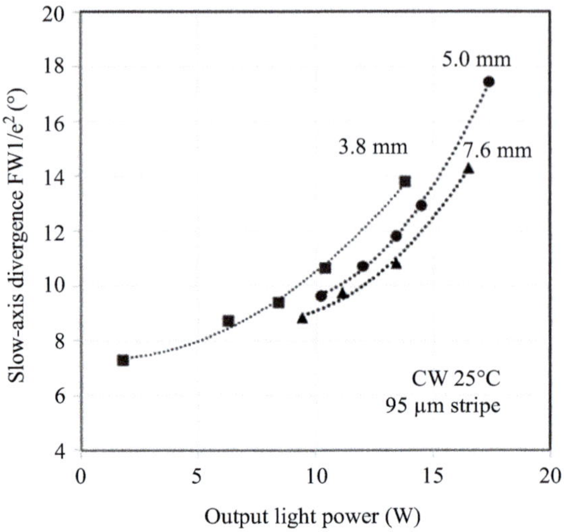

Figure 1.10 Plots showing slow-axis divergence as a function of output power for 3.8 mm, 5.0 mm, and 7.6 mm cavity length and 95 μm stripe-width broad area lasers

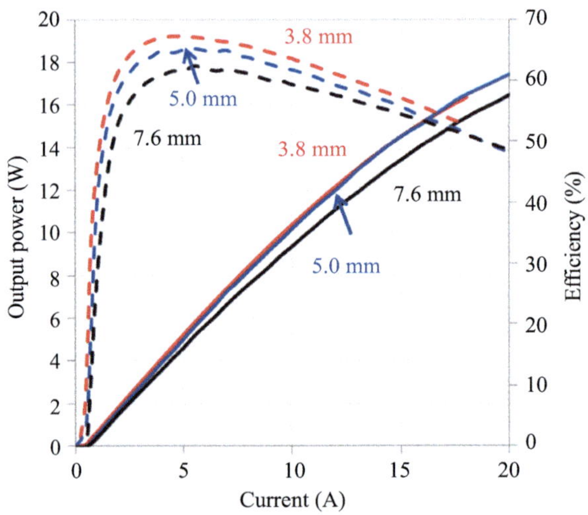

Figure 1.11 Plots showing output power and efficiency for 3.8 mm, 5.0 mm, and 7.6 mm cavity length and 95 μm stripe-width broad area lasers

1.5 mm, 3.8 mm and 7.6 mm cavity length and 100 μm stripe-width devices mounted on AlN substrate are shown in Figure 1.10. Corresponding L-I and PCE for these devices are shown in Figure 1.11. At the same divergence angle, the 7.6 mm diode laser has the highest output laser power followed by the 5.0 mm and then the 3.8 mm diode laser. As an example, at a divergence angle of 11°, the 3.8 mm, 5.0 mm, and 7.6 mm diode lasers have an output power of 10.5 W, 12.5, and 13.5 W, respectively. Since their far-field divergence angles in the fast axis direction are kept at the same levels as discussed before, the numbers show that the 5.0 mm and 7.6 mm lasers are about ~19 percent and ~29 percent brighter than the 3.8 mm devices. However, there is a tradeoff. Additional loss incurs from increase in the cavity length leading to reduced efficiency. The results demonstrate that increasing the cavity length can offer a path to improving slow-axis brightness. Again, this is because longer cavity diode lasers yield lower thermal resistance, resulting in reduced thermal lensing and thus improved slow-axis divergence. There is a limit to the extent that diode laser cavity can be increased. State-of-the-art diode lasers have a residual total internal loss of ~0.5 cm^{-1}. This implies that when cavity length is longer than 5-mm, it ensues in a severe slope-efficiency penalty.

1.3.3.2 Reducing thermal gradient across the stripe-width

The primary reason for the thermal gradient across the stripe-width is due to one-dimensional heat flow primarily in the lateral direction which leads to increase in temperature at the center of the stripe. This is well understood and the exact thermal profile across the stripe-width depends on the size and thermal resistance of the substrate and heatsink. Therefore, reducing the lateral heat flow by engineering microscopic thermal path was proposed and it was referred to as the pedestal heat sinking technique. As the name suggests, the heat generated within the stripe-width of the laser diode was forced to conduct through a thermally conducting path that is sized to match the heat source, that is, the stripe-width. To take the pedestal heat-sinking idea a step further, thermal contact can be made only onto the stripe—leaving all the surrounding region with an air-gap. The pedestal heatsinking concept uses an identical width thermal path as the heat source right underneath the active region to force the heat to be transferred only transversally down for some distance before it can laterally spread out into the heatsink. The prevention of lateral heat spreading in the thermal path just underneath the laser emitter can dramatically reduce the thermal gradient. The one-dimensional heat flow in the engineered thermal path is illustrated in Figure 1.12. The cross-sectional view of the heat source indicating the junction inside the laser diode resides directly above the pedestal of the same lateral dimension and finite height that is connected to the submount. Figure 1.13 shows the lateral temperature profile across the device when the pedestal height is increased from zero up to a thickness of 100 μm. Compared to the baseline, the lateral temperature rise reduces by more than 50 percent at a 10-μm thick pedestal height; while the number drops by approximately 75 percent at a 20-μm thickness and there is essentially no lateral thermal gradient as the pedestal height reaches a thickness of 100 μm. Therefore, microscopic thermal path engineering in abroad area diode laser is a viable method for mitigating thermally induced slow-axis lensing effect but it comes at the expense of higher thermal resistance as shown in Figure 1.14.

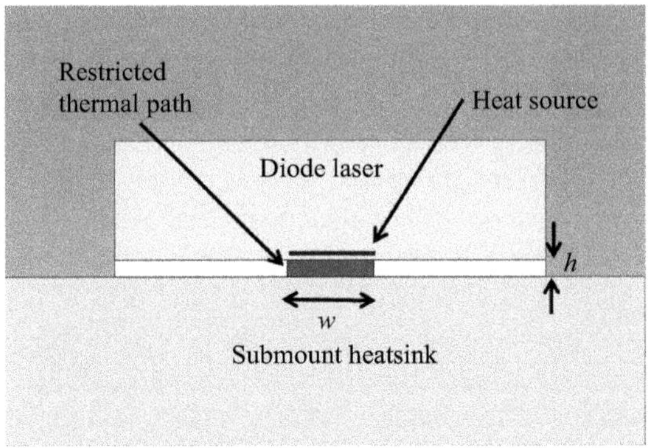

Figure 1.12 *Schematic showing a cross-sectional view of a laser diode with a heat-source located at the junction at a distance, h, away from the heatsink. This heat source is engineered to flow only along a restricted thermal path with width, w*

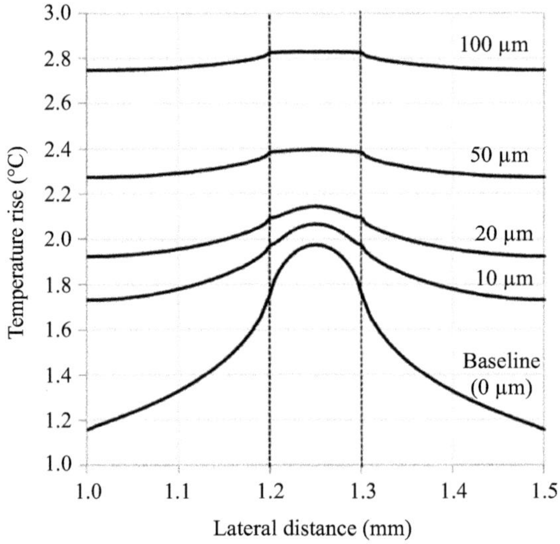

Figure 1.13 *Plot showing temperature rise in the lateral direction of a 95-μm wide diode laser with various thickness of lateral thermal diffusion constriction*

1.3.3.3 Reducing the divergence by suppressing higher-order modes

Historically, techniques discussed in Sections 1.3.3.1 and 1.3.3.2 such as reducing lateral index contrast, inverse thermally induced lens using complex heatsink designs [35,36] or on-chip phase corrections [37] have been employed to overcome

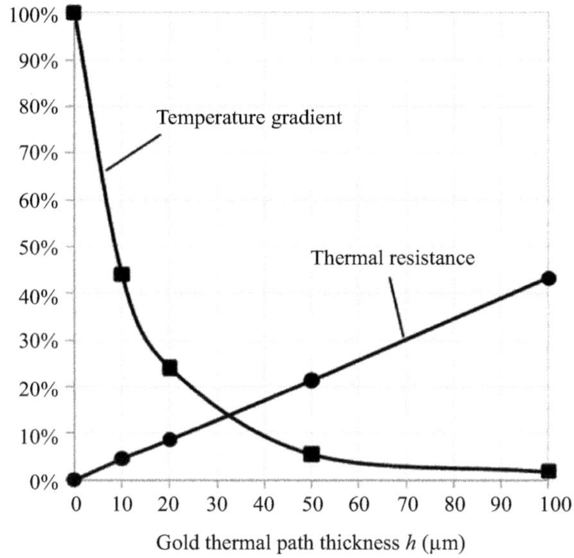

Figure 1.14 Plot showing reduction in temperature gradient across the diode laser emitter width as a function of increase in thickness of the restricted heat flow path and the resulting increase in thermal resistance

this problem. These past efforts produced some modest improvement, but not significant enough to drive down the cost of laser diodes. Most significant slow-axis brightness improvement, to date, has resulted from suppressing higher order modes in the lateral direction; whereby, improvement in slow-axis brightness has translated to increase in output power for a given BPP.

Advancements have been made in slow-axis brightness using BAL design known as flared laser oscillator waveguide (FLOW) and reduced mode diode (REM-diode) technologies [38]. In these devices, the ratio of the partial reflector (PR) to high reflector (HR) emitter width is always greater than unity. Light propagating in the flared current injection region can form a thermal waveguide that is closer to the width of the narrower HR-side causing a beam output at the front facet to have a substantially narrower beam width than the actual partial reflector facet width. The narrower HR reflector also helps to suppress higher order modes upon reflection thus reducing the slow-axis divergence. So, the slow-axis BPP is smaller for FLOW and REM devices compared to conventional broad area devices. A range of designs with various PR to HR emitter width ratios and various current-injection patterns invariably displayed lower slow-axis divergence compared to conventional BALs. This allows REM-diodes to operate with larger emission widths resulting in larger pumped areas. For the same BPP, the thermal footprint of these devices can be designed to be larger making the thermal and the series resistances lower.

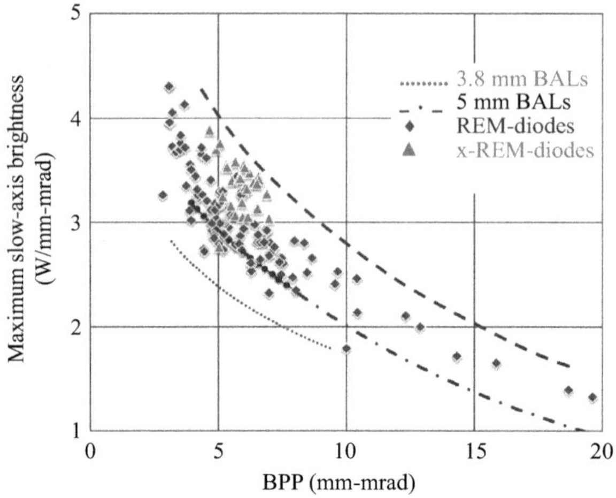

Figure 1.15 Plot of maximum slow-axis brightness for 3.8 mm BAL, 5 mm BAL, and REM and x-REM diodes for a range of BPP

Consequently, the rollover power is higher for these devices and the efficiency does not drop off as quickly as the standard BALs when operating beyond the peak efficiency drive currents. Therefore, the REM-diodes are more efficient at the same operating powers. Figure 1.15 shows maximum slow-axis brightness as a function of BPP for a wide range of REM-diodes in the range of 920 to 976 nm. The symbols denote the maximum slow-axis brightness as a function of BPP for 5 mm long REM and x-REM-diodes of various designs with BPP in the range of 3 to 20 mm-mrad. The dashed line shows the boundary of the maximum slow-axis brightness that is achievable at any chosen BPP for a given REM architecture. Most of the REM-diode designs show higher brightness compared to BALs. It shows that, in absolute terms, REM-diodes have improved brightness more than when increasing just the cavity length from 3.8 mm to 5 mm for a fixed BPP. To date, REM-diodes with brightness as high as 4.3 W/mm-mrad has been demonstrated at a BPP of 3 mm-mrad.

REM-diode architecture is a burgeoning versatile technology. This architecture allows augmenting additional means of stripping higher order modes. As the new suppression techniques for inducing absorption or scattering preferentially to the higher order modes are supplemented to the REM architecture, slow-axis divergence remains relatively flat at even higher operating currents further increasing the slow-axis brightness. For example, Figure 1.16 shows slow-axis divergence as low as 11 degrees even at ~26 A operating current. When additional higher order mode suppression techniques are augmented to REM-diode technology, this new class of BAL is generally referred to as the x-REM diode technology. Based on these results and the trends of past findings, a new family of curve with higher brightness as a function of BPP is expected for the x-REM technology.

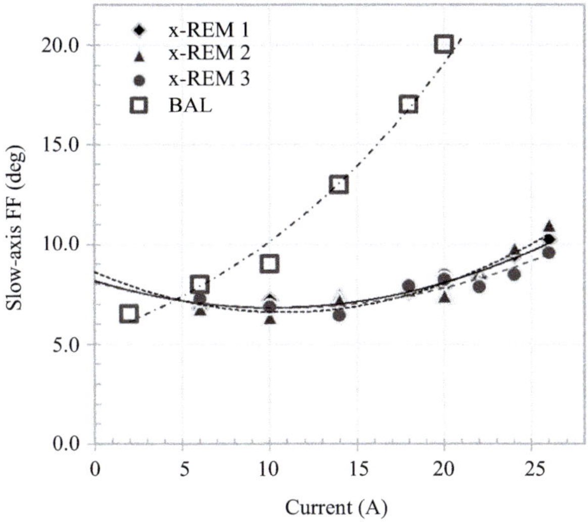

Figure 1.16 Plot of slow-axis divergence of BAL and three different x-REM diode designs

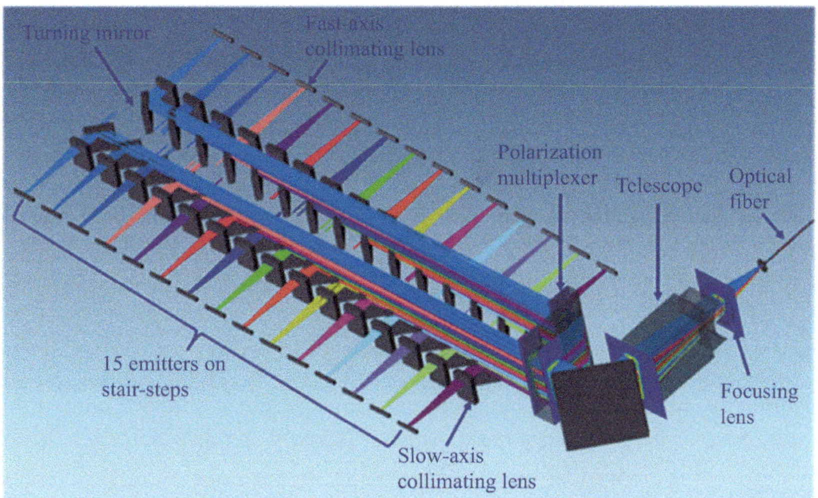

Figure 1.17 ZEMAX optical model of fiber-coupled diode laser pump showing two rows of fifteen diodes that are polarization multiplexed

1.3.4 Fiber-coupled multi-single-emitter diode lasers

Figure 1.17 shows an example of a fiber-coupling scheme that comprises two rows of fifteen diode laser chip-on-submounts (CoS). These CoSs are mounted on a stair-step configuration. Each diode is collimated by using fast-axis and slow-axis

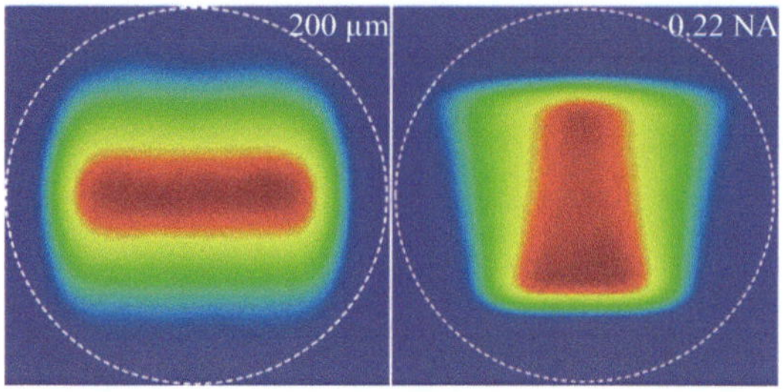

Figure 1.18 The intensity and excitation NA profile at the fiber input as modeled by ZEMAX (dashed line indicates the 200 μm fiber core and 0.22 NA) for the design shown in Figure 1.17

collimating lenses. These fifteen beamlets are then directed toward the polarization multiplexer using a turning mirror for each beam. After the turning mirrors, each row of 15 beams are vertically stacked with nearly 100 percent fill-factor and directed toward polarization multiplexer with the aid of 90° reflecting mirrors. Subsequently, the two rows of 15 compacted-beamlets are combined using a polarization multiplexer to achieve 2× brightness and these beams are then focused into a fiber using a combination of telescope and focusing lens. Since each beamlet is individually lensed, this beam-combining architecture achieves high optical-to-optical efficiency of greater than 90 percent. Since every diode is actively aligned, this architecture achieves beam-pointing accuracy typically not possible with bar technology. Bars have inherent "smile" issues which result in "rogue" uncorrected beams that are not unidirectional thus degrading beam quality and introducing high numerical-aperture cladding-light when coupled into fibers. This is significantly different for this particular optical assembly. It can achieve under-filled numerical aperture and spot size (see Figure 1.18) when focused into a fiber thus eliminating the need for cooling delivery fibers. Additionally, this optical topology conserves usage of volume since no large opto-mechanical components are required to arrange the beamlets into a compact and nearly 100 percent fill-factor collimated beam format—a requirement before efficient coupling into fibers (Figure 1.17).

Using such fiber-coupling schemes, very high brightness fiber-coupled diode laser pumps have been demonstrated with output power more than 300 W from 105 μm and 0.15NA beam as shown in Figure 1.19.

1.3.5 Reliability of fiber-coupled multi-single-emitter diodes

Reliability of laser diodes can be assessed with good accuracy. The acceleration model that is assumed for power and temperature acceleration is a combination of a power law (which describes the effective acceleration due to increasing optical power of the laser) and the Arrhenius law (which describes the effective acceleration due to increasing junction temperature of the laser). From (1.14) the acceleration of unreliability as a function of optical power, P, current, I, and junction temperature, T_j can be derived. In (1.14), m is the acceleration parameter of drive

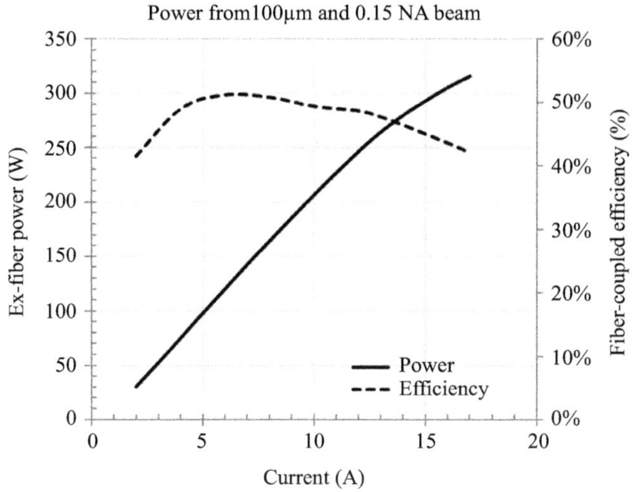

Figure 1.19 Power generated from 105 μm and 0.15NA beam using 30 single-emitter fiber-coupled package

current, n is the acceleration parameter of optical power, E_a is the activation energy, and k_B is Boltzmann's constant:

$$\text{Acceleration Factor} \propto I^m P^n \exp\left(\frac{-E_a}{k_B \cdot T_J}\right) \tag{1.14}$$

For InGaAs quantum-well lasers on GaAs, nominally $m = 2$, $n = 2$, and $E_a = 0.45$ eV. During laser diode reliability assessment, both random and wear out failures must be assessed. The mixed Weibull distribution is a common model for analyzing the case of more than one failure mode, as each failure mode can be described with a distinct Weibull distribution.

One of the additional advantages of using single-emitter-based high power packages is that the reliability of the fiber-coupled module is also further improved from the already highly reliable single emitter diode lasers by effective redundancy. For example, in a 16-emitter module, when the 4th sudden failure of laser happens inside the module, the module power drops by over 20 percent of the initial power at a fixed current. At this condition, the module reliability R_{module} can be described as a function of laser reliability R_{laser} with a binomial equation shown in (1.8):

$$R_{\text{module}} = \sum_{i=0}^{3} \frac{16!}{(16-i)!i!} \cdot {R_{\text{laser}}}^{16-i} \cdot (1 - R_{\text{laser}})^i \tag{1.15}$$

Laser FIT at 1,000 means laser reliability is greater than 98 percent by the end of 2 years of uninterrupted operation. Using (1.8), module reliability can be calculated as 99.98 percent by the end of 2 years of uninterrupted operation. This indicates that the reliability of modules based on single emitter diode lasers is also greatly improved and it gets even better for packages with higher number of emitters.

Typically, single emitters are electrically connected in a series in fiber-coupled packages. The operating current is only in the range of 15 to 25 amperes compared to 60 to 80 amperes in the case of bar-based packages due to parallel connection of 4 to 19 emitters per single bar at 0.5 mm spacing. When defects occur in the semiconductor epitaxial layer, many tens of amperes of current can preferentially go through these defects and create catastrophic damage to the entire bar deeming the entire package unusable. Low operating current in conjunction with series connection prevent such catastrophic reliability issues for single emitter packages. Typically, when a single emitter fails, it becomes short circuited leaving the rest of the population unaffected. Failure in single emitter packages lead to graceful degradation with fractional power loss which is highly attractive for applications that need very high reliability.

1.4 Diode laser applications

1.4.1 Diode-pumped solid-state lasers and fiber lasers

High-power diode laser improvements have enabled growth in a wide range of other laser technologies where the diode laser is used to excite a doped crystal (DPSSLs) or doped fiber (fiber lasers). While diode lasers offer a highly efficient, compact, low-cost source of laser energy, it has two key limitations. They do not store energy and their brightness is many orders of magnitude lower. Essentially diode lasers need to be in tandem with a brightness converting medium for it to be useful for many applications.

In the late 1980s, the use of diode lasers to pump solid-state lasers (DPSSLs) started to gain popularity for commercial deployment. DPSSLs dramatically reduced the size and complexity of the thermal management systems (primarily recirculating chillers) and gain blocks that historically had incorporated arc lamps to pump the solid-state laser crystal. Diode laser wavelengths were chosen based on their overlap with spectral absorption features of the solid-state laser gain medium, significantly reducing the heat load when compared to the broad band emission spectrum of arc lamps. Given the popularity of neodymium-based lasers emitting at 1,064 nm, a pump wavelength of 808 nm was the highest volume wavelength in production for diode lasers for over 20 years.

As the brightness of multimode diode lasers improved and the ability to stabilize narrow emission linewidths with volume Bragg gratings (VBGs) became possible in the mid-2000s, a second generation of improved diode pumping efficiency was achieved. Weaker and spectrally narrow absorption features around 880 nm attracted great interest for high brightness diode lasers that could be spectrally stabilized. These higher performance lasers enabled direct pumping of the upper laser level $4F_{3/2}$ in neodymium, reducing the quantum defect, thus improving fundamental mode operation at higher average powers that would otherwise be limited by thermal lensing. By the early 2010s, a trend grew for significant power scaling of single transverse mode 1,064 nm lasers and their higher harmonics operating in the visible and ultraviolet wavelengths. Given the long upper-state lifetime of Nd:YAG and Nd:YVO4, Q-switched operation of these DPSSLs offered high pulse energies and peak powers well suited for ablative material processing and high precision micromachining applications.

Fiber lasers, perhaps is the most cost-effective means to convert the brilliance of diode lasers. While wavelength multiplexing optics can convert a relatively low brightness diode laser to a brighter one, this comes at the expense of lower spectral brightness and optomechanical complexity and the beam quality is still unmatched. Fiber lasers have proven particularly effective at brilliance conversion. For instance, compare the following best-in-class laser where ~5 kW of 976 nm power from 400 μm and 0.46 NA beam (95 mm-mrad of beam quality), which is equivalent to 0.55 W/(mm-mrad)^2 of pump brilliance. That pump light can be easily converted by a fiber laser to ~4 kW of 1,064 nm from a 20 μm and 0.06 NA beam, which has a brilliance of 11,098 W/(mm-mrad)^2. The brilliance enhancement provided by the fiber laser is greater than 20,000 times or over four orders of magnitude!

In general, as diode laser brilliance advances, so does the power scaling of the lasers they pump. Brilliance improvements in diode lasers tend to promote more efficient brightness conversion. In the future, both spatial and spectral brightness will become even more demanding for systems that will enable low quantum-defect pumping of narrow absorption features in SSLs. It will also become indispensable for dense wavelength multiplexing for direct diode laser applications.

1.4.2 Markets and applications

These advancements in high-power diode lasers have enabled numerous important applications. These lasers have displaced older technologies and enabled a new class of products as the performance improved exponentially in dollars per bright-watt. With an improvement of cost and performance of over a factor of ten every 8 years, high-power diode lasers have disrupted markets that were unforeseen. While it is impossible to precisely forecast future applications, it is instructive to look back over the last three decades to frame the prospects in the coming decade.

1.4.2.1 1980s: optical storage and initial niche applications

Optical storage was the first large application for diode lasers. Soon after Hall's initial demonstration of an infrared semiconductor laser, General Electric demonstrated the first visible red diode laser. Twenty years later the compact disc (CD) was introduced to the market and the optical storage market flourished. Continued innovations in diode laser technologies brought optical storage evolutions such as the digital versatile disc (DVD) and then the Blu-ray disc (BD). This was the first large market for diode lasers, but modest power levels limited proliferation to other applications. These devices remained applicable to relatively small niche markets such as thermal printing, medical applications, and select aerospace and defense applications.

1.4.2.2 1990s: optical networking boom

In the 1990s diode lasers became the cornerstone of communications networks. Semiconductor lasers were used to transmit signals down fiber optic networks, but the higher power single mode pump lasers used in optical amplifiers were critical to expanding optical networks to economically viable scale and ultimately led to fully support the growth in data over the Internet.

1.4.2.3 2000s: laser as a tool

As both performance and cost improved, diode lasers started to increasingly displace legacy gas lasers or other sources of energy transfer in a broad variety of processes. Today diode-pumped lasers have become ubiquitous tools used in a wide range of applications. Industrial applications range from traditional manufacturing processes such as cutting and welding to new advanced manufacturing techniques such as additive manufacturing for 3D printing metal parts. Microfabrication applications are even more diverse as key products such as smartphones have been made commercially viable by the precise power and energy delivery of these lasers. Aerospace and defense applications span a wide range of mission-critical applications and in the future will likely include next-generation of directed energy systems. For this application, SWaP will become exceedingly important. Therefore, there will be a great demand for lowering the specific mass and specific volume of fiber-coupled diode laser packages to well below 1 g per watt and 0.5 cm^3 per watt, respectively.

1.5 The future prospects of diode lasers

Figure 1.20 shows Moore's law for transistors along with the trend for laser diode brilliance. It shows number of transistors per chip increase ten times every 7 years. Similarly, laser diode brilliance has been increasing ten times every 8 years.

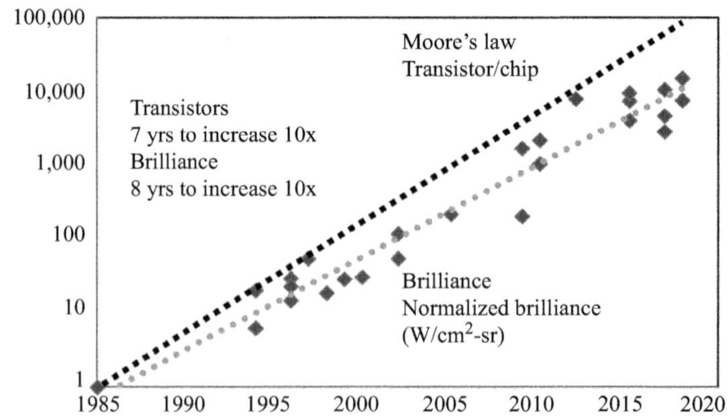

Figure 1.20 Plots comparing Moore's Law for density of transistors per chip with brilliance of diode lasers improve over time

The improvement in brilliance was the result of a wide range of unforeseen technology advancements. While further new innovations will be required to continue this trend, there is reason for optimism because innovation in diode laser technology has not yet been fully exhausted. Well-understood physics can further enhance the performance of diode lasers with continued engineering development. For example, quantum dot gain medium offers significant improvements in efficiency and brightness over current quantum wells. Diode laser slow-axis brightness can also be improved by more than a factor of ten. New higher thermal conductivity materials offer significant enhancements to lowering the diode junction temperature. These and other innovations are some of the critical areas of development that offer a glimpse into the path for continued development in the coming decades.

There are four primary areas of development that will be required to continue to improve fiber-coupled laser diode brightness.

1.5.1 Increasing power and efficiency

Using quantum dot gain medium can reduce the residual free-carrier absorption by tenfold compared to what is possible with quantum-well gain medium. This will allow even longer cavity devices which will improve carrier confinement, differential efficiency, series resistance, and thermal resistance and therefore improve total ex-facet efficiency from the current 73 percent to 85 percent. Achieving quantum dot gain medium will require significant nontrivial material development effort.

1.5.2 Reducing slow-axis BPP and increasing fast-axis brightness

Currently, BALs developed for coupling into 105 μm and 0.15 NA have a BPP of about 4.5 mm-mrad. Theoretically, one could achieve 0.31 mm-mradian (at 976 nm wavelength). This is an improvement of about 14-fold in BPP which translates to 14 times higher slow-axis brightness that is yet to be harnessed. Coherent double-junction lasers can have, in principle, two times higher brightness in the fast-axis direction. So, in total, one can envision, at least, 28 times higher brightness.

1.5.3 Increasing submount thermal conductivity

Carbon and graphene-based material research is beginning to show encouraging results paving the path toward higher thermal conductivity submount technology. With these next generation submounts, one can envision thermal conductivity improvement of approximately three to four times compared to copper and six to eight times higher compared to currently used AlN/CuW types of submounts. For instance, at 20W output power and 16 W heat dissipation from diodes at 2.5 C/W, one can expect about 7 percent improvement in diode efficiency.

1.5.4 Improving optical coupling scheme

Current ex-facet to ex-fiber fiber-coupling efficiency for commercial products is about 85 percent. This can be improved by another 5 percent to 10 percent in the future by lower loss coatings and use of higher fill-factor, compact beamlets.

When all these improvements come to fruition, one can envision almost 30 times higher brightness and about 30 percent higher efficiency compared to current state-of-the-art before reaching a combination of theoretical and practical constraints posed by material properties.

More than 50 years ago Gordon Moore noted the exponential improvement in density of integrated circuits as shown in Figure 1.20. His predictions continued to hold for decades and enabled a series of disruptive innovations that could not have been envisioned in the 1960s. When Robert Hall demonstrated the diode laser over 50 years ago he launched another technology revolution. Like Moore's Law, no one could have predicted the subsequent exponential advances in the brilliance of high-power diode lasers that have been enabled by a wide range of different innovations. It is likely that continued technological advancements will sustain this exponential improvement in brilliance because there are numerous diode laser attributes that are yet to be harnessed. This will continue the displacement of legacy technologies and will further change the way that things are made. Even more fundamental to economic growth, high-power diode lasers will also revolutionize what things can be made.

Acknowledgments

The author would like to thank nLIGHT's Management, Advanced Technology, Device Development and Product Development groups for their valuable input. All the figures are courtesy of nLIGHT.

References

[1] Frenkel, Y. I., and Ioffe, A. F., "On the electrical and photoelectric properties of contacts between a metal and a semiconductor," *Phys. Z. Sowjetunion*, **1**, 60, 1932.
[2] Gross, E. F., and Kariev, N. A., "Light absorption by cuprous oxide crystal in infrared and visible spectrum," *Dokl. Akad. Nauk SSSR*, **84**, 26, 1952.
[3] Hall, R. N., Fenner, G. E., Kingsley, J., Soltys, T. J., and Carlson, R. O., "Coherent light emission from GaAs junctions," *Phys. Rev. Lett.*, **9**, 366, 1962.
[4] Kroemer, H., "A proposed class of heterojunction lasers," *Proc. IEEE*, **51**, 1782–83, 1963.
[5] Alferov, Z. I., and Kazarinov, R. F. Semiconductor laser with electrical pumping, USSR Patent 181737, 1963.
[6] Panish, M. B., Hayashi, I., and Sumski, S., "Double-heterostructure injection lasers with room temperature thresholds as low as 2300 A/cm^2," *Appl. Phys. Lett.*, **16**, 326–7, 1970.
[7] Alferov, Z. I., Andreev, V. M., Garbuzov, D. Z., *et al.*, "Investigation of the influence of the AlAs-GaAs heterostructure parameters on the laser threshold current and realization of continuous emission at room temperature," *Semiconductor*, **4**, 1826, 1970.

[8] Dingle, R., Wiegmann, W., and Henry, C. H., "Quantized state of confined carriers in very thin AlGaAs-GaAs-AlGaAs hetero-structures," *Phys. Rev. Lett.*, **33**, 827–30, 1974.

[9] Chang, L. L., Esaki, I., and Tsu, R., "Resonant tunneling in semiconductor double barriers," *Appl. Phys. Lett.*, **24**, 593–5, 1974.

[10] Van der Ziel, J. P., Dingle, R., Miller, R. C., Wiegmann, W., and Nordland, W. A., Jr. "Laser oscillations from quantum states in very thin $Al_{0.2}$Gas-GaAs- multilayer structures," *Appl. Phys. Lett*, **26**, 463–5, 1975.

[11] Dupuis, R. D., and Dapkus, P. D., "Room temperature operation of GaAlAs/GaAs double-hetero-structure lasers grown by metal organic chemical vapor deposition," *Appl. Phys. Lett.*, **31**, 466–8.

[12] Dupuis, R. D., Dapkus, P. D., Holonyak, N., Jr., Rezek, E. A., and Chin, R., "Room temperature operation of quantum-well GaAlAs-GaAs laser diodes grown by metalorganic chemical vapor deposition," *Appl. Phys. Lett.*, **32**, 295–7.

[13] Mawst, L. J., Bhattacharya, A., Lopez, J., *et al.*, "8 W continuous wave front-facet power from broad-waveguide Al-free 980 nm diode lasers," *Appl. Phys. Lett.*, **69**, 1532–35, 1996.

[14] Emanuel, M. A., Carlson, N. W., and Skidmore, J. A., "High efficiency AlGaAs-based laser diode at 808 nm with large transverse spot size," *IEEE Photon. Technol. Lett.*, **8**, 1291–1293, 1996.

[15] Kanskar, M., Earles, T., Goodnough, T. J., Stiers, E., Botez, D., and Mawst, L. J., "73 percent CW power conversion efficiency at 50 W from 970 nm diode laser bars," *Electron. Lett.*, **41**, 245–7, 2005.

[16] Kanskar, M., He, Y., Cai, J., *et al.*, "53% wallplug efficiency 975 nm distributed feedback broad area laser," *Electron. Lett.*, **42**, 1455–57, 2006.

[17] Bour, D. P. and Rosen, A., "Optimum cavity length for high conversion efficiency quantum well diode lasers," *J. Appl. Phys.*, **66**, 2813 (1989).

[18] Kanskar, M., Earles, T., Goodnough, T., Stiers, E., Botez, D., and Mawst, L. J., "Novel in-plane semiconductor lasers IV," *Proc. SPIE*, **5738**, 47, 2005.

[19] Kanskar, M., Earles, T., Goodnough, T. J., Stiers, E., Botez, D., and Mawst, L. J., "73% CW power conversion efficiency at 50W from 970 nm diode laser bars," *Electron. Lett.*, **41**(5), 245–7, 2005.

[20] Mawst, L. J., Bhattacharya, A., Lopez, J., *et al.*, "8 W continuous wave front-facet power from broad-waveguide Al-free 980 nm diode lasers," *Appl. Phys. Lett.*, **69**, 1532–5, 1996.

[21] Emanuel, M. A., Carlson, N. W., and Skidmore, J. A., "High efficiency AlGaAs-based laser diode at 808 nm with large transverse spot size," *IEEE Photon. Technol. Lett.*, **8**, 1291–1293, 1996.

[22] Al-Muhanna, A., Mawst, L. J., Botez, D., Garbuzov, D. Z., Martinelli, R. U., and Connolly, J., "14.3 W quasi-continuous wave front-facet power from broad-waveguide Al-free 970 nm diode lasers," *Appl. Phys. Lett.*, **71**, 1142–4, 1997.

[23] Buda, M., van der Vleuten, W. C., Iordache, Gh., *et al.*, "Low-loss low-confinement GaAs–AlGaAs DQW laser diode with optical trap layer for high-power operation," *IEEE Photon. Tech. Lett.*, **11**, 161–3, 1999.

[24] Kanskar, M., Earles, T., Goodnough, T., Stiers, E., Botez, D., and Mawst, L. J., "73% CW power conversion efficiency at 50 W from 970 nm diode laser bars," *Electron. Lett.* **41**, 245–7, 2005.

[25] Kanskar, M., Earles, T., Goodnough, T., Stiers, E., Botez, D., and Mawst, L. J., "High-power conversion efficiency Al-free diode lasers for pumping high-power solid-state laser systems," *Proc. SPIE*, **5738**, 47–56, 2005.

[26] Kanskar, M., Bao, L., Bai, J., *et al.*, "High reliability of high power and high brightness diode lasers," *Proceedings of SPIE*, **8965**, 896508–18, 2014.

[27] Zucker, E., Zou, D., Zavala, L., *et al.*, "Advancements in laser diode chip and packaging technologies for application in kW-class fiber laser pumping," *Proc. SPIE*, **8965**, 896507–21, 2014.

[28] Karlsen, S. R., Price, R. K., Reynolds, M., *et al.*, "100-W, 105-μm, 0.15 NA fiber coupled laser diode module," *Proc. SPIE*, **7198**, 71980T, 2009.

[29] Price, K., Karlsen, S., Leisher, P., and Martinsen, R., "High brightness fiber coupled pump laser development," *Proc. SPIE*, **7583**, 758308, 2010.

[30] Dawson, J. W., Messerly, M. J., Beach, R. J., *et al.*, "Analysis of the scalability of diffraction limited fiber lasers and amplifiers to high average power," *Optics Express*, **16**, 13240–66, 2008.

[31] Tang, X., van der Ziel, J. P., and Chin, A. K., "Characterization of the array modes of high-power gain-guided GaAs single-quantum-well laser arrays," *IEEE J. Quantum Electron.* **32**, 1417–26, 1996.

[32] Lengyel, G., Wolf, H. D., and Zschauer, K. D., "A study of the lateral modes in wide double-heterostructure GaAs-AlGaAs laser diodes," *J. Appl. Phys.* **49**, 1047–53, 1978.

[33] Mailhot, S., Champagne, Y., and McCarthy, N., "Single-mode operation of a broad-area semiconductor laser with an anamorphic external cavity: experimental and numerical results," *Appl. Opt.* **39**, 6806–13, 2000.

[34] Stelmakh, N., "External-to-cavity lateral-mode harnessing devices for high-brightness broad-area laser diodes: concept, realizations and perspectives," *Proc. SPIE* **7230**, 72301B, 2009.

[35] Bai, J. G., Leisher, P., Shiguo Zhang, *et al.*, "Mitigation of thermal lensing effect as a brightness limitation of high-power broad area diode lasers," *Proc. SPIE*, **7953**, 79531F, 2011.

[36] Piprek, J., "Inverse thermal lens effects on the far-field blooming of broad area laser diodes," *IEEE Photon. Tech. Lett.*, **25**, 958–60, 2013.

[37] Eckstein, H., Zeitner, U. D., Tünnermann, A., Schmid, W., Strauss, U., and Lauer, C., "Mode shaping in semiconductor broad area lasers by monolithically integrated phase structures," *Opt. Lett.*, **38**, 4480, 2013.

[38] Kanskar, M., Flared laser oscillator waveguide, US Patent 9,166, 369, 2015.

Chapter 2

Coherent beam combining architectures for high-power laser diodes

Gaëlle Lucas-Leclin[1], Philipp Albrodt[1], David Pabœuf[1], Guillaume Schimmel[1] and Patrick Georges[1]

2.1 Introduction

The power and brightness of individual laser diodes are continuously improving over the years, not only in the near infrared range but also in the visible. As the cost per watt decreases accordingly, high-power direct diode laser systems are becoming more and more attractive for industrial applications where they are not limited anymore to the pumping of solid-state lasers and fibres [1–4]. Nonetheless, the highest power emitted from a single laser chip is typically below 20 W for a broad area element in the near infrared, and even below in other spectral ranges. Thus the high powers required for cutting, welding and other advanced manufacturing techniques are only available by combining multiple emitters in a single beam – strength lies in numbers in this matter too! This is indeed very general for any kind of lasers, and beam combining approaches are by far not used with diode lasers only [5,6]. Still, there are specifics when it comes to diode lasers, and this chapter will solely be devoted to them.

The *brightness* of a laser beam is defined as the power per unit area in a solid angle of divergence. It is evaluated as the optical power divided by the product of the mode area S at the beam waist and the solid angle Ω in the far field [2,5]. These quantities are linked through the beam quality parameter M^2. With elliptical beam profiles of diode lasers, the brightness can be expressed as

$$B = \frac{P}{S \times \Omega} = \frac{P}{\pi w_x w_y \times \pi \theta_x \theta_y} = \frac{P}{\lambda^2 \times M_x^2 M_y^2}$$

[1]Laboratoire Charles Fabry, Institut d'Optique Graduate School, CNRS, Université Paris-Saclay, Palaiseau, France

Beam combining techniques are classified in incoherent combining, spectral beam combining and coherent beam combining approaches. The first one, the simplest and most straightforward, is shown in Figure 2.1 and it consists of the side-by-side addition of laser beams. 1D-bars of laser diodes actually fall in this category, which takes benefit of the compactness and parallel manufacturing process of semiconductor technology. A single package containing tens of adjacent emitters remains extremely compact and easy to handle. The available optical power P_{bar} is then increased up to hundreds of watts but the beam quality is strongly deteriorated. Indeed, the total emission width w_{bar} is enlarged whereas the angular divergence θ_{bar} remains similar to an individual element. 2D stacks of bars scale further the available power, although the spatial brightness of those stacks is even less than of a single element ($B_{bar} < B_1$) [7]. Furthermore, the spectrum of the laser bar or stack is wide due to device-to-device disparities and temperature differences in the packaging mount. Beam transformation optics, based on step mirrors and cylindrical lenses, are commonly used to circularise the output beam of a laser bar and optimise its coupling into an optical fibre [8,9].

Spectral beam combining, on the other hand, consists in the wavelength multiplexing of diode lasers with spectrally selective passive components as shown in Figure 2.2. Both a serial configuration (a) using dichroic mirrors and (b) a parallel configuration using diffraction gratings, prisms or volume Bragg gratings (VBGs) exist [5,6,10]. Then the different beams emitted by the laser diodes overlap perfectly, and the spatial brightness of the overall system is increased. The spectrum of the laser emission consists of multiple distinct lines, and the total spectral width is of course significantly larger than that of a single device. This approach is simple for coarse wavelength multiplexing with laser emitters or bars in slightly different spectral ranges, when the laser lines are distant of a few tens of nanometers. When a narrower spectral width is required, each diode laser has to be spectrally stabilised in order to precisely maintain its central line with regard to the combining elements. This dense wavelength division multiplexing is far more demanding on the wavelength stability and spectral purity of the individual laser elements, which also impact the beam quality of the laser system [6]. Either internal or external spectral stabilisation have been demonstrated, with single-mode emitters or broad area lasers, individual elements or bars [11–15]. The wide variety of setups evidences the versatility of this approach for power scaling of lasers. Commercial kW-class diode laser systems with extremely high brightness in the GW.cm^{-2}.sr^{-1} range are nowadays available.

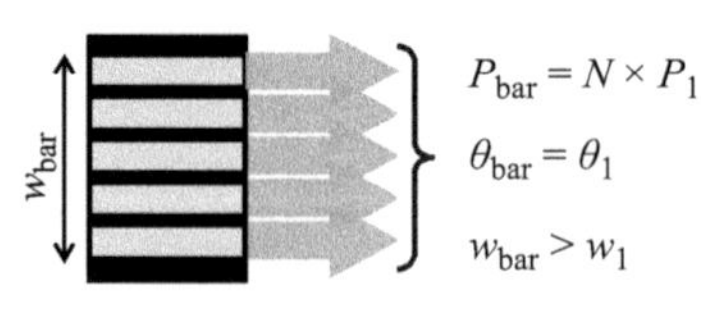

Figure 2.1 Concept of incoherent, side by side, addition of diode lasers

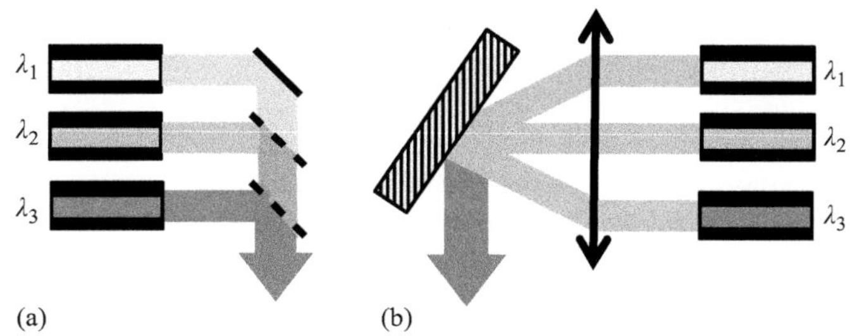

Figure 2.2 Concept of spectral beam combining in (a) serial configuration using dichroic mirrors and (b) parallel configuration using gratings

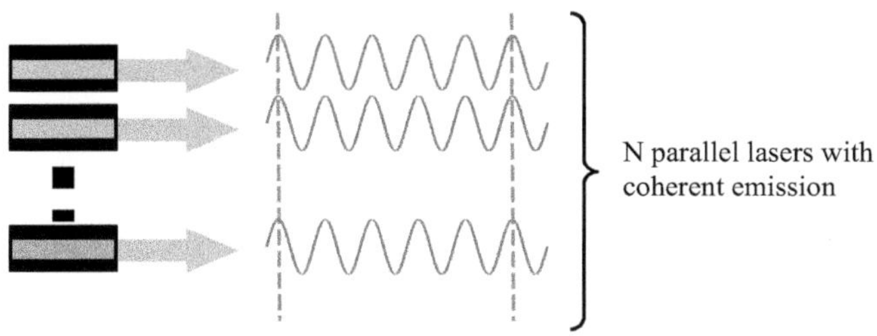

Figure 2.3 Concept of coherent beam combining

In comparison with the two previous combining approaches, coherent beam combining is the most demanding. It is based on constructive interference between phase-locked beams as shown in Figure 2.3, which result in a single high-power beam. Then the laser diodes have to be mutually coherent, operating at the same wavelength and with a stable phase relationship [5,6,16]. It implies that the devices to be combined are single-mode, diffraction-limited, emitters. Coherent beam combining of laser beams preserve, or even improve, the beam quality and spectral purity of the individual laser elements, and for this reason has been thoroughly investigated. Different configurations were implemented, with active or passive phase stabilisation, extended-cavity or parallel amplifier architectures. We propose in this chapter to present different realisations of coherently combined laser systems with laser diodes, and to comment on their limitations. Coherent beam combining architectures have been extensively studied for fibre laser systems to scale the available power/energy while maintaining high beam quality. Though the same general principles apply for laser diodes, there are still a few specificities that are worth mentioning. We shall describe more extensively the experiments made at Laboratoire Charles Fabry (Palaiseau, France) with high-brightness tapered laser diodes for the past 10 years; indeed, those emitters provide the highest power in a nearly diffraction-limited beam among laser diodes, and appear thus as a good starting point for coherent combination.

We begin this chapter with a summary of the properties of high power laser diodes and amplifiers, with a specific attention to the single-mode devices compatible with coherent beam combining approaches. Then we present the general principles behind them, by making a distinction between the phase-locking stage and the coherent superposition one. The majority of this chapter, in the last two sections, is devoted to the description and analysis of a few representative experimental setups to force a coherent and phase-locked operation of multiple amplifiers and laser emitters (Sections 2.4 and 2.5).

2.2 High-power semiconductor lasers and amplifiers used for coherent beam combining

The content of this chapter is focused on edge-emitting laser diodes. The power available from those devices is strongly dependent on the spectral range, and the lateral geometry of the waveguide. In the near infrared where the performance of semiconductor materials are the most favourable, an optical power in the range of 1 W can be obtained from a truly single-mode ridge emitter with a beam quality factor M^2 equal to 1 in both slow and fast axis [17]. But broad area edge-emitters achieve ~20 W in a highly multimode beam with M^2 above 20 in the slow axis [18]. Then the spatial brightness of the laser emission remains limited to typically 100 $MW.cm^{-2}.sr^{-1}$. Tapered lasers and amplifiers are edge-emitting devices consisting of a narrow ridge section followed by a wide flared section as shown in Figure 2.4. They have been proven to provide multi-Watt output powers and a slow-axis beam quality factor typically below 5 – though exact evaluation of this parameter should be considered cautiously with the complex beam profiles of those tapered laser diodes [19]. The spatial brightness reaches then 1 $GW.cm^{-2}.sr^{-1}$.

Coherent beam combining approaches are intended to further increase this brightness above the intrinsic limitations of single devices. Nevertheless, the laser beams to be coherently combined should be in the fundamental transverse mode. Indeed, no stable phase relationship exists between the transverse modes of a

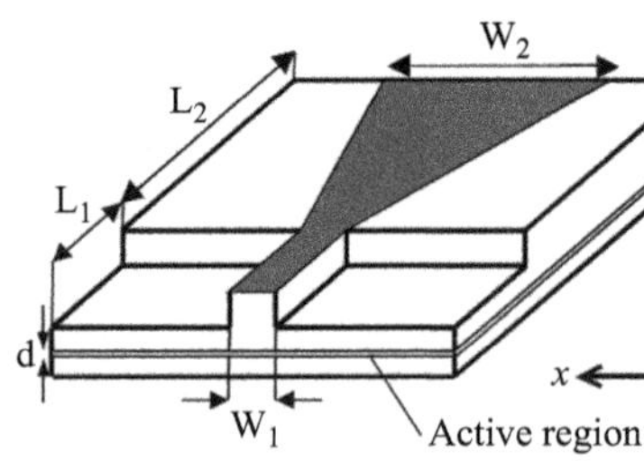

Figure 2.4 Schematic of a tapered laser/tapered amplifier consisting of a single mode ridge waveguide (width w_1 and length L_1) followed by a flared gain-guided tapered section (length L_2) resulting in a broad output facet $w_2 >> w_1$. Reprinted, with permission, from [20]

broad area laser diode, thus only one transverse mode could possibly be phase-locked at the same time. As compared to incoherent combining approaches that have been applied to single-mode ridge emitters as well as broad area devices, coherent beam combining is thus restricted to single-mode devices. The residual power within the high-order transverse modes is not effectively transferred to the combined beam and therefore limits the overall combining efficiency. This is a major limitation to the total optical power achieved by coherent beam combining of diode lasers: the single-mode power is much lower than in other types of lasers used in coherent beam combining, notably fibre lasers. On the bright side, coherent superposition of slightly multimode laser beams acts as a spatial filter and improves the beam quality of the combined beam by rejecting the high-order modal content.

We shall emphasise, and this will be further illustrated in this chapter, that controlling the phase experienced during propagation in an edge-emitting laser diode is much easier than in any other kind of laser devices. It is actually straightforwardly realised by changing the driving current, through the induced temperature change of the active waveguide. Phase noise in high power semi-conductor amplifiers is discussed in detail in Section 2.4.2.

Arrays versus individual emitters

Linear arrays of edge-emitting devices are very common, compact and convenient devices: the emitters are side by side in the slow-axis direction, with a distance p that depends of the emitter width $2w$ for obvious reasons. The number of emitters is typically between 5 and 20 on the same bar – up to 100 in some examples – and the fast-axis collimation of the entire bar can be realised with a single high numerical aperture acylindrical lens. In order to avoid crosstalk and evanescent coupling between adjacent emitters as well as to limit the thermal load, the pitch p is significantly larger than $2w$, typically a hundred micrometres up to 1 mm. The fill-factor $2w/p$ of the laser arrays is thus low, of the order of 1% typically. It may be increased by individual collimation of the emitters with a microlens array. It is even possible to further increase the number of emitters by stacking linear arrays on top of each other in the fast-axis direction, similarly to high power laser diode modules. The fill-factor is even lower in this direction though.

There are a few drawbacks of using arrays of laser devices for further coherent combination. First the collimation of the single-mode elements in the array should be excellent to preserve their beam quality. Minimising the bar smile is critical to avoid pointing errors and ensure a good overlap of the laser beams. Furthermore, it requires specific mounting and contacting processes to allow the individual addressability of the emitters for fine phase adjustments [21,22]. Since the emitters all share the same thermally stabilised mount, we may expect that individual control of the emitters would be unnecessary as phase fluctuations between emitters on the same array should be small. There have been, indeed, some successful demonstrations of phase-locking and coherent superposition for collectively controlled laser arrays, as will be demonstrated later in this chapter. But still, the optimisation of the phase-locked operation and the stabilisation of the combined power are improved if each emitter can be separately driven. Alternatively, individual laser devices on distinct mounts can be coherently combined, at the expense of somewhat stronger phase fluctuations. Nevertheless, the thermal management is improved as compared to a laser bar, so the power per emitter is increased. Each component is also separately collimated and aligned, which simplifies the overall experimental arrangement and allows fine optimisation of the beam superposition. This configuration is particularly well suited for the coherent combination of a few high-power emitters as discussed in Section 2.4.2.

2.3 Principles of coherent beam combining architectures

In contrast to the incoherent beam combining and spectral multiplexing approaches mentioned in the introduction, coherent beam combining consists of the addition of laser fields $E_i(t)$ – not only powers [5,6,16]. All the laser characteristics are preserved regarding beam quality, spectral purity or polarisation. Indeed, coherent beam combining of multiple beams results from constructive interferences between the optical fields, regardless of the technique used to superpose them. It implies that those optical fields are mutually coherent.

Coherent beam combining is divided into two distinct stages: firstly, the relative phases of the individual laser devices should be *locked* and *stabilised*, and secondly, the phase-locked beams are *superposed* so that they propagate as a single diffraction-limited beam.

2.3.1 Phase locking

Phase-locking approaches are usually classified into three categories, depending on how the mutual coherence of the laser devices is imposed:

- Self-organisation by coupling through evanescent or leaky waves,
- Self-organisation in a common extended laser cavity, and
- Parallel amplification of an external seed laser.

The first option relies on a monolithic linear array of closely packed ridge emitters, whose transverse modes are laterally coupled [23]. Different phase-locked array designs have been proposed, but, despite their inherent simplicity, the overall output power remains limited and the coherent operation of the whole array is not maintained at high operating currents. This approach will not be further discussed in this chapter, as the potential for high-power emission appears limited. In the second scenario, the phase-locked devices operate as coupled laser elements in an extended cavity (see Figure 2.5). On the contrary, in the third case, the devices are simple amplifiers, and their mutual coherence results from the seed laser. In both last cases, the relative phases of the beams through the devices might be either maintained passively, or stabilised with an external control. These two approaches are detailed in Sections 2.4 and 2.5.

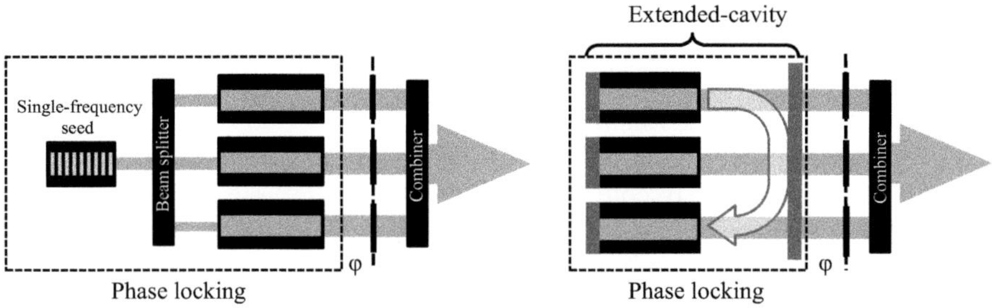

Figure 2.5 Comparison of two approaches for the phase-locking of laser devices: parallel amplification of a single-frequency source (left), or extended-cavity operation of laser emitters (right)

2.3.2 Coherent superposition

Two combining arrangements are possible for coherent superposition: either tiled-aperture implementations or filled-aperture ones [5,6,16]. In both cases, the goal is to concentrate the total power from the phase-locked emitters in a single lobe that can be directly used for applications. For tiled-aperture superposition, the phase-locked beams are simply positioned side by side and overlap in the far field. The coherent superposition is then equivalent to near-field aperture synthesis. In contrast, filled-aperture superposition requires additional optical elements such as beam splitters to overlap the phase-locked beams in the near field. In this section, we briefly summarise the fundamental principles of these combining approaches, and emphasise their respective constraints for the coherent combining of individual laser diodes and laser diode arrays.

2.3.2.1 Tiled aperture approach: far-field superposition

Tiled-aperture superposition of phase-locked laser beams in the far field is the simplest approach. The laser beams are individually collimated and positioned closely adjacent to each other, with a pitch p. They interfere in the far field without additional combining elements as shown in Figure 2.6.

We consider N in-phase Gaussian beams distant of pitch p, each of them with a near-field full-width measured at $1/e^2$ $2w$. The far-field profile of the phase-locked array is given by the Fourier transform of the near field. It consists in contrasted fringes with angular separation λ/p within a wide Gaussian envelope $\Delta\theta = 2\lambda/\pi w$ corresponding to the incoherent far-field profile of the whole set. The number of fringes is directly related to the fill-factor $ff = 2w/p$: the higher ff, the lower the number of fringes. It is highly desirable to get the largest fill-factor so that the combined power is within a single narrow lobe. Indeed, the superposition efficiency is in this case the ratio between the power contained in the far-field central lobe and the total power. For example, a fill-factor ff of 50% would result in only 45% within the central lobe with perfect Gaussian beams [6,16].

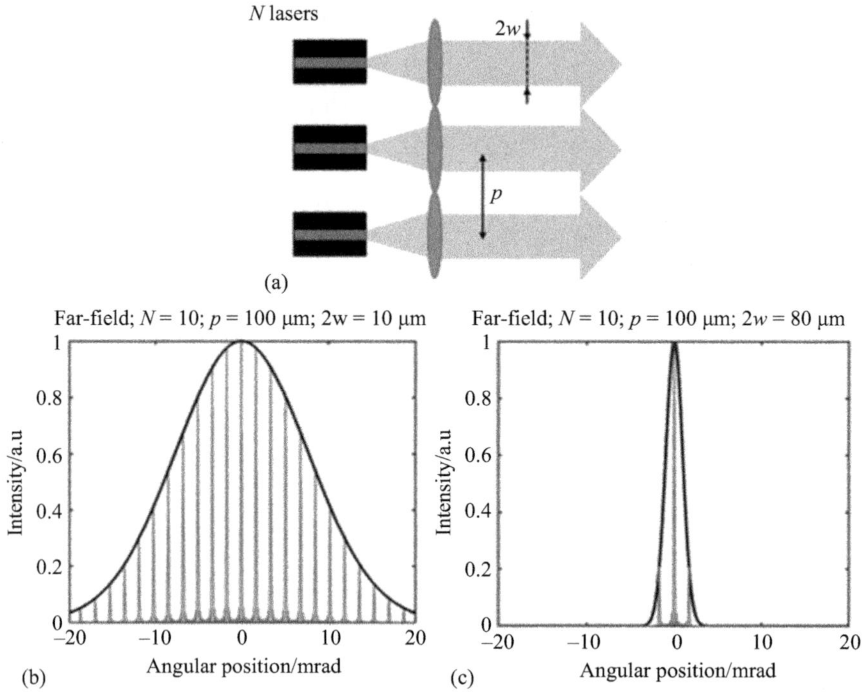

Figure 2.6 (a) Concept of coherent superposition in the far field, (b) corresponding far field for a fill factor of ff = 10% and (c) corresponding far field for a fill factor of ff = 80%

The tiled-aperture superposition has several advantages: it is suitable for beam steering and beam shaping, atmospheric turbulence compensation and the combined power is not supported by a single optical element. It is commonly used with fibre amplifiers, whose beams are arranged in closely packed 2D-arrays in order to maximise the fill-factor [24,25]. Nonetheless, this approach is not practical with laser diodes, as the fill-factor of single-mode emitter arrays is low in the slow axis. The calculated far field for coherent superposition for a laser bar with $N = 10$ emitters and a fill-factor of only 10% is shown in Figure 2.6(b). A large number of fringes limit the power content in the central lobe to about 5%. Individual collimation of the emitters would increase the fill factor, but the near-field intensity profile remains non-uniform, and side-lobes would still limit the power content in the central lobe [26,27]. Figure 2.6(c) shows the calculated far-field for the same laser bar but with a higher fill factor of 80% (after individual collimation). The far-field is narrower in this case but still shows a few fringes and the power content in the central lobe is in the range of 70%. In most cases, the far field of an array of phase-locked laser diodes exhibit multiple peaks within a wide envelope, corresponding to the interference fringes between the different beams. To illustrate this, Figure 2.7 shows the experimental far-field for the coherent superposition of 218 amplifiers, arranged in a stack of eleven arrays on top of each other in the fast axis (low fill-factor), each array consisting of 21 individually collimated emitters side-by-side in the slow axis (high fill-factor) [26].

2.3.2.2 Filled aperture approach: near-field superposition

The general principle of the coherent superposition of phase-locked beams in the filled-aperture approach is shown in Figure 2.8(a) and (b). It uses one or more combining elements, such as beam splitters or phase gratings, on which the high-power beams are superposed and interfere constructively into a unique beam.

The achievable combining efficiency for coherent superposition in the filled aperture approach is in both cases limited by the spatial overlap of the incident beams. The combined optical power is $P_{\text{comb}} = \frac{1}{N}\left|\sum_1^N E_n\right|^2$, and the combining efficiency is $\eta_{\text{comb}} = P_{\text{comb}}/\sum_1^N P_n$. In the case of imperfect spatial overlap of the

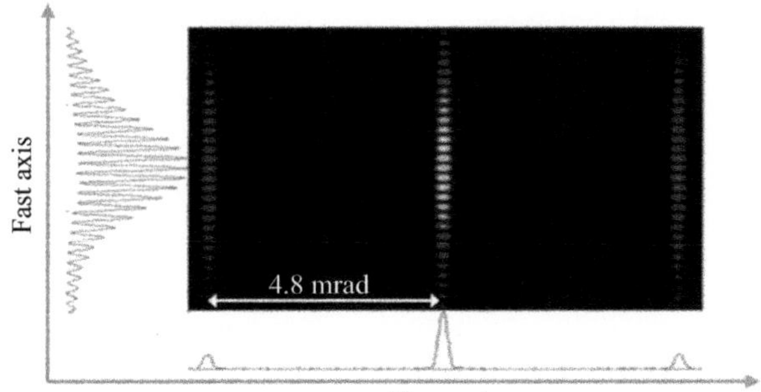

Figure 2.7 Experimental far-field of eleven 21-element SCOWA arrays coherently combined. Reprinted, with permission, from [26]

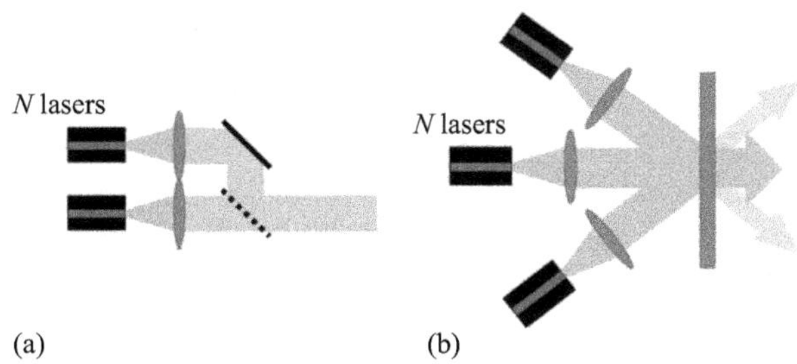

Figure 2.8 Concept of coherent superposition in the near field with (a) beam splitters and (b) phase gratings as combining elements

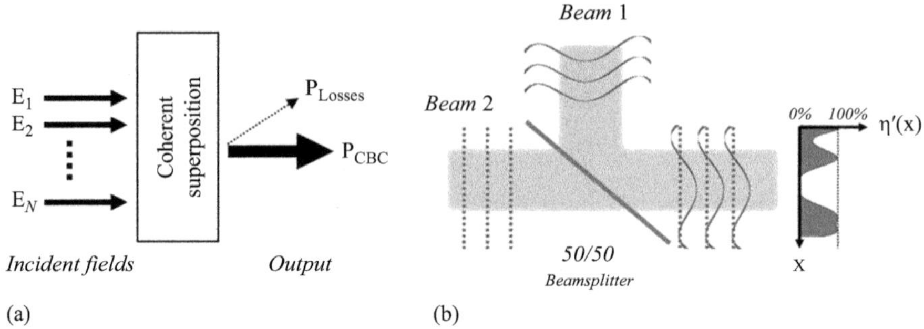

Figure 2.9 (a) Schematic of coherent superposition in the filled aperture approach. (b) Schematic of a simplified case for the interference of two spatially mismatched beams

complex incident fields, P_{comb} is lower than the sum of incident powers and a part of the optical energy is lost during coherent superposition (see Figure 2.9(a)) [28]. A simplified case for the interference of two imperfect fields, for example with a mismatch in the phase front, is shown in Figure 2.9(b). The beams interfere with a local combining efficiency of:

$$\eta'(x,y) = \frac{1}{N}\frac{\left|\sum_n E_n(x,y)\right|^2}{\sum_n |E_n(x,y)|^2} \tag{2.1}$$

which is equal to unity only where the incident complex fields $E_n(x,y)$ perfectly match in amplitude and phase. The global combining efficiency can then be rewritten in the more general form:

$$\eta_{\text{comb}} = \frac{\iint \eta'(x,y)\sum_n I_n(x,y)dx\,dy}{\iint \sum_n I_n(x,y)dx\,dy} \tag{2.2}$$

with $I_n(x,y) = |E_n(x,y)|^2$ the transverse profile of the nth beam. While perfectly single mode beams should have small mismatches and can therefore be efficiently combined, there are other types of beams, such as slightly multimode beams from tapered lasers, that usually have strong mismatches in the higher angle side lobes. On the one hand, this leads to combining losses, but on the other hand it is also the reason for the beam clean-up inherent to coherent beam combining.

Coherent superposition on beam splitters

The simplest combining element is a 50/50 beam splitter, which combines two incident beams into one if they are in phase. It is a successful approach for the combination of individual devices; combining efficiencies higher than 90% are commonly demonstrated, only limited by residual discrepancies between the spatial beam profiles [29,30]. Interestingly, the combining efficiency η_{comb} is not as sensitive as one might guess to misbalanced powers of the incident beams. A significant difference in power, for example, $P_2 = 2P_1$, would lead to minor combining losses in the range of only 3%. Combining more than two elements with beam splitters is straightforward by cascading superposition stages, either in a binary tree configuration or a linear chain one, but may become cumbersome. Furthermore, it is not suitable for diode laser mounted on an array, as the elements are usually too close to each other to position individual optical elements.

Coherent superposition on phase gratings

For large number of elements, a specifically designed phase-grating can be used. The phase profile of the combiner is optimised to diffract one beam into N with known phases $\{\Phi_i, 1 \leq i \leq N\}$ as shown in Figure 2.10. Used backwards with N beams of proper phases and amplitudes, it acts as an efficient combiner. Such a diffractive optical element (DOE) was used as early as the late 1980s [31,32]. The phase profile consists in periodic pattern with a period Λ, related to the diffraction angle $\mathrm{asin}(\lambda/\Lambda)$ between the beams. The simplest design adopts a binary profile with only two phase levels 0 and π, but at the expense of a low diffraction efficiency typically below 80%. Multilevel phase profiles provide higher diffraction efficiencies, but the highest ones are achieved with continuous phase profiles [33,34]. Sub-λ phase profiles have also been investigated [35]. Those phase gratings are used either as the coherent combiner for a set of phase-locked beams, or in extended-cavity configurations to induce a coupling between emitters and superpose their beams, as we shall show in Sections 2.4.1 and 2.5.2.

The diffraction efficiency η_{DOE} of a DOE used as a splitter ($1 \rightarrow N$) is measured as the proportion of power in the N orders of interest, with $|q_n|^2$ the diffraction

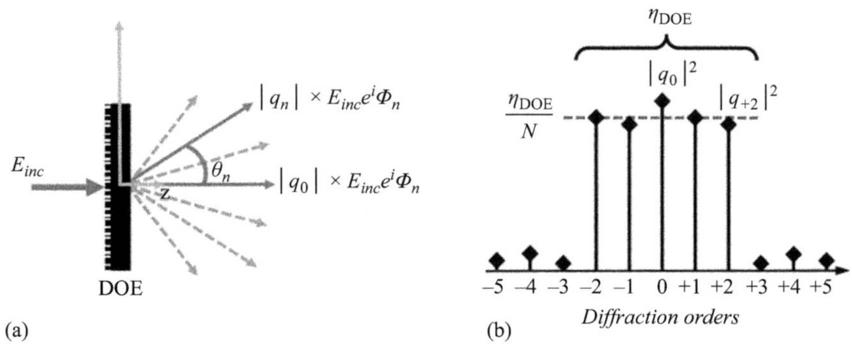

Figure 2.10 Definition of the combining efficiency from the amplitudes of the diffracted orders

efficiency in the nth order. The variance σ^2_{DOE} of the $\{|q_n|^2\}$ ensemble is a measurement of the imbalance between the different orders with regard to the average efficiency $\frac{\eta_{\mathrm{DOE}}}{N}$:

$$\eta_{\mathrm{DOE}} = \sum_{n=-\frac{N}{2}}^{\frac{N}{2}} |q_n|^2, \tag{2.3}$$

$$\sigma^2_{\mathrm{DOE}} = \frac{1}{N} \sum_{n=-\frac{N}{2}}^{\frac{N}{2}} \left(|q_n|^2 - \frac{\eta_{\mathrm{DOE}}}{N} \right). \tag{2.4}$$

Considering an incident plane wave E_0, the diffracted optical fields are:

$$E_n(\vec{r}) = |q_n| E_0 e^{i\Phi_n} e^{i\ \vec{k_n}\cdot\vec{r}}, \tag{2.5}$$

where $\mathbf{\Phi}_n$ is the relative phase and $\overrightarrow{k_n}$ the wavevector which coordinate along the grating is $\frac{2\pi}{\lambda}\sin\theta_n = n\frac{2\pi}{\Lambda}$. Inversely the same DOE illuminated with N beams coinciding to the complex conjugate $E_n{}^*(\vec{r})$ – same intensity distribution $\{|q_n|^2, n = -\frac{N}{2}\ldots\frac{N}{2}\}$, opposite phases $\{-\mathbf{\Phi}_n\}$ and wavevectors $\left\{\overrightarrow{-k_n}\right\}$ – would combine the beams into one with the combining efficiency η_{DOE}. For the coherent superposition of N beams of equal powers, the ideal phase grating should thus provide $\eta_{\mathrm{DOE}} = 100\%$ and $\sigma^2_{\mathrm{DOE}} = 0$.

Within the scalar approximation of diffraction, and as long as the pitch Λ remains much larger than the wavelength λ, the efficiency distribution $\{|q_n|^2, n \in \mathbb{Z}\}$ does not depend on Λ but only on the exact profile *m(x)* of an elementary phase profile [31,36]. The amplitude of the diffraction orders are given by the Fourier Transform of the grating profile:

$$q_n = FT\left[e^{im(x)}\right]\Big|_{\frac{n}{\Lambda}}, \ n \in \mathbb{Z}. \tag{2.6}$$

Consequently, for a given phase grating, the diffraction profile $\{q_n, n \in \mathbb{Z}\}$ can easily be computed and thus the quantities η_{DOE} and σ^2_{DOE} following (2.3) and (2.4). On the contrary, the calculation of a grating profile from the desired diffraction distribution is not straightforward, as the Fourier Transform of any diffraction distribution may not correspond to a phase grating – not to mention a binary one. Various numerical algorithms exist for the optimisation of the elementary pattern, depending on its complexity [36–38]. The cost function should take into account both the overall efficiency η_{DOE} in the N orders of interest and the uniformity of the diffraction distribution through the variance σ^2_{DOE}. It is noteworthy that a perfect balance between the diffracted orders ($\sigma^2_{\mathrm{DOE}} = 0$) is achieved at the expense of a lower diffraction efficiency η_{DOE}, as is demonstrated in Figure 2.11, which compares the results obtained for $N = 5$ orders and different phase profiles.

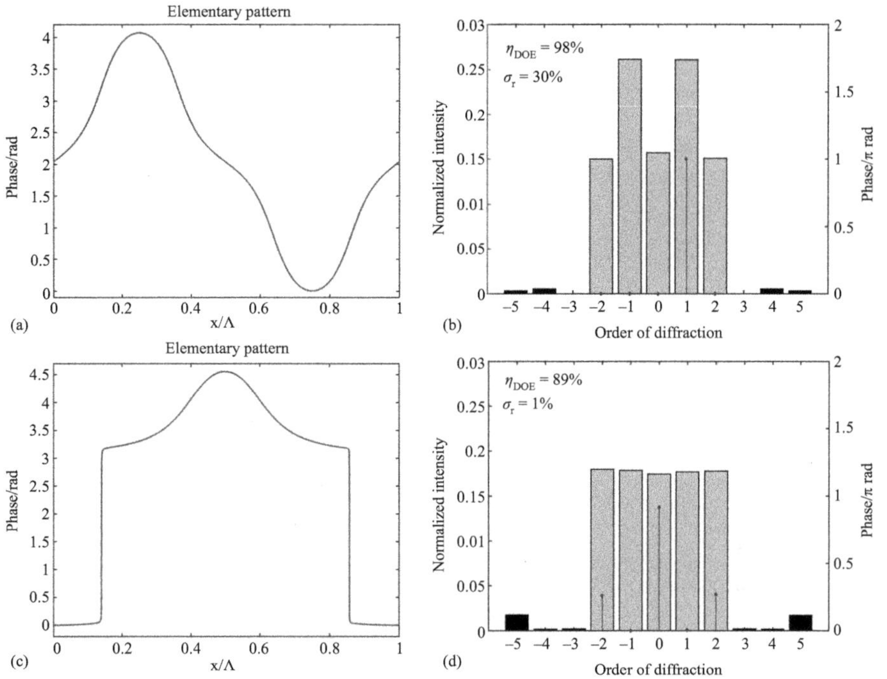

Figure 2.11 *Two examples for continuous phase patterns for a 1 → 5 beam splitter. The elementary phase patterns are shown in (a) and (c) and the corresponding diffraction orders are given in (b) and (d). The pattern in the first line is optimised for high diffraction efficiency whereas the pattern in the second line is optimised for a homogenous distribution.* σ_r *is the relative standard deviation between the orders, defined as* $\boldsymbol{\sigma_{DOE}}$ *divided by* $\boldsymbol{\eta_{DOE}/N}$

2.4 Master-oscillator power amplification architectures

The master-oscillator power amplification (MOPA) configuration is well established in high-power laser diode architectures and is the standard approach to increase the power available from single frequency laser diodes. It allows to preserve the high temporal and spatial coherence of a low power seed source in the powerful amplified laser beam [39–42]. The master oscillator (MO) is either a monolithic laser diode with an integrated distributed Bragg reflector (DBR or DFB laser), or an extended-cavity laser diode. In both cases, the spectral purity of the laser line is excellent, characterised by a narrow linewidth (≤ 10 MHz) and a high side-mode suppression ratio (>40 dB). In most cases the power amplifier is a tapered amplifier in order to reach multi-Watt powers and maintain decent spatial beam quality.

The parallel amplification of a single seed source in multiple amplifiers is the direct extension of the MOPA approach in order to further increase the available output power above the limits of the individual elements. The seed beam is divided into the *N* amplifiers by means of cascaded beam splitters or DOEs and is injected into the parallel amplifiers (see Figure 2.5). The whole arrangement follows a multi-arm Mach-Zehnder interferometer. The injection of the seed laser is usually done in an external optical setup; however, it is also possible to monolithically integrate the master oscillator, the beam splitter and the amplifiers on one chip [43]. The coherence between the N amplified beams is thus ensured. Nonetheless, the phase pistons experienced in each amplifier may fluctuate and need to be stabilised for an efficient superposition of the N beams. This is usually achieved by small adjustments of the drive currents into the amplifiers [44] or by an monolithically integrated section for phase modulation [43]. In contrast to coherent beam combining of fibre amplifiers in MOPA configuration [45,46], no external phase modulators are required. Additionally, the phase fluctuations in semiconductor amplifiers are very low and dominated by low frequencies (see Section 2.4.2.1). Different phase-stabilisation methods have been proposed for the active coherent combining of multiple beams, from heterodyne detection to hill-climbing algorithms – see [16] for an overview of the most popular approaches. The best choice concerning the phase control for a certain experimental setup depends on the number of amplifiers. While simple sequential hill-climbing algorithms are sufficient for a limited number of amplifiers [47], more sophisticated phase control algorithms have to be used for coherent combining of a large number of amplifiers [44].

Simple and linear in its very principle, the parallel MOPA architecture requires however some precautions in its implementation. First, the seed power injected into each amplifier should be larger than their saturation power, in order to limit the contribution of amplified spontaneous emission (ASE) and to maximise the extracted power per amplifier. As the number *N* of amplifiers grows, it may be necessary to use a pre-amplifier to increase the single-frequency power available from the MO to the required level [44]. Second, strong optical isolation ($>$ 40 dB) of the MO from the amplifiers is crucial to maintain a stable operation of the seed laser diode. Finally, parasitic oscillation of the amplifiers needs to be prevented, as it would introduce mode competition against the amplified line, reduce the coherence of the amplified beam and might result in random phase jumps. Very high-quality anti-reflection coatings and/or angle-cleaved facets are required.

2.4.1 Coherent beam combining of amplifier arrays

In most experimental demonstrations of coherent combining of semiconductor amplifiers, the active devices were used in arrays, as they allow to use a large number of devices simultaneously in a compact packaging. We summarise in the following the most relevant works with arrays of single-mode ridge waveguide amplifiers (Section 2.4.1.1) and arrays of tapered amplifiers (Section 2.4.1.2).

2.4.1.1 Ridge waveguide amplifiers

Parallel MOPA architectures have been investigated experimentally with arrays of ridge amplifiers since the 1990s, with up to 900 phase-locked amplifiers in an arrangement of nine arrays of 100 elements [48]. More recently, the phase-locking and coherent combining of an array of 47 high-power single-mode elements (slab-coupled optical waveguide amplifiers, or SCOWA) were demonstrated with a record combined power of 40 W in a diffraction-limited beam [44]. The experimental setup is shown in Figure 2.12.

Here, the MO was a DFB laser diode at 1,064 nm amplified in a ytterbium-doped fibre amplifier. The array pitch was 250 μm, and each amplifier was individually addressed with a separate electrical contact. The seed beam was split and injected into the SCOWA through a butt-coupled planar lightwave circuit consisting of cascaded y-branch waveguide splitters. On the front side of the laser array, the 47 amplified beams were combined into one with a high-efficiency (>99%) DOE. The phase control loop was based on a stochastic parallel gradient descent (SPGD) algorithm, which was intended to maximise the combined power on a single photodetector using random dithering of the amplifier currents. The combining efficiency was independent of the operating current and reached 87%. Though this experiment has demonstrated, to date, the highest coherently combined

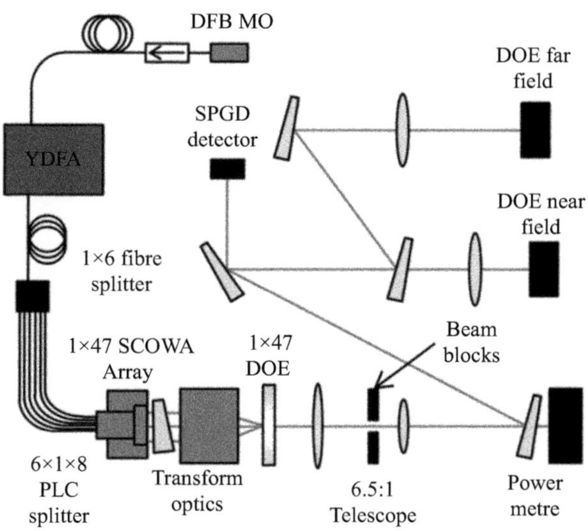

Figure 2.12 Experimental setup for the coherent combining of an array of 47 slab-coupled amplifiers (SCOWA). A single-frequency source is amplified in a Yb-doped fibre and split into 48 beams, which are butt-coupled to the array. The SCOWA output beams are combined using a single Fourier lens and DOE, and the output is sent to diagnostics. Only three SCOWA beams are shown for clarity. Reprinted, with permission, from [44]

power from a semiconductor-based MOPA architecture, it remains a complex arrangement, with a combined power per amplifier P_{CBC}/N of only 0.85 W. The latter is indeed limited by the single elements themselves, neither by the combining configuration nor the beam quality of the amplified beams.

Further increase of the total power could in principle be achieved by increasing the total number of elements with parallel arrays in a 2D-stack as attempted in [26], at the cost of an increased complexity of the setup.

2.4.1.2 Tapered waveguide amplifiers

We describe in the following the coherent combining of five tapered waveguide amplifiers achieved at Laboratoire Charles Fabry (Palaiseau, France) [22]. As they reach high powers in nearly diffraction-limited beams [19,20,49], tapered amplifiers are the perfect building block for simple coherent beam combining architectures relying on a limited number of devices only.

The laser array used in this experiment was a custom laser mini-bar with five tapered emitters (2.5 mm-long ridge section, 3 mm-long tapered section, 6° tapered angle) and a pitch $p = 1$ mm. An extremely low vertical divergence design ("ELoD"), with 95% power content within a divergence angle $\Theta_V^{95\%} = 25°$, was chosen for the epitaxial structure of the amplifiers [50]. The rear (ridge section) facet has a $R < 0.1\%$ coating, and the front (tapered section) facet has a $R = 0.5\%$ residual reflectivity. The amplifiers were soldered *p*-junction-up on a copper heat sink to allow separate electrical contact to ridge and taper for all devices, and external optical access to both facets while ensuring a good heat extraction. A photo of the mini-array is shown in Figure 2.13.

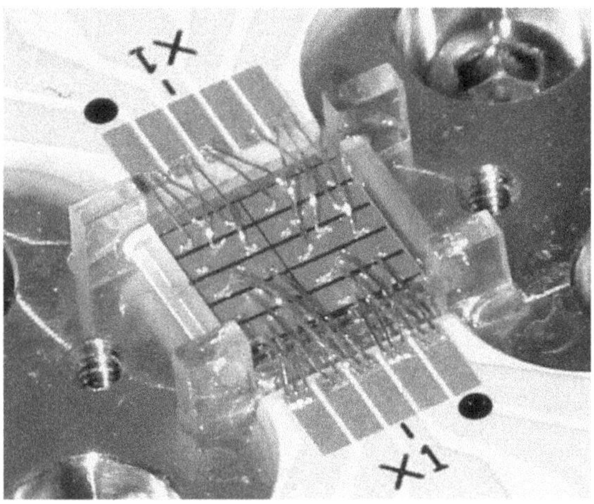

Figure 2.13 Picture of the five-emitter tapered laser bar: the front (tapered) section is located on the right side of the picture while the rear (ridge) section is located on the left side, and the individual electrical contacts are clearly visible on both sides. Two fast-axis cylindrical lenses are glued on the mount on both sides

Each element can be individually addressed with separate electrical contacts for the ridge and tapered sections.

The seed source was a narrow-linewidth distributed feedback (DFB) laser diode at $\lambda = 975.5$ nm. The DFB laser emission was protected from parasitic optical feedback with −60 dB isolation (Figure 2.14). Its beam was re-shaped with a cylindrical afocal system to match the mode profile of the amplifiers ridge section. The seed beam was divided by a diffractive optical element (DOE_1), with a total diffraction efficiency of 98% in five orders and a power distribution $\{P_i\} = (14\%, 28\%, 16\%, 27\%, 14\%)$ following the design described in Figure 2.11(b). With a DOE pitch $\Lambda = 78$ μm, the five diffractive beams were imaged into the amplifiers in a *f–f* Fourier conjugation with a $f' = 80$ mm doublet, which ensured that $p = \lambda f'/\Lambda$. The power extracted per element is between 2.3 W and 3.4 W at the operating currents $I_{RW} = 350$ mA and $I_{TP} = 6$ A in the ridge and tapered sections, respectively [22]. The maximum power emitted by the array was 14.7 W, limited by insufficient heat evacuation of the mounting and incomplete saturation of the amplifiers. The relative phases of the amplified beams were controlled by properly adjusting the current in each ridge section through the current-induced heating of the amplifiers. A current change $\Delta I_{RW} \cong 55$ mA corresponded to a π phase-shift in a single amplifier. As the devices are almost saturated, such a current change in the ridge section of the amplifiers has no significant impact on the extracted power. The front far-field profile in Figure 2.15 was measured on the front side of the array, while the phases were equalised for in-phase operation. The high visibility ($V \geq 93\%$) of the fringe pattern confirms the strong mutual coherence of the five amplified beams.

The coherent superposition of the five beams was realised with a second DOE identical to the first one positioned in the back focal plane of another $f' = 80$ mm doublet (Figure 2.14). The combined beam resulted from the constructive interferences of all five beams in the on-axis order behind the DOE, when the phases of each beam matched the phase distribution $\{\phi_i\} = (0, 0, 0, \pi, 0)$ imposed by the DOE design. The currents in the ridge sections were sequentially adjusted with a hill-climbing algorithm targeting to maximise the combined power P_{CBC} measured on a photodiode. As the relative phase fluctuations were extremely small and the

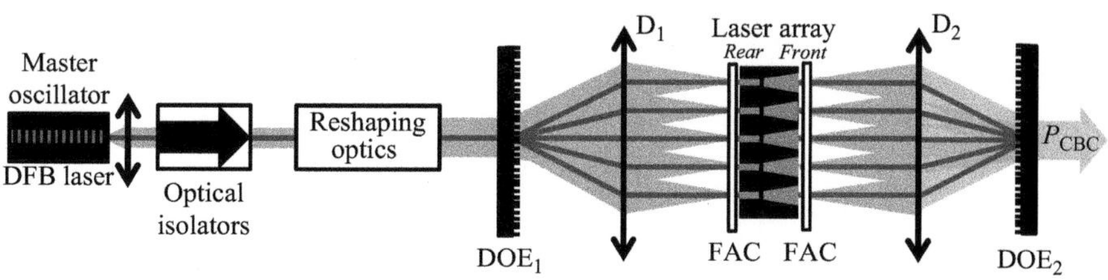

Figure 2.14 Experimental setup used for the coherent beam combining of an array of five tapered-section edge-emitting devices in the master-oscillator power-amplifier configuration; D_1, D_2: 80 mm doublets; FAC: 0.6 mm fast-axis collimators; $DOE_{1,2}$: diffractive optical elements (see [22])

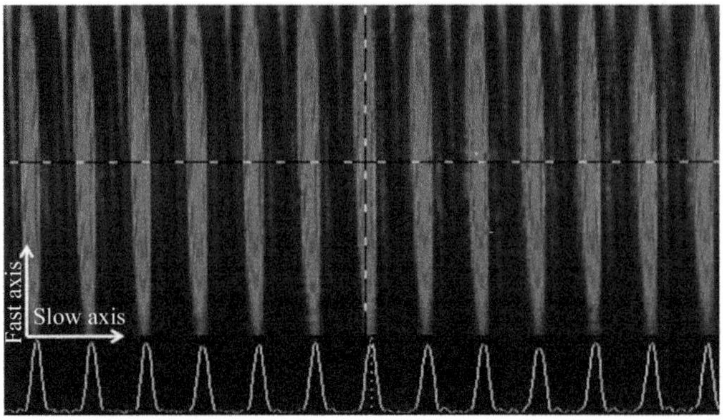

Figure 2.15 Far-field interference profile observed on the front side of the amplifier array under in-phase operation of the five devices, at $I_T = 2$ A in each emitter

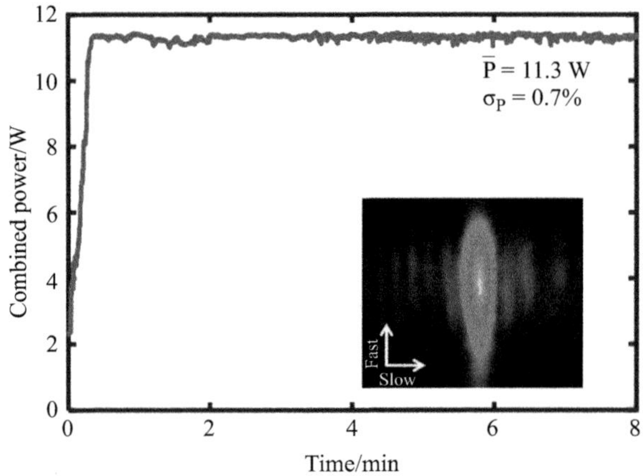

Figure 2.16 Evolution of the combined power at $I_{TP} = 6$ A in the tapered section of each device; the servo-loop optimises the ridge currents in typically 20 s, then maintains the power at its maximum. Inset: far-field profile of the combined beam

number of elements in the array was low, this optimisation approach was the simplest and yet very effective. For low operating currents ($I_T \leq 3$ A), the relative phase fluctuations in the array were so small that the active stabilisation was only necessary once per hour at moderate drive currents. Nevertheless at the highest operating currents, the servo loop was continuously operating to maintain the combined power at its maximum. Then the combined power remained very stable as shown in Figure 2.16, with a standard deviation below 1%. This corresponds to phase fluctuations within elements in the laser array below 0.1 rad ($\lambda/30$)

over 10 min. The combining efficiency, defined as the zeroth-order power behind DOE_2 (P_{CBC}) divided by the total power extracted from the five-element laser array, reached 82% at $I_T = 3$ A, and decreased to 78% at the maximum operating current $I_T = 6$ A [22]. It was mostly limited by the single-mode power content in the beam profiles of the individual tapered-section amplifiers. The maximum combined power reached 11.5 W, corresponding to a combined power per element $P_{CBC}/N = 2.3$ W. The beam quality of the coherently combined beam was improved with respect to the beam quality of one individual tapered amplifier. The power content in the diffraction limited central lobe was higher than 80%. The M^2_{1/e^2} beam quality factor was 1.1 in fast axis and below 1.3 in slow axis. The standard M^2_σ factor (second moment beam widths) in slow axis is significantly higher (≈3) as the higher angle side lobes still contain about 20% of the total energy.

2.4.1.3 Limitations related to coherently combined arrays

Comparison of the two previous experiments highlights common features and limitations. The coherent superposition of the amplified beams relies on a free-space phase grating, which is specifically designed for the number of elements in the array – respectively 47 in the MIT Lincoln Laboratory experiment and 5 in ours at LCF. The pitch of the phase grating being typically tens of micrometres, the focal length of the Fourier lens that matches the laser array far-field profile to the diffraction orders of the grating is in the range of 100 mm. It significantly contributes to the overall size of those setups. Furthermore, it is noteworthy that the combined power per element in the array is below the maximum power demonstrated from single devices amplifiers, because of the overall heating of the array which degrades the performance of the individual elements. Additionally, access to both facets of the array makes the heat management more challenging. Regarding the mounting of the array itself, special attention has to be attributed to achieve a low smile in between the elements. All these challenges require a high level of technical expertise with diode laser bars, and may discourage in the long run the development of commercial products based on coherently combined arrays of amplifiers. Individual semiconductor amplifiers may be considered as an alternative to arrays.

2.4.2 Coherent combining of individual amplifiers

In contrast to Section 2.4.1, we describe here the coherent combination of individual tapered waveguide amplifiers, which are mounted on separate, individually cooled, heatsinks. While the general approach of the coherent beam combining remains similar to configurations with arrays, there are some differences in the characteristics of the devices. For example, the heat management of each device is optimised leading to a higher extracted power. The coupling and collimation of each device become easier and can be optimised individually. If one device is behaving badly, for example, due to damages to the facet, one can simply replace this device without any influence on the other amplifiers in the interferometer. Beam splitting and combining can be done with standard beam splitters. However, the relative phase fluctuations in between the parallel amplifiers are higher than for amplifiers on one bar, as the temperature fluctuations in each device are different. Furthermore, this approach seems not to be scalable to a very high number of devices with reasonable footprint and complexity of the setup. All these points considered, one can say that there may be a potential for simple architectures based on a limited number coherently combined high power tapered amplifiers for applications where an increase of the available power is needed.

2.4.2.1 Coherent beam combining of three high power tapered amplifiers

A first demonstration of coherent beam combining with three individual high-power tapered amplifiers is shown in Figure 2.17. This was a laboratory scale experiment, with typical arm lengths of about 0.5 m [47,51]. The seed source was a DFB laser ($\lambda = 976$ nm). We used a double stage Faraday isolator with more than -50 dB of isolation. Standard nonpolarising 50:50 beam splitters were used for splitting and recombining the beams. The design of the amplifiers was identical to the DBR tapered lasers described in reference [49] but with the DBR grating replaced by a longer ridge waveguide. Each section of the amplifier was driven independently, allowing decoupled control of the phase (via the ridge section current I_{RW}) and the power (via the tapered section current I_{TP}). The maximum extracted optical power from a single amplifier reached 6.5 W at $I_{TP} = 10$ A and $I_{RW} \geq 300$ mA for an injected seed power ≥ 10 mW [51].

Phase noise of high-power tapered amplifiers

Both mechanical and thermal fluctuations of the free-space interferometer, as well as temperature fluctuations inside the amplifier devices, contribute to the overall phase fluctuations between the parallel arms. We compared the phase fluctuations of a *passive arm* (the amplifier was removed: free space propagation only) and of an *active arm* with an amplifier, in our noise-protected laboratory conditions. The phase fluctuations were measured with two fast photodiodes to obtain simultaneously in-phase and quadrature signals. The measured phase drift (see Figure 2.18(a)) was only slightly stronger in the active arm, with a peak-to-peak value of 0.4 rad over 40 s. The corresponding power spectral densities (PSD) $S_\varphi(f)$ are given in

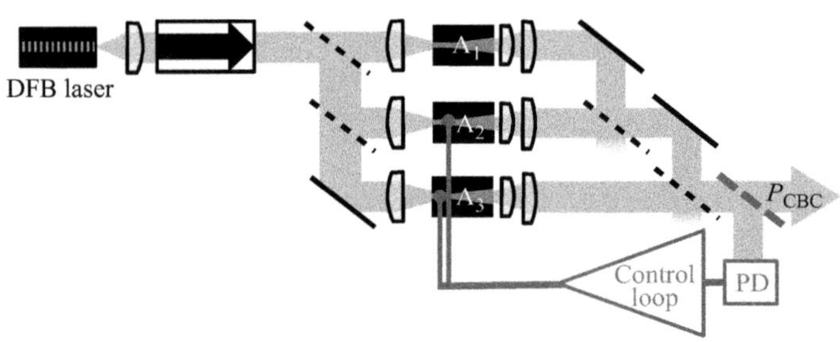

Figure 2.17 Experimental setup of the coherent beam combining interferometer with three individual amplifiers A_i; PD: photodiode

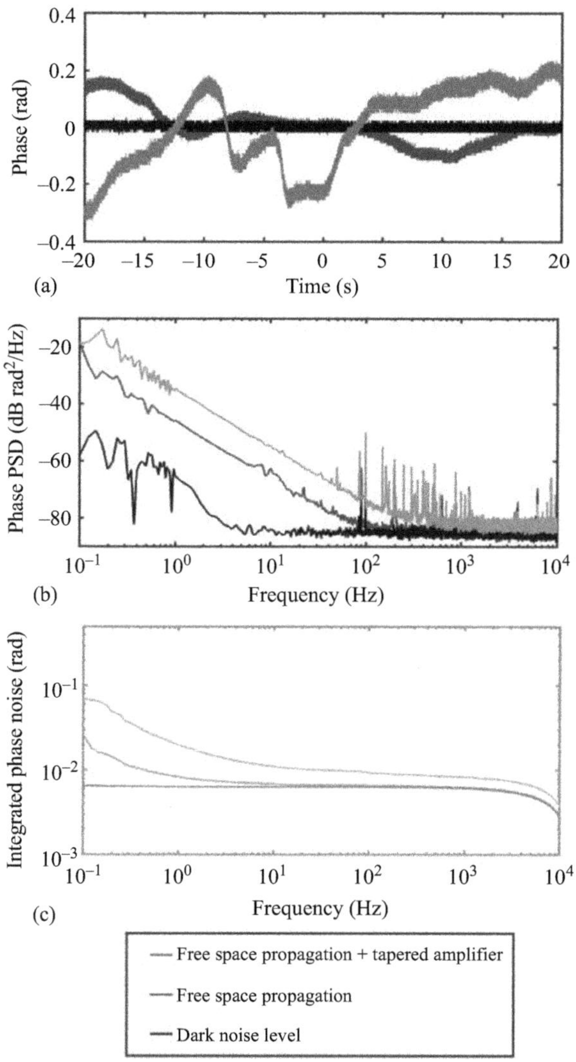

Figure 2.18 (a) Measured relative phase fluctuations in the interferometer, (b) corresponding phase noise spectral density and (c) corresponding integrated phase noise. The current into the tapered amplifier were I_{RW} = 400 mA and I_{TP} = 10 A. The seed power was P_{in} = 10 mW. A moving average filter was applied to the PSD plot from 1 to 10 kHz for better visibility

Figure 2.18(b). Obviously, the PSDs of the passive and active arms are similar. The noise inherent to the amplifier is deduced from their comparison: a minor increase in low frequency noise is attributed to thermal fluctuations in the amplifier and some peaks from 100 Hz to 1 kHz are related to electrical noise in the laser diode drivers.

The integrated phase noise, defined as

$$S_{\text{int}}(F) = \int_{F}^{+\infty} S_{\varphi}(f)df, \tag{2.7}$$

is shown in Figure 2.18(c). For the active arm of the interferometer, $S_{\text{int}}(0.1\text{Hz})$ was below $\pi/30$, which implied that correction of the amplifier currents every 10 s should be sufficient to maintain combining losses attributed to phase mismatch below 1% [28]. The phase fluctuations of those tapered amplifiers were thus extremely slow and low, and could easily be corrected.

Coherent combining power and efficiency

The stabilised power for coherent superposition of three tapered amplifiers is shown in Figure 2.19(a). The rise time after closing the loop for active phase

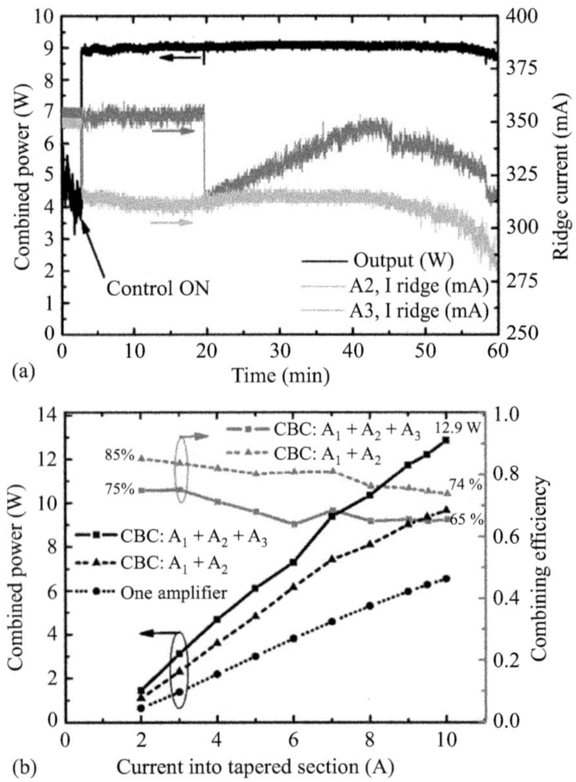

Figure 2.19 (a) Stability of the combined power and actively regulated ridge currents of A_2 and A_3. (b) Combined power at output and total combining efficiency as a function of the currents in the tapered section. The ridge currents were actively controlled in the range of 250 to 400 mA. The astigmatism of the amplifiers was corrected at each operating current (see [47,51])

control was below one second. The standard deviation of the stabilised power was better than 0.6%. Meanwhile, the changes of the ridge currents remain below 35 mA, demonstrating that the phase of the whole setup is quite stable. It corresponds to overall phase fluctuations $\Delta\varphi$ below π over 30 min. The combined power reached 12.9 W at the maximum operating currents (Figure 2.19(b)), corresponding to a combined power per element $P_{CBC}/N = 4.3$ W. This latter value emphasises the advantage of using individual amplifiers as compared to amplifiers in an array, as the extracted power per element at thermal roll-over is higher, increasing the combined power as compared to a similar setup with a mini-array (see Section 2.4.1.2). The overall combining efficiency of the three amplifiers reached 75% at low currents and decreased to 65% at 10 A. Figure 2.19(b) also shows the achieved power for the coherent combining (CBC) of only two tapered amplifiers. In this case, the achievable combining efficiencies were higher as there were less combining steps in a more symmetric setup [52].

The strongest impact on the combining losses is related to spatial mismatches between the beams. We analysed thoroughly the beam profiles at each stage of the setup in order to evaluate the different contributions to the combining efficiency, following (2.1) and (2.2) in Section 2.3.2.2. The amplified beam extracted from each individual device is not purely single-mode in the slow axis, with only $\rho_0 \cong$ 70% of the total power within the fundamental mode at the maximum operating current. Coherent superposition of the beams automatically leads to a clean-up of the spatial profile, thanks to the rejection of the incoherent high-order modes inherent to the combining process. This effect is clearly visible in beam profiles at waist shown in Figure 2.20. Considering the power in the central lobes only, the

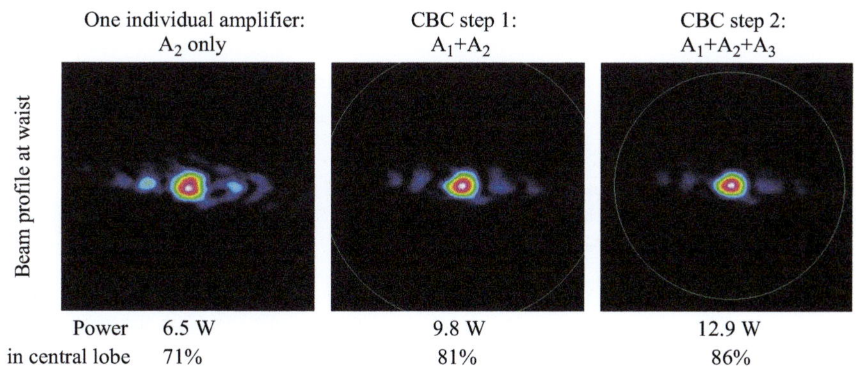

Figure 2.20 Beam profiles at waist for one individual amplifier (A_2), CBC of two amplifiers (A_{1+2}) and the final CBC step of all three amplifiers (A_{1+2+3}). Measurement for $I_{tp} = 10$ A, $T = 20$ °C and actively controlled ridge currents in the range of 250 to 400 mA. The FA is in vertical and the SA is in horizontal direction. The given central lobe power content was calculated by fitting a 2D Gaussian beam profile to the central lobe of the beam and calculating the power content within the fitted profile. Reprinted, with permission, from [47]

combining efficiency of the whole system was evaluated to be higher than 82% [47]. The lower experimental combining efficiency of about 65% for the whole beam was thus clearly related to the power content in the fundamental mode of the individual amplifier beams. The spatial brightness of the combined beam was significantly improved as compared to the individual amplifier beam. The optical power in the fundamental mode was increased by a factor of 2.4, which demonstrated the increased spatial brightness of the combined beam. Furthermore, the use of improved amplifiers with a higher beam quality did increase the combining efficiencies [53].

2.4.2.2 Coherent beam combining module based on commercially available amplifiers

The experiments described in Sections 2.4.1 and 2.4.2.1 used state-of-the-art research devices in customized packaging. Here we describe some additional experiments using commercial tapered amplifiers and off-the-shelf optical elements only. Such amplifiers are now available over a wide coverage of different wavelengths, most commonly from 650 nm to 1,060 nm. Usually, those devices are mounted *p*-side down with only one common electrical contact for the ridge and tapered sections. This makes it difficult to control the phase and the power of the device independently and an external element for phase control is required.

Figure 2.21(a) shows the experimental setup of a four-arm Mach-Zehnder interferometer. The seed source was a narrow linewidth DFB laser ($\lambda = 976$ nm). The 5-mm long amplifiers were mounted on a so-called DHP-heatsink that ensures a good heat evacuation, and reached an optical power above 3 W at 5 A. The amplifiers were mounted two by two on a common thermally stabilised baseplate in order to limit their relative phase fluctuations. Each device was separately

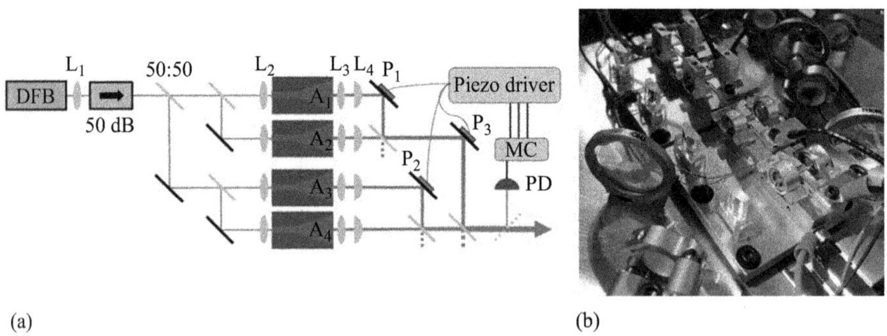

Figure 2.21 (a) Experimental setup using four commercially available tapered amplifiers (DILAS TA-0976-3000) in a compact experimental setup. The dimensions of the interferometer including the four amplifiers, beam splitters, mirrors and glued optics are below 300 mm × 200 mm. (b) Photo of the mounted amplifiers. The whole module consists of two submodules with two amplifiers each

collimated in the fast and slow axes at its maximum operating current. On the output side of the amplifiers, three mirrors were glued on piezoelectric elements (P1, P2, P3), which allow adjusting the phase without impacting the output power (Figure 2.21(b)). The feedback optimisation loop was implemented on a standard miniaturised microcontroller.

The combined power was measured over 160 mn (Figure 2.22(a)). The average power at the output of the interferometer was 9.25 W, with an RMS stability better than 0.6%. The overall combining efficiency of the setup was 66%. The amplitude of the corrections applied to the piezo elements P1 and P2 was below 1 V during the whole measurement, which corresponded to phase fluctuations smaller than 0.5 π between amplifiers on the same submodule (Figure 2.22(b)). As expected, the phase fluctuations between the two submodules, corrected by P3, were slightly higher. Finally, the beam profiles of one amplifier and of the coherently combined module (Figure 2.22(c) and (d)) evidenced the beam clean-up resulting from the

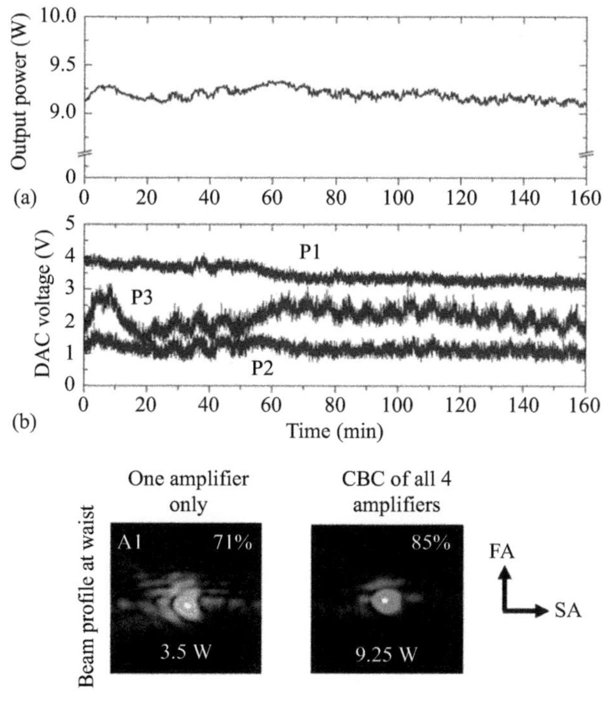

Figure 2.22 (a) Power stability of the coherently combined beam over 160 min. The average power was 9.25 W. (b) Signals sent by the microcontroller to the piezo driver. A change of about 2 V corresponds to a phase shift of π. (c) Beam profile at waist of one individual amplifier. (d) Beam profile at waist of the final coherently combined beam (FA: fast axis; SA: slow axis)

rejection of mismatched high-order modes. The power content in the central lobe was significantly increased and reached more than 86%; the beam quality parameter of the combined beam was $M^2_{2\text{nd moment}} < 1.8 \times 2.7$, which was significantly better than for one individual amplifier of this type.

These results demonstrate how to develop small footprint modules of coherently combined tapered amplifiers based on off-the-shelf elements only. Such modules could significantly improve the performance in all applications, such as spectroscopy or atom cooling, relying on tapered amplifiers.

2.5 Extended-cavity architectures

2.5.1 Principles of operation

As compared to the previous Master Oscillator Power Amplifier architecture that requires an external single-frequency laser source to ensure the mutual coherence of the combined amplifiers, extended-cavity configurations rely solely on optical coupling between the laser elements for that purpose. The active devices share the same laser cavity that is designed in order to enforce their co-operation, and they act thus altogether as a sampled active medium. Depending on the design of the extended-cavity, the coherent operation of the different elements is achieved passively thanks to the self-organisation of the cavity, or some active control of their relative phases is necessary. This approach has been widely investigated in the literature with diode lasers since the 1990s [54]. Most of the time, it is implemented with an array of equally spaced laser elements, and the extended cavity is set on the front side of the emitters (Figure 2.23).

Phase-locking the gain elements with an extended cavity consists in favouring the cavity operation on a coherent mode (spatial and longitudinal) of the whole set of active devices. The modal analysis of phase-locking extended cavities is thoroughly explained in [55,56]. The authors propose a general description of the optical cavity as a linear system, which couples N individual active elements operating on a single spatial mode. There are then N independent sets of intensity

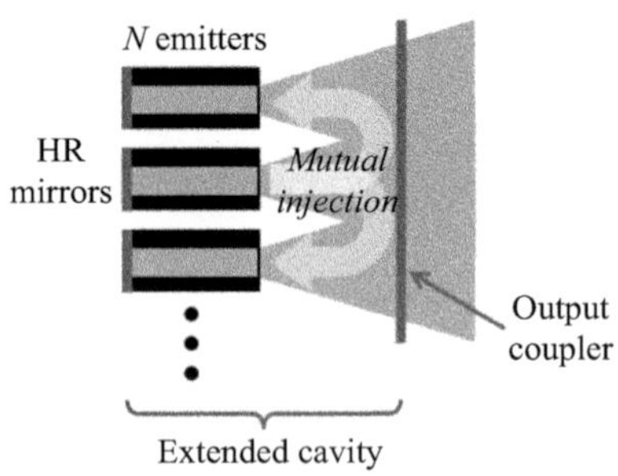

Figure 2.23 General scheme of a common extended cavity for the phase-locking of multiple emitters, built classically between the rear side of the laser emitters (HR mirrors) and the front output coupler

and phase distribution for the elements, the so-called cavity supermodes, which are the eigenmodes of the passive extended-cavity. Phase-locked operation of the N elements is achieved when the coupled-cavity laser array runs on one of those supermodes. This is achieved if this supermode experiences minimum losses on a round-trip inside the cavity – as compared to other supermodes and to the individual incoherent operation of the emitters. Consequently, the requirements on the design of the extended cavity are:

1. to ensure maximum sharing of photons between them,
2. to forbid individual and incoherent operation of the elements,
3. to provide that a resonant supermode does exist in the cavity and
4. to limit competition between multiple supermodes.

An important result of the linear analysis of the extended cavity is that the existing supermode satisfies the resonance condition: its phase-shift upon one round-trip in the cavity is an multiple integer of 2π. The coherent state lases at some specific frequency ν_i matching this phase condition. It might occur that no solution exists within the gain bandwidth of the devices, which prevents the passive self-organisation of the cavity. Then some active control on the path lengths is required to adjust the round-trip phase shift. With laser diodes it is easily achieved by changing the injection current. It appears that the usual distinction between active master-oscillator power-amplifier architectures and passive extended-cavity ones is not accurate, as we shall illustrate in the following.

We describe in this chapter different extended-cavity architectures investigated in the literature, emphasising their specificities and illustrating some of their performance with results obtained at Laboratoire Charles Fabry (Palaiseau, France). We adopt the distinction between resonators based on beam superposition using intracavity combiners in multi-arm interferometric designs and those using parallel coupled cavities [55].

2.5.2 Cavity architectures based on beam superposition

2.5.2.1 General description

We usually describe extended-cavities based on the superposition of laser beams as interferometric resonators, as they follow a multi-arm interferometer design. The combiner elements are either beam splitters or DOE, arranged as for the filled-aperture coherent superposition of phase-locked beams described in Section 2.3.2.2. Such cavity setups are actually very similar to MOPA architectures, except for the seed source (see Section 2.4). The simplest configuration with only two lasers consists in a Michelson-interferometer-like cavity, with a beam splitter as the combiner. This setup can in principle be extended to combine more than two gain elements, either in a binary tree architecture which superposes two by two beams of equal powers, in a linear chain of beam splitters with adequate transmissions to maximise the cavity transmission or with a specifically designed DOE (see Figure 2.24). Interestingly in most practical cases, the extended-cavity is implemented on the front side of the gain elements, and realises simultaneously the phase-locking and the

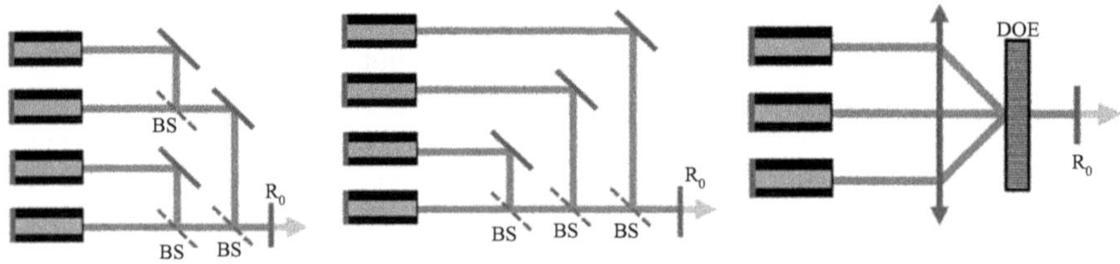

Figure 2.24 (Left) Binary tree extended-cavity, consisting in parallel superposition of two by two beams; (Centre) linear chain extended-cavity, using beam splitters with optimised splitting ratios; (Right) DOE extended-cavity with $N \rightarrow 1$ combiner; BS, beam splitter; R_0, output coupler

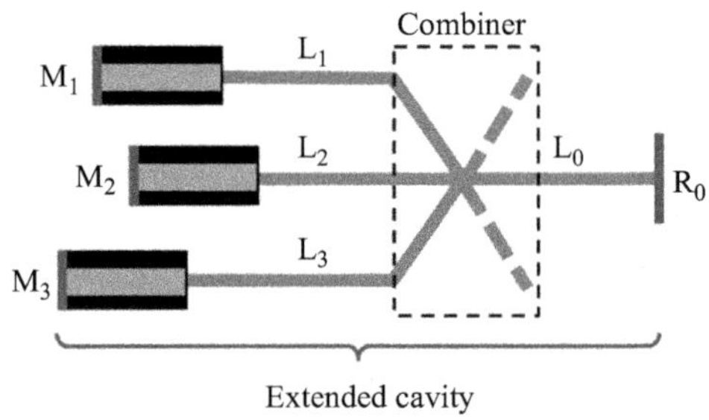

Figure 2.25 General scheme of an interferometric cavity with three arms of different optical path lengths L_1, L_2, L_3; the optical path length of the common arm is L_0; each emitter is closed on its back side by a high-reflection mirror M_i; R_0 is the output coupler of the extended-cavity

coherent superposition of all the emitters. Indeed, the output beam consists of a single diffraction-limited lobe, in contrast to parallel coupled cavities (see Section 2.5.3).

For the sake of simplicity, we shall describe those different cavities with the simplified scheme presented in Figure 2.25: N identical emitters are superposed by means of a $N \rightarrow 1$ interferometric combiner, with splitting/combining efficiency is η_c; each emitter supports only the fundamental transverse mode; the N arms of the cavity, from the back side mirrors to the combiner, may have different optical path lengths L_n; the extended-cavity is closed on the common arm by an output coupler (R_0). Those interferometric cavities do meet all the criteria established earlier (Section 2.5.1). On one hand, mutual injection of all the emitters is ensured by the combiner element, as is obvious following the optical paths in the laser cavity. The laser light reflected by the extended-cavity end mirror (R_0) is equally shared between the different active elements. The coupling ratio between the emitters is equal to the combiner splitting coefficient. Furthermore, the intracavity losses are minimum for a proper phase and intensity distribution (namely the cavity supermode) that results in constructive interferences between the laser beams in the common arm closed by the end mirror. Any other phase or intensity distribution would not be self-consistent: such extended-cavity supports one and only one supermode [56]. The phase-locked operation of the gain elements is favoured by the cavity design, as compared to their incoherent emission. Indeed the effective reflectivity of the cavity for a single emitter under incoherent operation would be in the range of $(\eta_c/N)^2 \times R_0$, corresponding to the two-way splitting of the light on the combiner on a cavity roundtrip. On the contrary, the effective reflectivity for the coherent supermode is $\eta_c^2 \times R_0$.

2.5.2.2 Self-organisation and spectral filtering

In addition to the requirements on the intensity and phase distribution of the cavity supermode, its frequency ν_i should match a resonance of the overall cavity, and simultaneously corresponds to a transmission maximum of the N-arm interferometric cavity. Indeed, the interferometric resonator acts as a selective spectral filter whose free spectral range is related to the path length differences between the N cavity arms. Different approaches are used to describe this effect, either a modal analysis of the resonator and its supermodes [55,56] or the equivalent reflectivity of the N-arm resonator using the circulating field theory [29,57,58]. They are both equivalent and provide a general constraint on the optical path-length differences ΔL_i with respect to the gain bandwidth Δk_g (expressed in wave numbers) for a resonant mode to exist with minimum intracavity losses. Typically, the product $\Delta k_g \times \Delta L$ should be much larger than π to ensure that multiple transmission peaks of the interferometric cavity exist within the gain bandwidth, in order to minimise the cavity roundtrip losses (see Figure 2.26). Depending on the active gain bandwidth and acceptable values of ΔL, the maximum number of emitters that can be *passively* phase-locked can be estimated [55,59,60]. For laser diodes, we shall consider a gain bandwidth $2\Delta k_g$ in the order of 600 cm^{-1}, equivalent to 10 nm at $\lambda = 1$ µm. Then if we expect two hundred transmission peaks within the gain

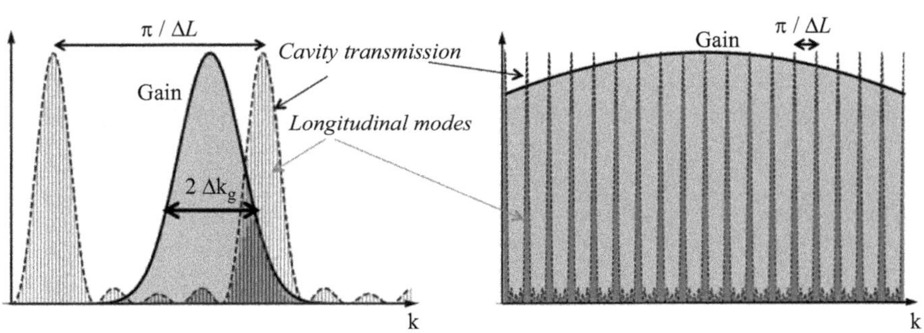

Figure 2.26 Transmission of the interferometric cavity ($N = 5$) as a function of the wavenumber k compared to the gain bandwidth, for two extreme cases $\Delta k_g \times \Delta L \ll \pi$ (left) and $\Delta k_g \times \Delta L \gg \pi$ as described in the text

bandwidth, the path length difference ΔL between adjacent arms should be typically of one centimetre; with $N = 50$, it roughly represents an interferometric cavity length of more than $N \times \Delta L = 500$ mm.

Experimentally, with a few individual laser diodes combined within a Michelson-type cavity as in Figure 2.24 (left), the interferometric arms may have significantly different path lengths: ΔL is of a few centimetres, and the period of the cavity transmission modulation is much smaller than the gain bandwidth as in Figure 2.26 (right). Then the odds are high to have a peak of the N-arm cavity transmission close to the maximum optical gain. If the arm lengths are changed, because of a random mechanical or thermal perturbation of the setup, then the optical supermode automatically shifts to another resonant mode k_i at the maximum of another transmission peak. In this "longitudinal-mode dominated regime" as defined by M. Khajavikhan and J. Leger in [55], the extended-cavity operation self-adapts to maintain minimum roundtrip cavity losses, without any specific control of the emitter phases. It corresponds to a truly passive phase-locking of the emitters, which uses the wide mode diversity of the cavity for its self-organisation. This effect was experimentally evidenced in [29,58], for example, with two laser diodes.

On the other hand, in an extended-cavity configuration with an array of laser diodes and a DOE as the combiner (see Figure 2.24 (right)), the optical path lengths of the interferometer arms are essentially comparable. Then $\Delta k_g \times \Delta L \ll \pi$, and the transmission peaks are distant from much more than the gain bandwidth, as in Figure 2.26 (left). No resonant supermode might get enough gain in the extended cavity to compensate for the transmission losses. The coherent operation of the laser array in the extended cavity is thus highly unlikely, and even so, the laser cavity cannot self-adjust to perturbations of the optical paths by longitudinal mode hops. Nevertheless, this description does not consider the possibility to finely tune the position of the transmission peaks by slightly adjusting the cavity arm lengths and optimise the phases – similarly to a Vernier effect. This is of course straightforward with laser diodes by changing their driving currents, with some active control to maximise the intracavity power likewise MOPA architectures. As the selected supermode undergoes minimum roundtrip losses, the lasers are phase-locked, but such an extended-cavity is not capable of self-organisation and its operation might be unstable. Nonetheless, the active control of the relative phases allows to overcome the limit in the number N of elements that could be mutually coherent and phase-locked in an extended-cavity architecture.

2.5.2.3 Multi-arm interferometric cavities: experimental results

A few experimental demonstrations of the two-arm Michelson resonator are reported in the literature with laser diodes. The first published demonstration by Sabourdy *et al.* [29] used fibre-coupled single-mode laser diodes operating at 975 nm. The beam splitter was a 50/50 fibre coupler, and the output coupler was one of the cleaved-fibre port with a 4% reflectivity. The combining efficiency reached 85%. In order to combine more than two active elements, the Michelson-type interferometer is commonly extended to a tree or linear chain configuration with fibres and solid-state crystals. But it requires individual devices, and is thus not suitable for arrays of laser diodes.

Actually most interferometric cavities reported in the literature with laser diodes use a phase grating (DOE) as the intracavity splitter/combiner (see Section 2.3.2.2). The extended-cavity is based on a 4 f optical arrangement with two consecutive Fourier conjugations: the DOE is positioned in the Fourier plane of a collimation lens where the beams overlapped in the slow axis, and a slit selects the on-axis diffracted order in the Fourier plane of another lens, corresponding to the geometrical image of the laser array (see Figure 2.27). A mirror closes the laser cavity. The focal length of the collimation lens is chosen to match the emitter beams with the DOE orders, so that $f_1 = p\Lambda/\lambda$ with p and Λ the array and the DOE pitches respectively, and λ the laser wavelength. As Λ/λ is typically above 100 and the array pitch p is in the range 0.1 to 2 mm with high power laser arrays, the cavity is usually quite long. Furthermore, the Fourier lens should be corrected from field geometrical aberrations in order to maintain a good overlap of the N beams onto the DOE in its focal plane. The roundtrip intracavity losses are minimised when the active elements are mutually coherent with the proper phase relationship corresponding to the maximum diffraction efficiency on the phase grating. Nonetheless, as explained above in Section 2.5.2.2, this extended-cavity setup precisely corresponds to the situation where the optical path lengths of the multi-arm interferometric cavity, between the emitters and the DOE, are equal ($\Delta L \simeq 0$). Consequently, the intracavity losses might not be at the lowest theoretical value, unless the optical path lengths are adjusted. The coherent operation of the laser array might not be favoured as compared to their incoherent emission.

A first demonstration of this architecture was described with $N = 6$ individual laser diodes and a binary phase grating in 1986 [31,32]. The experimental combining efficiency reached 68%, below the theoretical 81% efficiency. Nonetheless, it seems that there was no adjustment of the currents in the individual devices to optimise the path length differences. More recently the same architecture was implemented with an array of $N = 21$ high-power single-mode laser diodes (SCOWA design, as in Figure 2.12) and a continuous phase grating [61]. A diffraction grating at

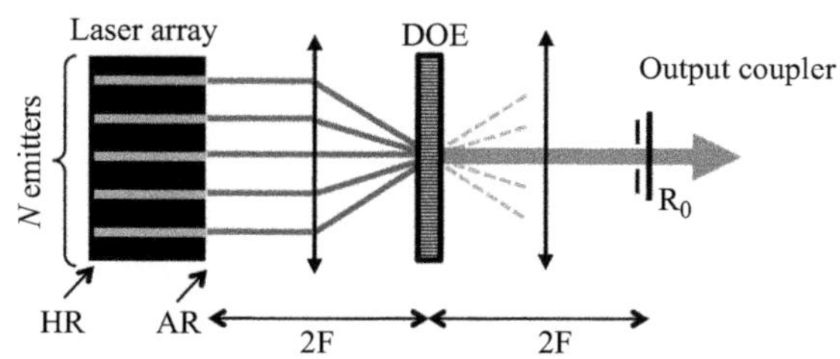

Figure 2.27 Extended-cavity setup with a diffractive optical element as the splitter/combiner element; HR, AR: high-reflection and anti-reflection coatings, respectively; F is the focal length of the Fourier lenses; R_0 is the output coupler reflectivity

Littrow incidence was used as the end mirror to stabilise the laser line, and the output beam was extracted on an intracavity beam splitter (Figure 2.28). Active control of the emitter currents was required in order to maximise the cavity transmission and observe a coherent operation of the laser array in the extended-cavity, with a standard SPGD optimisation servo-loop. The intracavity combining efficiency reached 81%, slightly below the DOE diffraction efficiency of 90%. The setup was not optimised for high power extraction, though, and the useful output power in a single beam was only 1.2 W (Figure 2.28).

A similar experiment was used to coherently combine $N = 5$ individual quantum cascade lasers (QCLs) operating at $\lambda = 4.6$ μm [58,62]. Each laser device was separately collimated, and the beams were orientated to match the DOE orders. The driving currents in every laser were adjusted to equalise the powers onto the DOE. Two configurations were compared, either with nearly equal cavity lengths ($\Delta L \cong 0$) or with one arm longer than the others ($\Delta L \cong 10$cm), in order to demonstrate experimentally the impact of the path-length differences on the combining efficiency in a passive interferometric resonator. As expected, the combining efficiency slightly increased in the long-arm cavity, which was attributed to the higher probability to find a transmission peak of the interferometric cavity within the gain bandwidth. With a $R_0 = 30\%$ output coupler and binary DOE which diffraction efficiency was $\eta_{DOE} = 75\%$, the intracavity combining efficiency reached 70%, close to the theoretical limit. Nonetheless, the extraction efficiency, which measures the combined output power with respect to the power available from the individual emitters, was only 40%.

In all those experimental demonstrations, the intracavity combining efficiency is high, close to the theoretical diffraction efficiency of the combiner element. The availability of optimised phase gratings with a continuous profile rather than a binary one limits the intracavity roundtrip losses, and allows operating the laser

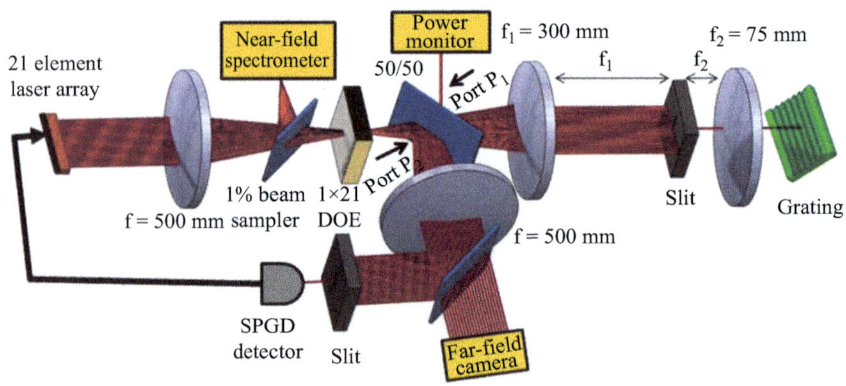

Figure 2.28 Experimental setup for the interferometric extended-cavity with a $N = 21$ SCOWA array and an intracavity phase grating; the diffraction grating closes the cavity and provides wavelength selectivity. Reprinted, with permission, from [61]

diodes at high currents. The output beam consists in a single diffraction-limited lobe, as the extended-cavity ensures simultaneously the phase-locking and coherent superposition of the emitters. Nevertheless, the combined power extracted from the laser cavity is typically below the values reached with MOPA architectures and similar laser devices (see Section 2.4). This is closely related to the need for the extended cavities to ensure a strong coupling between the emitters in order to maintain their mutual coherence. It is indeed crucial to favour the phase-locked operation of the active elements in the extended-cavity. Consequently, the reflectivity R_0 of the output coupler that closes the extended cavity is high. It reduces the total extracted power as compared to the maximum optimal power available from the free-running laser diodes. Experimentally, insufficient end mirror reflectivity R_0 prevents phase-locked operation far above the extended-cavity laser threshold, as the individual elements start to run independently. Obviously, in those superposition cavities, R_0 results from a careful compromise between optical feedback and power extraction.

2.5.2.4 Back-side resonator configurations

In a quest for improvement of the extended cavity scheme and its overall efficiency, one may wonder whether it is mandatory for the extended-cavity to ensure simultaneously the phase-locking of the individual emitters and the coherent combining of the laser beams into one. This is clearly not the case with the MOPA architectures described earlier in Section 2.4, neither with parallel coupled cavities as we shall show in Section 2.5.3. We have proposed and demonstrated an alternative configuration based on a common extended cavity on the back side of the diode lasers for phase locking, while the coherent beam superposition was implemented on the front side similarly to MOPA architectures (Figure 2.29). In this specific arrangement, the extended cavity guarantees a strong optical feedback and a high mutual injection between the emitters, while maintaining maximum power extraction efficiency on the front facet of the laser array.

The rear-side interferometric cavity was first demonstrated in a two-arm Michelson interferometer with individual laser devices [63], and then implemented

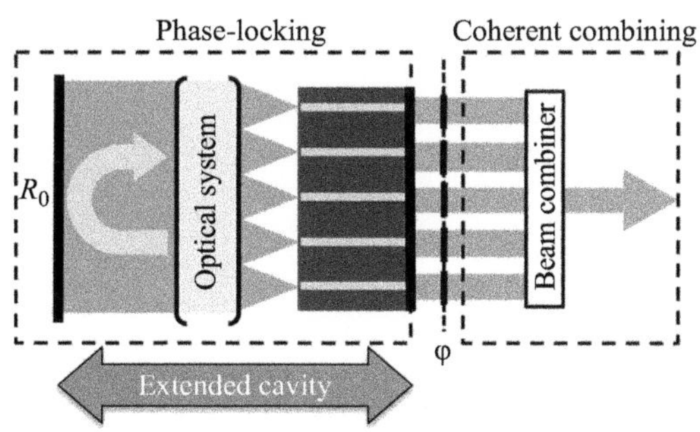

Figure 2.29 General scheme of the extended-cavity configuration with rear-side phase-locking and coherent combining on the front side

in a multi-arm cavity including a DOE with an array of five lasers with a pitch $p = 1$ mm [22]. The latter demonstration was based on the very same array of five tapered-waveguide elements as for the MOPA experiment described in Figure 2.14 (Section 2.4.1.2). The experimental arrangement was very similar too, except that there was no seed laser source here (Figure 2.30). Instead, a high-reflection (HR) mirror closed the cavity in the back focal plane of a 80-mm doublet (D_3, identical to D_1 and D_2). The laser extended-cavity was thus built between this rear mirror and the $R = 0.5\%$-coated front facet of the device, in a 4 f-long optical conjugation involving two successive Fourier transformations similarly to the standard front-side configuration. A narrow slit selected the on-axis beam to favour the phase-locked operation of the laser bar on the array supermode which maximised the on-axis diffracted power, corresponding to the phase state $\{\phi\}_{in} = (0, 0, 0, \pi, 0)$ of the laser array.

It is noteworthy that the longitudinal positioning of the intracavity DOE was critical, since it acted as a narrow-bandwidth (FWHM $\leq$ 2 nm) spectral filter in the cavity: it was adjusted experimentally to match the maximum-gain wavelength of the laser array at high current. On the front-side of the laser bar, the optical setup for the coherent superposition is identical to the MOPA's (Figure 2.30).

In its very principle our extended-cavity laser was a Mach-Zehnder interferometer consisting of five arms of comparable optical path lengths ($\Delta L \simeq 0$). As discussed in the previous Sections 2.5.2.2 and 2.5.2.3, action on the cavity arms was required to finely tune their path lengths and adjust the position of the interferometric cavity transmission peaks. It was straightforward to optimise the currents in the ridge sections in order to minimise the intracavity losses, as for the MOPA architectures, with a sequential hill-climbing algorithm using the intracavity combined power as the merit function [22]. The highest power corresponded to the phase-locked operation of the laser array on the correct phase state $\{\phi\}_{in} = (0, 0, 0, \pi, 0)$. The intracavity combining efficiency reached $\eta_{in} = 78\%$ at low operating current (2 A per emitter) and decreased to 50% at the maximum operating current of 5 A. This was far below the theoretical 98% diffraction

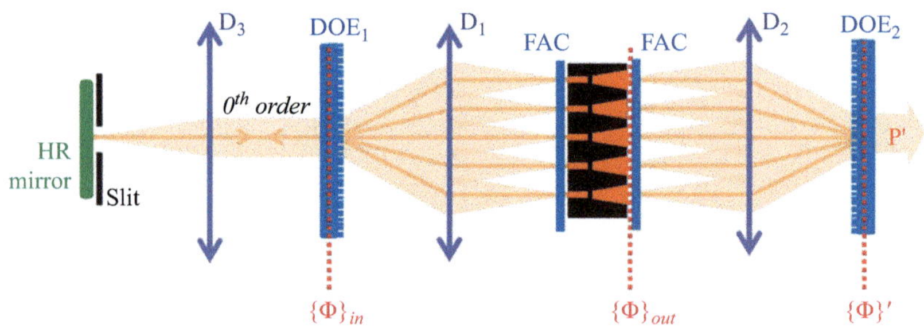

Figure 2.30 Experimental setup used for the coherent beam combining of an array of five tapered-section edge-emitting devices in the rear-side extended-cavity configuration

efficiency of the DOE. It was attributed to imperfect overlap of the beams on the DOE due to the doublet aberrations, residual incoherent operation of the emitters and parasitic feedback into the amplifiers that disturbs the stability of the extended-cavity array. Nonetheless, the phase-locked operation of the array is still favoured as compared to its incoherent operation: the roundtrip intracavity losses remain below $1 - \eta_{in} \cong 50\%$ for the coherent operation, whereas they are of $1 - 1/N \cong 80\%$ under incoherent emission.

On the front side of the laser array, the coherent superposition of the beams was realised in the same manner as in a MOPA architecture, with an additional DOE identical to the intracavity one. Though the phase state $\{\phi\}_{in}$ was known on the rear DOE, the phase distribution $\{\phi\}_{\text{out}}$ among the beams on the front side could be any combination of 0 and π, as the phase of each laser beam was equal to $\mathbf{\Phi}_{\text{in}}$ modulo π because of the half roundtrip of the light in the cavity. This is a typical property of this rear-side extended-cavity architecture. The adequate phase state $(0, 0, 0, \pi, 0)$ for the maximum combining efficiency on the front DOE was obtained by adding a constant current $\Delta I_{\pi} \approx 55$ mA in the ridge sections of the emitters, corresponding to a π phase shift for a single device. This phase correction had no impact on the phase-locked operation of the extended-cavity. Finally, the combining efficiency η_{comb} on the front side remained between 62% and 78% on the whole operating range. We estimated that it was limited by the power content in the fundamental mode and the residual incoherent operation of the emitters. The combined extracted power reached 7.5 W, corresponding to $P_{\text{CBC}}/N = 1.5$ W [22]. To the best of our knowledge, this is the highest power extracted to date in a single beam from an array of diode lasers in a phase-locking extended-cavity (see Figure 2.40). The rear-side extended-cavity architecture did allow circumventing the conventional compromise between optical feedback and extraction efficiency. As compared to the MOPA setup implemented with the same array, though, it is noteworthy that the coherence of the laser beams was reduced at the highest operating currents, resulting in the somewhat reduced combining efficiency.

2.5.3 Parallel coupled cavities

We describe in this section extended-cavity configurations that consist in the parallel operation of coupled cavities in a spatial-filtering resonator. These cavities are based on the free-space diffraction of a set of phase-locked laser beams. The diffraction pattern of the same beams without mutual coherence would simply result in their incoherent superposition – a wide and highly divergent beam. A spatial filter is inserted in the extended-cavity that will induce large losses for the incoherent operation of the emitters, or for any other phase-locked superposition on a coherent supermode than the one selected by the filter. Such architectures are only adapted to arrays of closely spaced emitters, as their beams must overlap inside the resonator.

In the examples commented in this section, the spatial filter may be a physical component inserted in the resonator, but even more conveniently the extended-cavity may be designed for the laser array to be the spatial filter of its own emission after one roundtrip in the cavity: this is the case in both the architectures based on the Talbot effect (Section 2.5.3.1) and the self-Fourier cavities (Section 2.5.3.2). The modal description of such cavities was extensively studied by M. Khajavikhan and J. Leger [55], we shall focus here on the experimental demonstrations.

2.5.3.1 Cavities using near-field spatial filtering: the Talbot cavity

Principles of operation

A well-established extended-cavity configuration is the Talbot resonator. It is based on the near-field self-imaging Talbot effect that was evidenced in 1836 by H. F. Talbot and explained later by Lord Rayleigh [64,65]: the propagation of an infinite coherent 1D periodic field results in the apparition of self-images (amplitude and phase) of the initial field at distances equal to integer multiple of $Z_T = 2p^2/\lambda$. Z_T is the so-called Talbot distance, proportional to the square of the array pitch p, and image-like patterns also appear at fractional Talbot distances Z_T/n (see Figure 2.31). For arrays with finite dimensions, the Talbot effect is also observed, but edge losses appear that degrade the imaging quality [66]. Talbot laser cavities take benefit of this effect to force the laser array into a phase-locked emission. On a roundtrip cavity length $2L_{\text{ext}} = Z_T/2$, the in-phase supermode (corresponding to identical phases for all emitters) is imaged with a $p/2$ lateral shift which can be compensated by a tilt of the cavity mirror (see Figure 2.31). Each self-image in the near-field propagation corresponds to constructive interferences between the adjacent emitters, and is thus the coherent superposition of light from the whole array.

The roundtrip propagation of the optical field in the Talbot cavity induces mutual injection between adjacent emitters. The laser array acts as a spatial filter located at half the Talbot distance, which favours the self-replicating supermode as compared to incoherent individual operation of the emitters or to any other supermode phase distribution [67]. Nonetheless for such phase-locking extended-cavity, the requirements exposed in Section 2.5.1 for the extended-cavity are only partially verified. Indeed, the modal analysis of the Talbot cavity demonstrates that N spatial supermodes – N being the number of emitters in the array – co-exist, and they are

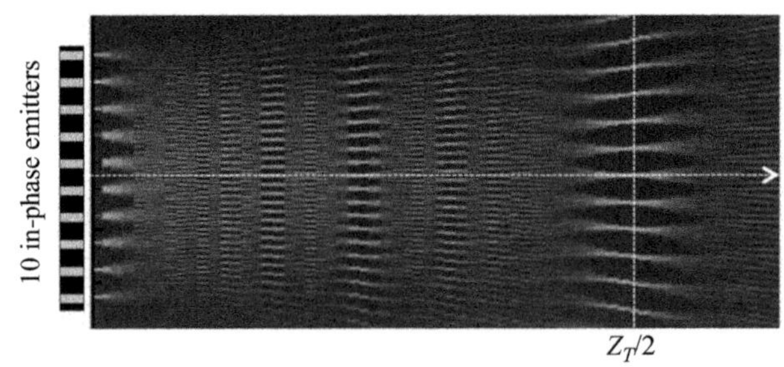

Figure 2.31 Simulation of the near-field propagation of an array of N = 10 in-phase laser beams with Gaussian transverse profile; fill-factor = 4%

poorly discriminated [55,67]. Furthermore, the strength of the coupling between emitters is decreasing with the distance between them. This may prevent the global phase-locked operation of a large number of emitters.

There have been multiple successful demonstrations of Talbot cavities for the phase-locking of arrays of high-power laser diodes [68–71]. It is indeed a simple and compact configuration, which does not require any specific intracavity element. Output powers up to 12 W have been demonstrated [69,70], but the actual amount of phase-locked power is unknown in those experiments where the far-field profiles consist of multiple peaks because of the low fill-factor of the laser arrays. Indeed, the characterisation of the degree of coherence of the phase-locked emission is not straightforward with such far-field profiles. A first indication may be given by the visibility of the interference pattern, defined as $V = (I_{\max} - I_{\min})/(I_{\max} + I_{\min})$. But the relationship between V and η_{coh} is not simple except for $N = 2$, and depends on the number of phase-locked beams (Figure 2.32). Operation on residual high-order supermodes may also degrade the coherence of the laser array with limited impact on V [72]. The width of the interference peaks is another information on the number of lasers that interfere constructively with regard to the diffraction limit λ/Np. Thus comparison of phase-locked powers in different experiments should be made cautiously, and only the coherent combining of the phase-locked beams would provide a robust and quantitative assessment of the degree of coherence of the extended-cavity laser array.

Wavelength-stabilised Talbot cavities

We have proposed to use a VBG as the extended-cavity mirror to stabilise the spectrum and discriminate the phase-locked operation against the free-running emission [71]. The cavity setup is shown in Figure 2.33. We used an array of ten index-guided flared section emitters driven in parallel, with a $p = 100$ µm pitch. The rear ridge facet of the array was HR coated, while the front flared facet was anti-reflection coated ($R < 10^{-3}$) to prevent free-running emission of the individual emitters. The Talbot distance was then $Z_T = 20$ mm ($\lambda \simeq 1$ µm), and the extended-cavity length was 5 mm. The VBG was tilted by $\lambda/2p$ (5 mrad) to compensate for the lateral shift of the in-phase mode on a $Z_T/2$ propagation. With a VBG reflectivity of 40%, the maximum power extracted from the extended-cavity was 1.7 W at the operating current of 4 A. The phase-locked operation was stable, and required no active feedback on the active devices. In this demonstration, the far-field profile exhibited highly contrasted narrow interference peaks on top of a low incoherent pedestal (Figure 2.33). The peak width (FWHM) was below 1.4 times the diffraction limit λ/Np, which indicates that most of the emitters were actually phase-locked and contributed to the coherent emission of the laser array.

Experimentally we observed a significant increase of the incoherent background content with decreasing VBG reflectivities (40%–30%–20%) under the same operating conditions, corresponding to a lower proportion of the extracted power within the in-phase supermode. This further confirms that the optical feedback from the extended-cavity should be strong to enforce the phase-locked

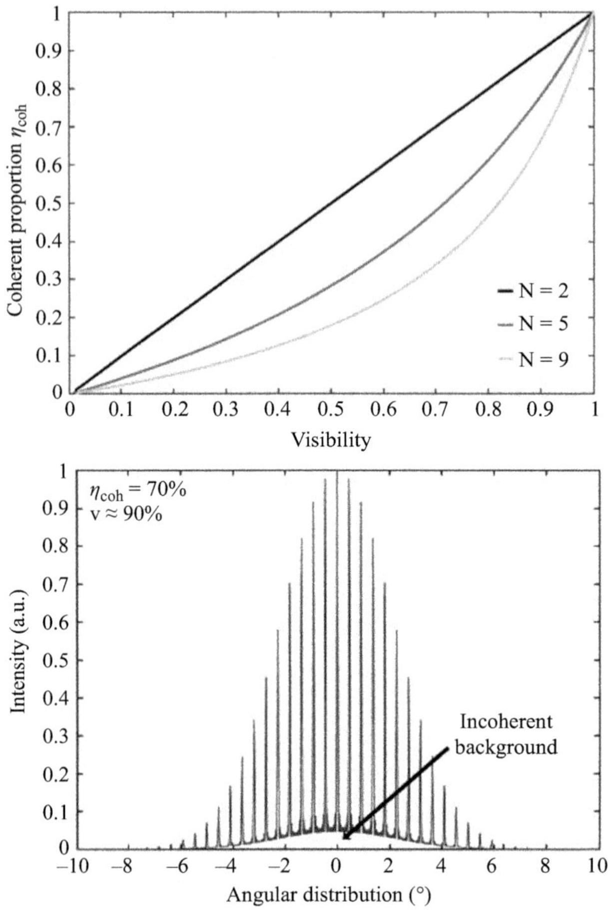

Figure 2.32 (Top) Simulation of the relationship between the proportion of coherent power and the visibility of the interference pattern in the far-field profile, for a single supermode and N phase-locked beams; (Bottom) Theoretical far-field profile of the in-phase mode for $N = 9$ emitters ($p = 120$ μm, fill-factor $= 3\%$), considering 30% incoherent power

operation of the whole array and makes it prevail over the incoherent operation of the individual emitters, at the expense of the electrical-to-optical power conversion efficiency as was also observed in [68]. This is a common feature of extended-cavity laser setups, and was already mentioned in Section 2.5.2.3 for other cavity architectures. Similarly, there is a stronger contribution of the broad pedestal in the far-field profile of the phase-locked array while using a simple dielectric mirror as the output coupler of the Talbot cavity instead of a VBG [73,74], as was also evidenced in [75]. It appears then that the spectral stabilisation of the laser elements

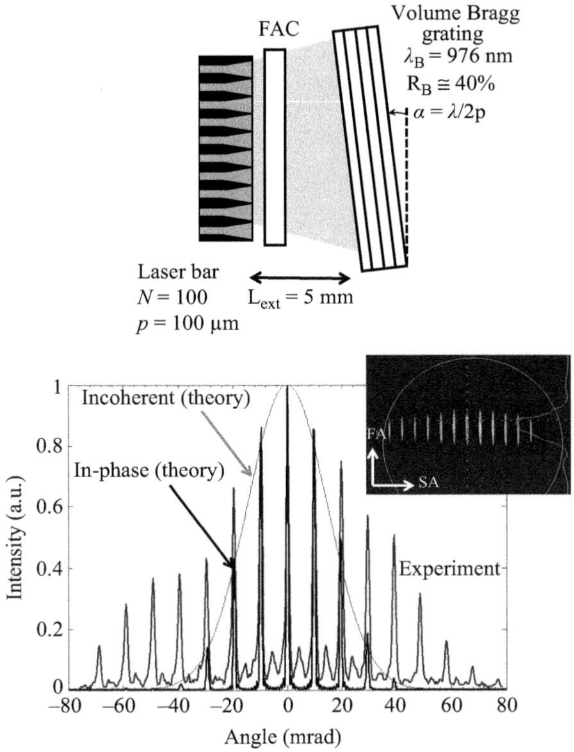

Figure 2.33 (Top) Extended-cavity setup for the phase-locking of an array of $N = 10$ index-guided tapered amplifiers, with a volume Bragg grating as the output coupler; (bottom) experimental far-field profile in the slow axis compared to the theoretical ones (incoherent and phase-locked)

limits their parasitic incoherent operation and thus enhances the coherence of the extended-cavity phase-locked array.

Back side resonators

Similarly, to the discussion in Section 2.5.2.4, it is possible to implement the phase-locking extended-cavity on the back side of the laser array, and circumvent the compromise between strong optical feedback and high power conversion efficiency. With these parallel coupled-cavities, the setup is not much more complicated, as an additional coherent superposition stage is already needed. We implemented a simple 7-mm-long Talbot cavity on the back facet of an array of $N = 9$ single-mode ridge waveguide devices (Figure 2.34). The reflectivity of the back facets was below 1%, and ~3% on the front facets. The array pitch was $p = 120$ µm, and the driven current in each active element could be independently controlled. In this preliminary experiment, a high-reflectivity mirror was used to close the cavity. After proper alignment of the end mirror, the passive phase-locked

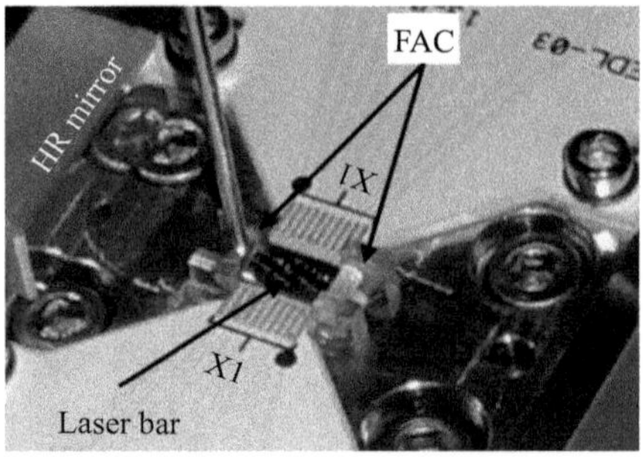

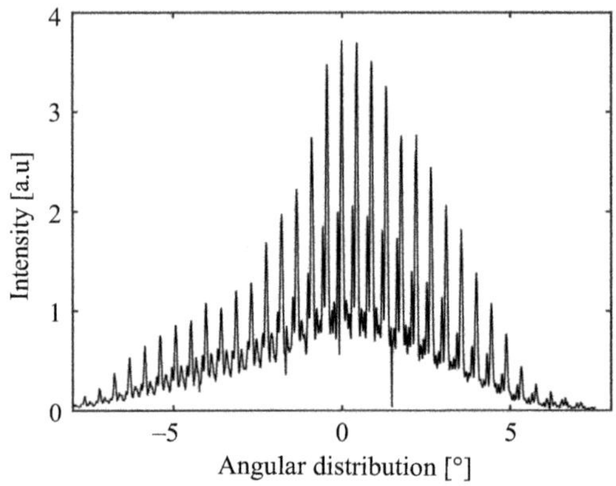

Figure 2.34 (Top) Experimental setup for the phase-locking of an array of N = 9 ridge waveguide devices in a rear-side Talbot scheme; (bottom) Experimental far-field profile in the slow axis at the maximum operating current in each emitter (0.4 A) demonstrating the strong residual incoherent background

operation of the whole array was evidenced by stable and highly contrasted ($68\% \leq V \leq 97\%$) fringes in the far-field pattern on its front side. Fine adjustment of the currents in the individual emitters allowed improving the laser array coherence by optimisation of the different optical path lengths (see Section 2.5.2.2). Furthermore, as mentioned for the rear-side multi-arm cavity architecture described above in Section 2.5.2.4, we could adjust the phase distribution on the front side of the laser array by changing the currents of some of the emitters to apply a π-phase shift without disturbing the phase-locking on the back side. Nonetheless, the extracted power on the front side was limited to 1.45 W, corresponding to 70% of the total power from the same laser array under incoherent operation. From the far-field profile we estimated that the proportion of coherent power was decreasing from 88% at threshold to only 32% at the maximum operating currents. The strong incoherent background is attributed to an insufficient antireflection coating of the rear facet of the laser array (Figure 2.34).

Interestingly, Liu and Braiman have demonstrated that an array of high-power broad area lasers could be phase-locked in a Talbot cavity [70,75]. They have developed a V-shape cavity, which forces simultaneously the transverse single-mode operation of the individual broad area emitters and the in-phase operation of the fundamental modes. The extended cavity is implemented on one off-axis arm on the front side of the laser array while the output power is extracted in the symmetric off-axis direction (Figure 2.35). This folded two-side scheme follows the same principle as our rear-side extended-cavity configuration, and results in high output powers and nearly diffraction-limited far-field profiles with narrow and contrasted fringes [75]. Still, the optical power per element is not higher than what could be achieved from truly single-mode emitters, and the proportion of the output power being actually in a single phase-locked supermode cannot be assessed with certainty.

Coherent superposition of an array of phase-locked emitters

Because of the low fill-factor of the laser arrays used in the experiments described here, the far-field profiles consist of numerous peaks, and the beam profile cannot be directly used for applications. Coherent superposition of the phase-locked beams is then further required by diffraction on a phase grating, as was described in Section 2.3.2.2. The combining of the N beams into one provides also a direct measurement of the coherent power extracted from the laser array. Specific DOEs

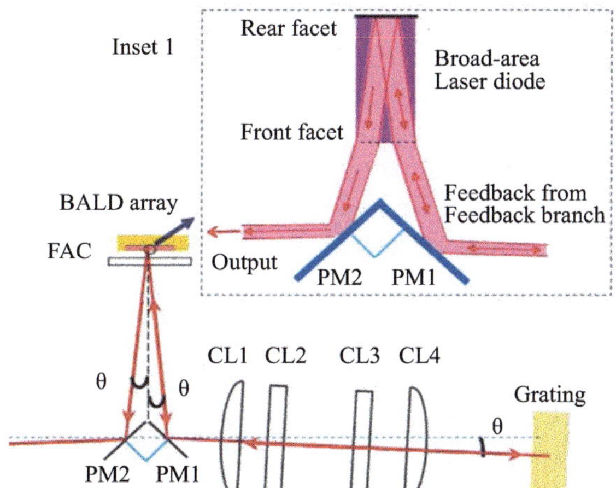

Figure 2.35 Schematic design of the V-shaped external Talbot cavity array with broad area laser diodes (top view – slow axis). FAC: fast-axis collimator. Two prism mirrors (PM1, PM2) form right angle and separate the laser beams into two branches (feedback branch and output branch). Feedback branch consists of cylindrical lenses CL1-CL4 and grating. Inset 1 shows an individual broad area laser in V-shaped external cavity (from [76])

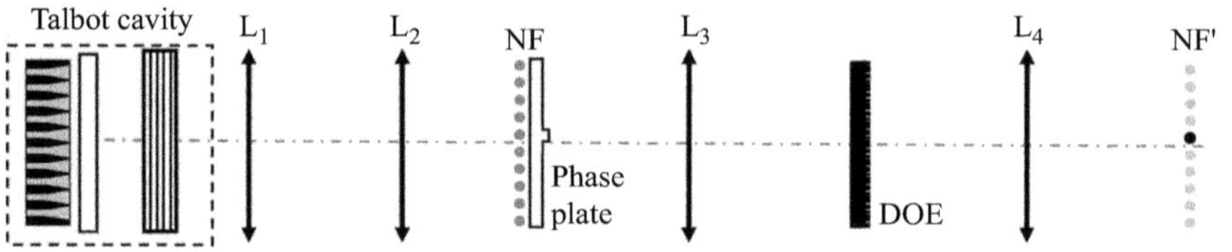

Figure 2.36 Experimental setup for the coherent superposition of the phase-locked laser array with an external binary DOE; L1–L4 are cylindrical lenses; NF, NF′: near-field images. The Talbot cavity was aligned to favour the out-of-phase mode {0,π,0,π,0,π,0,π,0,π} and a π-phase step was added on a central beam to adjust the phase distribution (see [72])

should be used, as the phases of the emitters in the two favoured supermodes (in-phase or out-of-phase) are either 0 or π. It places additional constraints on the design of the phase grating.

The coherent superposition of the ten phase-locked laser beams from the Talbot extended-cavity described above in Figure 2.33 was implemented with a binary phase grating [72]. A phase plate which adds a π phase-shift on one beam was necessary to adjust the phase distribution of the phase-locked emitters to the optimal phases of the grating (Figure 2.36). It required then an additional imaging of the near-field profile in a 4f arrangement to position the phase plate, followed by a Fourier conjugation to put the DOE in the far-field plane. The diffraction efficiency of the phase grating η_{DOE} was 83%. Disappointingly the proportion η_{coh} of power that was efficiently combined was no more than 50% of the power extracted from the phase-locked array, even when the visibility V of the far-field pattern was estimated to be as high as 95%. It further demonstrates that the estimation of the phase-locked power from the far-field profile does not provide an accurate evaluation of the coherence of the laser array.

2.5.3.2 Cavities using far-field angular filtering

We describe in the section a few examples of extended-cavity configurations based on the a priori knowledge of the far-field interference profile of the phase-locked laser array. Such a profile being significantly different from the incoherent one, a spatial filter is introduced in the extended-cavity in order to favour it. The extended-cavity is thus designed to realise a spatial Fourier transformation of the array output in a plane where a filter can be introduced.

A well-known implementation of this principle was demonstrated by C. Chang-Hasnain and her colleagues in an extremely compact cavity with an array of $N = 10$ emitters [77]. As the fill-factor of the array was high (60%), the out-of-phase array supermode consisted of only two lobes. A narrow off-axis mirror was used in the Fourier plane of a gradient-index lens to selectively feedback one lobe while the symmetric one was taken as the output of the extended cavity. This is actually an approach very similar to the V-shaped resonator used to improve the beam quality of broad area lasers already mentioned (see Figure 2.35) [78,79]. Variants of this architecture with an off-axis concave mirror [80] or a VBG used as an angular filter [73] were also proposed (see Figures 2.37 and 2.38). In any case, the fill-factor of the laser array is high for the far-field profile of the out-of-phase supermode to consist of only to two lobes. This increases the thermal load of the array, which is why this approach did not demonstrate very high powers so far. Nonetheless, no additional coherent superposition of the phase-locked beams is necessary, as the power is already concentrated in a single diffraction-limited lobe. It is also noteworthy that the optical feedback is intrinsically high, as the extended-cavity mirror is not positioned on the output beam. Actually, this phase-locking cavity architecture is a folded version of the rear-side configuration described in Figure 2.29.

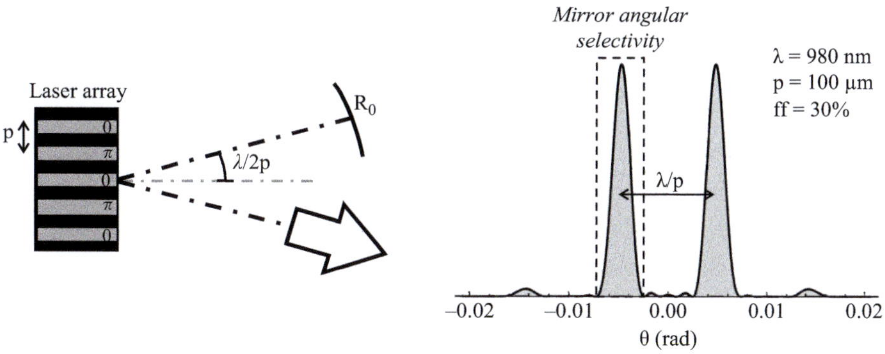

Figure 2.37 (Left) Angular filtering cavity; (Right) theoretical far-field profile of a N = 5 phase-locked laser array with a fill-factor ff = 30% in the out-of-phase mode

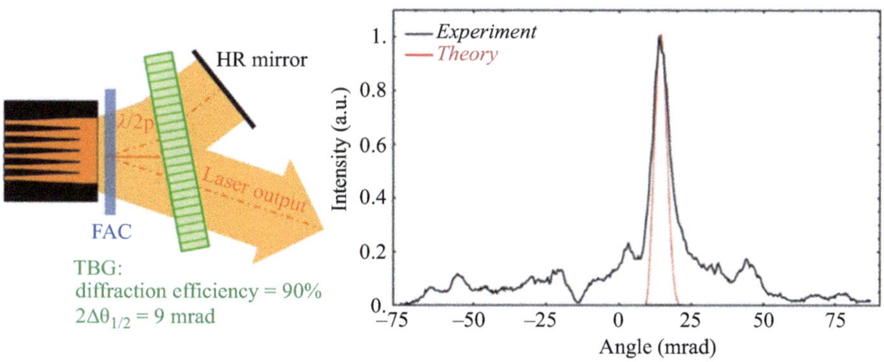

Figure 2.38 (Left) Experimental setup of an angular filtering cavity with a transmission volume Bragg grating (TBG); (Right) far-field profile of the N = 6 phase-locked laser diodes at the output power of 971 mW (from [73])

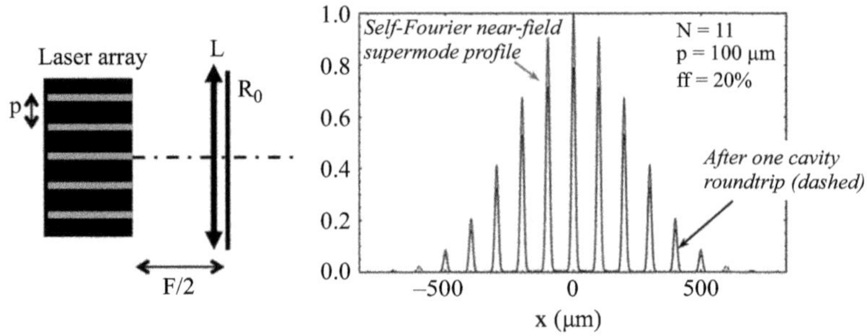

Figure 2.39 (Left) Scheme of a self-Fourier extended-cavity; (Right) simulated near-field profile of the in-phase self-Fourier supermode for N = 11 Gaussian beams

As an example, Figure 2.38 shows the far-field pattern of an array of $N = 6$ narrow tapered lasers, with a pitch $p = 30$ μm and a filling factor close to 100% [73]. The out-of-phase operation of this array would correspond to a far-field profile of two lobes emitted in directions $\theta_1 = \lambda/2p = 16$ mrad, and an angular width $\delta\theta = \lambda/Np = 5$ mrad. The extended-cavity used a transmissive volume Bragg grating, with a diffraction efficiency of 90% at the design wavelength and an angular acceptance of 9 mrad. The far-field profile of the extended-cavity evidences one lobe with a width about 1.4 times the diffraction limit $\delta\theta$, over a broad incoherent pedestal. The maximum output power was 1.3 W ($P_{\mathrm{CBC}}/N = 0.2$ W) for an operating current of 3 A, but we estimated that only 33% was in the narrow lobe.

The self-Fourier cavity is another very specific example of this far-field filtering scheme, in which the array of emitters acts itself as a sampled spatial filter. It requires though that the number of emitters, their pitch and their width are properly chosen with respect to each other, so that the spatial distribution of the beams after one Fourier transformation – corresponding to the far-field profile of the phase-locked array – is equal to the near-field spatial distribution of the original laser array [81]. The focal length of the Fourier lens is then $F = 2p^2/\lambda$. The only demonstration with laser diodes was made with an array of $N = 35$ single-mode emitters [82]. The coherence of the phase-locked array was limited, which was partly attributed to the aberrations of the Fourier lens. This configuration puts stringent constraints on the laser array parameters, and the far-field profile should be further converted into a single mode beam (Figure 2.39).

2.6 Conclusion

Beam combining approaches are a key-technology towards the future development of high-power and cost-effective high-brightness laser systems based on semiconductor lasers. While incoherent and spectral beam combining are already widely used in industrial laser diode systems, coherent beam combining of diode lasers remains an exploratory field.

In this chapter we described the general principles of coherent beam combining of diode lasers, for coherent superposition and phase locking. We described and commented a few experimental demonstrations, mostly with on high-brightness laser diodes emitters. The purpose of the phase-locking of multiple emitters is forcing their mutual coherent operation, while coherent combining consists in the superposition of the phase-locked beams into one single diffraction-limited beam, which can be directly used for applications. As the fill-factor of diode laser arrays is usually low, the favoured coherent combining architecture is based on the superposition of the emitters in the near-field, with beam splitters or phase gratings (see Section 2.3.2). Regarding phase-locking, we compared two architectures that differ in the way the mutual coherence of the laser emitters is achieved: either amplification of a single-frequency seed source in parallel amplifiers (MOPA approach – see Section 2.4), or laser operation of multiple active elements in a common extended-cavity that induces a mutual coupling (see Section 2.5). The latter approach has led to a wide variety of cavity configurations, with single emitters and arrays. Nonetheless, the behaviour of the phase-locking extended-cavities becomes unstable at currents far above threshold, and the phase-locked emission is then always deteriorated by incoherent emission of the laser emitters or the competition with other cavity supermodes. It clearly sets a limit in the coherent output power obtained with those configurations. Figure 2.40 compares the achieved output power and the used number of emitters in the different demonstrations discussed in this chapter, in both investigated architectures. Diverse emitter designs, cavity configurations and largely different numbers of emitters, from only 2 to 900

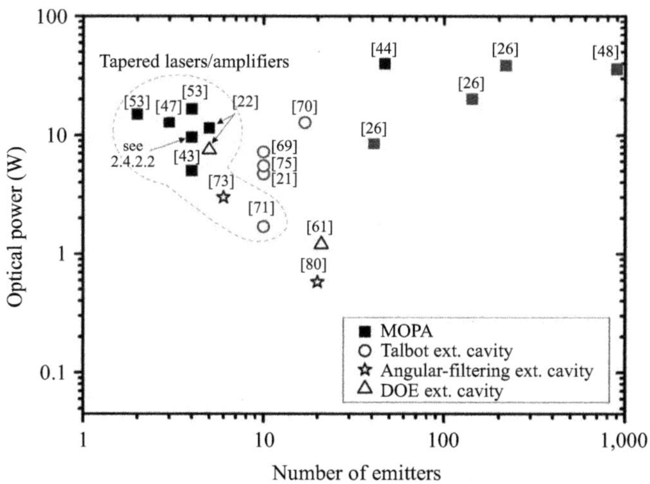

Figure 2.40 Optical power versus number of emitters for different phase-locked laser diode systems; the output power is not combined in a single mode in [21,69–71,75] and in [26,48], respectively, for extended-cavity and MOPA configurations

emitters, have been studied so far. Obviously, the highest powers have been achieved in a parallel MOPA configuration, and it is noteworthy that the highest powers mentioned with extended-cavities do not always correspond to a single combined beam, making the comparison somewhat meaningless. When similar devices were used in both configurations, as with slab-coupled optical waveguide emitters or tapered waveguides, the extracted powers and phase stability were higher in the MOPA architecture. The need for an active control of the relative phase experienced in each amplifier in the MOPA configuration should not be seen as a significant drawback. Indeed we have explained that it is also required in extended-cavity configurations with equal path lengths (see Section 2.5.2.2). This is straight-forward with laser diodes to control their driving current; additionally, the very low and slow phase fluctuations make the phase stabilisation easily manageable.

The highest power achieved by coherent beam combining is 40 W from 47 high-brightness amplifiers, corresponding to less than 1 W per emitter [44]. Achieving a similar or even higher power with few emitters only would greatly simplify the optical arrangement and the phase stabilisation, and would be even more promising for future applications. We believe that our work on coherent beam combining of a limited number of high-brightness tapered amplifiers discussed in Section 2.4.2 shows the highest potential in this respect. The output power of such systems could even be further increased by dense spectral beam combining of several similar CBC modules operating at slightly different wavelengths, offering a potential path to future CBC-based direct diode high-brightness material processing systems. Besides industrial applications, coherent beam combining is also a promising approach to improve the laser sources used in the field of biomedical applications, spectroscopy and remote sensing.

References

[1] Kanskar, M., Keeney, S., and Martinsen, R. 'Laser diodes: the power of brilliance – the past and future of high-power semiconductor lasers', *LaserFocusWorld*. 2018; pp. 1–8. Available from: https://www.laserfocusworld.com/lasers-sources/article/16555251/laser-diodes-the-power-of-brilliance-the-past-and-future-of-highpower-semiconductor-lasers, accessed on July 18, 2018

[2] Strohmaier, S., Tillkorn, C., Olschowsky, P., and Hostetler, J. 'High-power, high-brightness direct-diode lasers', *Optics and Photonics News*. 2010; **21**(4): p. 24

[3] Britten, S., Poprawe, R., Weisheit, A., *et al*. 'Brightness and average power as driver for advancements in diode lasers and their applications', *Proc. SPIE* **9348**; 2015: p. 93480B

[4] Patil, D.S. *Semiconductor Laser Diodes Technology and Applications*, London: IntechOpen; 2012

[5] Fan, T.Y. 'Laser beam combining for high-power, high-radiance sources', *IEEE Journal of Selected Topics in Quantum Electronics*. 2005; **11**(3): pp. 567–577

[6] Yu, C.X., and Fan, T.Y. 'Beam combining', in *High Power Laser Handbook'*, New York: McGraw Hill; 2011
[7] Brauch, U., Loosen, P., and Opower, H. 'High-power diode lasers for direct applications' in Diehl, R. (Ed.) *High-Power Diode Lasers*, Berlin/Heidelberg: Springer; 2000: pp. 303–368
[8] Treusch, H., Du, K., and Baumann, M. 'Fiber-coupling technique for high-power diode laser arrays', *Proc. SPIE* **3267**, 1998: pp. 98–106
[9] Kruschke, B., Fritsche, H., Kern, H., *et al.* 'Beam combining techniques for high-power high-brightness diode lasers', *Proc. SPIE, Components and Packaging for Laser Systems*; 2015: p. 934614
[10] Sanchez-Rubio, A., Fan, T.Y., Augst, S.J., *et al.* 'Wavelength beam combining for power and brightness scaling of laser systems', *Lincoln Laboratory Journal.* 2014; **20**(2): pp. 52–66
[11] Daneu, V., Sanchez, A., Fan, T.Y., Choi, H.K., Turner, G.W., and Cook, C.C. 'Spectral beam combining of a broad-stripe diode laser array in an external cavity', *Optics Letters.* 2000; **25**(6): pp. 405–407
[12] Huang, R.K., Chann, B., Missaggia, L.J., *et al.* 'High-brightness wavelength beam combined semiconductor laser diode arrays', *IEEE Photonics Technology Letters*. 2007; **19**(4): pp. 209–211
[13] Hengesbach, S., Klein, S., Traub, M., and Witte, U. 'Simultaneous frequency stabilization, wavelength multiplexing and improvement of beam quality using a self-optimizing external cavity diode laser', *Optics Letters.* 2016; **41**(3): pp. 595–598
[14] Sevian, A., Andrusyak, O., Ciapurin, I., Smirnov, V., Venus, G., and Glebov, L. 'Efficient power scaling of laser radiation by spectral beam combining', *Optics Letters.* 2008; **33**(4): pp. 384–386
[15] Chann, B., Kaiman, M., Tayebati, P., *et al.* 'Direct diode lasers with comparable beam quality to fiber, CO2 , and solid state lasers', *Proc. SPIE* **8241**; 2012: p. 824102
[16] Brignon, A. (Ed.). *Coherent Laser Beam Combining*, Hoboken: John Wiley & Sons; 2013
[17] Huang, R.K., Donnelly, J.P., Missaggia, L.J., *et al.* 'High-power nearly diffraction-limited AlGaAs-InGaAs semiconductor slab-coupled optical waveguide laser', *IEEE Photonics Technology Letters*. 2003; **15**(7): pp. 900–902
[18] Crump, P., Blume, G., Paschke, K., *et al.* '20W continuous wave reliable operation of 980nm broad-area single emitter diode lasers with an aperture of 96 μm', *Proc. SPIE* **7198**; 2009: p 719814
[19] Wang, X., Erbert, G., Wenzel, H., *et al.* '17-W near-diffraction-limited 970-nm output from a tapered semiconductor optical amplifier', *IEEE Photonics Technology Letters*. 2013; **25**(2): pp. 115–118
[20] Kelemen, M.T., Weber, J., Mikulla, M., and Weimann, G. 'High-power high-brightness tapered diode lasers and amplifiers', *Proc. SPIE* **5723**; 2005: p. 29
[21] Huang, R.K., Chann, B., Missaggia, L.J., and Al, E. 'Coherent Combination of Slab-Coupled Optical Waveguide Lasers', *Proc. SPIE* **7230**; 2009: p. 72301G

[22] Schimmel, G., Janicot, S., Hanna, M., *et al.* 'Coherent beam combining architectures for high power tapered laser arrays', *Proc. SPIE* **10086**; 2017: p. 100860–22
[23] Botez, D. 'Monolithic phase-locked semiconductor laser arrays', in Botez, D., Scifres, D.R. (Eds.), *Diode Laser Arrays*, Cambridge University Press, Cambridge, UK; 1994, pp. 1–71
[24] Bourderionnet, J., Bellanger, C., Primot, J., and Brignon, A. 'Collective coherent phase combining of 64 fibers', *Optics Express.* 2011; **19**(18): pp. 3931–3933
[25] Yu, C.X., Kansky, J.E., Shaw, S.E.J., Murphy, D.V., and Higgs, C. 'Coherent beam combining of large number of PM fibres in 2-D fibre array', *Electronics Letters.* 2006; **42**(18): p. 1
[26] Redmond, S.M., Creedon, K.J., Kansky, J.E., *et al.* 'Active coherent beam combining of diode lasers', *Optics Letters.* 2011; **36**(6): pp. 999–1001
[27] Huang, W.R., Montoya, J., Kansky, J.E., Redmond, S.M., Turner, G.W., and Sanchez-Rubio, A. 'High speed, high power one-dimensional beam steering from a 6-element optical phased array', *Optics Express.* 2012; **20**(16): p. 17311
[28] Goodno, G.D., Shih, C.-C., and Rothenberg, J.E. 'Perturbative analysis of coherent combining efficiency with mismatched lasers', *Optics Express.* 2010; **18**(24): pp. 25403–25414
[29] Sabourdy, D., Desfarges-Berthelemot, A., Kermène, V., and Barthélémy, A. '975-nm single-mode laser source: External coherent combining of two pigtailed laser diodes', *IEEE Journal on Selected Topics in Quantum Electronics*. 2004; **10**(5): pp. 1033–1038
[30] Bloom, G., Larat, C., Lallier, E., Carras, M., and Marcadet, X. 'Coherent combining of two quantum-cascade lasers in a Michelson cavity', *Optics Letters.* 2010; **35**(11): pp. 1917–1919
[31] Leger, J., Swanson, G., and Veldkamp, W. 'Coherent laser addition using binary phase gratings', *Applied Optics.* 1987; **26**(20): pp. 4391–4399
[32] Leger, J.R., Swanson, G.J., and Veldkamp, W.B. 'Coherent beam addition of GaAlAs lasers by binary phase gratings', *Applied Physics Letters.* 1986; **48**(14): pp. 888–890
[33] Romero, L.A., and Dickey, F.M. 'Theory of optimal beam splitting by phase gratings. I. One-dimensional gratings', *Journal of the Optical Society of America A* 2007; **24**(8): pp. 2280–2295
[34] Miklyaev, Y.V., Imgrunt, W., Pavelyev, V.S., *et al.* 'Novel continuously shaped diffractive optical elements enable high efficiency beam shaping', *Proc. SPIE* **7640**; 2010: p. 764024
[35] Bloom, G., Lallier, E., Lee-Bouhours, M.-S.L., Huignard, J., Larat, C., and Loiseaux, B. 'Design and optimization of a high-efficiency array generator in the mid-IR with binary subwavelength grooves', *Applied Optics* 2011; **50**(5): p. 701
[36] Sidick, E., Knoesen, A., and Mait, J.N. 'Design and rigorous analysis of high-efficiency array generators' *Applied Optics* 1993; **32**(14): pp. 2599–2605

[37] Zhou, C., and Liu, L. 'Numerical study of Dammann array illuminators', *Applied Optics.* 1995; **34**(26): pp. 5961–5969

[38] Walker, S.J., and Jahns, J. 'Array generation with multilevel phase gratings', *Journal of the Optical Society of America A.* 1990; **7**(8): pp. 1509–1513

[39] Spießberger, S., Schiemangk, M., Sahm, A., *et al.* 'Micro-integrated 1 Watt semiconductor laser system with a linewidth of 3.6 kHz', *Opics. Express.* 2011; **19**(8): pp. 7077–7083

[40] Blume, G., Pohl, J., Feise, D., *et al.* 'Single-mode master-oscillator power amplifier at 647 nm with more than 500 mW output power', *Optics Letters.* 2015; **40**(8): pp. 1757–1759

[41] Wandt, D., Laschek, M., v. Alvensleben, F., Tünnermann, A., and Welling, H. 'Continuously tunable 0.5 W single-frequency diode laser source', *Optics Communications.* 1998; **148**(4): pp. 261–264

[42] Wilson, A.C., Sharpe, J.C., McKenzie, C.R., Manson, P.J., and Warrington, D.M. 'Narrow-linewidth master-oscillator power amplifier based on a semiconductor tapered amplifier', *Applied Optics.* 1998; **37**(21): pp. 4871–4875

[43] Osinski, J.S., Mehuys, D., Welch, D.F., *et al.* 'Phased array of high power, coherent, monolithic flared amplifier master oscillator power amplifiers', *Applied Physics Letters.* 1995; **66**(5): pp. 556–558

[44] Creedon, K.J., Redmond, S.M., Smith, G.M., *et al.* 'High efficiency coherent beam combining of semiconductor optical amplifiers', *Optics Letters.* 2012; **37**(23): pp. 5006–5008

[45] Müller, M., Klenke, A., Steinkopff, A., Stark, H., Tünnermann, A., and Limpert, J. '3.5 kW coherently combined ultrafast fiber laser', *Optics Letters.* 2018; **43**(24): p. 6037

[46] Augst, S.J., Ranka, J.K., Fan, T.Y., and Sanchez, A. 'Beam combining of ytterbium fiber amplifiers', *Journal of Optical Society of America B.* 2007; **24**(8): p. 1707

[47] Albrodt, P., Jamal, M.T., Hansen, A.K., *et al.* 'Coherent combining of high brightness tapered amplifiers for efficient non-linear conversion', *Optics Express.* 2019; **27**(2): pp. 928–937

[48] Levy, J.L., and Roh, K. 'Coherent array of 900 semiconductor laser amplifiers', *Proc. SPIE* **2382**; 1998: pp. 58–70

[49] Fiebig, C., Blume, G., Kaspari, C., *et al.* '12 W high-brightness single-frequency DBR tapered diode laser', *Electronics Letters.* 2008; **44**(21): pp. 1253–1255

[50] Crump, P., Decker, J., Winterfeldt, M., *et al.*: 'Development of high-power diode lasers with beam parameter product below 2 mm x mrad within the BRIDLE project', *Proc. SPIE* **9348**; 2015; p. 93480D

[51] Albrodt, P., Hanna, M., Moron, F., *et al.* 'Coherent combining of high brightness tapered lasers in master oscillator power amplifier configuration', *Proc. SPIE* **1051**; 2018: p. 105140T

[52] Fan, T.Y. 'The effect of amplitude (power) variations on beam combining efficiency for phased arrays', *IEEE Journal of Selected Topics in Quantum Electronics.* 2009; **15**(2): pp. 2008–2010

[53] Albrodt, P., Jamal, M.T., Hansen, A.K., *et al.* 'Recent progress in brightness scaling by coherent beam combining of tapered amplifiers for efficient high power frequency doubling', *Proc. SPIE* **1090**; 2019: p. 109000O
[54] Leger, J.R. 'External methods of phase-locking and coherent beam addition of diode lasers', in *Surface Emitting Semiconductor Lasers and Arrays*, Cambridge: Academic Press, 1993. pp. 379–443
[55] Khajavikhan, M., and Leger, J.R. 'Modal theory of coupled resonators for external cavity beam combining', in Brignon, A. (Ed.) *Coherent Laser Beam Combining*, Hoboken: John Wiley&Sons; 2013
[56] Khajavikhan, M., and Leger, J.R. 'Modal analysis of path length sensitivity in superposition architectures for coherent laser beam combining', *IEEE Journal on Selected Toptics in Quantum Electronics*. 2009; **15**(2): pp. 281–290
[57] Pedersen, C., and Skettrup, T. 'Laser modes and threshold conditions in N-mirror resonators', *Journal of the Optical Society of America B*. 2008; **13**(5), p. 926
[58] Bloom, G., Larat, C., Lallier, É., Carras, M., and Marcadet, X. 'Intracavity combining of quantum cascade lasers', in Brignon, A. (Ed.), *Coherent Laser Beam Combining*, Hoboken: John Wiley & Sons: pp. 401–425
[59] Corcoran, C.J., and Rediker, R.H. 'The dependence of the output of an external-cavity laser on the relative phases of inputs from five gain elements', *IEEE Photonics Technology Letters*. 1992; **4**(1): pp. 1197–1200
[60] Kouznetsov, D., Bisson, J.-F., Shirakawa, A., and Ueda, K. 'Limits of coherent addition of lasers: simple estimate', *Optical Review*. 2005; **12**(6): pp. 445–447
[61] Montoya, J., Augst, S.J., Creedon, K., Kansky, J., Fan, T.Y., and Sanchez-rubio, A. 'External cavity beam combining of 21 semiconductor lasers using SPGD', *Applied Optics*. 2012; **51**(11): p. 1724
[62] Bloom, G., Larat, C., Lallier, E., *et al.* 'Passive coherent beam combining of quantum-cascade lasers with a Dammann grating', *Optics Letters*. 2011; **36**(19): p. 3810
[63] Schimmel, G., Doyen Moldovan, I., Janicot, S., *et al.* 'Rear-side resonator architecture for the passive coherent combining of high-brightness laser diodes', *Optics Letters*. 2016; **41**(5), pp. 950–953
[64] Talbot, H.F. 'LXXVI. Facts relating to optical science. No. IV', *The London, Edinburgh, and Dublin Philosophical Magazine and Journal of Science*. 1836, **9**(56): pp. 401–407
[65] Rayleigh, L. 'XXV. 'On copying diffraction-gratings, and on some phenomena connected therewith', *The London, Edinburgh, and Dublin Philosophical Magazine and Journal of Science*. 1881; **11**(67): pp. 196–205
[66] Paboeuf, D., Lucas-leclin, G., Georges, P., *et al.* 'Wavelength-stabilized tapered laser diodes in an external Talbot cavity: simulations and experiments', *Proc. SPIE* **7198**; 2009: p.71981L
[67] Mehuys, D., Streifer, W., Waarts, R.G., and Welch, D.F. 'Modal analysis of linear Talbot-cavity semiconductor lasers', *Optics Letters*. 1991; **16**(11): pp. 823–825
[68] Waarts, R., Mehuys, D., Nam, D., Welch, D., Streifer, W., and Scifres, D. 'High-power, CW, diffraction-limited, GaAlAs laser diode-array in an external Talbot cavity', *Applied Physics Letters*. 1991; **58**(23): pp. 2586–2588

[69] Huang, R.K., Chann, B., Missaggia, L.J., Augst, S.J., Swint, R.B., and Donnelly, J.P. 'High-power coherent beam combination of semiconductor laser arrays', *Preceedings of the Conference on Lasers and Electro-Optics*, Optical Society of America. 2008; **961**: pp. 2007–2008
[70] Liu, B., Liu, Y., and Braiman, Y. 'Coherent beam combining of high power broad- area laser diode array with a closed-V-shape external Talbot cavity', *Optics Express* 2010; **18**(7): pp. 7361–7368
[71] Paboeuf, D., Lucas-Leclin, G., Georges, P., *et al.* 'Narrow-line coherently combined tapered laser diodes in a Talbot external cavity with a volume Bragg grating', *Applied Physics Letters*. 2008; **93**(21): p. 211102
[72] Pabœuf, D., Emaury, F., Rossi, S. De, Mercier, R., Lucas-leclin, G., and Georges, P. 'Coherent beam superposition of ten diode lasers with a Dammann grating', *Optics Letters*. 2010; **35**(10): pp. 1515–1517
[73] Paboeuf, D., Vijayakumar, D., Jensen, O.B., *et al.* 'Volume Bragg grating external cavities for the passive phase locking of high-brightness diode laser arrays: Theoretical and experimental study', *Journal of the Optical Society of America B*. 2011; **28**(5): pp. 1289–1299
[74] Hassiaoui, I., Michel, N., Bourdet, G., *et al.* 'Very compact external cavity diffraction-coupled tapered laser diodes', *Applied Optics*. 2008; **47**(6): pp. 3–7
[75] Liu, B., and Braiman, Y. 'Coherent addition of high power broad-area laser diodes with a compact VBG V-shaped external Talbot cavity', *Optics Communications*. 2018; **414**: pp. 202–206
[76] Liu, B., and Braiman, Y. 'Coherent beam combining of high power broad-area laser diode array with near diffraction limited beam quality and high power conversion efficiency', *Optics Express*. 2013; **21**(25): pp. 1771–1773
[77] Chang-Hasnain, C.J., Berger, J., Scifres, D., Streifer, W., Whinnery, J., and Dienes, A. 'High-power with high-efficiency in a narrow single-lobed beam from a diode-laser array in an external cavity', *Applied Physics Letters*. 1987; **50**(21): pp. 1465–1467
[78] Raab, V., and Menzel, R. 'External resonator design for high-power laser diodes that yields 400 mW of TEM00 power', *Optics Letters*. 2002; **2**(3): pp. 167–169
[79] Yiou, S., Balembois, F., Georges, P., and Huignard, J.-P. 'Improvement of spatial beam quality of laser sources with an intracavity Bragg grating', *Optics Letters*. 2003; **28**(4): pp. 242–244
[80] Barthelemy, A., Louradour, F., and Couderc, V. 'Wavelength-tunable diffraction-limited operation of a standard high-power diode-laser array using an off-centred extended cavity', *Electronics Letters*. 1992; **28**(22): pp. 2038–2040
[81] Corcoran, C.J., and Pasch, K.A. 'Modal analysis of a self-Fourier laser cavity', *Journal of Optics A: Pure and Applied Optics*. 2005; **7**(5): pp. L1–L7
[82] Corcoran, C.J., and Durville, F. 'Passive coherent combination of a diode laser array with 35 elements', *Optics Express*. 2014; **22**(7): pp. 8420–8425

Chapter 3

High-power laser diodes for direct applications and laser pumping

Paul Crump[1], Olaf Brox[1], Jörg Fricke[1], Hans Wenzel[1], Andreas Klehr[1], Gunnar Blume[1], Katrin Paschke[1], Andrea Knigge[1] and Günther Tränkle[1]

3.1 Introduction

GaAs-based high power diode lasers (HPDLs) are the most efficient technology for converting electrical energy into optical energy, with HPDLs defined here as devices that emit operational output power $P_{op} \geq 1$ W, as described in [1–3]. HPDLs are in strong demand in many applications, for example, in laser material processing, where they are deployed either directly or as pumps for other lasers [4]. An overview is presented here of latest research in this field, summarizing recent performance data as well as briefly describing the state of understanding of physical and material effects that limit further performance scaling. We focus here on single packaged devices, either single emitters or laser bars. We do not discuss how the emitted beams from larger arrays of HPDLs can be combined [5,6], or progress in lasers grown on other substrates such as InP, GaN or GaSb [7–9].

This chapter focuses primarily on progress in HPDLS that are grown on GaAs substrates and emit in the operating wavelength range $\lambda_{op} = 900 \ldots 1000$ nm, where the highest performance is reported, and the majority of studies are performed. Our discussion starts with a review of progress in broad area BA-HPDLs that reach the very highest powers and efficiencies, at the cost of degraded beam quality. Laterally single mode SM-HPDLs are briefly discussed next, and these offer diffraction limited beam quality, but at cost of lower power and efficiency. Next a review is given on technological progress in the implementation of monolithically integrated wavelength stabilization in HPDLs. Pulsed solid-state laser systems require high performance, flexible seed lasers and progress in developing customized diode lasers for this specialist application are also presented. We conclude with a discussion on studies into GaAs-based HPDLs at short ($\lambda \sim 635$ nm) and long ($\lambda \sim 1200$ nm) wavelengths. In all cases, the most optimal combination of

[1]Ferdinand-Braun-Institut, Leibniz-Institut für Höchstfrequenztechnik (FBH), Berlin, Germany

P_{op}, power conversion efficiency η_E and beam quality M^2 is sought. Progress is illustrated throughout using recent results from studies by the authors and their co-workers, supported by citations and discussion of results from other groups.

3.2 High power broad area lasers

3.2.1 Motivation

Single semiconductor devices that emit optical output powers of $P_{op} \geq 1$ W enable many applications, from sensing to cutting. Broad area BA-HPDLs primarily target the highest P_{op} and $\eta_E(P_{op}) = P_{op}/I_{op}U_{op}$, where I_{op} and U_{op} are the electrical current and bias voltage at the operating point, respectively. Larger P_{op} at higher η_E is constantly sought, on economic and performance grounds. BA-HPDLs are edge-emitting devices, with P_{op} delivered from an aperture of width $W = 20 \ldots 1200$ μm, as shown schematically in Figure 3.1. The lateral (in-plane) emission angle is broad, quantified via the beam parameter product $\mathrm{BPP}_{95\%} = 0.25 \times W_{95\%} \times \Theta_{95\%}$ for a beam with measured width and angle at 95% power content of $W_{95\%}$ and $\Theta_{95\%}$, respectively. The beam quality factor M^2 quantifies the proportional difference between measured $\mathrm{BPP}_{95\%}$ and values for a Gaussian beam at the same wavelength, with lateral $M^2 = \mathrm{BPP}_{95\%}/\mathrm{BPP}_{Gauss} = 2 \ldots 20$ for a BA-HPDL. In the vertical direction, HPDLs are limited by design to emission with $M^2 \sim 1 \ldots 2$. Many research efforts seek highest P_{op} and η_E while maintaining $\mathrm{BPP}_{95\%}$ in a range suitable for the application, for example, to

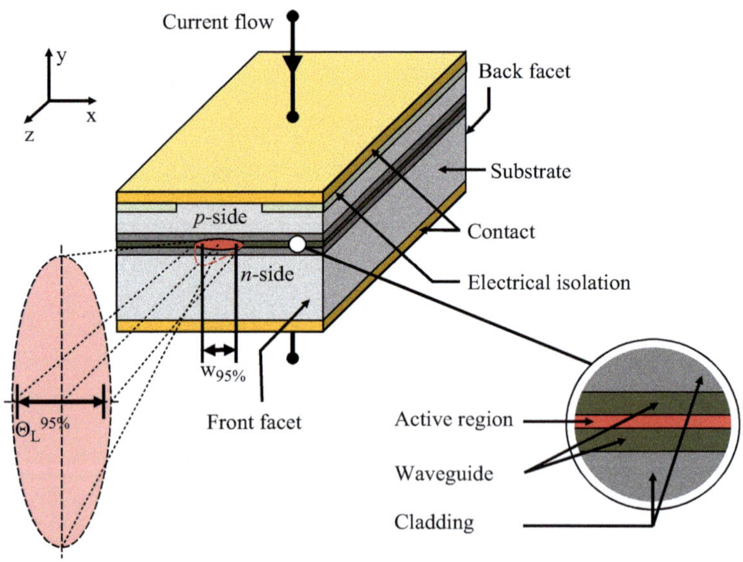

Figure 3.1 Schematic depiction of a BA-HPDL

increase the power coupled into a 100 μm core glass fiber, for fiber laser pumping. Other studies target lower $BPP_{95\%}$, while minimizing losses in P_{op} and η_E.

Alterations in HPDL design and technology are also needed for peak performance at various specialist operating conditions, which span the range from short pulse $\tau \sim$ 1–100 ns (e.g., for ranging and sensing), through long pulse $\tau \sim$ 0.1....100 ms (pumping of pulsed solid-state lasers) to continuous wave (CW) operation (direct use, or CW pumping).

Finally, attention must be paid to the emission spectrum. BA-HPDLs that (for example) operate at $\lambda_{op} \sim$ 976 nm emit P_{op} within a spectral window of $\Delta\lambda_{95\%} \sim$ 7... 10 nm (width with 95% power content). This is too large for some applications, (e.g., pumping the narrow absorption line in Yb:doped glass, which is <1 nm), so wavelength stabilization is needed, using either an external or a monolithically integrated grating.

3.2.2 Device configurations and performance comparison

There are two main BA-HPDL configurations in application. First, single emitters with $W_{95\%}$ = 90... 120 μm (see Figure 3.1 schematic), which are primarily used as pumps for fiber lasers or for direct application. Second, 1 cm wide bars that contain many emitters in a single semiconductor component are in wide use for pumping larger lasers such as solid state disk, slab, and gas-based amplifiers. Resonator lengths in the range L = 1... 6 mm are typical, although up to L = 9... 10 mm has been reported. Custom configurations are also possible such as devices with especially narrow (W = 20... 30 μm for lower M^2) or wide stripes (W = 0.2... 1.2 mm, for larger P_{op}) or custom mini-arrays (such as "T-bars" with 5 emitters each with W = 100 μm with 1 mm lateral spacing, for economic beam combining). An overview of recent peak power performance of BA-HPDLs in the 930... 980 nm wavelength range is given in Table 3.1, for the two main device configurations, and current topics in research are discussed in Sections 3.2.3–3.2.5.

3.2.3 Challenge 1: efficiency and power

The effects limiting efficiency and power in BA-HPDLs have been studied for many years and were discussed in detail in [1,3,15,16]. We briefly summarize the conclusions here. Peak reported $\eta_E(P)$ in a BA-HPDL is around 75%–76%, with values >70% reported from λ_{op} = 800–1060 nm [1,17]. Comparison between groups is challenging, due to limits on the accuracy of power measurement (±3... 5%) [1]. Laser threshold current I_{th}, the slope S of the light-current curve and U_{op} are all important limits. In a well-optimized structure, peak $\eta_E(P)$ is typically mainly limited by the differential internal efficiency (main limit to S), that is, by how effectively injected carriers are converted into stimulated photons above threshold. The electrical resistance and optical loss from the device p-side also play a key role (loss and resistance from the n-side layers is a factor of at least two smaller than for the p-layers). Current research studies seek improved S by addressing carrier recombination effects (eliminate leakage paths and other non-radiative losses), and design and technology efforts seek lower resistance and loss

Table 3.1 Summary of peak powers obtained by BA-HPDLs single emitters and bars

P_{op} (W)	**η_E**	**Test condition**	**Wavelength (nm)**	**Device configuration**	**Reference**
2000	57% @ 203 K	200 μs 10 Hz	940	1-cm bar, 69% fill factor, $L = 4$ mm, passive cooling	[10]
1000	70% @ 203 K 62% @ 288 K	1.2 ms 10 Hz	940	1-cm bar, 69% fill factor, $L = 4$ mm, passive cooling	[11]
800	63% @ 15 °C	CW	930	1-cm bar, 70% fill factor, $L = 6$ mm, active cooling	[12]
940	57% @ 20 °C	CW	925	1-cm bar, 77% fill factor, $L = 5$ mm, active cooling	[13]
30	52% @ 25 °C	CW	976	W not noted (~100 μm ?) $L = 5.1$ mm, passive	[14]
25	15 °C	CW	975	$W = 100$ μm $L = 4.0$ mm, active cooling	[15]
35	22% @ 15 °C	100 μs 100 Hz	975	$W = 90$ μm $L = 9.0$ mm, active cooling	[15]

through the use of highly asymmetric vertical structures (less p-side loss), termed here EDAS (extreme double asymmetric) designs. In EDAS vertical structures, there is (in general) low overlap between the optical field and the active region, leading to low gain and increased threshold current. In the most recent work, it was shown that adding further levels of asymmetry to the structure (extreme triple asymmetric ETAS) can retain the benefits of a thin p-side while simultaneously increasing the gain for lower threshold and hence increased peak efficiency [18].

At high bias $P_{op} > 10$ W in devices with $W_{95\%} = 90 \ldots 100$ μm, $\eta_E(P)$ is limited by electrical resistance (mostly from the p-side) and power saturation effects. In this operating range the BA-HPDL is operated under extreme conditions, with high carrier density, temperature, photon flux, electrical field, and current density. This leads to an interacting mix of power saturation terms arising, including current-driven self-heating (heating leading to degraded threshold and slope at increased temperatures, quantified via the temperature parameters T_0, T_1), bias-driven leakage (BDL), gain saturation (also termed spectral hole burning), longitudinal or lateral spatial hole burning, two-photon absorption and waveguide collapse. Other spatial nonuniformities and external effects (e.g., packaging-induced effects) can also play a role. Separating the contributions is challenging and is the topic of intensive current research, not least because CW power from emitters with $W = 90 \ldots 100$ μm has not increased above $P_{op} = 25$ W for the last 10 years [15,19]. Thermal effects can be minimized by operating with short-pulse currents (<1 μs), which enables higher peak powers in the range $P_{op} = 100 \ldots 145$ W to be achieved from devices with $W = 90 \ldots 100$ μm [16,20].

Performance progress in η_E is summarized in Figure 3.2 for single emitter BA-HPDLs grown using MOVPE and fabricated with $W = 90$ μm and $L = 4$ mm, tested

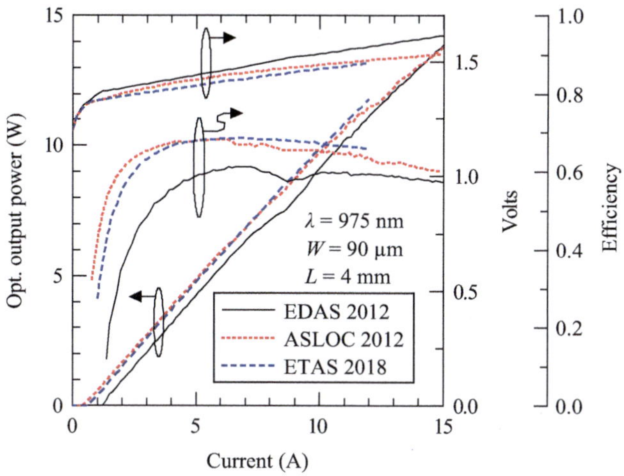

Figure 3.2 Optical output power, operating voltage and conversion efficiency as a function of current for BA-HPDLs using three epitaxial designs

in CW mode at heatsink temperature T_{HS} = 25 °C. The devices are mounted junction-down with hard solder on passively cooled CuW (10:90) carriers. Figure 3.2 compares the performance of lasers with operating wavelength close to λ_{op} = 975 nm with well-optimized conventional (close to) symmetric waveguides (ASLOC [1]) and highly vertically asymmetric designs (EDAS, ETAS). The most recent ETAS designs have comparable modal gain parameter ($\Gamma g_0 \sim 9$ cm^{-1}) to reference ASLOC designs ($\Gamma g_0 \sim 11$ cm^{-1}) and show higher *S*, lower U_{op} and lower power saturation for overall higher η_E(12 W, ETAS) = 66% cf. η_E(12 W, ASLOC) = 62%.

3.2.4 *Challenge 2: beam quality*

As noted above, BA-HPDLs operate with lateral $M^2 = 2\ldots20$, which corresponds to $BPP_{95\%} = 0.6\ldots6$ mm-mrad for λ_{op} = 976 nm. In the most recent studies, the optical field in these large volume devices is found (both experimentally and theoretically) to be well described in terms of a defined series of guided optical modes [21,22]. The waveguide supporting these modes develops due to refractive index changes arising from heating in the laser stripe and from the presence of high carrier densities (and associated gain-loss profile) in the device. Current spreading into the device edges leads to extended regions where gain does not clamp, and these provide additional gain and hence prefer higher-order optical modes with wide near fields (and hence high M^2).

Process- and packaging-induced waveguiding can also play a role, with intentional guides stabilizing but increasing $BPP_{95\%}$. In addition, longitudinal or lateral nonuniformity and associated non-clamping within the stripe due to (for example) spatial hole burning can lead to a significant (1 mm-mrad) increase in $BPP_{95\%}$ [23]. In general, increasing bias leads to higher $BPP_{95\%}$ due to stronger waveguiding from increased temperatures and higher carrier densities, with density continuing to rise above threshold in local regions where gain does not clamp. Local regions of high (or low) intensity can also be induced due to external optical or thermal feedback or the presence of material defects, further increasing $BPP_{95\%}$.

For a given device geometry (e.g., $L = 4$ mm, $W = 90 \ldots 100$ μm), successful technological methods to reduce $BPP_{95\%}$ include measures to minimize the impact of process- and package-induced waveguiding (suppress mounting stress) and current spreading (suppress current spreading) [21]. In addition, efforts to minimize the impact of temperature profiles [24] have been beneficial, for example by making changes in the vertical layer design [21]. Improved $BPP_{95\%}$ can also be achieved by adding structures that introduce in-chip losses for higher order modes [25]. It is however technologically challenging to add loss for higher order modes without compromising performance elsewhere.

For any given device technology, lower $BPP_{95\%}$ is achievable by reducing stripe width (narrower resonator leads to higher order modes being cut off, at the expense of power and efficiency) [21], by increasing resonator length or improving cooling (lower temperatures) [21,24,26]. Although well-motivated, attempts to minimize the impact of temperature-induced waveguiding by structuring the metallization or heatsink or using more thermally conductive heatsink materials have to date brought only minimal benefit [27], which is likely due to the presence of significant thermal barriers between the *p*-side of the GaAs chip and the metallization [28]. Results from recent studies are collected in Table 3.2 and Figure 3.3. Wider stripes enable higher power per component (lower cost in €/W) at penalty of larger $BPP_{95\%}$ and lower peak brightness. It should be noted that more complicated resonators such as tapered lasers or externally optically stabilized semiconductor amplifiers can operate with high powers in a beam with low $BPP_{95\%}$ (for example, tapered amplifiers can deliver $P_{op} = 17$ W with $M^2 < 1.5$ but only at

Table 3.2 Summary of linear brightness as a function of stripe width obtained in BA-HPDLs

P_{op} (W)	$BPP_{95\%}$ (mm-mrad)	$P_{op}/BPP_{95\%}$	η_E	λ_{op} (nm)	Device configuration	Reference
4	0.7	5.6	53	970	$L = 6$ mm $W = 20$ μm Passively cooled, CW	[21]
10.5	11.5	4.8	~60	976	$L = 5.2$ mm $W = 100$ μm 55 W from 5 emitter T-Bar Actively cooled, CW	[24]
55	26,5	2.1	~60	940	$L = 4$ mm $W = 400$ μm Passively cooled, 1 ms 10 Hz	[30]
110	63	1.7	63	940	$L = 6$ mm, $W = 1200$ μm Passively cooled, 1 ms 100 Hz	[31,32]

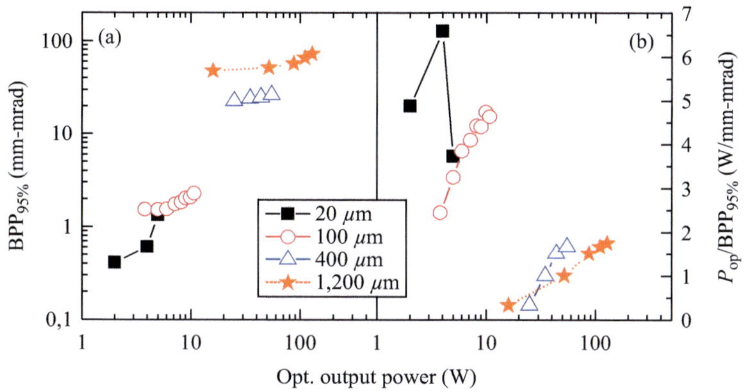

Figure 3.3 *(a) Beam parameter product and (b) lateral linear brightness (lines) as a function of optical output power for BA-HPDLs, with $W = 20\ldots1200$ μm and $L = 4000\ldots6000$ μm [21,24,30,31]*

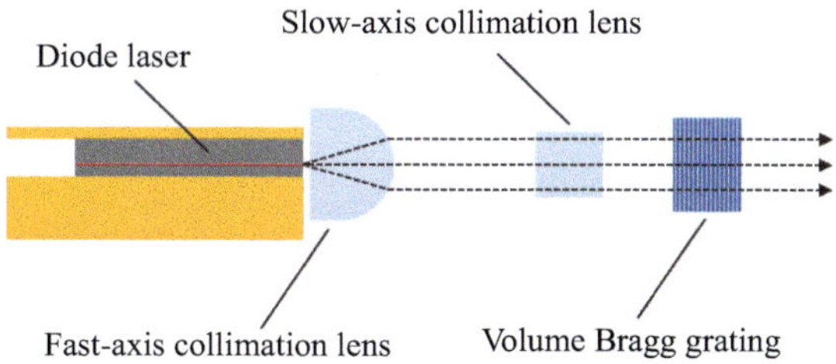

Figure 3.4 *Schematic configuration (cross section) of inverse-self-imaging configuration typically used for externally optically wavelength stabilized BA-HPDLs*

$\eta_E \sim 27\%$ [29]), at the cost of further reduced conversion efficiency, but these are dealt with elsewhere.

3.2.5 Challenge 3: external stabilization

As noted in the introduction, the spectral width at the operation power of BA-HPDLs is relatively broad, at $\Delta\lambda_{95\%}(P_{op}) = 5\ldots10$ nm. The spectrum can be stabilized and narrowed using an external spectrally selective element, with a narrow, temperature stable reflectivity. Many configurations are possible, reviewed in [33], but the most common approach is inverse self-imaging, using a volume Bragg grating, as depicted schematically in Figure 3.4 and discussed in [34,35]. In these systems, there are two optical resonators: (a) one between the diode laser front and rear facets with reflectivities R_f and R_r, respectively and (b) one between the diode laser rear facet and the grating (with reflectivity R_{VBG}). The wavelength for peak optical gain of the diode laser material shifts rapidly with temperature at around

0.3–0.4 nm/K, and gain and grating wavelength (which shifts very little with P_{op}) are typically selected to coincide at P_{op}. High powers with $\Delta\lambda_{95\%} < 0.1$ nm are required for pumping very narrow lines (for example in gas-lasers [36]) and this can be achieved when the diode laser resonator is much weaker than the external grating (e.g. $R_{VBG} = 20\%$, $R_f \sim 0.1\%$), but this typically leads to >10% reduction in η_E due to optical losses in the external resonator. High powers with $\Delta\lambda_{95\%} \sim$ 1 nm (suitable for example for pumping the narrow absorption line at $\lambda = 976$ nm in Yb-doped glass) and reasonable efficiencies (<5% reduction in η_E) can be achieved with weaker external feedback (e.g., $R_{VBG} = 15\%$, $R_f = 2\%$). However, for weak feedback, when gain and grating wavelength do not coincide (away from P_{op}), lasing can occur at the gain wavelength with $\Delta\lambda_{95\%} = 5 \ldots 10$ nm, limiting the useful operating range. A precise and mechanically stable optical alignment is needed and this must be tolerant against non-idealities in the optical path, for example from mechanical nonuniformity such as "bar smile" (few-micron class vertical offset in emitter position within a laser bar).

The largest current challenge however is that external optical feedback typically leads to a strong reduction in reliable power level [37]. The reasons for this failure are currently researched, and depend on operating wavelength and the nature of the feedback. Known lifetime limiting effects include those generated by feedback into the diode laser, within the optical waveguide (leading to increased local optical intensities), feedback into the diode laser, outside the waveguide (for increased local heating, carrier generation, stimulation of pathological optical modes) and feedback into the packaging (for increased local heating, or defect generation).

Lower stabilization losses and more tolerant alignment can be achieved when diode lasers with low vertical far-field angles $\Theta_V^{95\%} \leq 25°$ (at 95% power content) are used. The best-optimized devices to date show just 5%–10% reduction in η_E (P_{op}) compared to conventional diode lasers with $\Theta_V^{95\%} = 40 \ldots 60°$ [35]. Improved reliability is observed when higher front facet reflectivity is used, or when feedback is reduced (better system design, higher quality optical elements).

3.3 High power laterally single mode lasers

In contrast to BA-HPDLs, laterally single mode high power diode lasers SM-HPDLs are forced by design to operate with a (close-to) single lateral mode, and deliver their power in a beam with $M^2 \sim 1$. Their schematic layout is similar to that shown in Figure 3.1, with typically $W < 20$ μm. Also in contrast to BA-HPDLs, a large lateral refractive index step is introduced technologically, sufficient to overcome guiding from thermal or carrier effects. The index step can be introduced by etching grooves into the top surface of the device, by etching a mesa and burying this in epitaxially regrown material or by using a variety of diffusion and implantation steps. The strength of the index step is normally quantified in terms of the difference in effective refractive index between the fundamental vertical mode in stripe and edge regions, Δn_{eff} (although this does not apply when the vertical waveguide is fully removed, as in some buried mesa designs). Values of W and

Δn_{eff} are selected to cut-off higher-order laterally guided modes both for cold-cavity case and considering any additional guiding from current-induced heating, injected carriers, or stress effects in normal operation.

High power SM-HPDLs typically use wide $W \sim 5 \ldots 10$ μm and large $L = 4 \ldots 6$ mm for low thermal and electrical resistance. Carrier losses due to current spreading can be severe for these narrow stripes, so measures to suppress this are critical, either through well-chosen trench depth (no current flows in the trenches), or by using current blocking structures defined by regrowth, implantation, or diffusion steps. Optical (scattering) losses in the ridge must also be suppressed by minimizing the roughness of the ridge side walls.

Power and efficiency are limited by the same mechanisms as in BA-HPDLS, with an additional limit being the onset of higher order lateral modes as temperature and carrier density in the ridge section slowly increase with bias, leading to abrupt steps (kinks) in the light-current curve, shifts in wavelength, and alteration in the beam profile, such as beam steering. The kink power is typically much lower than the peak power (e.g., in [38], kink at $P_{op} = 1$ W, peak $P_{op} = 2.5$ W). High order lateral modes can be suppressed and kink power increased by adding additional lateral losses for higher order modes via a variety of techniques [25,39–42].

Thick vertical waveguides enable narrow vertical far fields in the $\Theta_V^{95\%} = 17 \ldots 25°$ range for peak $\eta_E < 45\%$ at $P_{op} = 0.1 \ldots 0.5$ W [38,39,43]. Thinner vertical guides with wider vertical far fields $\Theta_V^{95\%} \sim 50°$ ($\sim$25° FWHM) enable higher power and efficiency $\eta_E > 60\%$ at $P_{op} = 1$ W to be achieved (kink power > 2 W), when combined with lateral loss techniques [42]. The EDAS and ETAS designs used in BA-HPDLs also show benefit for SM-HPDLs, as their highly asymmetric vertical profile and thin *p*-side waveguide suppresses higher order lateral modes by design, as discussed in [44] and shown in Figure 3.5. In addition, trenches can be etched deeply, with etching halted close to the active region ($\sim$ 200 nm) cutting off lateral current spreading for higher η_E while maintaining low $\Delta n_{eff} \sim 10^{-3}$. In recent studies, $\eta_E \sim 60\%$ at $P_{op} = 1$ W could also be achieved in SM-HPDLs using EDAS designs with $L = 4$ mm and $W = 5 \ldots 15$ μm, when mounted *p*-up on CuW (10:90) submounts and tested CW, with results shown in Figure 3.5. Kink power must still be improved, by implementing lateral mode filtering.

3.4 High power lasers with monolithic grating stabilization

Various high power diode laser applications require narrow linewidths and precisely defined wavelengths of the optical emission. Application examples include the pumping of narrow lines in solid-state lasers and the use of wavelength beam combining for power scaling in direct diode material processing systems. Moreover, a stabilized wavelength with a reduced thermal shift is requested, which is caused by current-induced internal heating or by variations of the environment conditions.

Linewidths narrowing and wavelength stabilization can be obtained with Bragg gratings incorporated into the diode laser cavity. Studies have therefore been performed to monolithically integrate optical gratings into high power diode lasers

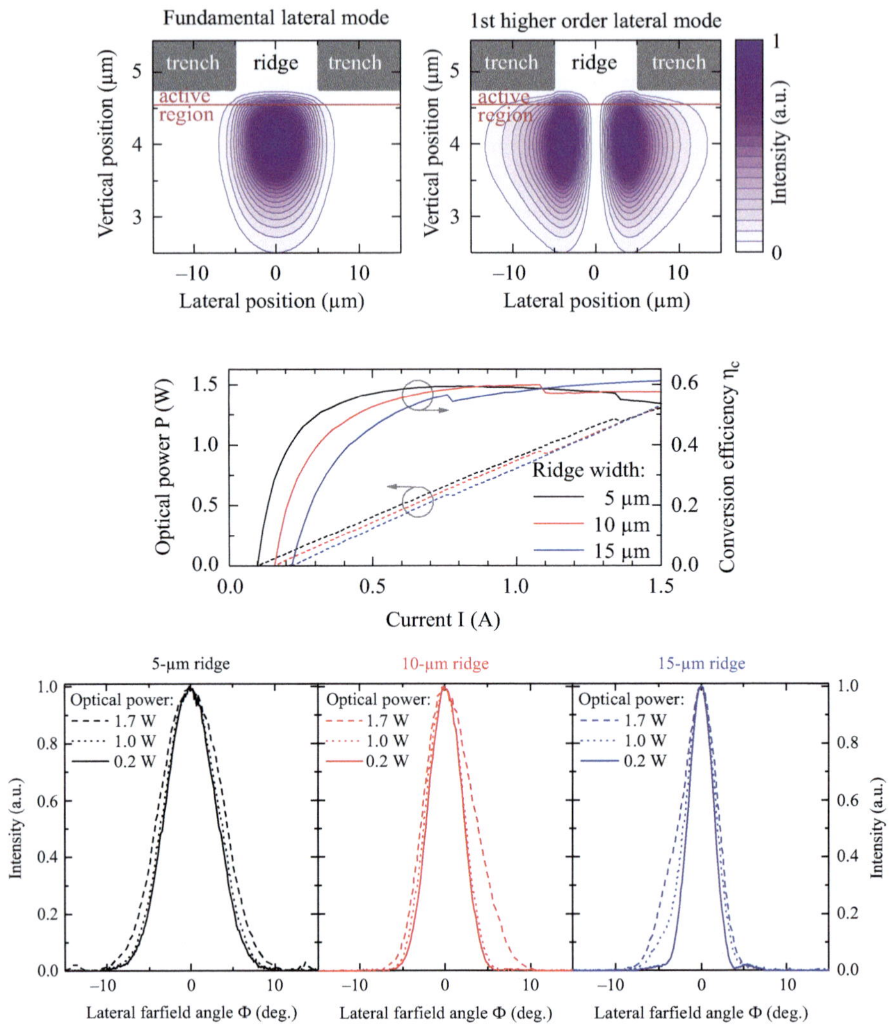

Figure 3.5 CW performance of SM-HPDL that make use of EDAS vertical designs [44]. Top: Simulated intensity contour plots of the fundamental mode (left) and 1st higher order lateral mode (right) for a 10 µm wide ridge. Center: Optical output power and efficiency as a function of current for different values of stripe widths. Bottom: Intensity as a function of lateral angle for different stripe widths and P_{op}

without degrading efficiency and power. While design requirements for efficiency-optimized wavelength stabilized laser diodes are summarized in [45–47] we review here technological approaches that allow the monolithic integration of gratings.

In one approach the vertical laser structure is partially grown, then the wafer surface is patterned with shallow grating grooves with periods and widths to form low order gratings. Subsequently it follows an epitaxial overgrowth step which completes the vertical laser structure. Challenges of the overgrown gratings (OG) are the surface preparation prior to overgrowth and the adjustment of growth conditions to avoid oxygen at the interface, which acts as fast carrier trap and strongly reduces efficiency [48]. Another technological approach applies high order gratings deeply etched into the surface of the *p*-cladding and waveguide layers of the complete vertical structure. The surface etched grating (SE) approach is challenging in terms of etch mask preparation and stability during the etch step.

In the following the technological approaches will be described and examples of fabricated high power diode lasers with distributed feedback (DFB) as well as with distributed Bragg reflectors (DBR) will be discussed.

3.4.1 Overgrown gratings

In contrast to the InP material system, where overgrowth as well as selective growth techniques are well established [49,50], the overgrowth of GaAs devices needs special attention due to the higher propagation velocity of crystal defects. Vertical structures are typically designed with gratings in the *p*-side waveguide to not disturb the active region during the overgrowth. Among different fabrication techniques reviewed in [45] preferably a sequence of GaAsP and InGaP epitaxial layers is grown at the end of the first growth step. After patterning the resist wet-chemical etching is applied because it produces surfaces with better crystal quality when compared to dry-etching. The high wet-chemical etch contrast between the GaAsP and InGaP grating layers enables the stable production of highly defined grating dimensions for lasers across a wavelength range from 760 to 1120 nm. Appropriate dimensioning of the grating layer thickness and the offset from the active region allows the precise control of the coupling coefficient κ. DFB lasers ($\kappa = 0.2 \ldots 2\ \text{cm}^{-1}$) as well as DBR lasers ($\kappa > 60\ \text{cm}^{-1}$) have been fabricated using this technique [51,52].

During grating preparation and MOVPE overgrowth the contamination with oxygen has to be avoided, since the resulting deep-level trap states promote non-radiative recombination and hence degrade the laser performance. Besides careful surface cleaning procedures before overgrowth different strategies have been proposed to suppress oxygen in the vertical structure (Figure 3.6). (a) The first epitaxial step ends with a *p*-InGaP protection layer on top of the partially grown *p*-AlGaAs confinement layer. The protection prevents the oxygen susceptible Al to be exposed to air. (b) A preferred strategy avoids the protection layer and applies in-situ etching with CBr_4 in the MOVPE reactor [53]. It has been applied to fabricate high power DFB lasers and loss free grating integration was demonstrated [54]. Figure 3.7 illustrates the full technological process to fabricate OG grating with in-situ etching in the MOVPE reactor.

However, when in-situ etching techniques are used, oxygen contamination can still arise during the overgrowth process. The effect is illustrated in Figure 4 in [55], for diode lasers with $L = 3$ mm, $W = 90$ μm and $R_f \sim 0.1\%$ that use OG gratings with $\kappa L \sim 0.2$ realized with in-situ etching. The slope close to threshold is

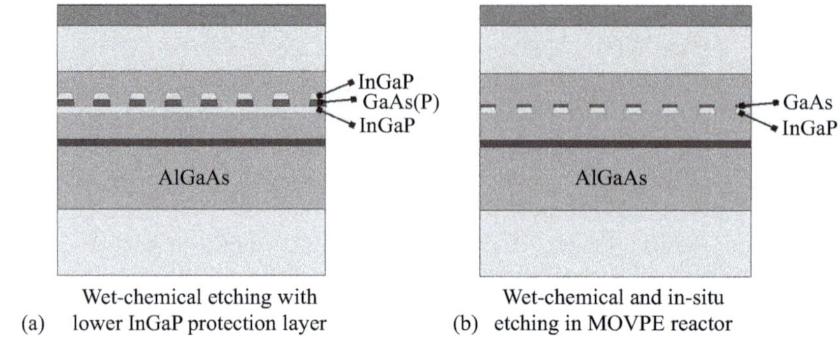

Figure 3.6 Vertical laser structure with OG gratings (a) with and (b) without InGaP protection layer

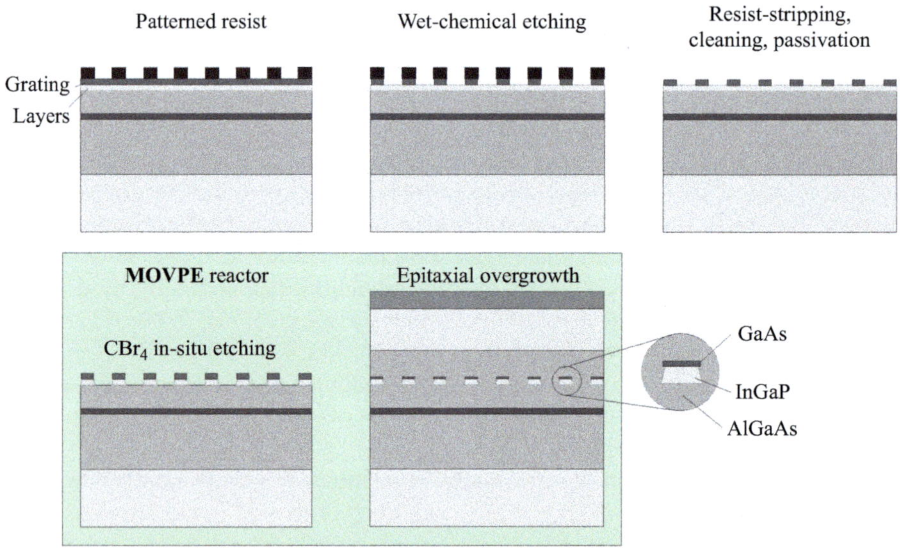

Figure 3.7 Grating fabrication process with in-situ CBr_4 etching in the MOVPE reactor

measured on *p*-down mounted devices and can be reduced by 0.2 to 0.4 W/A, directly correlated to the oxygen level at the grating layer measured using SIMS.

Current research focuses on the improvement and the stabilization of the OG process with in-situ CBr_4 etching (Figure 3.7).

3.4.2 Surface gratings

When the OG grating approach is too costly or not possible for a given material system or device configuration SE gratings can be applied. The advantage of grating patterning by anisotropic reactive ion etching (RIE) is the independency of the process from orientation of the grating section with respect to the crystallographic orientation of the wafer whereas for the manufacturing of OG gratings the growth and material composition during growth is influenced by the crystal orientation and the grating alignment.

Important for the choice of the surface grating layout and the preparation of the etch mask is the material composition of the vertical structure and the required etch depth of the grating to achieve the desired reflectivity. Usually numerical simulations are used to define the necessary etch depth and estimate fabrication tolerances. Whereas a DFB laser requires only a weak grating (low reflectivity) DBR laser gratings usually are designed to achieve high reflectivity, which in most cases means that these gratings have to be etched deeper.

The material of the etch mask must resist the whole etch process. The necessary aspect ratio and the crystallographic orientation of the material will in almost all cases require RIE or the faster inductively coupled plasma (ICP) etching. The optimization of the etch mask system must fulfil three requirements. The first is a sufficient selectivity, meaning the mask material should etch slower than the semiconductor material. The second is that the etch mask itself can be patterned with sufficient resolution, meaning that there is another process available that is able to etch the mask fast while preserving a high aspect ratio. The third requirement affects the size of the grating openings. Due to the micro loading effect, which describes the dependence of etch velocity from aspect ratio, especially in sub-μm structures (Figure 3.8), there exists a minimum size of the grating structures. A too high aspect ratio will almost stop the etching.

The etch masks described by different research groups normally consists of a multilevel system [56–59], combining silicon nitride or silicon oxide with a thin metal layer. Patterning can be realized by different techniques such as conventional photo-, holographic- or wafer-stepper-lithography, [58,60] electron beam or nanoimprint lithography [61]. In a first step the high resolution lithography is carried out. The pattern of the resist layer is then transferred into the metal layer which serves as a sacrificial layer to further transfer the pattern into the underlying

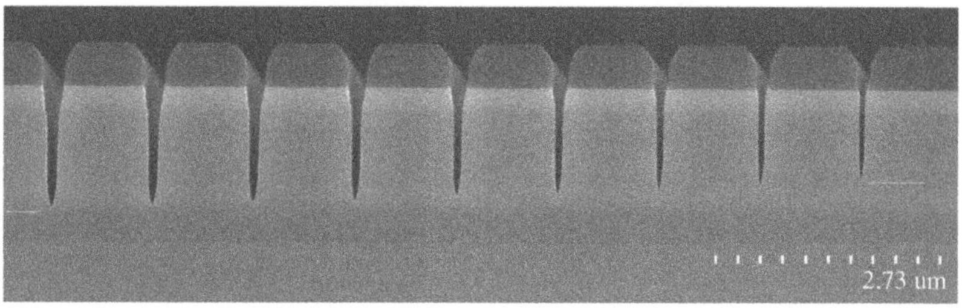

Figure 3.8 SEM image of an apodized surface grating. The residual baked resist is still present on top of the grating structure. Apodization was achieved by varying the width of the etch window from 100 nm (right) to 200 nm (left). The etch rate increases with increasing width of the resist opening

silicon nitride or silicon oxide. The silicon nitride can for example also be substituted by a polymer with good selectivity (Figure 3.9). A single photoresist layer can be sufficient, when the size of the structures will be increased. This technique will for instance be applied when i-line stepper lithography is used.

The layout and arrangement of the gratings depend on the laser type. Highly reflective DBR Bragg gratings can be deeply etched through the whole vertical laser structure [58,59,62–64]. Due to the high semiconductor-air-contrast the grating section can be kept short. However to ensure low scattering losses the vertical etch walls must be close to 90° with only minimum sidewall roughness. When DBR gratings are etched only into the *p*-waveguide layer the DBR section typically must be 300 μm to 1000 μm long but the demands on the mask selectivity are weaker.

DFB laser gratings can be arranged similar to the DBR gratings, that is, directly into the resonator axis (within laser stripe) or as laterally coupled gratings beside the resonator axis [65,66].

Simulations clarify that for high reflectivity the duty cycle of the grating (unetched area/grating-period) must be high (close to 1) [67]. This however is in contradiction to the limits of the aspect ratio with sufficiently high etch rate but also to the resolution limits of the lithography tool used. A solution is the tailoring of the used RIE process which is able to form V-shaped profiles for the gratings only etched into the *p*-waveguide layer. In the laser resonator the optical mode mainly overlaps with the bottom of the V-groove and therefore leads to a higher duty cycle compared to the original one produced by the lithography process.

Once all aspects of the fabrication process are clarified the gratings can be implemented into ridge waveguides in DBR-RW lasers [63], broad area lasers [62,68], tapered lasers [69,70], or Y-branch lasers [71,72].

Applications that require power-current characteristics with fewer or even no spectral mode hops are best served by DFB-Bragg-gratings instead of DBR-Bragg-gratings. Two principal concepts with (a) laterally coupled gratings nearby the

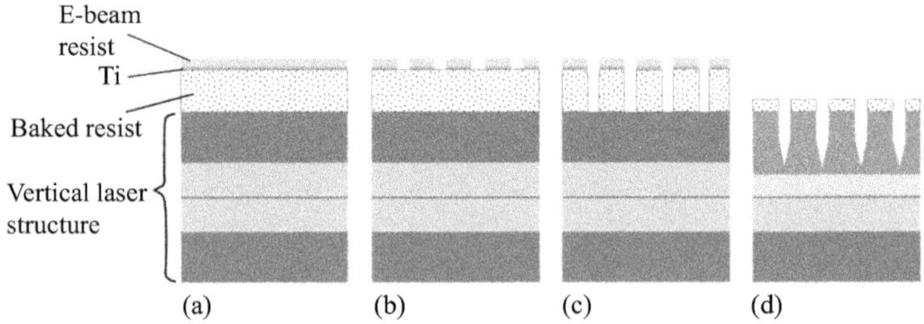

Figure 3.9 Schematics of surface grating fabrication process: (a) The vertical laser structure is covered with a three-level mask stack. (b) The upper resist is exposed and developed. (c) RIE of the Ti-layer before the etch gas is changed to oxygen and the baked resist is patterned while Ti acts as a sacrificial mask. (d) RIE of the p-cladding and a part of the p-waveguide layer. The gas chemistry must be chosen for high selectivity between baked resist and semiconductor

ridge waveguide and (b) gratings implemented into the ridge waveguide are known. The fabrication of laterally coupled gratings can be combined with the waveguide etching. Laterally coupled gratings however have a strong overlap with higher order modes. In comparison to DFB lasers with gratings arranged along the resonator axis, their multimode emission can start at lower injection currents. In broad area DFB lasers the gratings can only be arranged along the resonator axis. Such devices have been fabricated successfully for high power applications (Figure 3.10), for example, as sources for spectral beam combining systems.

3.4.3 Comparison

Especially high power (several W to kW) lasers or laser bars require high efficiency in order to limit the effort for cooling and power supply and to ensure sufficient life time. In contrast to DBR lasers a properly designed DFB laser requires a low reflectivity grating which is at best suitable for low losses. This chapter is therefore focused on this device type.

Despite the low reflectivity in DFB laser gratings optical losses occur especially in (SE) DFBs due to the use of high index contrast grooves. Scattering losses can also be influenced by the shape of the etched grooves [47]. This is for instance the case, when the grating groove is not V-shaped or any other etching anomaly widens the groove. Further the intensity of the optical mode is high close the quantum well (QW) and falls rapidly with vertical offset between grating and active region d_{res}. Small values of d_{res} are therefore coupled with high reflectivity of the grating but simultaneously also losses increase. A deviation from the optimum of d_{res} can therefore significantly reduce the internal efficiency of the device. The exact control of the grating etch depth is consequently essential for (SE) DFB-laser fabrication.

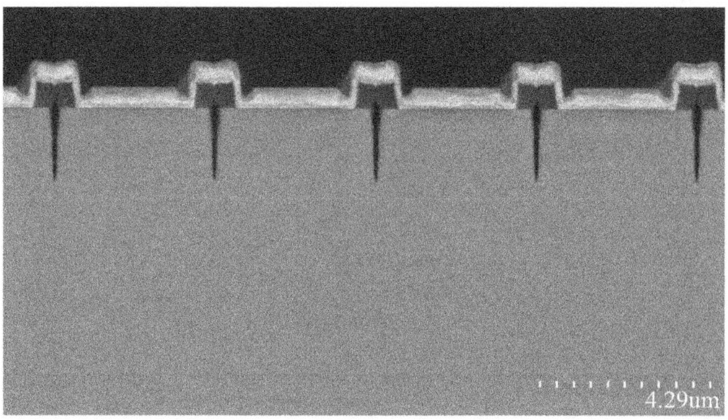

Figure 3.10 SEM image of a 25th order DFB-laser surface grating etched into AlGaAs p-cladding and confinement layers. The grating period is 4 µm. Etch grooves are encapsulated with 700 nm SiN_x. The current injection takes place between the grooves via Ti/Pt/Au metallization layer

Figure 3.11 shows the comparison of power current characteristic (cw) and conversion efficiency of $L = 6$ mm long devices with and without grating. For efficient coupling into 35 μm core multimode fiber a stripe width $W = 30$ μm is used in this example. Nominally identical lasers with no grating have been fabricated on the same wafer for direct comparison. These lasers are coated at the rear facet with $R_r = 96\%$ and at the front facet with $R_f = 0.8\%$. Laser diodes without grating achieve 56% conversion efficiency at 5 W output power. This is 1.17x larger than SE DBR-lasers from the same wafer, which achieve 48% at 5 W due to lower slope and early roll-over.

Well-optimized OG DFB lasers (see Figure 3.11(b)) have also been demonstrated to be able to achieve a conversion efficiency closer to reference lasers without gratings. In Figure 3.11(b) a 3 mm long, 90 μm wide OG-DFB laser achieves 56% conversion efficiency whereas a nominally identical broad area laser without grating achieves 60%, indicating only low scattering loss.

The spectra of these two BA-DFB laser types have been recorded (Figure 3.12). The implemented gratings reduce the spectral width at 10 W output power to $\Delta\lambda_{95\%} = 0.49$ nm (OG) DFB compared to $\Delta\lambda_{95\%} = 4.7$ nm spectral width of a BA laser. (SE) DFB lasers achieve $\Delta\lambda_{95\%} = 0.8$ nm at $P_{op} = 5$ W whereas the corresponding BA laser hat $\Delta\lambda_{95\%} = 5.3$ nm.

These results confirm the high potential of the SE-DFB as well as OG-DFB lasers for industrial applications

Further gains in terms of loss reduction and fabrication tolerances can be realized, when the grating depth is adapted to the distribution of the optical intensity along the resonator axis. With the knowledge of the intensity distribution along the resonator the aim is to continuously reduce the groove etch depth from the rear facet

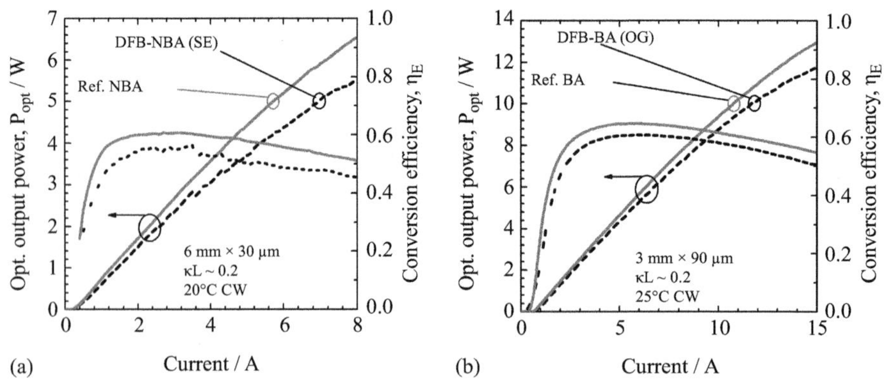

Figure 3.11 CW optical output power and power conversion efficiency dependency on injection current of DFB-lasers compared to nominally identical reference lasers. (a) 6 mm × 30 μm SE-DFB-lasers; (b) 3 mm × 90 μm broad area DFB lasers with regrown gratings. The vertical structures of both devices are identical

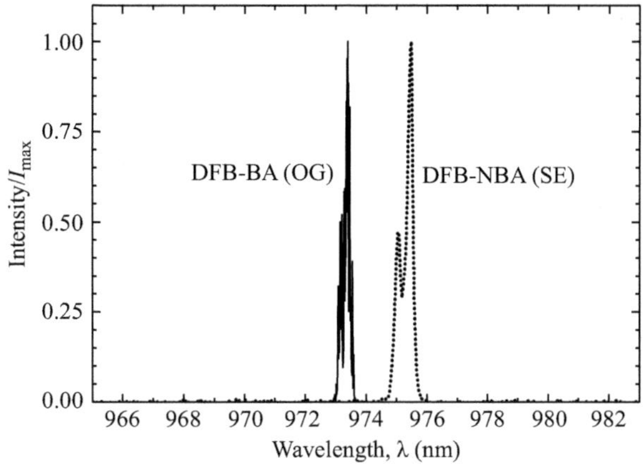

Figure 3.12 (SE, $W = 30$ μm) and (OG, $W = 90$ μm) DFB laser spectra, measured at $P_{op} = 5W$ and $10W$, respectively. The reduction of the spectral width is in both configurations approximately by factor 10 compared to devices without gratings

(low optical intensity, small d_{res}, and high feedback) to the front facet (high optical intensity, large d_{res}, and low feedback). Figure 3.8 shows such an apodized grating. For better visibility the grating period is reduced in both examples.

The performance of such devices could successfully be demonstrated at around $\lambda_{op} = 975$ nm. In particular it was shown in [60,73] that by applying apodized SE gratings the peak conversion efficiency could be increased from 57% to 60% compared to uniform gratings with identical spectral properties (Figure 3.13). By using apodized gratings instead of uniform ones the fabrication tolerances (i.e., the change of device performance with change of d_{res}) have also been significantly increased.

A compilation of fabrication processes and device properties is given in Table 3.3. The choice of the best fabrication strategy is strongly influenced by the available equipment and the vertical epitaxial structure used for the laser.

When a proper vertical structure is used both SE and OG DFBs can achieve high performance. Technological challenges as reduction of scattering losses in surface etched gratings or the surface preparation prior to the overgrowth require careful fabrication control.

The combination of high output power, high conversion efficiency with precise wavelength control as well as wavelength stabilization makes laser diodes with implemented surface gratings more and more interesting as a replacement for fiber or gas lasers. In the field of material processing currently they are mostly used as pump sources, but the aim for the future is the direct use as stacks for material processing. This requires high reliability and improved beam quality. Another important future application with increasing interest lies in the automotive field, for instance, light detection and ranging (LIDAR). These future applications will

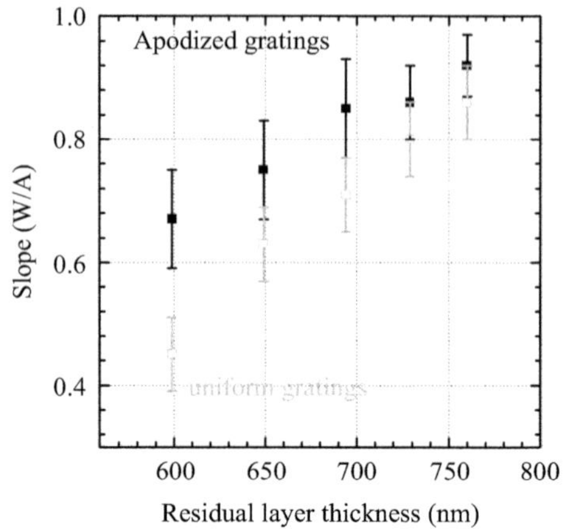

Figure 3.13 Slope of SE-DFB lasers as a function of residual layer thickness d_{res} for uniform grating (gray, open symbols) and apodized gratings (black, closed symbols). The d_{res} of apodized gratings is a mean value along the resonator axis

Table 3.3 Compilation of advantages/disadvantages, requirements and parameters of grating overgrowth and surface grating technology

Grating integration technology	Control of coupling strengths κ	Grating order m	Min. structure size	Pros	Cons	References
Surface etched grooves (SE)	Depths and shape of etched grooves	>2	Resist typ. >100 nm decreased toward active layer	Single epitaxial step	Control of etch depth	[62] [56] [57] [58] [59] [60] [61]
Epitaxial overgrowth (OG)	Epitaxial layer thicknesses and duty cycle	Typ. 1 or 2		Dimensions defined by epilayers	Oxygen incorporation, growth depends on crystal orientation	[49] [53] [50] [45] [51] [52]

require a high number of devices or laser bars. The semiconductor fabrication technology is best suited to solve this challenge.

3.5 Seed lasers

Diode lasers are ideally suited for the generation of optical pulses in the nanoseconds and picoseconds ranges by gain-switching, Q-switching, mode-locking, or pulse-gating. However, many applications (such as seeding) require spectrally stabilized and nearly diffraction limited pulses for which special designs have to be implemented. Diffraction limited emission is achieved by a several microns wide ridge waveguide (RW), so that only the fundamental lateral mode is supported. Spectral stabilization is realized with a Bragg grating integrated into the semiconductor chip, resulting in DFB or DBR types of lasers. Whereas single DFB or DBR lasers can emit peak pulse powers up to several watts, higher peak powers of several tens of watts can be reached by an amplification of the pulses with semiconductor optical amplifiers, which can be either monolithically or hybrid integrated with the master oscillators (MOs). In order to diminish self-heating and generation of amplified spontaneous emission between the pulses, the amplifier should be operated in pulsed mode, too. Whereas the current pulses injected into the power amplifier (PA) can be longer than those injected into the MO, the delay between both pulse trains has to be carefully chosen.

To inject current pulses in the nanoseconds and picoseconds ranges with high amplitudes into the lasers, the final amplifier stages of the driving electronics must be placed as close as possible to the lasers and amplifiers to minimize parasitic conductance. In general, the smaller the pulse width the more difficult is the generation of high peak currents. By using micro-optics for beam-shaping and submicron positioning technology, MO, PA, and the high-frequency electronic can be hybrid integrated in small modules [74].

In this chapter, we give a survey of wavelength stabilized diode lasers suitable as pulsed seed sources in the wavelength range between 900 and 1100 nm.

3.5.1 Gain switching

Gain switching, that is, modulating the laser gain by turning on and off the current injected into the cavity, offers a simple, cost effective and power-efficient possibility to generate optical pulses with arbitrary repetition rates. Typically, one can distinguish two different schemes of operation: for current pulses longer than ~500 ps the optical pulse follows the electrical pulse, except during turn-on and turn-off. If the current pulses are sufficiently short, only the first relaxation oscillation peak with duration of <100 ps is excited. For both modes of operation a single-electrode device such a DFB or DBR RW laser (Figure 3.14) can be used. However, for the generation of pulses shorter than 100 ps the utilization of a multi-electrode DBR laser with a reverse biased saturable absorber (SAB) (similar to those used for Q switching) is advantageous to suppress the trailing part of the pulses. Table 3.4 gives an overview of published peak powers. Detailed discussions of the temporal, spectral, and spatial properties of gain-switched pulses have been published in [84–87]. Figure 3.15(a) shows the temporal behavior of the optical pulses of a gain-switched DFB laser measured using a 70 GHz sampling scope NRO 9000 (LeCroy) and a fast 25 GHz photo diode 1434 (New Focus) for a variety of pulse currents up to 6 A [74]. For small currents slightly above threshold, the first relaxation oscillation dominates. With increasing pulse current the optical pulse follows more and more the electrical pulse. The rise time of 1.2 ns excluding

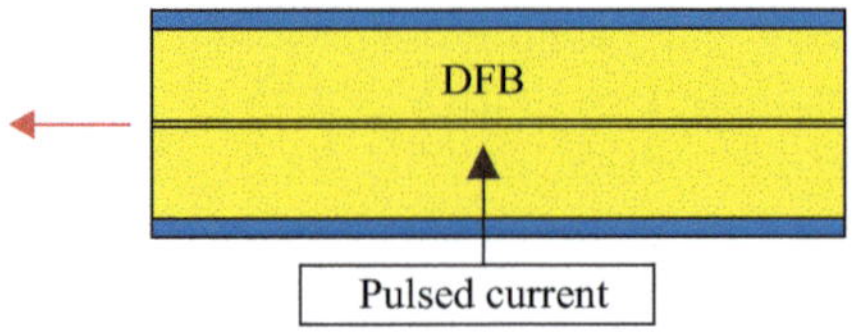

Figure 3.14 Scheme of gain switching

Table 3.4 Summary of peak powers obtained by gain switching of DFB or DBR lasers

Peak power (W)	Pulse width (ns)	Repetition rate (MHz)	Wavelength (nm)	Configuration	Reference
1.6	50	1	976	DFB laser	[75]
0.7	4	0.02	976	DBR laser	[76]
0.6	4	0.02	1063	DBR laser	[76]
6.3	0.07	1000	1064	DBR tapered laser	[77]
28	0.03	5 ... 80	1062	DFB laser + tapered PA	[78]
1	0.07	10	1065	DFB laser	[79]
2	500	1	1062	DFB Laser	[80]
42	0.08	1	1064	Monolithically integrated DFB laser + tapered PA	[81]
3.8	4	0.25	1063	DFB laser	[74]
10	4	0.25	1063	DFB laser + tapered PA	[74]
4	0.065	1	1063	DBR laser	[82]
65	0.035	1	1063	DBR laser with absorber section + tapered PA	[82]
0.1	0.05	0.1	1064	DFB laser	[83]
> 0.6	1–10	0.1	1064	DFB laser	[83]

the first relaxation time corresponds to that of the current pulse (4 ns). The fall time has the same magnitude. Time-averaged optical spectra were measured using an optical spectrum analyzer Q8384 (Advantest) with a resolution of 10 pm. Figure 3.15(b) shows a color-scale mapping of the spectra on a logarithmic scale versus pulse current. The spectra reveal that over the whole current range the gain-switched DFB laser operates in a single mode. The peak wavelength decreases slightly from 1062.65 nm slightly above threshold to 1062.62 nm at a power of 2.7 W probably caused by an increase of the carrier density in the waveguide. This behavior is accompanied by a broadening of the spectrum at the longer wavelength side which indicates a thermally induced chirp during the pulse due to starting self-heating.

3.5.2 Q-switching

Pulses having duration of 100 ps and less can be obtained with an arbitrary repetition rate by the modulation of the intra-cavity loss of a three-section DBR laser consisting of gain, SAB and DBR sections (Figure 3.16). The gain section can be operated in

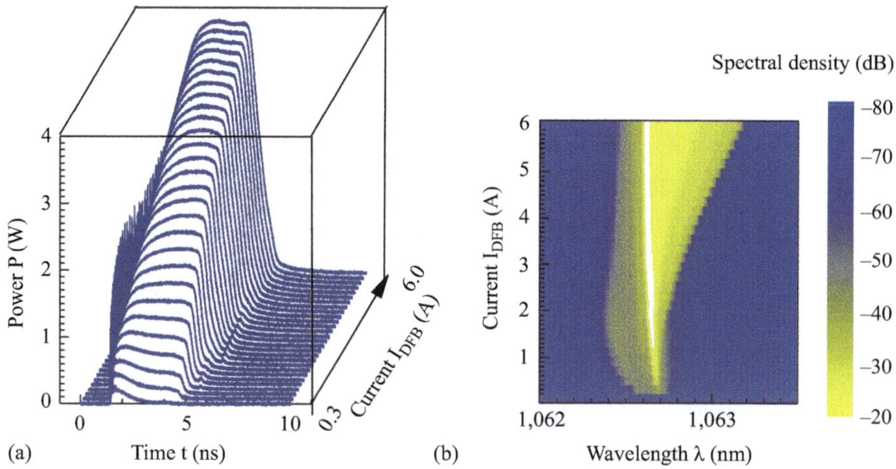

Figure 3.15 (a) Optical power of a gain-switched DFB laser versus time. Parameter is the pulse current. (b) Corresponding pseudo-color mapping of the time-averaged optical spectra on a logarithmic scale

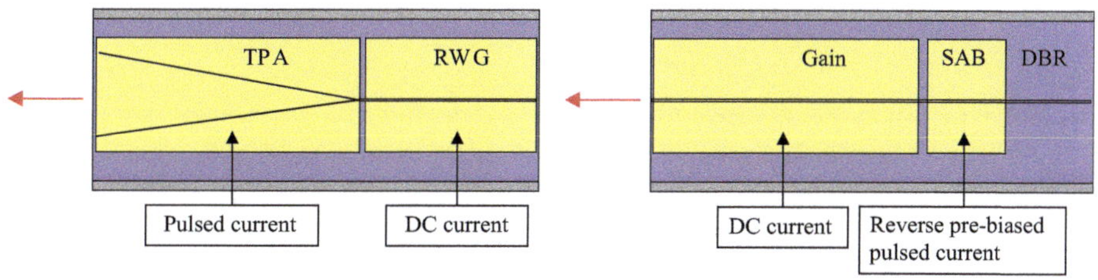

Figure 3.16 Scheme of Q-switching and subsequent amplification

Table 3.5 Summary of peak powers obtained by Q-switching of DBR lasers

Peak power (W)	Pulse width (ps)	Repetition rate (MHz)	Wavelength (nm)	Configuration	Reference
1	83	20	1066	DBR laser + tapered PA	[88]
50	80	1.25	1066	DBR laser + tapered PA	[88]
3.8	66	40	1063	DBR laser	[74]
35	80	1.25	1067	DBR laser + tapered PA	[74]

CW or in pulsed mode to minimize the self-heating of the laser at high currents. Table 3.5 gives an overview of published peak powers. Figure 3.17 shows the temporal and spectral behavior of an amplified optical pulse generated by Q-switching [74]. With a compact assembly having a footprint 5 cm × 4 cm, combining DBR

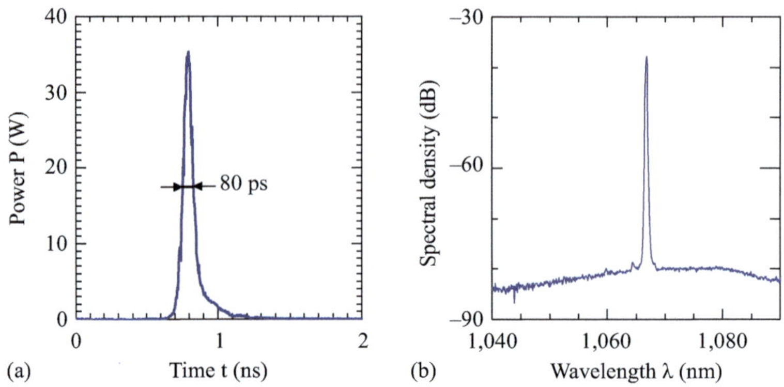

Figure 3.17 (a) Optical power of an amplified pulse generated with a Q-switched DBR laser versus time. (b) Corresponding optical spectrum on a logarithmic scale

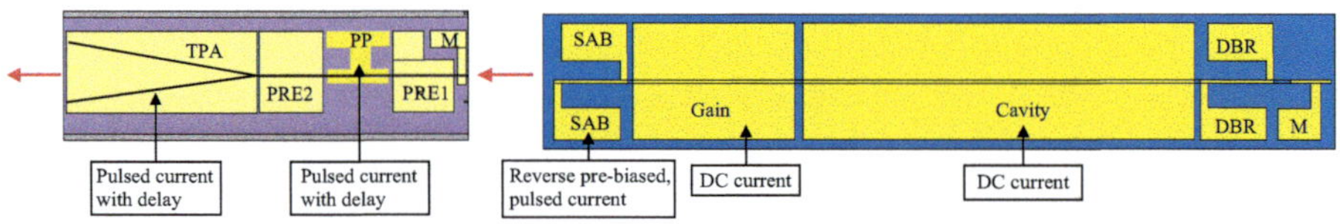

Figure 3.18 Scheme of mode locking and subsequent pulse picking and amplification

laser, tapered PA consisting of ridge-waveguide (RWG) and tapered section (TPA) (Figure 3.16), micro-optics for beam-shaping and driver electronics, a maximum peak power of 35 W with a pulse width of 80 ps (FWHM) has been obtained. The time-averaged optical spectrum shows that the wavelength is locked by the Bragg grating. The spectral width (FWHM) is 0.3 nm and the suppression of the ASE is better than 40 dB.

3.5.3 Mode locking and pulse picking

Mode-locking, where the phases of the longitudinal modes of the cavity are synchronized, generally allows for shorter pulse durations down to less than 10 ps. However, the pulse repetition rate is fixed to the cavity roundtrip frequency. Thus pulse pickers (PPs) have to be used in order to reduce the repetition rate. Passive mode locking which yields the shortest pulses necessitates a SAB in the cavity. In order to reduce the timing jitter, the absorber can be modulated with a frequency equal to the cavity roundtrip frequency, thus combining active and passive to hybrid mode locking (Figure 3.18). In order to reduce the repetition frequency the optical pulses generated by the MO are coupled into a tapered PA with a ridge-waveguide entrance section containing a PP whose transparency is modulated by the injection of short current pulses (Figure 3.18). Thus the PP acts as an optical gate and selects the desired pulses. The width of the optical gate determined by the width of the current pulse must be smaller than the period of the pulse train. By utilizing fast GaN high-electron mobility transistors for gating, selective pulse picking with a free choice of the repetition frequency in the range from kHz up to 100 MHz can be achieved. Afterwards the pulses are amplified in the tapered section (TPA), which is driven with short current pulses of high amplitude (Figure 3.18). Table 3.6 gives an overview of published peak powers. A successful pulse picking is demonstrated in Figure 3.19(a) [91]. The pulse train emitted by the MO has a repetition frequency of 4.2 GHz corresponding to a period of 240 ps. The fluctuation of the peak power is caused by sampling effects of the oscilloscope. The corresponding time-averaged optical spectrum is shown on linear (Figure 3.19(b)) and logarithmic (Figure 3.19(c)) scales.

3.5.4 Pulse gating

In a MO-PA setup the amplifier or a part of it can be used as an optical gate. The MO (a DFB or DBR laser) is operated in CW mode emitting a single longitudinal mode with a small spectral line width. The amplifier or an electrically separated section normally unbiased or reverse biased absorbs the incoming beam emitted by the MO but becomes transparent if a current pulse is injected so that the beam can pass. An optional integrated tapered section amplifies the generated optical pulse. The delay time between the current pulses injected into the gate and into the tapered section have to be adjusted to minimize the amplified spontaneous emission. The generated and amplified optical pulse preserves to a large extend the spectral properties of the CW emission of the MO. Table 3.7 gives an overview of published peak powers. It is possible to use the same set up for pulse gating and for pulse picking described in Section 3.5.3 (Figure 3.20) [91]. Figure 3.21(a) shows that a continuous variation of the optical pulse width between 0.25 and 1 ns can be achieved. An advantage of pulse gating is the absence of a first relaxation peak, which might be an interesting feature for applications such as seeding of fiber lasers. Figure 3.21(b) showing the optical spectra corresponding to Figure 3.21(a) reveals that wavelength and spectral width are nearly independent of the pulse width.

Table 3.6 Summary of peak powers obtained by mode-locked DBR lasers

Peak power (W)	Pulse width (ps)	Repetition rate (MHz)	Wavelength (nm)	Configuration	Reference
3.6	7	4100	927	DBR laser	[89]
1	7 … 9	4300	1065	DBR laser	[90]
125	8	16.7	1065	DBR laser + pulse picker integrated with tapered PA	[90]
0.6	5	4300	1068	DBR laser	[74]
74	9	67	1065	DBR laser + pulse picker integrated with tapered PA	[74]
0.6	4	4200	1030	DBR laser	[91]
50	4 … 15	0.01 … 20	1030	DBR laser + pulse picker integrated with tapered PA	[91]

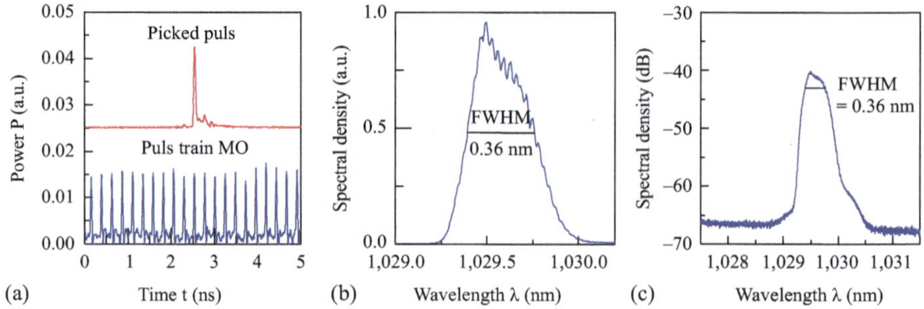

Figure 3.19 (a) Pulse train emitted by a mode locked DBR laser (bottom) and picked pulse (top). (b) and (c) Optical spectra on linear and logarithmic scales, respectively

Table 3.7 Summary of peak powers obtained by pulse gating

Peak power (W)	Pulse width (ns)	Repetition rate (MHz)	Wavelength (nm)	Configuration	Reference
0.6	0.4 … 1	0.1 … 0.8	1064	DFB laser + tapered PA	[92]
4	5	0.2	973	DBR laser + RW amplifier	[93] [94]
16	2	0.8	1064	DFB laser + pulse picker integrated with tapered PA	[95]
16	8	0.025	975	DFB laser + tapered PA	[96]
20	0.25 … 1	0.01 … 20	1030	DBR laser + pulse picker integrated with tapered PA	[69]

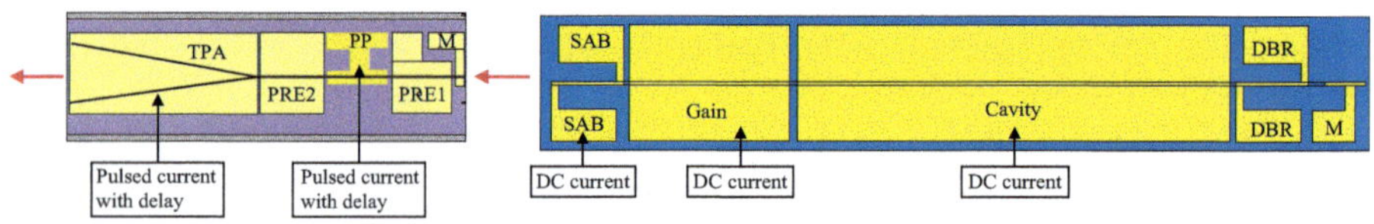

Figure 3.20 Scheme of pulse gating

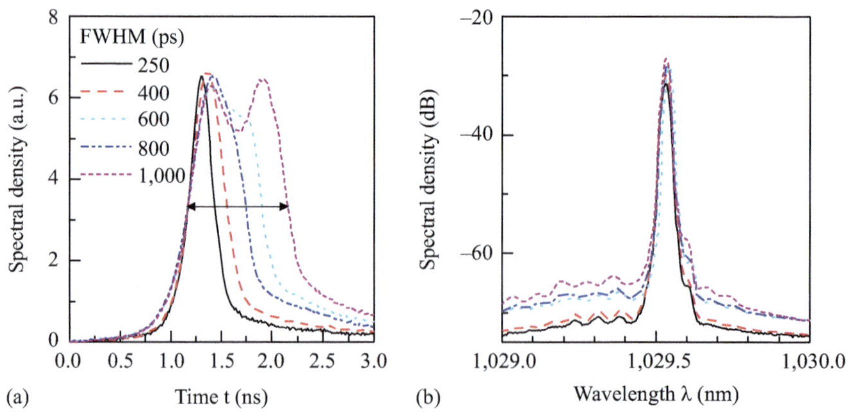

Figure 3.21 (a) Variation of optical pulse widths between 0.25 ns and 1 ns by pulse gating. (b) Corresponding optical spectra on a logarithmic scale

3.6 Wavelength limits on GaAs-based high radiance quantum well lasers

3.6.1 Introduction

Many other wavelengths outside the 9xx nm band are interesting for high power applications. Semiconductor material based on QWs grown on GaAs offers the ability to generate laser radiation between 630 and 1200 nm.

The shorter wavelengths have long drawn interest due to the visibility of the red light for large screen cinema projectors or for pumping exotic solid-state lasers such as the alexandrite laser. The longer wavelengths beyond 1100 nm are interesting for pumping of rare earth laser or for possible frequency conversion to the yellow spectral range.

3.6.2 Short wavelength limit

The short wavelength limit of QW lasers based on GaAs is defined by the barrier height possible with AlInP lattice matched to GaAs of about 2.4 eV. In order to confine the carriers to the QWs this practically limits the transition energies below 2.0 eV (i.e., >620 nm) for lasing applications at room temperature [97].

For high power applications, significant self-heating takes place, thus most works have been concentrating on the wavelength range between 630 and 640 nm (see Table 3.8). Here the optical output power of singe broad area devices is in the region below between 500 mW and 1 W. An example of characteristic curves of a broad area with a width of 60 μm and a length of 2 mm at a temperature of 15 °C is shown in Figure 3.22(a). The forward voltage is close to 2.4 V at threshold indicating good ohmic contacts (i.e., the absence of Schottky barriers). The lasing threshold is at about 900 mA and the slope efficiency at about 0.88 W/A. At 2 A the laser emits about 900 mW at a wavelength peaked at 639 nm (see inset).

The lifetime test data of seven diodes is shown in Figure 3.22(b). No diodes failed up to 3300 h when the aging test was halted. Assuming one failure at the switch off time the mean time to failure MTTF is higher than 20 000 h.

Increasing the wavelength drastically increases the device performance. At 670 nm the threshold decreases to 311 mA and the slope efficiency increases to 1.3 W/A leading to an optical output power of more than 2 W at 2 A (see Figure 3.23). Results for an aging test close to $P_{op} = 1$ W are also shown in Figure 3.23, where two devices with $L = 1.5$ mm and $W = 60$ μm, have operated to date for more than 4,000 h without failure, twice the reliable power of devices at 635 nm.

Table 3.8 GaAs-based broad area lasers at wavelength below 640 nm

Manufacturer	Aperture width/μm	Wavelength/ nm	Output power/ mW	Temperature range/°C
Intense [98]	200	635	700	15 °C
Mitsubishi [99]	40	638	500	−5 ... +40 °C
Modulight [100]	100	635	500	0 ... +25 °C
FBH [101]	60	635	500	15 °C

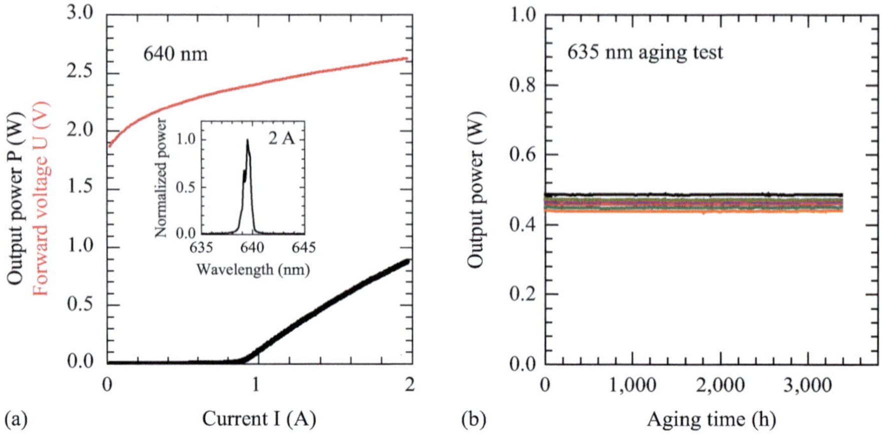

Figure 3.22 635 nm: (a) Optical power and forward voltage versus pump current at 15 °C. The inset shows the spectrum at 2 A. (b) Output power versus operation time of seven 635 nm LDs at 500 mW at 15 °C

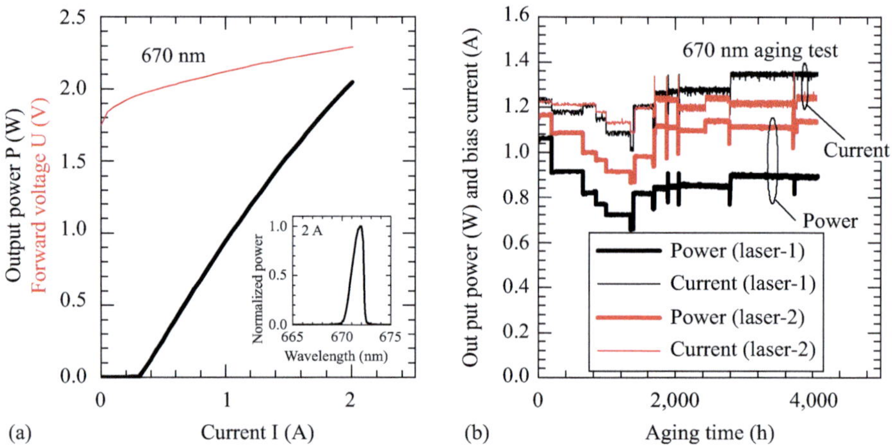

Figure 3.23 670 nm: (a) Optical power and forward voltage versus pump current at 15 °C. The inset shows the spectrum at 2 A. (b) Output power and operating current versus operation time of two 670 nm LDs at 15 °C. Notice that the pump current changes slightly, leading to steps in the power

3.6.3 Long wavelength limit

While the short wavelengths are limited by the barrier height, the long wavelength limit depends on the compressive strain of the QW. So far most publications featuring high power lasers with InGaAs QWs remained below 1,100 nm, preventing applications such as direct medical applications, pumping of holmium or thulium lasers [102], or even frequency doubling to the yellow spectral region. When a rigorous strain management is employed around the QW, it is possible to extend the emission wavelength beyond 1,100 nm, which so far have only been possible in semiconductor lasers using InAs quantum dots [103], dilute nitride N:InGaAs [104] or GaAsSb [105] as active material, each with its own advantages and drawbacks. Recent performance results are summarized in Table 3.9.

Strain management usually concentrates on compensating the compressive strain introduced by the InGaAs QW by adjacent tensile-strained layers such as GaAsP or AlGaAsP. However, the abrupt change in strain creates high stress potential leading to an increased defect formation, which in turn leads to a reduced lifetime. Hence an additional GaAs spacer layer can be employed to mitigate this effect and allow long lifetimes. This was attempted for broad area lasers with an aperture of 50 μm and a length of 3 mm that emit near 1120 nm (see Figure 3.24).

The lasers feature a threshold current of 370 mA, a slope efficiency of 0.83 W/A and reach an optical output power of 2.85 W at a pump current of 4 A. In a step stress test these lasers were operated with regularly increased output power. It was found that lasers without GaAs spacers between the QW for strain compensation started to fail at about 2 W, while the lasers with GaAs spacers operated reliably over 1000 h at 2.5 W and only failed when driven as high as 3.0 W.

At longer wavelengths of 1154 nm the results are similar. Here the efficiency is somewhat reduced. The broad area lasers feature a threshold current of 300 mA, a slope efficiency of 0.65 W/A and emit an output power of 2.35 W at 4 A (see Figure 3.25). They can be operated for a substantial period of time at 2 W output power without failure.

The same approach can also be applied to the even longer wavelength of 1,180 nm. Here the strain in the $In_xGa_{1-x}As$ QW is very close to the critical thickness. Therefore two different QWs were tested: one with a width of 9 nm and an In content of 0.365 resulting in a compressive material strain of −2.7% and one with 6 nm width with $x = 0.39$ with a strain of −2.54%. Both emitted at 1,180 nm with similar electro-optic performance characteristics, as shown in Figure 3.26(a), but the aging behavior was quite different (see Figure 3.26(b)). The devices with the 9 nm wide QW failed at power levels below 500 mW (black curves), while the 6 nm devices could be operated up to 1 W (red curves). The 6 nm devices were

Table 3.9 GaAs-based broad area lasers at wavelength above 1,100 nm

Manufacturer	Aperture width/μm	Wavelength/nm	Output power/W	Temperature range/°C
Innolume [106]	90	1120	6.0	25
FBH [107]	100	1120	4.0	25
Innolume [106]	90	1150	4.0	25
FBH [107]	50	1154	2.0	25
Innolume [106]	90	1180	4.0	25
FBH [107]	50	1180	1.0	25

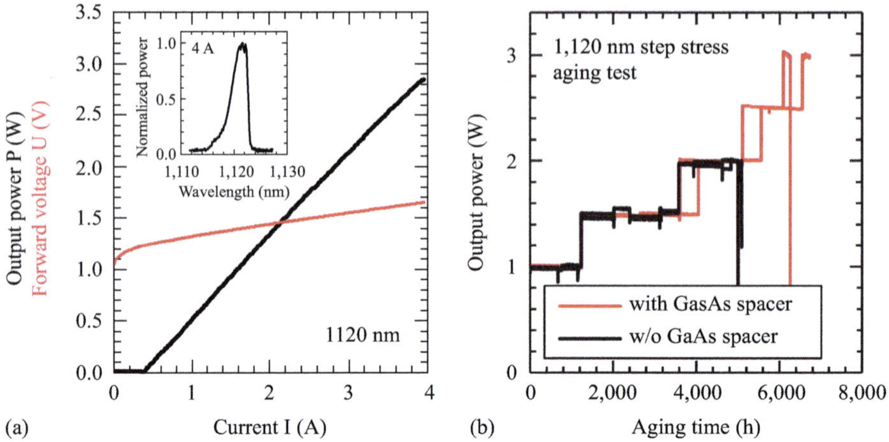

Figure 3.24 1,120 nm: (a) Optical power and forward voltage versus pump current at 25 °C. The inset shows the spectrum at 2 A. (b) Output power versus operation time of four 1,120 nm LDs at 25 °C [107]

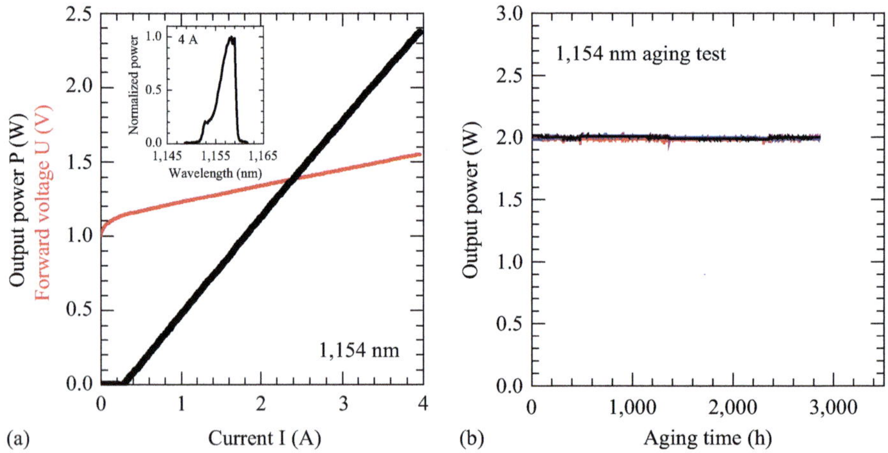

Figure 3.25 1,154 nm: (a) Optical power and forward voltage versus pump current at 25 °C. The inset shows the spectrum at 4 A. (b) Output power versus operation time of four 1,154 nm LDs at 25 °C

developed further by including the GaAs spacer also used for the shorter wavelengths. The laser featured a threshold of 0.3 A, a slope efficiency of 0.61 W/A, and emitted 2.25 W at 4 A (see Figure 3.2.3(a)).

The lifetime of the 6 nm devices with the GaAs spacers (blue curves) was significantly improved and the lasers could be operated up to 1.5 W without failure. At 1.7 W the laser started to fail, corresponding to an MTTF of about 1,500 h.

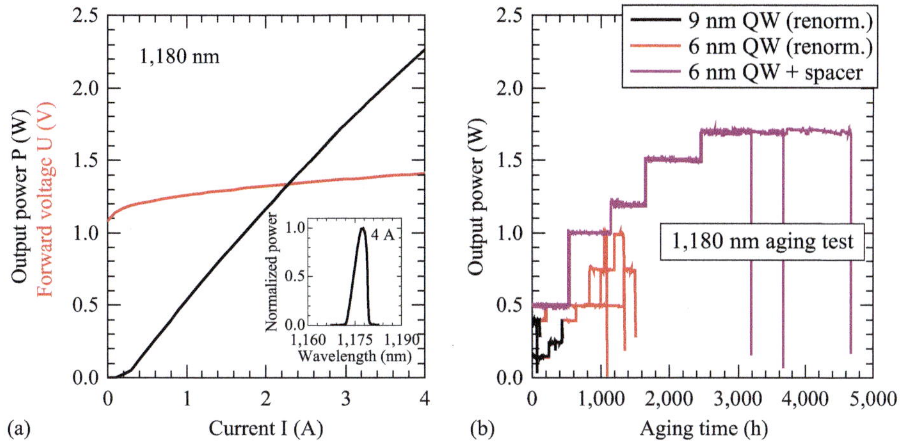

Figure 3.26 1,180 nm: (a) Optical power and forward voltage versus pump current at 25 °C. The inset shows the spectrum at 4 A. (b) Output power versus operation time of nine 1,180 nm LDs. Notice that the 9 nm and 6 nm w/o spacer devices had an aperture of 100 μm at 25 °C [107]. Their output power was renormalized (divided by two) to match the 50 μm devices with GaAs spacer

3.7 Conclusions and path forward

An overview was presented on the status of high power GaAs-based diode lasers for direct use and for pump applications, focusing on several key research areas. Research studies into the device physics and material science of diode lasers operating around λ_{op} = 940–975 nm continue to enable increases in power, efficiency, and beam quality, as needed for improved industrial systems. Advanced process technology and sophisticated device configurations also enable new functions, such as efficient emission with a narrow spectrum and the generation of very high power optical pulses, for use as seeds for solid-state lasers. In addition, the performance at longer and shorter wavelength is improving, by careful device engineering in the presence of high crystal strain and low barriers to carrier escape, respectively. The combination of steady performance progress in baseline configurations, and the enabling of new functions and wavelengths will continue to make new diode laser applications possible and enhance performance in established systems.

Acknowledgments

The authors thank Juliane Rieprich for preparing Figure 3.1, and Martin Wilkens for preparing the figures used in Section 3.3.

References

[1] Crump P., Erbert G., Wenzel H., *et al.* 'Efficient high-power laser diodes'. *IEEE J. Sel. Top. Quantum Electron.* 2013, vol. 19(4), p. 1501211.

[2] Erbert G., Bärwolff A., Sebastian J., Tomm J. 'High-power broad-area diode lasers and laser bars', in Diehl R. (ed.); *High-Power Diode Lasers: Fundamentals, Technology, Applications: With Contributions by Numerous Experts* (Berlin, Springer, 2000), pp. 173–223.

[3] Wenzel H. 'Basic aspects of high-power semiconductor laser simulation'. *IEEE J. Sel. Top. Quantum Electron.* 2013, vol. 19(5), pp. 1502913-1–1502913-13.

[4] Liu X., Zhao W., Xiong L., Liu H. 'Packaging process of high power semiconductor lasers', in Liu X., Zhao W., Xiong L., Liu H. (eds.); *Packaging of High Power Semiconductor Lasers* (New York, Springer, 2015), pp. 185–226.

[5] Fan T.Y. 'Laser beam combining for high-power, high-radiance sources'. *IEEE J. Sel. Top. Quantum Electron.* 2005, vol. 11(3), pp. 567–577.

[6] Haas M., Rauch S., Nagel S., Beißwanger R., Dekorsy T., Zimer H. 'Beam quality deterioration in dense wavelength beam-combined broad-area diode lasers'. *IEEE J. Quantum Electron.* 2017, vol. 53(3), pp. 2600111-1–2600111-11.

[7] Crump P. '9.3 Diode lasers in the near infra-red spectral range', in Weber H., Loosen P., Poprawe R. (eds.); *Landolt–Börnstein – Numerical Data and Functional Relationships in Science and Technology, Group VIII – Advanced Materials and Technologies, Laser Physics and Applications, Subvolume B (Laser Systems: Part 3)* (Berlin, Springer, 2011), pp. 54–73.

[8] Crump P., Patterson S., Dong W., *et al.* 'Room temperature high power mid-IR diode laser bars for atmospheric sensing applications'. *Proc. SPIE* 2007, vol. 6552, pp. 655216-1–655216-11.

[9] Kneissl M., Rass J. 'Blue and green-emitting laser diodes', in Weber H., Loosen P., Poprawe R. (eds.); *Landolt–Börnstein – Numerical Data and Functional Relationships in Science and Technology, Group VIII – Advanced Materials and Technologies, Laser Physics and Applications, Subvolume B (Laser Systems: Part 3)* (Berlin, Springer, 2011), pp. 27–40.

[10] Frevert C., Crump P., Bugge F., Knigge S., Ginolas A., Erbert G. 'Low-temperature optimized 940 nm diode laser bars with 1.98 kW peak power at 203 K'. *Proc. CLEO/QELS* San Jose, USA, May 2015 (IEEE, 2015), p. SM3F.8.

[11] Frevert C., Bugge F., Knigge S., Ginolas A., Erbert G., Crump P. '940nm QCW diode laser bars with 70% efficiency at 1 kW output power at 203K: analysis of remaining limits and path to higher efficiency and power at 200K and 300K'. *Proc. SPIE* 2016, vol. 9733, pp. 97330L-1–97330L-13.

[12] Strohmaier S.G., Erbert G., Rataj T., *et al.* 'Forward development of kW-class power diode laser bars'. *Proc. SPIE* 2018, vol. 10514, p. 1051409.

[13] Knapczyk M.T., Jacob J.H., Eppich H., *et al.* '70% efficient near 1 kW single 1-cm laser-diode bar at 20°C'. *Proc. SPIE* 2011, vol. 7918, p. 79180F.

[14] Gapontsev V., Moshegov N., Berezin I., *et al.* 'Highly-efficient high-power pumps for fiber lasers'. *Proc. SPIE* 2017, vol. 10086, p. 1008604.

[15] Crump P., Wenzel H., Erbert G., Tränkle G. 'Progress in increasing the maximum achievable output power of broad area diode lasers'. *Proc. SPIE* 2012, vol. 8241, p. 82410U.

[16] Crump P., Wenzel H., Kaul T., *et al.* 'Experimental and theoretical studies into the limits to peak power in GaAs-based diode lasers'. *Proc. 25th IEEE Int. Semicon. Las. Conf.* Kobe, Japan, September 2016 (IEEE, 2016), pp. 1–2.

[17] Crump P., Wang J., Patterson S., *et al.* 'Diode laser efficiency increases enable >400-W peak power from 1-cm bars and show a clear path to peak powers in excess of 1-kW'. *Proc. SPIE* 2006, vol. 6104, pp. 64560M-1–64560M–10.

[18] Kaul T., Erbert G., Maaßdorf A., Martin D., Crump P. 'Extreme triple asymmetric (ETAS) epitaxial designs for increased efficiency at high powers in 9xx-nm diode lasers'. *Proc. SPIE* 2018, vol. 10514, pp. 105140A-1–64560M-7.

[19] Petrescu-Prahova I.B., Modak P., Goutain E., *et al.* '253 mW/μm Maximum power density from 9xx nm epitaxial laser structures with d/Γ greater than 1 μm' *Proc. 21st IEEE Int. Semicon. Las. Conf.* Sorrento, Italy, 2008 (IEEE, 2008), pp. 135–136.

[20] Slipchenko S.O., Sokolova Z.N., Pikhtin N.A., Borschev K.S., Vinokurov D.A., Tarasov I.S. 'Finite time of carrier energy relaxation as a cause of optical-power limitation in semiconductor lasers'. *Semiconductors* 2006, vol. 40(8), pp. 990–995.

[21] Crump P., Winterfeldt M., Decker J., *et al.* 'Novel approaches to increasing the brightness of broad area lasers'. *Proc. SPIE* 2016, vol. 9767, p. 97671L–1.

[22] Zeghuzi A., Radziunas M., Wenzel H., Wünsche H.-J., Bandelow U., Knigge A., 'Modeling of current spreading in high-power broad-area lasers and its impact on the lateral far field divergence' *Proc. SPIE* 2018, vol. 10526, pp. 105261H.

[23] Rauch S., Modak P., Holly C., Zimer H. 'Beam quality improvement of broad-area laser diodes by symmetric facet reflectivities'. *Proc. IEEE Int. Semicon. Las. Conf. ISLC* Santa Fe, USA, September 2018 (IEEE, 2018), pp. 1–2.

[24] Bachmann A., Lauer C., Furitsch M., König H., Müller M., Strauß U. 'Recent brightness improvements of 976 nm high power laser bars'. *Proc. SPIE* 2017, vol. 10086, pp. 100861-1–100861-9.

[25] Wenzel H., Crump P., Fricke J., Ressel P., Erbert G. 'Suppression of higher-order lateral modes in broad-area diode lasers by resonant anti-guiding'. *IEEE J. Quantum Electron.* 2013, vol. 49(12), pp. 1102–1108.

[26] Crump P., Frevert C., Hösler H., *et al.* 'Cryogenic ultra-high power infrared diode laser bars'. *Proc. SPIE* 2014, vol. 9002, pp. 90021I-1–90021I-11.

[27] Bai J.G., Leisher P., Zhang S., *et al.* 'Mitigation of thermal lensing effect as a brightness limitation of high-power broad area diode lasers'. *Proc. SPIE* 2011, vol. 7953, p. 79531F.

[28] Rieprich J., Winterfeldt M., Kernke R., Tomm J.W., Crump P. 'Chip-carrier thermal barrier and its impact on lateral thermal lens profile and beam parameter product in high power broad area lasers'. *J. Appl. Phys.* 2018, vol. 123(12), p. 125703.

[29] Wang X., Erbert G., Wenzel H., *et al.* '17-W near-diffraction-limited 970-nm output from a tapered semiconductor optical amplifier'. *IEEE Photonics Technol. Lett.* 2013, vol. 25(2), pp. 115–118.

[30] Platz R., Eppich B., Crump P., *et al.* '940-nm broad area diode lasers optimized for high pulse-power fiber coupled applications'. *IEEE Photonics Technol. Lett.* 2014, vol. 26(6), pp. 625–628.

[31] Pittroff W., Eppich B., Erbert G., Platz R., Tyralla D., Tränkle G. 'Simple design for fiber coupled 9xx nm kW-QCW pump module with high duty cycle based on customized chips and lateral heat removal'. *Proc. SPIE* 2014, vol. 8965, pp. 896515-1–896515-8.

[32] Platz R., Frevert C., Eppich B., *et al.* 'Progress in high duty cycle, highly efficient fiber coupled 940-nm pump modules for high-energy class solid state lasers'. *Proc. SPIE* 2018, vol. 10513, p. 1051319.

[33] Chi M., Petersen P.M. '15.1 Edge emitters with external cavities', in Weber H., Loosen P., Poprawe R. (eds.); *Landolt–Börnstein – Numerical Data and Functional Relationships in Science and Technology, Group VIII –Advanced Materials and Technologies, Laser Physics and Applications, Subvolume B (Laser Systems: Part 3)* (Berlin, Springer, 2011), pp. 223–235.

[34] Crump P., Hengesbach S., Witte U., Hoffmann H.D., Erbert G., Tränkle G. 'High-power diode lasers optimized for low-loss smile-insensitive external spectral stabilization'. *IEEE Photonics Technol. Lett.* 2012, vol. 24(8), pp. 703–705.

[35] Crump P., Knigge S., Maassdorf A., *et al.* 'Low-loss, smile-insensitive external frequency-stabilization of high power diode lasers enabled by vertical designs with extremely low divergence angle and high efficiency'. *Proc. SPIE* 2013, vol. 8605, pp. 86050T1–86050T13.

[36] Koenning T., McCormick D., Irvin D., Stapleton D., Guiney T., Patterson S. 'Narrow-line fiber-coupled modules for DPAL pumping'. *Proc. SPIE* 2015, vol. 9348, pp. 934805-1–934805-9.

[37] Kissel H., Leonhäuser B., Tomm J.W., Hempel M., Biesenbach J. 'Accelerated degradation of high power diode lasers caused by external optical feedback operation'. *Proc. 26th IEEE Int. Semicon. Las. Conf. ISLC* Santa Fe, USA, September 2018 (IEEE, 2018), pp. 1–2.

[38] Pietrzak A., Wenzel H., Crump P., *et al.* '1060-nm ridge waveguide lasers based on extremely wide waveguides for 1.3-W continuous-wave emission into a single mode with FWHM divergence angle of $9° \times 6°$'. *IEEE J. Quantum Electron.* 2012, vol. 48(5), pp. 568–575.

[39] Wenzel H., Bugge F., Dallmer M., *et al.* 'Fundamental-lateral mode stabilized high-power ridge-waveguide lasers with a low beam divergence'. *IEEE Photonics Technol. Lett.* 2008, vol. 20(3), pp. 214–216.

[40] Bettiati M., Cargemel V., Pagnod P., *et al.* 'Reaching 1 watt reliable output power on single-mode 980 nm pump lasers'. *Proc. SPIE* 2009, vol. 7198, p. 71981D.

[41] Guermache A., Voiriot V., Capella R.-M., Gallion P., Jacquet J. 'New design rules and experimental study of slightly flared 1480-nm pump lasers'. *IEEE Photonics Technol. Lett.* 2006, vol. 18(6), pp. 782–784.

[42] Sverdlov B., Pfeiffer H.-U., Zibik E., *et al.* 'Optimization of fiber coupling in ultra-high power pump modules at 1 = 980 nm'. *Proc. SPIE* 2013, vol. 8605, pp. 860508-1–860508-10.

[43] Zhao S., Wang Y., Qu H., *et al.* '2-W high-efficiency ridge-waveguide lasers with single transverse mode and low vertical divergence'. *IEEE Photonics Technol. Lett.* 2017, vol. 29(23), pp. 2005–2007.

[44] Wilkens M., Wenzel H., Fricke J., *et al.* 'High-efficiency broad-ridge waveguide lasers'. *IEEE Photonics Technol. Lett.* 2018, vol. 30(6), pp. 545–548.

[45] Crump P., Brox O., Bugge F., *et al.* 'Chapter 2 – High-power, high-efficiency monolithic edge-emitting GaAs-based lasers with narrow spectral widths', in Coleman J.J., Bryce A.C., Jagadish C. (eds.); *Semiconductors and Semimetals 86* (New York, Elsevier, 2012), pp. 49–91.

[46] Crump P., Schultz C.M., Wenzel H., Erbert G., Tränkle G. 'Efficiency-optimized monolithic frequency stabilization of high-power diode lasers'. *J. Phys. Appl. Phys.* 2013, vol. 46(1), p. 013001.

[47] Decker J., Crump P., Fricke J., *et al.* '25-W monolithic spectrally stabilized 975-nm minibars for dense spectral beam combining'. *IEEE Photonics Technol. Lett.* 2015, vol. 27(15), pp. 1675–1678.

[48] Bugge F., Mogilatenko A., Zeimer U., *et al.* 'Characterization and optimization of 2-step MOVPE growth for single-mode DFB or DBR laser diodes'. *J. Cryst. Growth* 2011, vol. 315(1), pp. 74–77.

[49] Chu S.N.G., Tanbun-Ek T., Logan A., *et al.* 'Grating overgrowth and defect structures in distributed-feedback-buried heterostructure laser diodes'. *IEEE J. Sel. Top. Quantum Electron.* 1997, vol. 3(3), pp. 862–873.

[50] Kulkova I.V., Kadkhodazadeh S., Kuznetsova N., Huck A., Semenova E.S., Yvind K. 'High-quality MOVPE butt-joint integration of InP/AlGaInAs/InGaAsP-based all-active optical components'. *J. Cryst. Growth* 2014, vol. 402, pp. 243–248.

[51] Brox O., Bugge F., Mogilatenko A., *et al.* 'Distributed feedback lasers in the 760 to 810 nm range and epitaxial grating design'. *Semicond. Sci. Technol.* 2014, vol. 29(9), p. 095018.

[52] Brox O., Tawfieq M., Casa P.D., *et al.* 'Realisation of a widely tuneable sampled grating DBR laser emitting around 970 nm'. *Electron. Lett.* 2017, vol. 53(11), pp. 744–746.

[53] Maaßdorf A., Weyers M. 'In-situ etching of GaAs/Al_x $Ga_{1-x}As$ by CBr_4'. *J. Cryst. Growth* 2008, vol. 310(23), pp. 4754–4756.

[54] Schultz C.M., Crump P., Maaßdorf A., *et al.* 'In situ etched gratings embedded in AlGaAs for efficient high power 970 nm distributed feedback broad-area lasers'. *Appl. Phys. Lett.* 2012, vol. 100(20), p. 201115.

[55] Crump P., Decker J., Maaßdorf A., *et al.* 'Monolithically wavelength-stabilized high power diode lasers'. *Proc. IEEE High Power Diode Lasers and Systems Conference (HPD)* Coventry, UK, October 2017 (IEEE, 2017), pp. 11–12.

[56] Hofling E., Schafer F., Reithmaier J.P., Forchel A. 'Edge-emitting GaInAs-AlGaAs microlasers'. *IEEE Photonics Technol. Lett.* 1999, vol. 11(8), pp. 943–945.

[57] Lim M.H., Murphy T.E., Ferrera J., Damask J.N., Smith H.I. 'Fabrication techniques for grating-based optical devices'. *J. Vac. Sci. Technol., B: Microelectron. Nanometer Struct. Process. Meas. Phenom.* 1999, vol. 17(6), pp. 3208–3211.

[58] Docter B., Geluk E.J., Sander-Jochem M.J.H., Karouta F., Smit M.K. 'Deep etching of DBR gratings in InP using Cl_2 based ICP processes'. *Proc. Symp. IEEE/LEOS* Eindhoven, The Netherlands, 2006 (2006), pp. 97–100.

[59] Maximov M.V., Ramushina E.M., Skopina V.I., *et al.* 'Edge-emitting InGaAs/GaAs lasers with deeply etched semiconductor/air distributed Bragg reflector mirrors'. *Semicond. Sci. Technol.* 2002, vol. 17(11), pp. L69–L71.

[60] Fricke J., Decker J., Maaßdorf A., *et al.* 'DFB lasers with apodized surface gratings for wavelength stabilization and high efficiency'. *Semicond. Sci. Technol.* 2017, vol. 32(7), p. 075012.

[61] Higashiki T., Nakasugi T., Yoneda I. 'Nanoimprint lithography and future patterning for semiconductor devices'. *J. Micro/Nanolithogr. MEMS MOEMS* 2011, vol. 10(4), p. 043008.

[62] Krauss T.F., Painter O.J., Scherer A., Roberts J.S., Rue R.M.D.L. 'Photonic microstructures as laser mirrors'. *Opt. Eng.* 1998, vol. 37(4), pp. 1143–1149.

[63] Shutts S., Edwards G., Elliott S.N., Smowton P.M., Krysa A.B. 'Deep etched distributed Bragg reflector (DBR) InP/AlGaInP quantum dot lasers'. *Proc. SPIE* 2011, vol. 7953, p. 795308.

[64] Avary K., Rennon S., Klopf F., Reithmaier J.P., Forchel A. 'Reactive ion etching of deeply etched DBR-structures with reduced air-gaps for highly reflective monolithically integrated laser mirrors'. *Microelectron. Eng.* 2001, vol. 57–58, pp. 593–598.

[65] Dumitrescu M., Telkkälä J., Karinen J., *et al.* 'Narrow-linewidth distributed feedback lasers with laterally coupled ridge-waveguide surface gratings fabricated using nanoimprint lithography'. *Proc. SPIE* 2011, vol. 7953, p. 79530B.

[66] Achtenhagen M., Amarasinghe N.V., Jiang L., Threadgill J., Young P. 'Spectral properties of high-power distributed Bragg reflector lasers'. *J. Light. Technol.* 2009, vol. 27(16), pp. 3433–3437.

[67] Fricke J., Wenzel H., Matalla M., Klehr A., Erbert G. '980-nm DBR lasers using higher order gratings defined by i-line lithography'. *Semicond. Sci. Technol.* 2005, vol. 20(11), p. 1149.

[68] Fricke J., Bugge F., Ginolas A., *et al.* 'High-power 980-nm broad-area lasers spectrally stabilized by surface Bragg gratings'. *IEEE Photonics Technol. Lett.* 2010, vol. 22(5), pp. 284–286.

[69] Selmic S.R., Evans G.A., Chou T.M., *et al.* 'Single frequency 1550-nm AlGaInAs-InP tapered high-power laser with a distributed Bragg reflector'. *IEEE Photonics Technol. Lett.* 2002, vol. 14(7), pp. 890–892.

[70] Müller A., Zink C., Fricke J., *et al.* 'Efficient, high brightness 1030 nm DBR tapered diode lasers with optimized lateral layout'. *IEEE J. Sel. Top. Quantum Electron.* 2017, vol. 23(6), pp. 1–7.

[71] Tawfieq M., Müller A., Fricke J., *et al.* 'Extended 9.7 nm tuning range in a MOPA system with a tunable dual grating Y-branch laser'. *Opt. Lett.* 2017, vol. 42(20), pp. 4227–4230.

[72] Fricke J., Klehr A., Brox O., *et al.* 'Y-branch coupled DFB-lasers based on high-order Bragg gratings for wavelength stabilization'. *Semicond. Sci. Technol.* 2013, vol. 28(3), p. 035009.

[73] Decker J., Fricke J., Maaßdorf A., Erbert G., Tränkle G., Crump P. 'Non-uniform DFB-surface-etched gratings for enhanced performance high power, high brightness broad area lasers'. *Proc. SPIE* 2017, vol. 10086, p. 100860R.

[74] Wenzel H., Klehr A., Schwertfeger S., *et al.* 'Compact sources for the generation of high-peak power wavelength-stabilized laser pulses in the picoseconds and nanoseconds ranges'. *Proc. SPIE* 2012, vol. 8241, pp. 82410V1-1–82410V1-11.

[75] Chen S., Yoshita M., Sato A., Ito T., Akiyama H., Yokoyama H. 'Dynamics of short-pulse generation via spectral filtering from intensely excited gain-switched 1.55-μm distributed-feedback laser diodes'. *Opt. Express* 2013, vol. 21(9), pp. 10597–10605.

[76] Klehr A., Wenzel H., Brox O., Schwertfeger S., Staske R., Erbert G. 'Dynamics of a gain-switched distributed feedback ridge waveguide laser in nanoseconds time scale under very high current injection conditions'. *Opt. Express* 2013, vol. 21(3), pp. 2777–2786.

[77] Klehr A, Schwertfeger S, Wenzel H, *et al.* 'Dynamics of high power gain switched DFB RW laser under high current pulse excitation on a nanosecond time scale'. *Proc. SPIE* 2013, vol. 8840, pp. 88401N1–88401N19.

[78] Le Galès G., Joly S., Marcello G., *et al.* 'Seed laser diodes in pulsed operation: limitations and reliability'. *Proc. SPIE* 2017, vol. 10098, pp. 100980S1–100980S9.

[79] Zeller W., Kamp M., Koeth J., Worschech L. 'High-power pulsed 976-nm DFB laser diodes'. *Proc. SPIE* 2010, vol. 7682, p. 76820T.

[80] O'Daniel J.K., Achtenhagen M. 'High-power spectrally-stable DBR semiconductor lasers designed for pulsing in the nanosecond regime'. *Proc. SPIE* 2010, vol. 7616, pp. 76160W1–76160W11.

[81] Klehr A., Sumpf B., Hasler K., Fricke J., Liero A., Erbert G. 'High-power pulse generation in GHz range with 1064-nm DBR tapered laser'. *IEEE Photonics Technol. Lett.* 2010, vol. 22(11), pp. 832–834.

[82] Riecke S.M., Lauritsen K., Erdmann R., Übernickel M., Paschke K., Erbert G. 'Pulse-shape improvement during amplification and second-harmonic generation of picosecond pulses at 531 nm'. *Opt. Lett.* 2010, vol. 35(10), pp. 1500–1502.
[83] Riecke S.M., Wenzel H., Schwertfeger S., *et al.* 'Picosecond spectral dynamics of gain-switched DFB lasers'. *IEEE J. Quantum Electron.* 2011, vol. 47(5), pp. 715–722.
[84] Zibik E.A., Bertrand A., Kaiser W., *et al.* 'Laser diodes with distributed feedback for application as subnanosecond fiber laser seeder'. *Proc. SPIE* 2012, vol. 8277, pp. 82771O1–827710O10.
[85] Wenzel H., Schwertfeger S., Klehr A., Jedrzejczyk D., Hoffmann T., Erbert G. 'High peak power optical pulses generated with a monolithic master-oscillator power amplifier'. *Opt. Lett.* 2012, vol. 37(11), pp. 1826–1828.
[86] Schwertfeger S., Klehr A., Hoffmann T., Liero A., Wenzel H., Erbert G. 'Generation of sub-100 ps pulses with a peak power of 65 W by gain switching, pulse shortening, and pulse amplification using a semiconductor-based master oscillator-power amplifier system'. *Appl. Opt.* 2013, vol. 52(14), pp. 3364–3367.
[87] Yokoyama Y., Takada K., Kageyama T., *et al.* '1064-nm DFB laser diode modules applicable to seeder for pulse-on-demand fiber laser systems'. *Opt. Fiber Technol.* 2014, vol. 20(6), pp. 714–724.
[88] Schwertfeger S., Klehr A., Hoffmann T., Liero A., Wenzel H., Erbert G. 'Picosecond pulses with 50 W peak power and reduced ASE background from an all-semiconductor MOPA system'. *Appl. Phys. B.* 2011, vol. 103(3), pp. 603–607.
[89] Klehr A., Liero A., Wenzel H., *et al.* '1030-nm diode-laser-based light source delivering pulses with nanojoule energies and picosecond duration adjustable by mode locking or pulse gating operation'. *Proc. SPIE* 2017, vol. 10123, pp. 101230D1–101230D11.
[90] Schwertfeger S., Klehr A., Liero A., Erbert G., Trankle G. 'High-power picosecond pulse generation due to mode-locking with a monolithic 10-mm-long four-section DBR laser at 920 nm'. *IEEE Photonics Technol. Lett.* 2007, vol. 19(23), pp. 1889–1891.
[91] Klehr A., Liero A., Hoffmann T., *et al.* 'Compact ps-pulse laser source with free adjustable repetition rate and nJ pulse energy on microbench'. *Proc. SPIE* 2011, vol. 7953, pp. 79531D1–79531D10.
[92] Harth F., Ulm T., Lührmann M., *et al.* 'High power laser pulses with voltage controlled durations of 400–1000 ps'. *Opt. Express* 2012, vol. 20(7), pp. 7002–7007.
[93] Klehr A., Wenzel H., Fricke J., Bugge F., Erbert G. 'Generation of spectrally stable continuous-wave emission and ns pulses with a peak power of 4 W using a distributed Bragg reflector laser and a ridge-waveguide power amplifier'. *Opt. Express* 2014, vol. 22(20), pp. 23980–23989.

[94] Klehr A., Wenzel H., Fricke J., *et al.* 'Generation of spectrally-stable continuous-wave emission and ns pulses at 800 nm and 975 nm with a peak power of 4 W using a distributed Bragg reflector laser and a ridge-waveguide power amplifier'. *Proc. SPIE* 2015, vol. 9382, pp. 93821I1–93821I10.
[95] Vu T.N., Klehr A., Sumpf B., Wenzel H., Erbert G., Tränkle G. 'Wavelength stabilized ns-MOPA diode laser system with 16 W peak power and a spectral line width below 10 pm'. *Semicond. Sci. Technol.* 2014, vol. 29(3), p. 035012.
[96] Vu T.N., Klehr A., Sumpf B., Wenzel H., Erbert G., Tränkle G. 'Tunable 975 nm nanosecond diode-laser-based master-oscillator power-amplifier system with 16.3 W peak power and narrow spectral linewidth below 10 pm'. *Opt. Lett.* 2014, vol. 39(17), pp. 5138–5141.
[97] Coleman J.J., Bryce A.C., Jagadish C. 'Advances in Semiconductor *Lasers'*. Cambridge, USA: Academic Press; 2012.
[98] Intensco Corp *Intense Series 1200 Diode Laser Selection Chart* [online]. 2019. Available from http://www.intenseco.com/products/visible/series-1200-diode-laser-selection-chart/ [accessed January 2019].
[99] Nishida T., Shimada N., Ogawa T., Miyashita M., Yagi T. 'Short wavelength limitation in high power AlGaInP laser diodes'. *Proc. SPIE* 2011, vol. 7918, p. 791811.
[100] Modulight, Inc. *ChiliLase Chips 632...690 nm High-Power Laser Unmounted Chips for CW Applications* [online]. 2019. Available from https://www.modulight.com/datasheets/combined/ChiliLase%20Chips%20Datasheet.pdf [accessed January 2019].
[101] Pohl J., Bugge F., Blume G., *et al.* 'Combined Mg/Zn p-type doping for AlGaInP laser diodes'. *J. Cryst. Growth* 2015, vol. 414, pp. 215–218.
[102] Jackson S.D., Bugge F., Erbert G. 'Directly diode-pumped holmium fiber lasers'. *Opt. Lett.* 2007, vol. 32(17), pp. 2496–2498.
[103] Ribbat C., Sellin R., Grundmann M., Bimberg D. 'High power quantum dot lasers at 1160 nm'. *Phys. Status Solidi B.* 2001, vol. 224(3), pp. 819–822.
[104] Konttinen J., Korpijärvi V.-M. 'Frequency-converted dilute nitride laser diodes for mobile display applications'. *Nanoscale Res. Lett.* 2014, vol. 9(1), p. 82.
[105] Blum O., Klem J.F. 'Characteristics of GaAsSb single-quantum-well-lasers emitting near 1.3 μm'. *IEEE Photonics Technol. Lett.* 2000, vol. 12(7), pp. 771–773.
[106] Innolume GmbH *Products – Broad-Area Laser Diodes* [online]. 2019. Available from https://www.innolume.com/products/BA_HPLD.htm, [accessed January 2019].
[107] Bugge F., Bege R., Blume G., *et al.* 'Lifetime behavior of laser diodes with highly strained InGaAs QWs and emission wavelength between 1120 nm and 1180 nm'. *J. Cryst. Growth* 2018, vol. 491, pp. 31–35.

Chapter 4

Quantum cascade lasers

Arkadiy Lyakh[1–3]

4.1 Introduction

Quantum cascade lasers (QCLs) are majority carrier semiconductor devices that rely on radiative electron transitions between quantized energy states confined within the conduction band. Their development in the late 1980s and early 1990s was driven by a strong need for a compact and efficient laser source in the mid-wave infrared (MWIR; 3 μm to 5 μm) and long-wave infrared (LWIR; 8 μm to 12 μm) spectral regions. Compared to other infrared laser sources, such as gas lasers, fiber lasers, and optical parametric oscillators, QCLs offer a unique combination of a compact size, high efficiency, high optical power, and design flexibility in tailoring device characteristics to a specific infrared laser application. As such, QCLs have been of significant interest for various defense applications, such as standoff detection and free space optical communications, as well as for a number of industrial applications, including infrared imaging and spectroscopy. QCL technology has been commercialized by a number of companies, including Pranalytica, Daylight Solutions, Forward Photonics, EOS Photonics (now Pendar Technologies), Alpes Lasers, Thorlabs, AdTech Optics, Hamamatsu Photonics, and others.

QCL general structure is similar to that for diode lasers: QCL chips are typically 1 to 10 mm in length, several hundred microns in width, and approximately 200 microns in thickness. A photo of a QCL chip on a dime coin is shown in Figure 4.1(a). Electrical current is driven perpendicular to device surface through an approximately 10 μm-wide stripe, meaning that laser core footprint is 10 microns by several millimeters. This geometry results in a typical emission aperture for edge-emitting QCLs of 10 μm × 2 μm, as shown in Figure 4.1(b).

While QCLs are also semiconductor devices, their laser core structure is very different from that of diode lasers. QCL cores are typically comprised of over 1,000 nm-thick layers grouped into tens of stages. A QCL stage is a sequence of alternating quantum wells and barriers designed, in ideal case, for one electron to emit one photon. As electrical current is driven through a QCL structure, each electron

[1]NanoScience Technology Center, University of Central Florida, Orlando, FL, USA
[2]Department of Physics, University of Central Florida, Orlando, FL, USA
[3]CREOL, College of Optics and Photonics, University of Central Florida, Orlando, FL, USA

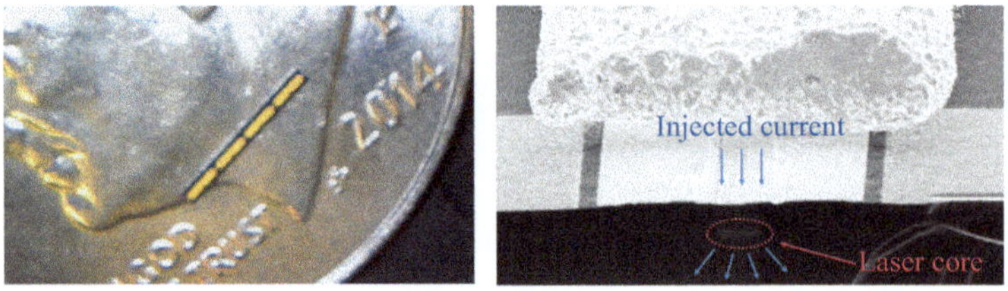

(a) (b)

Figure 4.1 (a) Photo of a QCL chip on a dime. (b) SEM photo of an edge-emitting buried heterostructure QCL

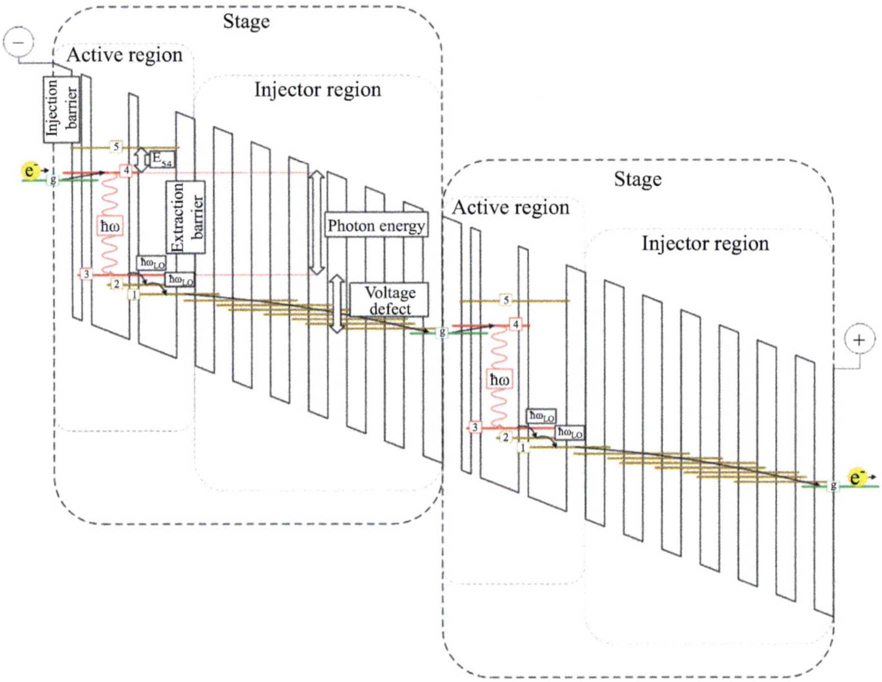

Figure 4.2 Schematic of two QCL core stages

propagates through multiple stages, emitting multiple photons. A schematic of two consecutive stages illustrating the cascading (carrier recycling) effect is shown in Figure 4.2.

Since QCL emission wavelength is largely independent of semiconductor material band gap, these devices can be designed to emit light in a very broad spectral range spanning from approximately 3 microns to hundreds of microns (terahertz region). For the same reason, QCLs emitting at the same wavelength can be realized by employing various compound semiconductor materials. One of the main requirements in this case is for laser core materials to provide a sufficient conduction band offset to prevent carrier leakage over the barriers. The rule of thumb for state of the art QCLs is for band offset (difference in bandgap) to be approximately three times larger than laser transition energy.

The large band offset achievable with AlInAs/InGaAs material composition strain-balanced relative to InP substrate, combined with an extensive knowledge acquired over the last several decades on near-infrared InP-based diode laser growth, processing, and packaging, allowed for rapid progress in InP-based QCL performance. Figure 4.3 demonstrates that QCL-pulsed efficiency exceeding 10% has been reported for various laser core designs with emission wavelength spanning from 4 to 11 μm [1–10], with the highest values of 27% reported at 4.9 μm [6] and 28% reported at 5.6 μm [7]. Also, continuous wave (CW) efficiency values exceeding 8% have been reported throughout the same spectral range, with the highest value of 21% achieved at 4.9 μm [6].

High laser efficiency directly translates to a high CW optical power as laser self-heating is reduced. The highest CW optical power achieved for various QCL designs is presented in Figure 4.4. The watt-level operation was reported for wavelengths as short as 3.76 μm [2] and as long as 10.7 μm [10], with 4.6 W reported at 4.7 μm [4] and a record-high 5 W reported at 4.9 μm [6].

In this chapter, we present a brief overview of the QCL technology with an emphasis on the commercially important 3 μm to 12 μm spectral region covering the infrared fingerprint spectral region and the two infrared atmospheric windows (3 to 5 μm and 8 to 12 μm). Specifically, laser core design, waveguide design, and various aspects of mode control for QCLs are discussed below.

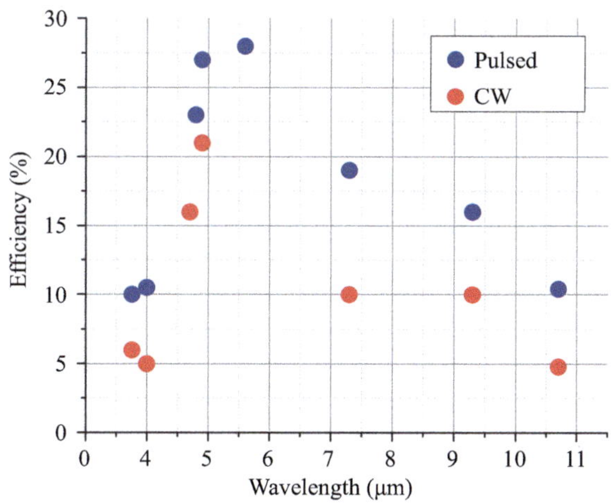

Figure 4.3 Highest pulsed (blue dots) and continuous wave (red dots) efficiencies for room temperature quantum cascade lasers reported in the wavelength range from 3 to 11 μm [1–10]

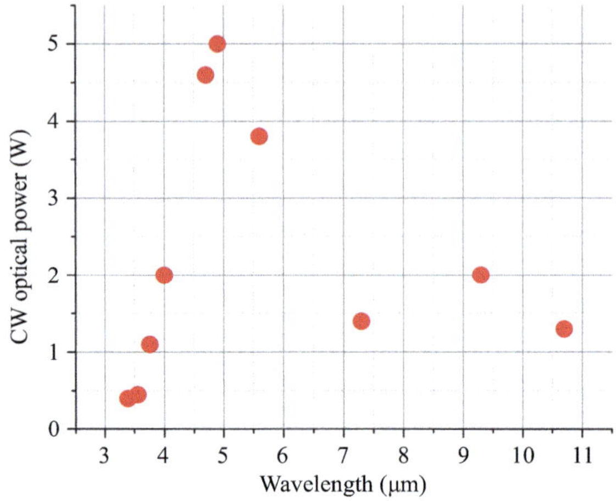

Figure 4.4 Highest continuous wave (CW) optical power for room temperature quantum cascade lasers reported in the wavelength range from 3 to 11 μm [1–10]

4.2 Governing equations for pulsed QCL operation

The flexibility in controlling energy levels supported by the multilayered laser core, as well as flexibility in controlling corresponding matrix elements and carrier lifetimes, opens up a broad optimization space for the key QCL parameters. Rigorous treatment of carrier transport through a multilayered intersubband structure is a very complex numerical problem (see [11,12]) and it is often impractical to carry out a full transport simulation in the QCL design optimization. Instead, a much simpler rate-equation model has been primarily used for guiding the development of high-performance QCLs.

Within the framework of the rate-equation model, carrier transport through the active and injector regions (see Figure 4.2) of each stage of the structure is described relying on numerically calculated scattering rates between eigenstates of a bandgap profile, neglecting coherent effects, while carrier tunneling from the lowest injector state to the upper laser level of the next stage is described using the density matrix approach that captures the effects of both coherent and incoherent tunneling [13,14]. The eigenstates supported by the structure and corresponding wavefunctions are typically determined using a simple shooting method for solving Schrodinger equation. A detailed description of the shooting method can be found in [15].

In its simplest form, the rate equations result in the following two equations for laser pulsed slope efficiency *dP/dI* and pulsed threshold current density J_{th}, the two main laser characteristics:

$$\frac{dP}{dI} = \frac{h\upsilon}{q} N_s \frac{\alpha_m}{\alpha_m + \alpha_w} \frac{1}{1 + \tau_3/\tau_4} \eta_i \tag{4.1}$$

$$J_{\text{th}} = J_{tr} + \frac{\alpha_m + \alpha_w}{\Gamma g} \tag{4.2}$$

where N_s is the number of stages, α_m is the mirror losses, α_w is the waveguide losses, τ_4 is the upper laser level lifetime, τ_3 is the lower laser level lifetime, η_i is the injection efficiency for the upper laser level, J_{tr} is the transparency current density, Γ is the mode overlap factor with the active region, and g is the differential gain.

For MWIR QCLs, the $\frac{\alpha_m}{\alpha_m + \alpha_w}$ term and $\frac{1}{1 + \tau_3/\tau_4}$ term on the right-hand side of (4.1) are both typically in the range of 0.8–0.9 and laser slope efficiency per stage is largely defined by injection efficiency for the upper laser level. The $\frac{\alpha_m}{\alpha_m + \alpha_w}$ term was estimated in this case assuming a 3 mm uncoated chip configuration and α_w was assumed to be in the range from 0.5 to 1.0 cm^{-1}. For LWIR QCLs, in addition to η_i, α_w and τ_4 need to be optimized to achieve a high slope efficiency.

The analysis of J_{th} in (4.2) requires J_{tr} and g that are given by:

$$J_{\text{tr}} = \frac{e}{\eta_i \tau_4} n_{3,\text{therm}} \tag{4.3}$$

$$g = \frac{4\pi e^2}{\varepsilon_0 n \lambda L_p} \eta_i \tau_4 \frac{\langle z_{43} \rangle^2}{\langle \gamma_{43} \rangle} \tag{4.4}$$

where $n_{3,\text{therm}}$ is the thermal backfilling of the lower laser level, L_p is the stage length, n is the effective refractive index, and z_{43} and γ_{43} are the matrix element and linewidth for the laser transition, respectively.

The $n_{3,\text{therm}}$ term can be estimated using:

$$n_{3,\text{therm}} = n_s e^{-\Delta_{inj}/kT} \tag{4.5}$$

where n_s is the sheet doping density of the laser core and Δ_{inj} is the voltage defect defined in Figure 4.2.

From (4.4) and (4.5), it is clear that η_{i} needs to be maximized to reduce threshold current density. In addition, the differential gain is inversely proportional to γ_{43}. The latter depends on interface scattering, which explains why QCLs are very sensitive to epi-growth quality.

To use (4.1)–(4.5) in QCL performance projections, it is necessary to calculate values of the matrix element for the laser transition, total lifetimes for the upper and lower laser levels, and the lifetime associated with carrier scattering from the level 4 to level 3 (τ_{43}). These parameters can be calculated with wavefunctions obtained using the shooting method:

$$z_{ij} = \langle \psi_i \mid z \mid \psi_j \rangle \tag{4.6}$$

$$\frac{1}{\tau_i} = \sum_f \frac{1}{\tau_{if}} = \frac{2\pi}{\hbar} \sum_f |\langle \psi_f | H^* | \psi_i \rangle|^2 \tag{4.7}$$

Equation (4.7) is Fermi's Golden Rule, where H^* is a scattering Hamiltonian. A list of possible scattering mechanisms in a multilayered semiconductor structure includes optical and acoustic phonon scattering, alloy scattering, interface scattering, electron-electron scattering, and others.

4.3 Laser core design

As for any laser, the primary goal for the QCL core design is to create an energy level arrangement conducive to population inversion between the upper laser level and the lower laser level. One of the main challenges in the QCL design is the ultra-short upper laser lifetime: in contrast to diode lasers, all the active region and injector states have approximately the same in-plane dispersion (Figure 4.5). As a consequence, carrier scattering between the subbands via emission and absorption of longitudinal optical (LO) phonons is allowed. This leads to an upper laser lifetime τ_4 on the order of one picosecond, which is three orders of magnitude shorter than that of diode lasers.

The first QCL operation at cryogenic temperatures was demonstrated utilizing single-phonon extraction of carriers from the lower laser level [16]. Specifically, by designing the energy separation between the lower laser level 3 and the level 2 underneath it to be equal to the LO phonon energy (E_{32} = 32 meV) the carrier lifetime for the lower laser level was reduced to approximately 0.2 ps, significantly shorter than the upper laser lifetime of approximately 1 ps. The difference in lifetimes led to population inversion between the two laser levels.

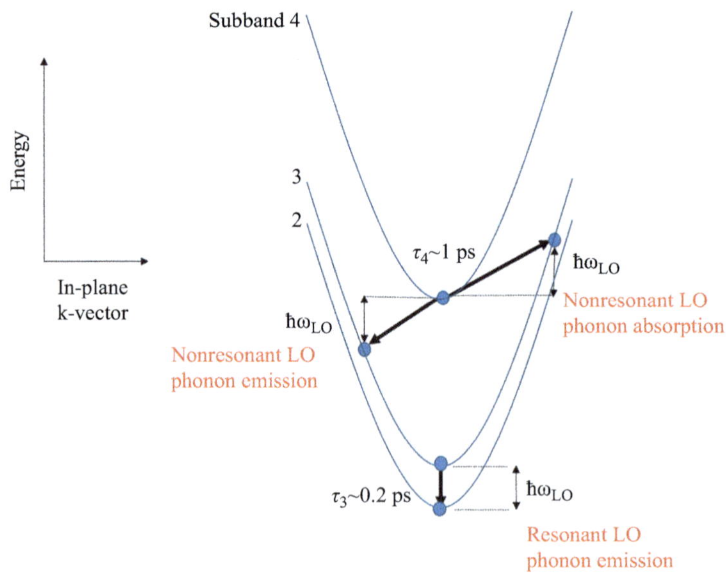

Figure 4.5 A diagram illustrating carrier scattering between subbands via emission and absorption of LO phonons

While the single phonon resonance design allowed for pulsed QCL operation at cryogenic temperatures, carrier backscattering from level 2 to level 3 prevented high-performance operation at higher temperatures. A good illustration of the backfilling effect has been recently provided in [17] where measured efficiency for a single-phonon structure, grown and processed using state-of-the-art techniques, reduced from 50% to 5% for laser temperature changing from 80K to 300K. To suppress carrier backscattering and improve high temperature QCL performance, bound-to-continuum, two-phonon ($E_{32} = E_{21} = 32$ meV), and three-phonon designs were subsequently explored [18,19]. The presence of several low energy levels located directly underneath the level 3 in these designs led to a suppressed backscattering and, therefore, to a significant improvement in laser performance at high temperatures. First room temperature CW operation was demonstrated for a two-phonon structure in 2002 [20]. The two-phonon resonance extraction is shown in Figure 4.2.

Further significant progress in QCL performance was achieved by several groups within the framework of the Efficient Mid-Infrared Laser (EMIL) program supported by Defense Advanced Research Projects Agency from 2007 to 2010. Watt-level CW room temperature operation was reported for 4.6 μm QCLs by the Northwestern University, Princeton University, and Pranalytica/Harvard University teams. CW room temperature efficiency was increased over the course of the program from approximately 5% to 21% at 4.9 μm [6] and 16% at 4.7 μm [7], whereas pulsed efficiency of 27% was reported at 4.9 μm [6] and 23% at 4.6 μm [5].

One of the main active region design goals set for the EMIL program was to improve carrier injection efficiency η_i for the upper laser level 4: the injection efficiency factors in laser slope efficiency (4.1), transparency current density (4.3), and differential gain (4.4). A straightforward (and intuitive) approach to increasing the injection efficiency term is to increase both energy and spatial separation between the upper laser level 4 and the nearest (parasitic) level 5 located in the active region above it (see Figure 4.2). Combination of a large E_{54} and a large spatial separation between the average coordinates for levels 4 and 5 isolates level 4 from other states in the active region and, therefore, leads to a more selective carrier injection into the upper laser level. A quantitative carrier leakage analysis that has recently been carried out based on a comprehensive rate-equation model indeed showed that carrier escape to level 5, followed by scattering to states other than level 4, is the main carrier leakage path for typical QCL designs [21] and that projected carrier performance can be significantly improved by increasing E_{54}.

It is difficult to significantly change E_{54} using either the two- or three-phonon relaxation approaches as active region epi-layers thicknesses have to be adjusted to provide an energy levels arrangement that is, on the one hand, yields a pre-determined laser transition energy (E_{43}) and, on the other hand, satisfies the resonance phonon-relaxation conditions. Meeting all these conditions in the same design does not leave enough flexibility to independently manipulate level 5 position. The Pranalytica/Harvard team circumvented the problem using nonresonant extraction (NRE) principles where the two-phonon resonance condition was relaxed [22,23]. Detuning from the resonance condition reduced scattering rates for corresponding carrier transitions. However, total lower laser level lifetime in the NRE design was still kept in the range of 0.2 ps to 0.3 ps by allowing for several parallel transitions from level 3. As a result, the lower laser lifetime for the NRE design was not very sensitive to variation in thicknesses of active region layers and this additional flexibility was used to manipulate the energy levels located above level 4. Specifically, E_{54} was increased from its typical value of 40–50 meV for the two-phonon design to 60–70 meV, which resulted in improved CW performance [22].

The Northwestern team, in turn, achieved the breakthrough performance employing a multimaterial active region composition. By constructing the active region bandgap profile using quantum wells/barriers with a variable quantum well depth/height, specific energy states could be individually manipulated, leaving other states largely unaffected. Using this approach, E_{54} exceeding 90 meV was achieved for the shallow-well design that was based on five different materials, resulting in 27% pulsed efficiency and 21% CW efficiency for MWIR room temperature QCLs. In addition, record-high injection efficiency of 86% at a wavelength of approximately 9 μm, close to its theoretical maximum of 90%, was demonstrated for a so-called step-taper active region design in [24].

The multimaterial composition offers an additional advantage for the QCL design: By varying barrier height in the active region, it is possible to locally control strength of interface scattering and, therefore, separately control the upper and lower laser lifetimes [25]. This additional design flexibility has not yet been fully explored and, when combined with other design improvements, will likely translate to further increase in laser performance.

It was recently demonstrated that $E_{54} \approx 90$ meV and performance comparable to that for designs based on multimaterial composition can be achieved for a conventional two-material composition. In the new approach, active region energy levels located above level 4 are pushed away from level 4 employing a combination of a two quantum well active region structure and a reduced coupling between the active region and injector in the same stage. An exemplary band diagram for this design is shown in Figure 4.6 with only principal states shown for simplicity. The curves in Figure 4.6 are squared wavefunctions anchored at energy levels for these states. In addition to the large E_{54}, the design also features a large spatial separation d_{54} of approximately 100 Å. Pulsed efficiency exceeding 28% at room temperature was demonstrated for a structure with this design [7]. Compared to the multimaterial approach, two-material designs can be grown essentially in any MOCVD or MBE reactor. In addition, growth is simplified as fewer number of materials needs to be calibrated.

In addition to improvement in injection efficiency, optimization of voltage defect (see Figure 4.2) has recently received considerable attention. Voltage defect is required to suppress carrier backscattering from the injector states to the lower laser level 3. The voltage defect range from 150 to 175 meV had been traditionally considered as the optimal one for reaching the highest laser efficiency [26,27]. This rule was derived assuming a single injector subband, which is a good approximation for cryogenic temperatures, but a poor one at room temperature. Presence of multiple

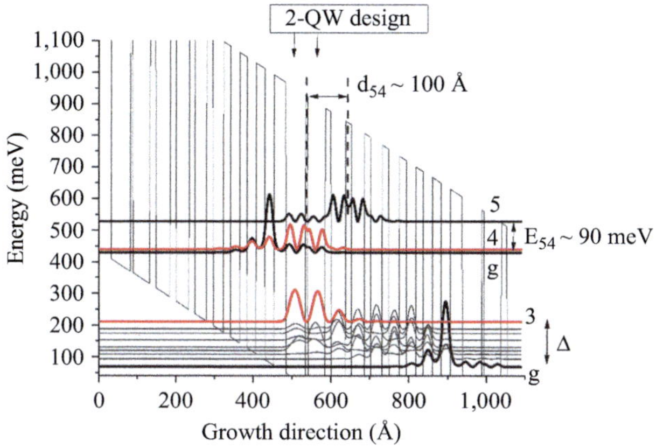

Figure 4.6 Exemplary band diagram of a two-quantum well design based on two-material composition with E_{54} ~ 90 meV and d_{54} ~ 100 Å

states in the injector was accounted for in [8]. The revised model showed that the optimal voltage defect was closer to 100 meV. Very high QCL performance was demonstrated for designs with a reduced voltage defect [6,8].

The QCL development under the EMIL program was dedicated to the 4.4 μm to 4.8 μm spectral region. Similar design concepts were later used to improve short wavelength (<4.2 μm) and long wavelength (>8 μm) QCL performance, both challenging for different reasons.

As evident from Figure 4.3, QCL performance progressively reduces as emission wavelength reduces below ~5 μm. The reduction in performance can be explained by increase in interface and alloy scattering (linewidth γ_{43} broadening; see (4.4)) [28] and reduced carrier confinement for the upper laser level (escape through continuum states, level 5, and indirect states) [29,30]. The best short-wavelength QCL results have been reported for strain-balanced structures with a very high strain in barrier layers resulting in a large band offset. The large band offset improves carrier confinement for the upper laser level. 2 W and 8% efficiency were reported in CW for 4.0 μm QCLs [3] and 1.1 W and 6% were reported for 3.76 μm QCLs [2] based on material composition with approximately −1.6% and −1.8% strain in the barriers, respectively. CW operation was also reported for wavelengths as short as 3.4 μm with strain in barrier layers increased to −2.3%, though reported power level was low [1].

As follows from Figure 4.3, QCL performance also progressively degrades when emission wavelength increases above 8 μm as free carrier losses are proportional to λ^2 and their relative contribution to overall losses becomes significant. From (4.1) and (4.2), higher waveguide losses reduce laser slope efficiency and increase threshold current density. In addition, nonresonant intersubband losses originating in the active region are also projected to increase as laser transition energy reduces, further reducing laser performance. Finally, LWIR QCLs suffer from a stronger Stark effect, which results into lower E_{54} at large bias: Thick quantum wells have to be employed in LWIR QCL design to reduce laser transition energy compared to that for MWIR QCLs. This, in turn, increases d_{54}, while still leaving a significant spatial overlap between states 4 and 5. Since Stark effect is larger for larger d_{54}, E_{54} is typically lower than 65 meV for LWIR QCLs and these devices exhibit sublinear optical power vs current (LI) behavior.

A CW optical power of 2 W and efficiency of 10% were reported for a 9.3 μm design with a thick and low-doped cladding layers to minimize free carrier losses [9]. In contrast to traditional long-wave infrared designs based on the lattice-matched composition, the structure was based on a strain-balanced composition to reduce carrier leakage via level 5 and escape to continuum states. However, achieved E_{54} of 60 meV was still insufficient to completely suppress the carrier leakage path and these LWIR QCLs exhibited a sublinear LI characteristic. Further increase in E_{54} will likely improve LWIR QCL performance.

4.4 Waveguide design

Under high duty cycle driving conditions, QCL core generates a large amount of heat waste. To suppress laser self-heating, the design of the waveguide and cladding layers surrounding the core should be optimized taking into account thermal transport.

AlInAs, InGaAs, and InP are the three main building blocks in the InP-based QCL waveguide design. Refractive index for InGaAs is ~3.5, for AlInAs ~ 3.2, and for InP ~3.1. The average refractive index for an AlInAs/InGaAs core typically lies in the range from 3.20 to 3.30, depending on Γ and ratio of total thickness for InAlAs barrier layers to that for InGaAs quantum well layers. Therefore, either the core itself or a combination of active region with InGaAs layers surrounding it can be used as a waveguide for the laser structure, while InP is typically used as a cladding material.

The core in early QCL designs was often sandwiched between InGaAs waveguide layers to maximize mode overlap factor with the core and, therefore, reduce threshold current density. However, improvement in overlap factor is moderate in this case: For example, COMSOL simulations for the waveguide design used in [7] show that calculated Γ for that design would only increase from 64% to 65% if 0.3 μm-thick InGaAs layers were added on both side of the core. At the same time, InGaAs is a ternary material with an inherently low thermal conductivity of ~5 W/(m·K), as opposed to ~70 W/(m·K) for InP. As a consequence, calculated thermal resistance for an epi-down mounted device from [7] with additional 0.3 μm-thick InGaAs waveguide layers would increase from 6.7 K/W (for all-InP device) to 8.4 K/W. To summarize, there is a relatively small penalty in threshold current density and a large improvement in thermal resistance for QCLs with an all-InP waveguide. As a consequence, the laser core for most state-of-the-art high power CW QCLs is encapsulated in low-doped InP layers and use of InGaAs waveguide or AlInAs cladding layers is generally avoided. Also, to further reduce core temperature, lateral mode confinement for high power QCLs is typically achieved utilizing semi-insulating InP overgrowth: presence of InP on each side of the waveguide reduces laser thermal resistance and, therefore, reduces core temperature rise under CW driving conditions [31,32]. A schematic of laser chip epi-down mounted on a submount with a high thermal conductivity is shown in Figure 4.7.

Mode overlap factor with the laser core Γ and waveguide losses αw both factor in (4.1) and (4.2). To accurately predict the two parameters, it is important to take into account a correction to refractive index and losses for each waveguide layer due to doping. First, dielectric function for undoped layers is taken from experiment. A Drude model is subsequently used to determine a correction due to doping using:

$$\epsilon(\omega) = \epsilon_r(\omega) + i\epsilon_i(\omega) = \epsilon_\infty \left[1 - \frac{N_d e^2}{\epsilon_\infty \epsilon_0 m^*} \frac{1}{\left(\omega^2 + i\omega\gamma_{pl}\right)} \right] \tag{4.8}$$

where N_d is free carrier density (doping level) and γ_{pl} is plasma damping frequency.

Real part of refractive index n and free carrier losses α_{fc} for a doped waveguide layer are calculated from (4.8) using:

$$n = \sqrt{\frac{\sqrt{\epsilon_r^2 + \epsilon_i^2} + \epsilon_r}{2}} \tag{4.9}$$

$$\alpha_{fc} = -\frac{4\pi}{\lambda} \sqrt{\frac{\sqrt{\epsilon_r^2 + \epsilon_i^2} - \epsilon_r}{2}} \tag{4.10}$$

Finally, a commercial 2D waveguide solver, such as COMSOL Multiphysics, can be used to calculate Γ and α_w using n and α_{fc} as input parameters for each waveguide layer in the structure.

TM-polarized light propagating in a QCL waveguide structure can couple to a lossy plasmonic mode at the metal/semiconductor interface. The dependence of the

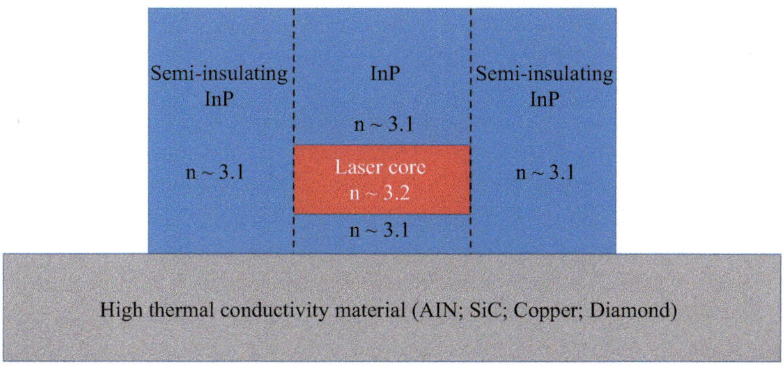

Figure 4.7 A schematic of a typical QCL waveguide

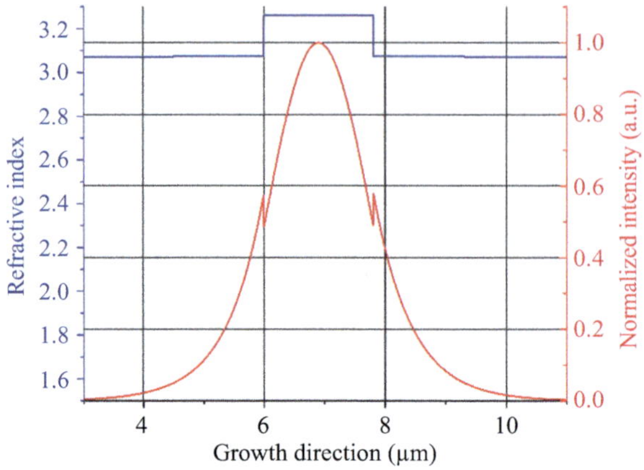

Figure 4.8 Intensity distribution in an exemplary 4.6 μm QCL structure

real part of refractive index on doping level (see (4.8) and (4.9)) can be used in the QCL waveguide design to suppress guided mode coupling to the plasmonic mode: the refractive index of the top InP layer located close to the metal/semiconductor interface is reduced by tuning the plasma frequency of that layer close to the optical frequency of emitted radiation. Typical doping level required for a significant reduction in refractive index is in the range of high 10^{18} cm^{-3}.

The selection rules for intersubband optical transitions force TM polarization for emitted light (in-plane magnetic field). The TM polarization, in turn, entails different boundary conditions compared to those for diode lasers: H_y and $\frac{1}{n^2}\frac{\partial H_y}{\partial x}$ should be continuous across the interfaces. As different layers have generally different refractive indices, near field intensity is discontinuous across the interfaces. This is illustrated in Figure 4.8 for an exemplary 4.6 μm waveguide design. This effect leads to a higher intensity outside the laser core and therefore to a somewhat lower Γ for TM polarization relative to that for TE polarization.

4.5 CW power scaling

The laser core and waveguide design principles discussed above were used in the development of multi-watt CW QCLs. Further improvement in efficiency will likely lead to a concurrent increase in optical power from a single emitter. However, it is unlikely that CW optical power exceeding 10 W from a single emitter required for a number of potential applications can be achieved in the traditional QCL configuration, that is, from devices with 5 mm × 10 μm × 2 μm laser core. CW optical power scaling with laser core dimensions can be explored for reaching output power level in the 10 to 20 W range.

The most straightforward approach to power scaling is increasing power through increase in cavity length of the device [33]. In this case, laser thermal resistance stays essentially unaffected by the change in cavity length as laser transverse dimensions stay the same. Mirror losses for variable cavity length can be kept at the point corresponding to the highest laser efficiency by applying controlled reflectivity coating to the front laser facet [33]. Using this approach, power scaling for a cavity length of up to 10 mm was demonstrated in [3] and [4]. However, the cavity length of 10 mm presents a practical limit for power scaling with cavity length as die handling and bonding become too difficult due to a large aspect ratio of a chip, significantly reducing yield for the laser fabrication. Further increase in CW optical power level requires an increase in lateral device dimensions, that is, by transitioning to the broad-area configuration shown in Figure 4.9 (right).

Power scaling with lateral laser core dimensions is straightforward when pulsed laser operation is required. For example, peak optical power exceeding 100 W was demonstrated in [34] for 400 μm-wide devices. However, power scaling is much more problematic in CW mode due to laser core overheating: The core is composed of ternary InGaAs and AlInAs materials that have inherently low thermal conductivity. In addition, the presence of the superlattice structure with numerous interfaces further hinders thermal transport in the growth direction. As a consequence, the laser core itself serves as a bottleneck for heat dissipation. Increase in its transverse dimensions further increases laser thermal resistance as the relative contribution of lateral heat escape path through semi-insulating InP layers is reduced. The effect is strong enough to completely prevent lasing for CW devices in a broad-area configuration [35,36].

An initial analysis and first demonstration of broad-area CW QCLs were presented in [37]. Assuming that laser core thickness d is much smaller than its width (1D case), an analytical formula for laser core temperature rise ΔT relative to surrounding bulk InP waveguide layers under CW driving conditions can be derived

$$\Delta T = \frac{Qd^2}{8k_\perp} \tag{4.11}$$

where Q is the heat waste density and $k_\perp$ is the effective thermal conductivity for the laser core in the growth direction. The formula clearly shows that laser core overheating can be suppressed by reducing its thickness. First CW operation for broad-area (30 μm-wide) QCLs was demonstrated for a design with ten stages (as opposed to traditional designs with 30 to 40 stages) [37]. Power level for these first devices was however low.

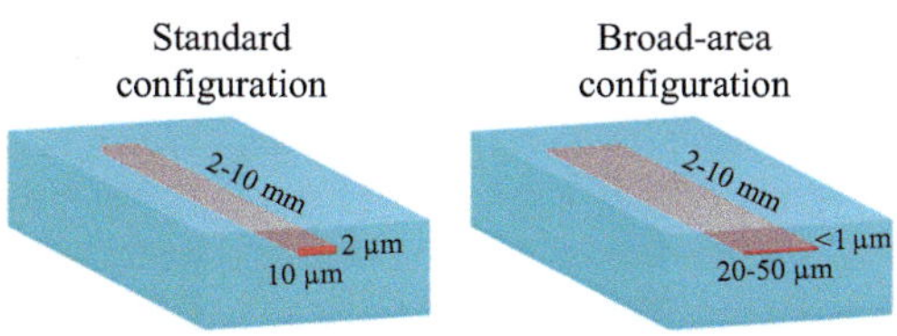

Figure 4.9 Schematic of the standard (left) and broad-area QCL configurations (right)

Optimization of broad-area QCLs requires a model capable of accurately predicting CW QCL performance in different laser configurations. A simple model predicting CW power for various device geometries based on *a known stage design* can be constructed using experimental pulsed data for the given design. The simple semiempirical model uses the measured pulsed values for transparency current, losses, differential gain, injection efficiency, and sublinearity of a pulsed LI characteristic as input model parameters to project pulsed LI characteristics for devices with an arbitrary configuration but the same layer sequence in the stage (variable cavity length, number of stages, doping, ridge width, etc.). A COMSOL waveguide simulation is used to account for changing mode overlap factor with the laser core for devices with different number of stages. The model is subsequently linked to a COMSOL thermal simulation to take into account laser core self-heating in CW mode and to project CW LI characteristics.

When in pulsed operation, a QCL will reach a threshold current defined by (4.2). Near threshold, pulsed power scales linearly with current as follows from (4.1). However, at higher bias slope efficiency has a sublinear dependence as injection efficiency for the upper laser level reduces. A rigorous theoretical treatment for injection efficiency would needlessly complicate the model, so an envelope is applied to the LI curve to replicate the curvature of the experimental results. This can be done with a parametrization of LI characteristic for a given design at large bias.

The portion of the semiempirical model discussed above is sufficient to project LI characteristics for devices with a given stage design in different configurations. Because short pulses (100s of nanoseconds) and low enough duty cycle prevent temperature buildup in the laser core, the core is taken to be at a fixed temperature for modeling-pulsed laser characteristics. However, CW operation allows for continual buildup of heat that both increases threshold current and decreases slope efficiency at threshold. This thermal behavior is typically described with characteristic temperatures T_0 and T_1:

$$I_{\mathrm{th,cw}} = I_{\mathrm{th,pulse}} e^{\frac{T-T_i}{T_0}} \tag{4.12}$$

$$\frac{\partial P}{\partial I} = \eta_{S,\mathrm{pulse}} e^{-\frac{T-T_i}{T_1}} \tag{4.13}$$

where $I_{\mathrm{th,cw}}$ is the CW threshold current, $\frac{\partial P}{\partial I}$ is the current dependent component of CW slope efficiency, T_i is the ambient temperature without heat buildup (i.e., the temperature set by a thermoelectric controller), T_0 is the characteristic temperature for threshold current density (taken from experiment), and T_1 is the characteristic temperature for slope efficiency (taken from experiment).

A projection of CW power above threshold requires consideration for thermal effects. By introducing a temperature varying with current, the derivative of power with respect to current is redefined:

$$\frac{dP}{dI} = \frac{\partial P}{\partial I} + \frac{\partial P}{\partial T}\frac{\partial T}{\partial I} \tag{4.14}$$

The first term is taken as the partial derivative with respect to current at threshold as defined in (4.13). For the second term, partial derivative of temperature with respect to current is taken from COMSOL thermal simulations and $\frac{\partial P}{\partial T}$ is found as:

$$\frac{\partial P}{\partial T} = \frac{\partial}{\partial T}\left[\frac{\partial P}{\partial I} * \left(I - I_{\text{th,cw}}(T)\right)\right] \tag{4.15}$$

In the thermal simulation, a linear model on current density approximates the linear portion of the voltage for the device. The linear model is given by a voltage offset V_0 and a differential resistance R_d that scale with a number of stages. Electrical power is deposited in the active region as the product of voltage and current, where the voltage is taken as:

$$V = N_s(V_0 + JR_d) \tag{4.16}$$

The active region is assigned with a cross plane thermal conductivity, $k_\perp$, and an in-plane thermal conductivity, $k_\parallel$. A stationary solution is found where the mean temperature in the active region is taken to be T.

For projections involving a change in the stage doping, it was assumed that transparency current (defined by backscattering), roll-over current density, and losses originating from the core (nonresonant intersubband losses) are all directly proportional to doping level.

The model can be used in CW power projections for both traditional and broad-area QCLs. Figure 4.10 shows exemplary CW projections obtained with this model for a fully characterized 4.6 μm QCL structure with 40 stages. The projections very well reproduce the experimental CW data, including thermal roll over,

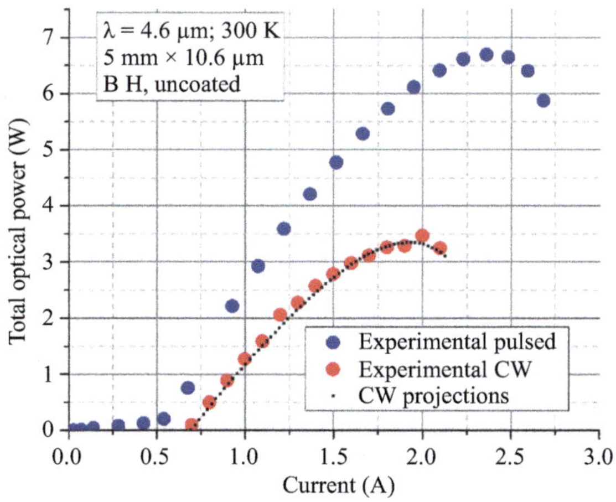

Figure 4.10 Comparison between experimental and model CW data for an exemplary 4.6 μm QCL with 40 stages

higher threshold current for CW operation, and lower slope efficiency for CW operation. These data show that the model correctly captures all main thermal and optical aspects of CW QCL operation. Also, a very good correspondence between experimental and model data was reported for broad-area devices with various ridge widths in [38].

To illustrate the approach, the 4.6 μm exemplary structure was optimized for CW operation in the broad-area configuration. The maximum optical power was achieved for 90 μm-wide laser core when the core was thinned to 14 stages, core doping was increased to 160% compared to that for the exemplary design, and InGaAs cladding layers were introduced with 200 nm on the substrate side and 100 nm on the episode to increase mode overlap with the core. The back facet was HR coated and the front facet left uncoated.

The optimized broad-area device projected 19.8 W in CW, with a wall-plug efficiency projected to be approximately 14% based on the linear voltage model. The model assumes that the device is actively cooled to 20 °C, keeping in mind the input power is ~140 W, corresponding to a heat removal rate of 7–8 $\frac{kW}{\mathrm{cm}^2}$. The output power can be increased to >20 W by increasing cavity length beyond 5 mm.

These model projection data show that CW optical power from a single broad-area QCL emitter of 10 to 20 W is achievable (Figure 4.11). Several research groups are actively working on the development of very high power broad-area QCLs. A parallel effort in this field is suppression of high-order mode operation for the broad-area configuration [39,40].

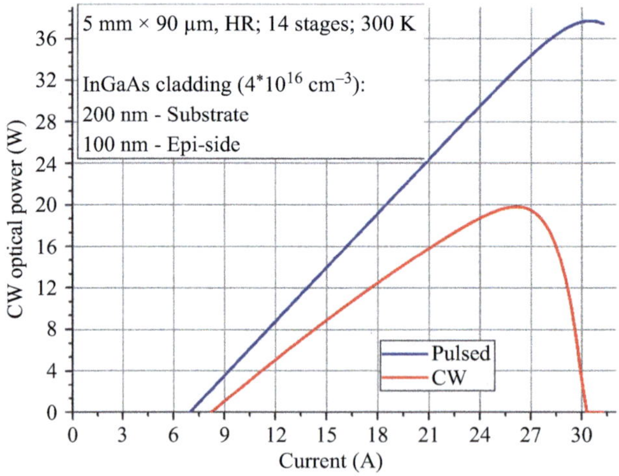

Figure 4.11 CW power projections for broad-area QCLs with the same laser core stage design of the exemplary 4.6 μm structure (Figure 4.11). Optimal configuration has 14 stages and includes InGaAs cladding layers to increase mode overlap factor with the laser core

4.6 Beam combining

Increase in power level to tens of watts with a high brightness output beam is pursued with various polarization, coherent, and spectral beam combining (SBC) schemes.

Output power from a laser module can be easily doubled without compromising beam quality using the polarization beam combining approach [41]. This configuration comprises two QCL emitters. Polarization for one of them is rotated by 90° with a half-wave plate and the two beams are subsequently combined with a polarizer that transmits one polarization and reflects the other one. Alternatively, the two beams can be combined using a Brewster's plate and by adjusting the angle of incidence for one of the two beams to be equal to the Brewster's angle [41]. System integration for polarization beam combining is simple and the configuration is not very sensitive to shock and vibration. Its main drawback is that it is limited to only two emitters.

One of the most promising approaches to reaching very high power levels is the SBC scheme utilizing the external cavity (EC) configuration [42]. The EC configuration is discussed in detail in Section 4.7. For SBC, an array of QCL emitters in a bar form is forced to have a single optical output with a combination of a transform lens, dispersive grating, and an output coupler. The EC selects a different wavelength for each element in the array, which ensures collinear and coaxial beam propagation for all emitters. Spectral brightness for this approach is the same as that for a single emitter. However, low spectral brightness is acceptable for many high power applications. One of the main advantages of the SBC is that it allows for using the emitters in their optimal configuration for CW operation, that is, BH chips mounted epi-down on a submount or directly mounted onto a heat sink. As a consequence, a high yield for laser material fabrication can be achieved, which makes it a practical solution for scaling up QCL power. The SBC approach has already been commercialized. For example, TerraDiode (now Forward Photonics) offers 20 W QCL systems with $M^2 < 1.5$.

A modified approach to SBC was explored in [43] where an array of distributed feedback (DFB) QCLs with each chip emitting at a different wavelength (different DFB grating period) was combined using a transform lens and an external grating (without output coupler). The DFB configuration is discussed in detail in Section 4.8. Since the DFB chips have preselected emission wavelengths, there is no need for the optical feedback, which simplifies optical alignment. This technology has been commercialized by EOS Photonics (now Pendar Technologies). SBC was also reported in [44] and [45].

Power scaling for monolithic arrays with a high output beam quality and a high spectral brightness can potentially be achieved with a coherent beam combining (CBC). The critical advantage of the monolithic approach is the inherently small size/weight and the capability to withstand strong vibration/shock profiles. Tree-arrays and leaky-mode waveguide arrays received most attention in the QCL CBC research.

In the tree-array configuration, a single-mode stem forces in-phase operation for all elements (branches) of the array [46,47]. Power scaling has been demonstrated for ridge waveguide devices in pulsed mode and BH devices both in pulsed and CW modes with up to eight-elements in the array. However, CW operation was limited to less than 1 W [48]. There are two main technical challenges for the tree-array CBC. First, the practical limit to the maximum number of elements is 16 as combined cavity length occupied by the stem section and coupling (Y-shaped) section is on the order of several millimeters as dictated by the requirement to achieve a high out-of-phase mode suppression (hence, long and narrow stem section is required), low bending losses (large bending radius), and minimal interelement thermal crosstalk (large interelement spacing). In addition, planar buried heterostructure is difficult to achieve in the coupling sections due to the formation of "rabbit ears" (nonplanar overgrowth), which significantly reduces array fabrication yield. As a consequence, tree arrays have not been transitioned to production yet.

Another monolithic CBC approach that is being actively pursued to scale up QCL power is based on the leaky-wave coupling. In contrast to the traditional QCLs, the gain region for these arrays is designed to have an effective refractive index significantly lower than that for the interelement regions. This leads to parallel optical coupling between the emitters in the array and to the formation of a "super-mode" spanning all the elements. Pulsed optical power of 5 W and a nearly diffraction limited, on-axis beam with 82% of power concentrated in the main lobe were reported at 8.3 μm for a five-element array [49]. These devices, however, require additional fabrication steps, which may affect device efficiency and fabrication yield. CW operation of coherently coupled arrays based on leaky-wave coupling is still to be demonstrated.

4.7 External cavity QCLs

Edge-emitting QCLs based on optical feedback from the two cleaved facets of the laser chip, that is, traditional Fabry–Perot QCLs described above, generally operate on numerous incoherent longitudinal modes with spectral envelope full width at half maximum of approximately 100 cm^{-1}. The intensity distribution throughout the longitudinal modes is difficult to control and it often unpredictably changes with a change in driving current and/or temperature. As a consequence, Fabry–Perot QCLs have very limited use in spectroscopic applications that often require single mode operation with tunable emission wavelength.

A single mode operation with a broad wavelength tunability (covering a substantial fraction of the QCL gain bandwidth) can be achieved employing the EC configuration [50]. A schematic of the EC QCL configuration is shown in Figure 4.12. The back facet of the laser chip is antireflection coated, preventing Fabry–Perot feedback. Collimated radiation from the back facet is incident upon an external (reflective) optical grating that provides angular separation for different wavelengths. The wavelength satisfying the Bragg condition, λ_G, couples back into the waveguide and therefore has the lowest threshold gain. Output beam can be collected from the other facet and emission wavelength can be controlled by rotating the grating, that is, by changing the Bragg condition.

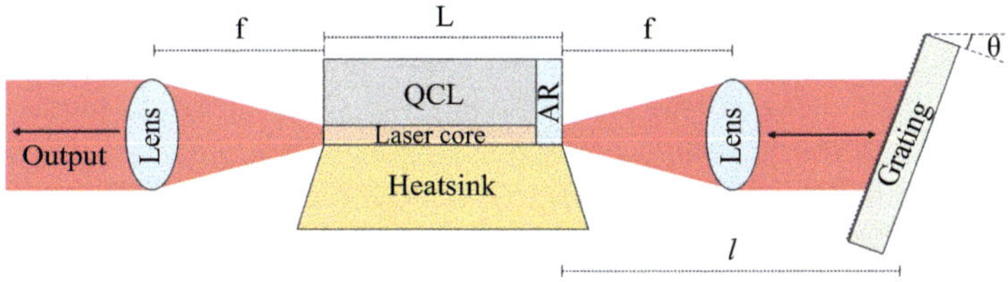

Figure 4.12 Schematic of EC QCL with a mechanically controlled grating

Similar to FP devices, threshold gain $G_{\text{th}}(\lambda)$ for EC QCLs is given by [51]:

$$G_{\text{th}}(\lambda) = \alpha_w + \frac{1}{L}\ln\frac{1}{|r_b(\Theta, \lambda)\cdot r_f|} \tag{4.17}$$

where L is the QCL chip length, r_f is the front facet reflectivity, and r_b is the effective reflectivity given by:

$$r_b(\Theta,\lambda) = \sqrt{R_G}\exp\left(-\frac{f^2(\lambda-\lambda_G)^2}{2W^2d^2\cos^2\Theta}\right)e^{2ikl} \tag{4.18}$$

where R_G is the first-order diffraction efficiency of the grating, f is the focal length of the collimating lens, W is the beam size at the back facet along the fast (vertical) axis, d is the grating period, and l is the distance between the back facet and the grating.

A set of longitudinal modes supported by the EC is determined by the steady-state condition:

$$arg[r_b(\Theta, \lambda)] + \frac{4\pi nL}{\lambda} = 2\pi N \tag{4.19}$$

Equations (4.17) and (4.18) can be used to determine which of the cavity modes defined by (4.19) lases for a given grating angle.

EC QCLs with a mechanically controlled grating have been demonstrated to operate in CW mode with a tuning range exceeding 150 cm^{-1} [52]. Tuning range can be increased employing QCL chips with heterogeneous laser core design where several stage designs emitting at different wavelengths share the same waveguide, which significantly broadens gain spectrum [53]. Lasing in the range from 6 to 11 μm from the same heterogeneous QCL material was demonstrated in [54].

EC QCLs based on a mechanically controlled grating suffer from two important disadvantages arising from the need to physically rotate the grating. These disadvantages are slow wavelength tuning and sensitivity to mechanical vibrations and shocks. The highest tuning rate reported for traditional EC QCLs with a mechanical grating is 5 kHz, achieved by tilting a small intracavity mirror [55]. However, the tuning range is limited. Also, using a micro opto-electro-mechanical systems scanning grating, a scanning rate of 1 kHz was achieved in [56]. The improvement in tuning speed demonstrated for these devices, while impressive, is still insufficient for many important laboratory applications, including combustion/

explosion diagnostics where a full spectral scan needs to be performed on the order of tens of microseconds. Applications involving optical detection of improvised explosive devices demand submillisecond spectral scan of a remote object. Sensitivity to vibrations and shocks, especially during the tunable laser operation, precludes the consideration of the traditional EC QCLs in applications where the tunable laser is deployed on moving platforms.

An alternative EC QCL configuration that does not have these two disadvantages has recently been explored in [57] and [58]. In contrast to the traditional EC QCLs with a mechanically controlled grating, the rapid wavelength tuning is achieved using an electrically controlled acousto-optic modulator (AOM), as shown in Figure 4.13. In the AOM, an acoustic wave at a radio frequency is generated by applying the electronic signal to a piezoelectric transducer attached to an optical crystal, such as Germanium, which is transparent at the wavelength at which the laser needs to operate. The acoustic wave (alternate compressions and rarefactions in the crystal) represents a phase transmission grating, from which the optical beam traversing the AOM crystal can be diffracted.

The response time of the change in the optical wavelength with a change in the AOM frequency is the main advantage of the new configuration. It is determined by the acoustic wave transit time across the optical beam. For a beam size of several millimeters, the response time can be less than 1 μs.

The AOM EC QCL configuration explored in [57] and [58] utilized an AOM with a traveling sound wave. The traveling wave incurs a frequency-shifted optical feedback as the laser beam propagates through the AOM crystal due to the Doppler shift equal to the sound wave frequency. The frequency shift prevents constructive interference and formation of axial optical modes in the laser (corresponding to the round trip time for the overall laser cavity) when zero-reflectivity AR coating is used on the laser facet facing the AOM. As a consequence, emission spectrum for the AOM controlled lasers in the traveling wave configuration is expected to be different from that for traditional Fabry–Perot and traditional EC devices.

Emission spectrum of lasers, including AOM-controlled QCLs, is seeded by spontaneous emission events in the gain medium. Therefore, phase consideration between these random events is not important. Each of the spontaneous emission events results into the formation of a broad spectral line as the beam propagates back and forth in the cavity. The resultant (integrated) spectrum is defined by the gain/loss spectrum, the AOM frequency response, and the frequency shift. FWHM

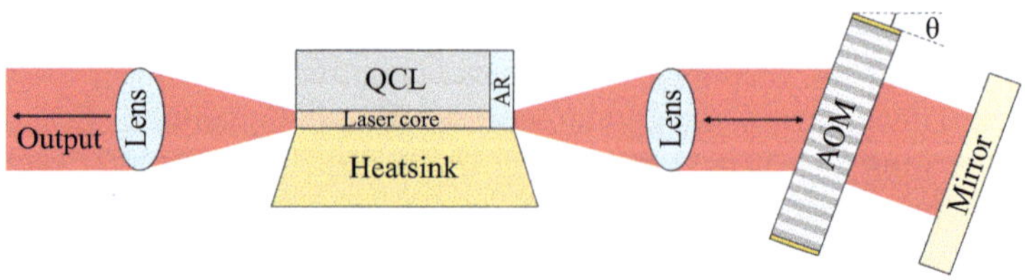

Figure 4.13 Schematic of an AOM-controlled EC QCL

for QCL gain spectrum is much larger than AOM frequency response. In this case, the integrated emission spectrum for an AOM-controlled laser, seeded by a broad spontaneous emission, has a nearly Gaussian shape with an estimated spectral full width at half maximum $\delta\vartheta$ given by [59]:

$$\delta\vartheta \approx \left(\Delta\cdot\Gamma_f^2\right)^{1/3} \tag{4.20}$$

where Δ is the roundtrip frequency shift and Γ_f is the spectral width of the AOM frequency response. For typical values of Δ and Γ_f, calculated value for $\delta\vartheta$ is approximately 1 cm^{-1}, acceptable for measuring species with broad spectral absorption features.

The EC QCLs with a mechanically controlled grating have been commercialized by a number of companies, including Daylight Solutions, Pranalytica, and Block Engineering. Pranalytica has also commercialized AOM-controlled EC QCLs. As of today, employment of EC QCLs in infrared spectroscopic measurements, mostly in laboratory environment, is the most important market niche for the QCL technology.

4.8 Distributed feedback QCLs

EC QCL systems described above require a labor-intensive EC alignment and they typically have relatively large overall dimensions to accommodate bulk optical elements. For handheld and ruggedized spectroscopic applications, a monolithic DFB QCL configuration discussed in this section can be used.

In DFB QCLs, a periodic grating is etched into the cladding and/or waveguide layers of the QCL structure, as shown in Figure 4.14. The grating creates a periodic perturbation in effective refractive index for the guided mode. As a consequence, the guided mode experiences weak reflections at each of the interfaces shown with dashed lines in Figure 4.14 as it propagates along the laser cavity, which is the source of wavelength-selective optical feedback. At the same time, both laser facets of the device are antireflection coated to prevent FP feedback.

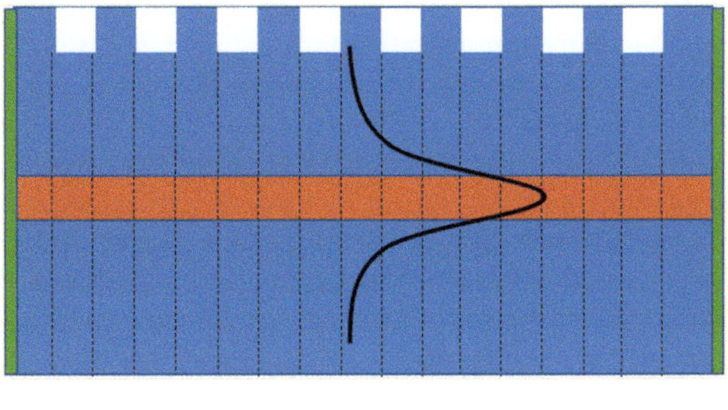

Figure 4.14 Schematic of a DFB QCL structure

The condition of constructive interference for reflected waves forces lasing on a spectral line approximately equal to the Bragg wavelength λ_B given by:

$$\lambda_B = 2n_{\text{eff}}\Lambda/N \tag{4.21}$$

where n_{eff} is the average effective refractive index along laser cavity, Λ is the grating period, and N is the grating order. By implementing the active stabilization loops controlled with low noise current and temperature controllers, a stable absolute DFB QCL operation with a linewidth in the range of KHz can be achieved [60,61].

A top-grating DFB QCL configuration, with grating, etched into the contact and top cladding layers, is the easiest one to fabricate [62]. Since the grating in this case is typically covered with gold for electrical current injection, the guided mode can interact with the surface plasmon mode and these devices often operate in a strong coupling regime.

A rigorous model for top-grating DFB QCLs is quite complex. For a TM-polarized beam propagating in the QCL waveguide structure, the governing equation for the magnetic field pointing in the y-direction (in-plane direction) is given by:

$$-\frac{\partial^2 H_y(x,z)}{\partial x^2} = \epsilon\left[\frac{\partial}{\partial z}\left(\frac{1}{\epsilon}\frac{\partial H_y(x,z)}{\partial z}\right) + k_0^2 H_y(x,z)\right] \tag{4.22}$$

The Floquet–Bloch theorem states that for a periodic perturbation along z-direction (beam propagation direction), the magnetic field can be expanded into partial waves as:

$$H_y(x,z) = \sum_{m=0,\pm1,\ldots} H_m(x)e^{i\beta_m z} \tag{4.23}$$

where $\beta_m = \beta_0 + mK$, β_0 is the propagation constant and $K = 2\pi/\Lambda$ is the grating vector. In addition, since the dielectric function is periodic, it can be expanded in Fourier series:

$$\epsilon(z) = \sum_{m=-\infty}^{+\infty} \tilde{\epsilon}_m^{(j)} e^{imKz} \tag{4.24}$$

For first-order DFB QCLs ($N = 1$ in (4.21)), the grating couples the guided mode propagating in $+z$-direction ($m = 0$ in (4.23)) to the guided mode propagating in $-z$-direction ($m = -1$) and laser power is collected from the edge of the laser [62], similar to that for FP devices. For second-order DFB QCLs ($N = 2$), the grating couples $m = 0$ and $m = -2$ guided modes, while H_{-1} is a partial wave radiating perpendicular to optical axis ($\beta_m = 0$). Useful optical output for second-order devices is collected from either epi-side [63] or substrate-side of the laser chip [64].

The model based on Floquet–Bloch approach can be used to accurately model DFB QCLs in various configurations even in the case of a large modulation in complex dielectric constant and/or a large grating height [65]. To project DFB QCL characteristics using this method, partial waves H_m need to be first determined by substituting (4.23) and (4.24) in (4.22) and by enforcing continuity conditions across interfaces between the waveguide layers (see Section 4.4). It was shown in [65] that up to 30 partial waves need to be taken into account in (4.23) for the model to converge, making this numerical approach difficult to implement. This explains why a systematic optimization of a metalized top-grating is rarely carried out.

Devices with "weak" gratings, on the other hand, can be analyzed using a much simpler approach described in [66]. In this approach, it is assumed that only two diffraction orders in (4.23) have significant amplitudes. For a sinusoidal grating shape, a pair of coupled equations for the amplitude of the +z-propagating wave *R(z)* and the amplitude of the −*z*-propagating wave *S(z)* can be derived:

$$-\frac{\partial R}{\partial z} + (\alpha - i\delta)R = i\kappa S \tag{4.25}$$

$$\frac{\partial S}{\partial z} + (\alpha - i\delta)S = i\kappa R \tag{4.26}$$

where δ is the wavelength detuning parameter from the resonance Bragg condition and coupling coefficient κ is given by:

$$\kappa = \frac{\pi n_1}{\lambda_B} + i\frac{\alpha_1}{2} \tag{4.27}$$

where n_1 is the amplitude of index modulation and α_1 is the amplitude of loss modulation.

The analysis of (4.25) and (4.26) shows that light intensity distribution is approximately uniform throughout the laser cavity when:

$$\kappa L \approx 1 \tag{4.28}$$

where L is the laser cavity length. In the design of DFB QCLs with a weak grating, first, effective refractive index and losses are calculated for the vertical cross sections at grating peaks and troughs, which provides n_1 and α_1. A coupling coefficient and an optimal cavity length are subsequently calculated using (4.27) and (4.28).

DFB QCLs with a weak index coupling can be realized using a shallow buried-heterostructure grating. Epi-growth of these devices is interrupted after growth of the upper InGaAs waveguide layer is completed and a shallow grating is then etched into the InGaAs layer. Epi-growth of the remaining top contact and cladding layers is subsequently carried out, followed by a standard QCL wafer processing. Since these devices rely only on index coupling (hence, low waveguide losses), a very high device performance can be achieved even in CW operation [67].

Wavelength tuning for DFB QCLs is achieved with thermal tuning of refractive index by either directly changing laser temperature and/or by changing laser driving current that heats up the laser core. A typical DFB QCL tuning range is on the order of several wavenumbers. The narrow tuning range is the main drawback of this technology. Monolithic DFB QCL arrays, with each emitter lasing at a slightly different wavelength (different grating period for each emitter), can be used to dramatically increase (combined) spectral coverage [68].

DFB QCLs have been commercialized by Alpes Lasers, Throlabs, Hamamatsu Photonics, and AdTech Optics. The devices are primarily used in spectroscopic applications measuring specific (narrow) absorption features.

4.9 Conclusion

Since their first demonstration in 1994, QCLs have experienced a rapid development cycle. A number of critical milestones, such as pulsed room temperature operation, the DFB configuration, the EC configuration, and CW room temperature operation, were all demonstrated for QCLs within the first 10 years. Subsequent improvement in room temperature performance made these devices the technology-of-choice for numerous high power and spectroscopic infrared applications.

One of the main factors currently limiting a QCL market size is a high price of $5,000 to $10,000 for a single laser chip. However, QCL cost can be significantly reduced through economy of scale: based on current pricing for wafer growth and processing, a price of $100 per laser chip is realistic in quantities exceeding thousand units. Therefore, QCLs will likely be soon adopted for low cost, large volume applications, such as sensors for greenhouse gasses monitoring.

The chapter covered only QCL configurations that have already been commercialized. Many interesting topics, such as QCL combs [69,70] and QCL integration on lattice-mismatched substrates [71,72], were left out due to chapter size constraints. In addition, only most important aspects of QCL operational principles were outlined. For a detailed description of the QCL basics, the reader is referred to an excellent textbook on Quantum Cascade Lasers by Jerome Faist [73].

References

[1] N. Bandyopadhaya, S. Slivken, Y. Bai, and M. Razeghi, "High power, continuous wave, room temperature operation of $\lambda \sim 3.4$ μm and $\lambda \sim 3.55$ μm InP-based quantum cascade lasers," Appl. Phys. Lett. 100, 212104 (2012)

[2] N. Bandyopadhaya, Y. Bai, B. Gokden, *et al.*, "Watt level performance of quantum cascade lasers in room temperature continuous wave operation at $\lambda \sim 3.76$ μm," Appl. Phys. Lett. 100, 212104 (2012)

[3] A. Lyakh, R. Maulini, A. Tsekoun, *et al.*, "High-performance continuous-wave room temperature 4.0 μm quantum cascade lasers with single-facet optical emission exceeding 2 W," PNAS 107, 18799 (2010)

[4] A. Lyakh, R. Maulini, A. Tsekoun, R. Go, and C. Kumar N. Patel, "Tapered 4.7 μm quantum cascade lasers with highly strained active region composition delivering over 4.5 watts of continuous wave optical power," Opt. Express 20, 4382 (2012)

[5] Y. Yao, X. Wang, J.-Y. Fan, and C. Gmachl, "High performance "continuum-to-continuum" quantum cascade lasers with a broad gain bandwidth of over 400 cm^{-1}," Appl. Phys. Lett. 97, 081115 (2010)

[6] Y. Bai, N. Bandyopadhyay, S. Tsau, S. Slivken, and M. Razeghi, "Room temperature quantum cascade lasers with 27% wall plug efficiency," Appl. Phys. Lett. 98, 181102 (2011)
[7] A. Lyakh, M. Suttinger, R. Go, P. Figueiredo, and A. Todi, "5.6 μm quantum cascade lasers based on a two-material active region composition with a room temperature wall-plug efficiency exceeding 28%," Appl. Phys. Lett. 109, 121109 (2016)
[8] R. Maulini, A. Lyakh, A. Tsekoun, and C. Kumar N. Patel, "$\lambda \sim 7.1$ μm quantum cascade lasers with 19% wall-plug efficiency at room temperature," Opt. Express 19, 17203 (2011)
[9] A. Lyakh, R. Maulini, A. Tsekoun, R. Go, and C. Kumar N. Patel, "Multiwatt long wavelength quantum cascade lasers based om high strain composition with 70% injection efficiency," Opt. Express 20, 24272 (2012)
[10] F. Xie, C. Caneau, H. Leblanc, *et al.*, "Watt-level room temperature continuous-wave operation of quantum cascade lasers with $\lambda > 10$ μm," IEEE J. Sel. Top. Quant. Electr. 19, 1200407 (2013)
[11] S. Lee, F. Banit, W. Worner, and A. Wacker, "Quantum mechanical wave-packet transport in quantum cascade laser structures," Phys. Rev. B 73, 245320 (2006)
[12] T. Kubis, C. Yeh, and P. Vogl, "Quantum theory of transport and optical gain in quantum cascade lasers," Phys. Stat. Sol. (c) 5, 232 (2008)
[13] R. Kazarinov and R. Suris, "Electric and electromagnetic properties of semiconductors with a superlattice," Sov. Phys. Semicond. 5, 707 (1971)
[14] J. Khurgin, Y. Dikmelik, P. Liu, *et al.*, "Role of interface roughness in the transport and lasing characteristics of quantum cascade lasers," Appl. Phys. Lett. 94, 091101 (2009)
[15] P. Harrison, *Numerical solutions in Quantum Wells, Wires, and Dots* (New York: Wiley, 2009), pp. 67–107
[16] J. Faist, F. Capasso, D. Sivco, C. Sirtori, A. Hutchinson, and A. Cho, "Quantum cascade laser," Science 264, 553 (1994)
[17] Y. Bai, S. Slivken, S. Kuboya, S. Darvish, and M. Razeghi, "Quantum cascade lasers that emit more light than heat," Nat. Photon. 4, 99 (2010)
[18] J. Faist, D. Hofstetter, M. Beck, T. Aellen, M. Rochat, and S. Blaser, "Bound-to-continuum and two-photon resonance quantum cascade lasers for high duty cycle, high-temperature operation," IEEE J. Quant, Electr. 38, 533 (2002)
[19] Q. Wang, C. Pflugl, L. Diehl, *et al.*, "High performance quantum cascade lasers based on three-phonon-resonance design," Appl. Phys. Lett. 94, 011103 (2009)
[20] M. Beck, D. Hofstetter, T. Aellen, *et al.*, "Continuous wave operation of a mid-infrared semiconductor laser at room temperature," Science 295, 301 (2002)
[21] D. Botez, C. Chang, and L. Mawst, "Temperature sensitivity of the electro-optical characteristics for mid-infrared (λ = 3–16 μm)-emitting quantum cascade lasers," J. Phys. D: Appl. Phys. 49, 043001 (2016)
[22] A. Lyakh, C. Pflugl, L. Diehl, *et al.*, "1.6 W high wall plug efficiency, continuous wave room temperature quantum cascade laser emitting at 4.6 μm," Appl. Phys. Lett. 92, 111110 (2008)

[23] A. Lyakh, R. Maulini, A. Tsekoun, *et al.*, "3 W continuous-wave room temperature single-facet emission from quantum cascade lasers based on non-resonant extraction design approach," Appl. Phys. Lett. 95, 141113 (2009)
[24] J. Kirch, C.-C. Chang, C. Boyle, *et al.*, "86% internal differential efficiency from 8 to 9 μm-emitting, step-taper active region quantum cascade lasers," Opt. Express 24, 24483 (2016)
[25] M. Semtsiv, Y. Flores, M. Chashnikova, G. Monastyrskyi, and W. Masselink, "Low threshold intersubband lasers based on interface scattering rate engineering," Appl. Phys. Lett. 100, 163502 (2012)
[26] J. Faist, "Wallplug efficiency of quantum cascade lasers: critical parameters and fundamental limits," Appl. Phys. Lett. 90, 253512 (2007)
[27] S. Howard, Z. Liu, D. Wasserman, A. Hoffman, T. Ko, and C. Gmachl, "High performance quantum cascade lasers: optimized design through waveguide and thermal modelling," IEEE J. Sel. Top. Quantum Electron. 13, 1054 (2007)
[28] R. Terazzi, "Transport in Quantum Cascade Lasers," Diss. ETH No. 20036, pp. 89–116 (2011)
[29] A. Adams, I. Marko, S. Sweeney, R. Teissier, A. Baranov, and S. Tomic, "The effect of hydrostatic pressure on the operation of quantum cascade lasers," Proc. SPIE 7222, Quantum Sensing and Nanophotonic Devices VI, 722203 (2009)
[30] H. Page, C. Becker, A. Robertson, G. Glastre, V. Ortiz, and C. Sirtori, "300 K operation of a GaAs-based quantum-well cascade laser at $\lambda \approx 9$ μm," Appl. Phys. Lett. 78, 3529 (2001)
[31] M. Beck, J. Fast, C. Gmachl, *et al.*, "Buried heterostructure quantum cascade lasers," Proc. SPIE 3284, In-Plane Semiconductor Lasers: from Ultraviolet to Mid-Infrared II, 231 (1998)
[32] S. Nida, B. Hinkov, E. Gini, and J. Faist, "Characterization of iron doped indium phosphide as a current blocking layer in buried heterostructure quantum cascade lasers," J. Appl. Phys. 121, 094502 (2017)
[33] R. Maulini, A. Lyakh, A. Tsekoun, *et al.*, "High power thermoelectrically cooled and uncooled quantum cascade lasers with optimized reflectivity coatings," Appl. Phys. Lett. 95, 151112 (2009)
[34] Y. Bai, S. Slivken, S. Darvish, A. Haddadi, B. Gokdenm, and M. Razeghi, "High power broad area quantum cascade lasers," Appl. Phys. Lett. 95, 221104 (2009)
[35] S. Slivken, J. Yu, A. Evans, J. David, L. Doris, and M. Razeghi, "Ridge-width dependence on high-temperature continuous-wave quantum-cascade laser operation," IEEE Photon. Technol. Lett. 16, 744 (2004)
[36] Y. Bai, S. Slivken, S. Darvish, and M. Razeghi, "Room temperature continuous wave operation of quantum cascade lasers with 12.5% wall plug efficiency," Appl. Phys. Lett. 93, 021103 (2008)
[37] M. Semtsiv and W. Masselink, "Above room temperature continuous wave operation of a broad-area quantum cascade laser," Appl. Phys. Lett. 109, 203502 (2016)

[38] M. Suttinger, R. Go, P. Figueiredo, A. Todi, J. Leshin, and A. Lyakh, "Power scaling and experimentally fitted model for broad area quantum cascade lasers in continuous wave operation," Opt. Eng. 57, 011011 (2017)
[39] R. Kaspi, S. Luong, C. Yang, C. Lu, T. Newell, and T. Bate, "Extracting fundamental transverse mode operation in broad area quantum cascade lasers," Appl. Phys. Lett. 109, 211102 (2016)
[40] R. Kaspi, S. Luong, T. Bate, C. Lu, T.C. Newell, and C. Yang, "Distributed loss method to suppress high order modes in broad area quantum cascade lasers," Appl. Phys. Lett. 111, 201109 (2017)
[41] J. Wagner, N. Schulz, B. Rosener, *et al.*, "Infrared semiconductor lasers for DIRCM applications," Proc. SPIE 7115, 71150A-1 (2008)
[42] R. Huang, B. Chann, J. Burgess, *et al.*, "Teradiode's high brightness semiconductor lasers," Proc. SPIE 9730, 97300C (2016)
[43] B. Lee, J. Kansky, A. Goyal, *et al.*, "Beam combining of quantum cascade laser arrays," Opt. Express 17, 16216 (2009)
[44] G. Bloom, C. Larat, E. Lallier, *et al.*, "Passive coherent beam combining of quantum cascade lasers with a Dammann grating," Opt. Lett. 36, 3810 (2011)
[45] R. Ostendorf, S. Hugger, R. Aidam, *et al.*, "External cavity spectral beam combining of 4.6 μm emitting quantum cascade laser arrays," 22nd IEEE International Semiconductor Laser Conference, 51 (2010)
[46] L. Hoffman, M. Klinkmuller, E. Mujagic, *et al.*, "Tree array quantum cascade laser," Opt. Express 17, 649 (2009)
[47] L. Hoffman, C. Hurni, S. Schartner, *et al.*, "Coherence in Y-coupled quantum cascade lasers," Appl. Phys. Lett. 91, 16106 (2007)
[48] A. Lyakh, R. Maulini, A. Tsekoun, R. Go, and C. Kumar N. Patel, "Continuous wave operation of buried heterostructure 4.6 μm quantum cascade laser Y-junctions and tree arrays," Opt. Express 22, 1203 (2014)
[49] J. Kirch, C.-C. Chang, C. Boyle, *et al.*, "5.5 W near-diffraction-limited power from resonant leaky-wavecoupled phase-locked arrays of quantum cascade lasers," Appl. Phys. Lett. 106, 061113 (2015)
[50] G.P. L, C. Peng, H.Q. Le, H.Q., *et al.*, "Grating-tuned external cavity quantum cascade semiconductor lasers," Appl. Phys. Lett. 78, 2834–36 (2001)
[51] R. Maulini, "Broadly tunable mid-infrared quantum cascade lasers for spectroscopic applications," Thèse de doctorat : Université de Neuchâtel, 2006
[52] R. Maulini, D.A. Yekha, J.-M. Bulliard, M. Giovannini, J. Faist, and E. Gini, "Continuous wave operation of a broadly tunable thermoelectrically cooled external cavity quantum cascade laser," Opt. Lett. 30, 19, 2584–86 (2005)
[53] C. Gmachl, D.L. Sivco, R. Colombelli, F. Capasso, and A.Y. Cho, "Ultra-broadband semiconductor laser," Nature 415, 883–87 (2002)
[54] N. Bandyopadhyay, M. Chen, S. Sengupta, S. Slivken, and M. Razeghi, "Ultra-broadband quantum cascade laser, tunable over 760 cm^{-1}, with balanced gain," Opt. Express 23, 21159–64 (2015)
[55] T. Tsai and G. Wocki, "External-cavity quantum cascade lasers with fast wavelength scanning," Appl. Phys. 100, 243–251 (2010)

[56] J. Grahmann, A. Mrten, A. Herrmann, *et al.*, "Large MOEMS diffraction grating results providing an EC-QCL wavelength scan of 20%." Proc. SPIE 2015, 9375, doi:10.1117/12.2079640

[57] A. Lyakh, R. Barron-Jenez, I. Dunayevskiy, R. Go, G. Tsvid, C. Kumar, and N. Patel, "Continuous wave operation of quantum cascade lasers with frequency shifted feedback," AIP Adv. 6, 29715 (2016)

[58] A. Lyakh, R. Barron-Jenez, I. Dunayevskiy, R. Go, G. Tsvid, C. Kumar, and N. Patel, "External cavity quantum cascade lasers with ultra-rapid acousto-optic tuning," Appl. Phys. Lett. 106, 2834 (2015)

[59] L.P. Yatsenko, B.W. Shore, and K. Bergmann, "Theory of a frequency-shifted feedback laser," Opt. Commun. 2004, 236, 183–202

[60] R. Williams, J. Kelly, J. Hartman, *et al.*, "Kilohertz linewidth from frequency-stabilized mid-infrared quantum cascade lasers," Opt. Lett. 24, 1844 (1999)

[61] F. Cappelli, I. Galli, S. Borri, *et al.*, "Subkilohertz linewidth room-temperature mid-infrared quantum cascade laser using a molecular sub-Doppler reference," Opt. Lett. 37, 4811 (2012)

[62] J. Faist, C. Gachl, F. Capasso, *et al.*, "Distributed feedback quantum cascade lasers," Appl. Phys. Lett. 70, 2670 (1997)

[63] D. Hofstetter, J. Faist, and M. Beck, "Surface-emitting 10.1 μm quantum-cascade distributed feedback lasers," Appl. Phys. Lett. 75, 3769 (1999)

[64] A. Lyakh and P. Zory, "Substrate-emitting, distributed feedback quantum cascade lasers," Appl. Phys. Lett. 91, 181116 (2007)

[65] N. Finger, W. Schrenk, and E. Gornik, "Analysis of TM-polarized DFB laser structures with metal surface gratings," IEEE J. Quantum Electron., 36, 7, 780–86 (2000)

[66] H. Kogelnik and C.V. Shank, "Coupled-wave theory of distributed feedback lasers," J. Appl. Phys. 43, 2327 (1972)

[67] Q.Y. Lu, Y. Bai, N. Bandyopadhyay, S. Slivken, and M. Razeghi, "2.4 W room temperature continuous wave operation of distributed feedback quantum cascade lasers," Appl. Phys. Lett. 98, 181106 (2011)

[68] B.G. Lee, M.A. Belkin, C. Pflügl, *et al.*, "DFB quantum cascade laser arrays," IEEE J. Quantum Electron., 45, 5 (2009)

[69] A. Hugi, G. Villares, S. Baser, H. Liu, and J. Faist, "Mid-infrared frequency comb based on a quantum cascade laser," Nature 492, 229 (2012)

[70] G. Villares, J. Wolf, D. Kazakov, *et al.*, "On-chip dual-comb based on quantum cascade laser frequency combs," Appl. Phys. Lett. 107, 251104 (2015)

[71] R. Go, H. Krysiak, M. Fetters, *et al.*, "Room temperature operation of quantum cascade lasers monolithically integrated onto a lattice-mismatched substrate," Appl. Phys. Lett. 112, 031103 (2018)

[72] H. Nguyen-Van, A. Baranov, Z. Loghmari, *et al.*, "Quantum cascade lasers grown on silicon," Sci. Rep. 8, 7206 (2018)

[73] J. Faist, Quantum Cascade Lasers. 1st edition (Oxford: Oxford University Press, 2018)

Chapter 5
Diode pumped high power lasers

Martin Smrž[1] and Martin Divoky[1]

Advances and price drop of diode lasers in recent years enabled construction of highly efficient high power and high energy diode pumped lasers. Diode pumped solid-state lasers became dominant in industrial applications and are slowly replacing gaseous or excimer high power lasers thank to small dimensions and high efficiency. Output power from thin disk-based systems emitting close to wavelength of 1 μm overcame 10 kW or 4 kW barrier from a single disk in continuous wave operation or in pulsed operation, respectively. The beam quality is given by parameter M^2 that equals to 8 or 1.38, respectively [1]. In regime of ultrashort pulse generation, thin-disk amplifiers approach average power of 2 kW with mJ energies in <2 picosecond long pulses [2,3] and are close to 1 kW barrier with 200 mJ picosecond pulses [4]. At repetition rate of 100 Hz pulse energy of 0.5 J from double disk amplifier was demonstrated [5]. Even diode pumped femtosecond thin-disk-based oscillators lately reached 270 W with 38 MW of peak power in a train of 210 fs long pulses [6]. Pulsed slab-based systems reached 100 J energy output with 1 kW average power [7]. Although employment of diode pumping significantly increased optical efficiency of all solid-state laser systems and efficiency approached limit given by the Stokes defect of gain media in some cases, the main issue of high average power lasers remains the heat extraction from a gain medium. Unlike in CW lasers, high small signal gain and nonlinear effects start to play role in pulsed lasers as the pulse energies increase. This leads to amplified spontaneous emission (ASE) and laser-induced damage threshold that also limit the laser operation.

In this chapter, we will guide the reader through development process of high average power diode pumped solid-state lasers, starting from material choice, through amplifier geometry and detailed description of the lasers at HiLASE [8,9].

Detailed insight into the problematic of solid-state lasers can be found in textbook by Koechner [10].

[1]HiLASE Centre, Institute of Physics of the Czech, Academy of Sciences, Dolní Břežany, Czech Republic

5.1 Material selection

Gain material of laser consists of transparent host material and active ions that participate in stimulated emission. Laser parameters are mostly given by active ions, while thermal parameters are usually given by the host. From laser perspective, the most important quantities are cross-section, fluorescence lifetime, and emission bandwidth for short pulse lasers. From thermal perspective, important parameters are thermal conductivity, thermo-optic coefficient (*dn/dT*), coefficient of thermal expansion, and thermal shock resistance.

5.1.1 Laser properties

Transition cross section defines probability that active ion absorbs or emits photon. The small signal gain g then be proportional to emission cross section σ_e and population inversion density ΔN. With higher emission cross-section, the gain material is more easily saturated and efficient amplification can be achieved with low intensities that reduce risk of optical damage. On the other hand, large emission cross-section increases losses due to ASE that lowers the gain and is detrimental mostly for high energy amplifiers.

Second important parameter is fluorescence lifetime τ. Fluorescence lifetime defines length of the spontaneous decay of electrons from upper laser level by photon emission. The longer the lifetime the longer the material can store energy absorbed from pump light. This is important for many diode pumped lasers, because diodes are not as bright as originally used flash lamps, so they need to emit light longer or their number must be largely increased. With long fluorescence lifetime, number of diodes and overall costs for the system can be therefore decreased. On the other hand, shortening of pump interval of pulsed lasers to fraction of the material lifetime by adopting highly bright pump diodes brings reduction of ASE losses [11].

5.1.2 Thermal properties

Thermal conductivity defines how easily the heat is conducted in the material. The higher heat conductivity, the smaller thermal gradients are induced in the gain media, and the more heat can be extracted from them. Consequent thermal lensing and mechanical stress originating from thermo-optic coefficient (*dn/dT*) and coefficient of thermal expansion generate optical path difference (OPD) and wavefront distortion, which cause drop of beam quality and finally can prevent lasing or lead to serious damage on the gain medium.

Thermal shock parameter indicates how much thermal loading can be tolerated before fracture appears:

$$R_S = \frac{K(1-\nu)}{\alpha E}\sigma_{\max}$$

where K is the thermal conductivity, ν is the Poisson ratio, E is the Young's modulus, α is the coefficient of thermal expansion, and $\sigma_{\max}$ is the maximum surface stress before fracture occurs.

5.1.3 Nonlinear properties

High peak intensity in optical materials induces nonlinear response of the material. Dielectric susceptibility, which describes material properties, is electric field-dependent. To simplify this relationship the material susceptibility is usually decomposed into Tailor series [12]. For centro-symmetric crystalline media, like YAG, or amorphous, like glass, the even orders of this series vanish. The lowest nonlinear odd order, the third order, is responsible for intensity-dependent refractive index of the material:

$$n(I) = n_0 + n_2 I$$

where n_0 is the standard refractive index of the material and n_2 is its intensity-dependent contribution. Propagation of laser beams and ultrashort pulses through nonlinear gain media causes self-focusing of laser beams and self-phase modulation of laser pulses. Although these phenomena are sometimes used for mode-locking of oscillators, or spectral broadening of ultrashort pulses, they are often unwanted and should be eliminated in order not to damage solid-state components of a laser system. Amount of accumulated nonlinear phase during the beam propagation is defined by so called B-integral [13]:

$$B = \frac{2\pi}{\lambda} \int I(z) dz$$

where λ is the radiation wavelength, I is the laser pulse intensity, and z is the spatial coordinate along propagation direction. B-integral value <1 is safe whereas value >1 is considered to be dangerous. In this case it is essential to adopt spatial filtering of the beam to improve the distorted wavefront, or to reduce peak intensity in order to avoid damages, for example, by expanding the beam size or stretching the pulse duration. Amplification of stretched pulses is then called chirped pulse amplification (CPA) [14], which is widely used nowadays for majority of high power ultrashort pulse amplifiers.

5.1.4 Gain bandwidth

The last mentioned quantity is gain bandwidth. The gain bandwidth defines the width of the amplified spectrum and therefore the achievable length of the amplified pulses. Gain bandwidth of the active medium is critical parameter for ultrashort pulse amplification. From the optical equivalent of uncertainty principle follows indirect proportionality between pulse bandwidth and pulse duration. The broader the bandwidth, the shorter pulses can be obtained after amplification. For picosecond pulse amplification, bandwidth of few nanometers is sufficient; however, femtosecond pulse amplification requires tens of nanometers broad spectra. Table 5.1 shows bandwidth of some active media.

5.1.5 Comparison

Both laser and thermal parameters do not change only with host material or active ions, but also with active ions concentration and ambient temperature. Since most

Table 5.1 Laser parameters of frequently used gain media [15–19]

Gain medium	Laser wavelength [nm]	Emission bandwidth [nm]	Fluorescence lifetime [μs]
Nd:YAG	1064.1	0.45	230
Yb:YAG	1030	6	960
$Ti:Al_2O_3$	800	320	3
Yb:S-FAP	1047	4	1100
$Yb:CaF_2$	1033	8	2400
Nd:glass	1054	28	300
$Yb:Lu_2O_3$	1032	13	820
Yb:LuAG	1031	7.5	1250
Cr:LiSAF	800–1050	257	67

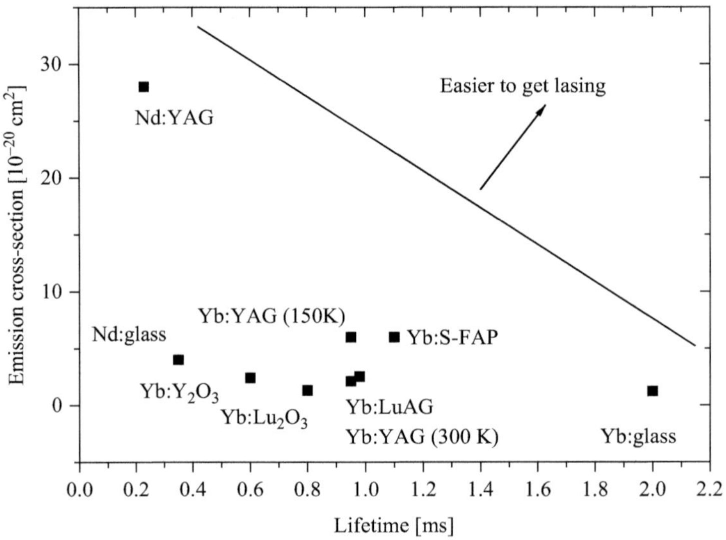

Figure 5.1 Emission cross-section versus lifetime for various host materials and different active ions

high power lasers rely on Yb^{3+} or Nd^{3+} doped materials, we will concentrate on comparison of those active ions.

Nd^{3+} doped materials are well established and their parameters can be found online. Thermal shock parameter was taken from [20], for Yb^{3+}:YAG from [20–22], for Yb^{3+}:LuAG from [22,23], and $Yb^{3+}:Y_2O_3$ from [23–28]. Thermal stress parameter R_T was assumed to scale linearly to R_S, $Yb^{3+}:Lu_2O_3$ from [26,28,29], Yb^{3+}:glass from [30], where thermal shock parameter was assumed to be similar to Nd^{3+} doped glasses, and Yb^{3+}:S-FAP from [15].

Usually used figure of merit is product of emission cross-section and lifetime [31]. From Figure 5.1, ideal materials seem to be Yb^{3+}:glass and Nd^{3+}:YAG, where this product is the highest and lasing is easiest to achieve. But when we take into account efficient energy extraction and resilience of the material, preferred materials change to Yb^{3+}:YAG, Nd^{3+}:glass or Yb^{3+}:S-FAP as can be seen in Figure 5.2. This reflects the reality, since high power operation was achieved on exactly those materials [7,32,33]. The preferred area marked with rectangle is arbitrary and can change depending on required output laser parameters.

Although maximum tolerable heat load is given by thermal shock parameter, the beam quality is usually proportional to thermal conductivity. Thermal conductivity is not constant, but it changes with active ion doping concentration and with temperature. This has to be taken into account when selecting host material for high power laser. For highly doped small thin-disk laser, sesquioxides can be better choice than YAG due to their higher thermal conductivity, see Figure 5.3; however, their melting point is too high and growing of good quality, large size single crystals for thin-disk manufacturing is difficult.

For low-doped high energy large slab amplifier, YAG is still the best option and its parameters can be even enhanced by cooling it down to cryogenic temperatures (Figures 5.4 and 5.5). By dropping the operating temperature from 300 to 150 K, thermal conductivity increases two times and emission cross-section three times. Other thermal properties benefit from low temperatures as well [22].

In addition, all Yb-doped materials can be pumped by affordable, well-developed, and efficient high brightness InGaAs pump diodes at wavelengths ranging approximately from 940 to 980 nm, depending on host material. This fact also contributes to expansion of Yb:YAG diode pumped solid-state lasers.

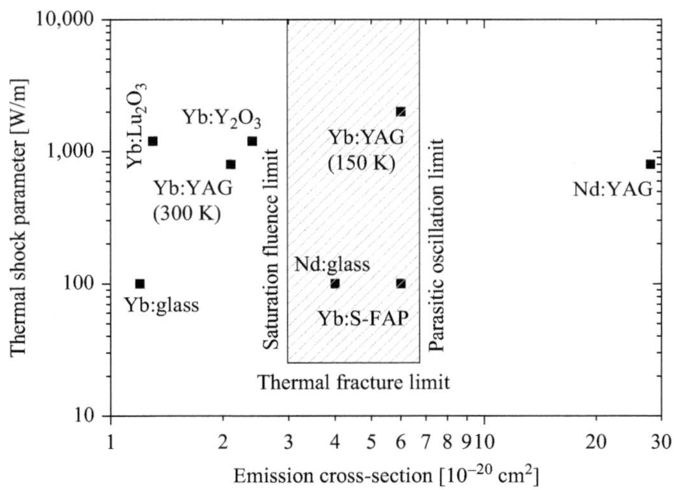

Figure 5.2 Thermal shock parameter versus emission cross-section

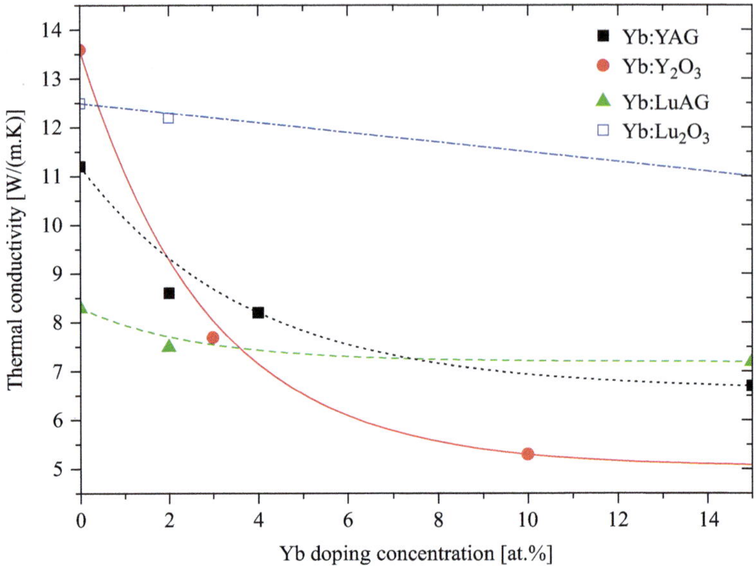

Figure 5.3 Thermal conductivity versus Yb^{3+} doping concentration in various hosts. Lines show simple exponential fits

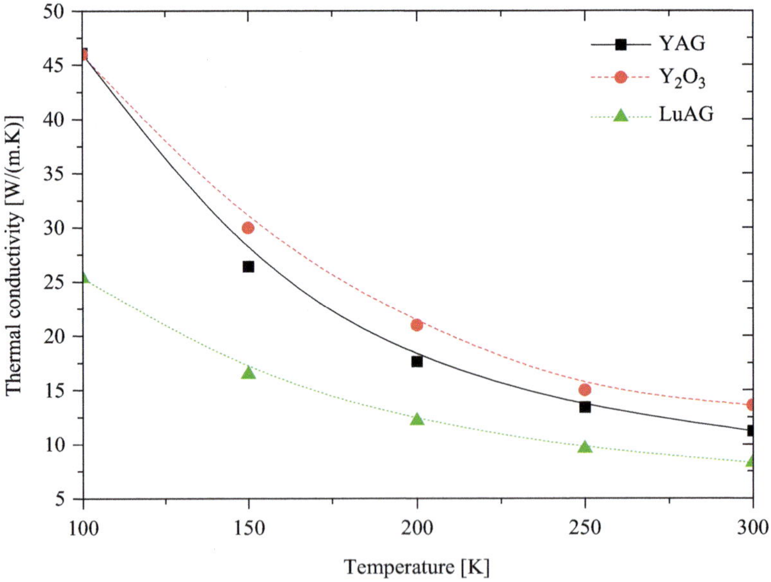

Figure 5.4 Thermal conductivity versus temperature for undoped crystals. Lines show simple exponential fits

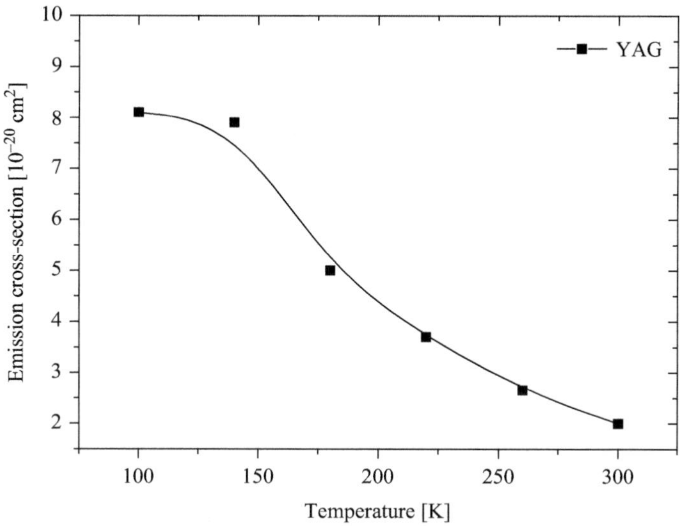

Figure 5.5 Emission cross-section versus temperature

5.2 Laser amplifiers

Most of the high-power-pulsed lasers work in MOPA scheme (Master Oscillator – Power Amplifier), where high quality seed beam is amplified in one or more power amplifiers. Those power amplifiers can be regenerative or multipass and both of them can be combined with CPA technique (chirped pulse amplification) to amplify ultra-short pulses. Behavior of the amplifiers can be predicted by Frantz–Nodvik equation [34]

$$E_{out} = E_{sat} ln\left\{ G_0 \left[\exp\left(\frac{E_{in}}{E_{sat}}\right) - 1 \right] + 1 \right\}$$

where E_{sat} is the saturation energy of the gain medium given by ratio of photon energy of the laser transition and its emission cross-section, G_0 is the small signal gain, and E_{out} is the pulse energy extracted from the amplifier. This equation implicates following facts:

- Gain for amplification of low fluence pulses compared to saturation fluence is $>>1$, but energy extraction efficiency is low.
- Gain for amplification of pulses with fluence comparable to or higher then saturation fluence approaches 1 and energy extraction efficiency approaches 100%.

To extract significant amount of stored energy and reach high efficiency of the amplifier is important to operate it in saturation. The saturation condition provides a clue for selection of an appropriate amplification system.

5.2.1 Regenerative amplifiers

Regenerative amplifiers [35–37] offer solution for amplification of pulses with energy $<<E_{sat}$. Although small signal gain in such an amplifier is high, single pass energy extraction is low and amplification is inefficient. Trapping of the pulse in an optical cavity allows for realization of tens of cavity round trips, increasing the accumulated gain and finally getting the pulse energy close to saturation one, leading to efficient energy extraction. Amplification undergone in optical resonator (Figure 5.6) does not degrade laser beam quality. Amplification process has following steps:

- Pumping of a gain medium—depending on pulse repetition rate, it can be pulsed or CW pumped amplifier;
- Injecting of a seed pulse to the amplifier—pulses are trapped in the cavity, usually by an acousto-optic switch, or more often by a Pockels cell and changing polarization state. Polarization locking by applying high voltage to a Pockels cell requires fast high voltage electronics and polarization optics in the cavity;
- Amplification of the trapped pulse—realization of tens or even hundreds of cavity round trips;
- Releasing of the amplified pulse from the cavity by switching of high voltage on the Pockels cell. Pulse is released after saturation of the gain medium. Efficiency of energy extraction is limited by losses—the higher the losses, the lower the extraction efficiency from the regeneration. Usual level of total losses is 4%–6%; and
- Input and output beam are separated by a Faraday rotator in standing wave cavity regenerative amplifiers usually. The input and output beams have identical polarization state but they have opposite propagation direction. In ring cavities [38], they are separated geometrically.

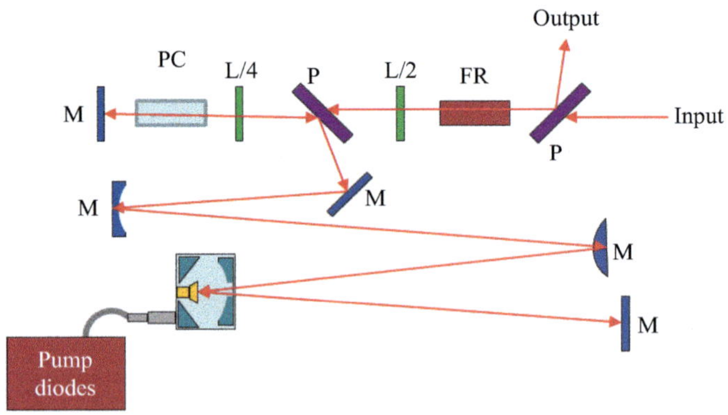

Figure 5.6 Regenerative amplifier with a standing-wave cavity (M, mirrors; PC, Pockels cell; FR, Faraday rotator; P, polarizer; L/2, half waveplate; L/4, quarter waveplate)

Regenerative amplifiers have several advantages:

- High net-gain, which can often reach several orders for weak seed pulses and
- Excellent beam quality because of presence of optical resonator filtering mode profile.

and drawbacks of regenerative amplifiers are:

- Gain narrowing [39], which is a consequence of high number of roundtrips and irregular spectral shape of gain profile. The effect narrows bandwidth of the pulse undergoing the amplification and so prolongs bandwidth-limited pulse duration;
- High complexity is given by presence of electro-optical or acousto-optical switches and optical resonator; and
- High repetition rate regenerative amplifiers can exhibit energy instability, so-called bifurcation [40,41], causing period multiplication or chaotic behavior. This typically occurs when pulse build-up time in the amplifier is comparable with upper laser level lifetime.

Because of their properties, regenerative amplifiers are often used like first stage amplifiers for short pulse amplification. They are very efficient in power amplifiers of thin-disk lasers, where they help to solve their main drawback—very low single-pass gain.

5.2.2 Multipass amplifiers

Unlike regenerative amplifiers, number of pulse round-trips in multipass amplifiers [36] is fixed and given by spatial multiplexing, that is, laser beam is folded by mirrors to multiple the number of roundtrips (Figure 5.7). The geometrical constraints are therefore the main limiting factor of roundtrip number. Thus, multipass amplifiers have to be able to saturate the gain medium in a few pulse roundtrips, and are normally used for power amplification in very last stages of a complex laser system, where net-gain lower than one order is sufficient. Frequently, a regenerative amplifier is followed by one or a series of multipass amplifiers in complex, high energy laser systems. Absence of resonator does not preserve excellent beam

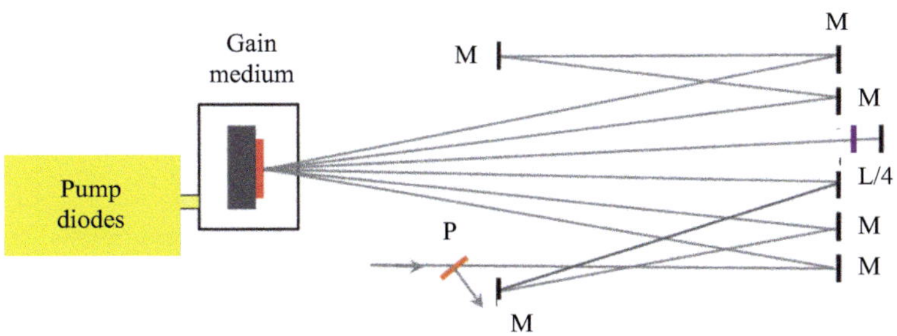

Figure 5.7 Multipass amplifier with a thin-disk gain medium (M, mirrors; P, polarizer; L/4, quarter waveplate)

quality; however, diffraction is often compensated by image relaying between the subsequent passes of the pulse through the gain medium. Imaging telescopes are often turned to spatial filters by installing pinhole in the common focal point of the telescope to increase the beam quality of the beam. Multipass amplifiers can be characterized as follows:

- Geometrical multiplexing enables simple setup without a need for complicated electro- or acousto-optic switches and high contrast polarizing optics;
- Resonator absence does not preserve beam quality and requires installation of spatial filters and image relaying but allows for increase of gain medium aperture (large mode resonators would be extremely long); and
- Low net-gain coming from geometrical constraints on alignment of pump beam and folding the signal beam.

5.2.3 Chirped pulse amplification

As mentioned in Section 5.1.3, amplification of ultrashort pulses brings complications with high intensity accumulating high B-integral as the pulse energy increases during the amplification process. Additionally, laser-induced damage threshold (LIDT) of materials drops with a square root of decreasing pulse length down to 2 ps [42]:

$$F_{th}(\tau_2) = F_{th}(\tau_1)\sqrt{\tau_2/\tau_1}$$

where F means the energy density (fluence) and τ is the pulse duration. Prediction of LIDT for femtosecond pulses is complicated due to many multiphoton effects participating in the process [43]. Crossing the damage threshold barrier of laser components and consequent damage on gain medium or dielectric coatings prevents direct amplification of femtosecond and picosecond pulses to high energy. High peak intensity important for laser-matter interaction is limiting factor for pulse amplification itself. Solution for such a drawback is amplification of broadband pulses stretched to nanosecond pulse duration, that is, CPA (Figure 5.8) [14,37,44].

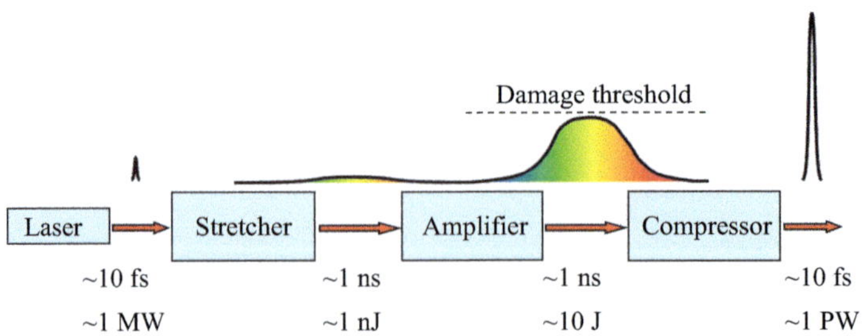

Figure 5.8 Scheme of ultrashort pulse amplification by Chirped Pulse Amplification Technique

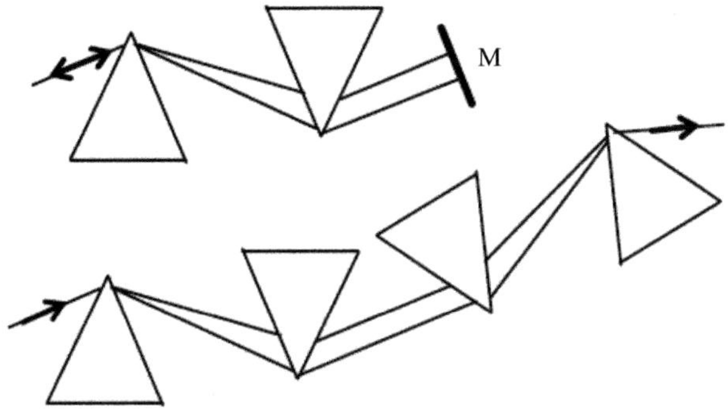

Figure 5.9 Compensation of spatial separation of spectral components in the stretched pulse in double-pass configuration using mirror (top) or another pair of dispersion elements (bottom)

Handling the pulse duration is possible due to dispersion effects [37,44]. Dephasing of spectral components of bandwidth-limited pulses generated by an oscillator stretches the pulses and generates frequency chirp. After amplification, the pulses are compressed again to original pulse duration by opposite dispersion compensating the actual chirp. Chromatic dispersion $n(\omega)$ in materials in near infrared spectral range generates so called normal (material) dispersion when higher frequency components propagate slower. Material dispersion is often insufficient or impractical for a CPA stretcher or compressor because chromatic dispersion doesn't have opposite counterpart that can fully compensate it. Therefore, OPD between spectral components is generated during propagation of the pulse in media exhibiting angular dispersion. Typically, a pair of dispersive prisms [45,46] or diffraction gratings [47] showing anomalous dispersion is used. Spatial separation of spectral components of the pulse generated after passing the second prism or grating can be compensated by a back-reflective mirror and double-pass configuration (Figure 5.9). Amount of generated frequency chirp is proportional to distance of prisms or gratings in the pair. Normal dispersion in near infrared spectral range can be generated by grating pair with negative distance of the gratings, that is, when the first grating is image relayed behind the second one by an imaging system between the gratings [48]. Such pair of Martinez stretcher and Tracy compressor generate opposite dispersions that compensate each other very well (Figure 5.9).

5.3 Geometry of the active medium in high power amplifier

High power amplifiers optimize geometry of the active medium to maximize heat extraction, minimize thermal lenses and wavefront aberrations. Construction of any of the following compact lasers would not be possible without availability of reliable and high brightness pump source like semiconductor laser diodes.

5.3.1 Rod-type amplifiers

Fundamental setup for solid-state amplifiers is a rod-type configuration [9] (Figure 5.10). The active medium is cylindrically shaped with a diameter of a few millimeters up to a few centimeters and length up to a few centimeters, depending on the actual doping, cooling strategy, pumping type, etc. There are two strategies for pump beam delivery. The rod can be end-pumped or side-pumped.

In case of the end pumping [49] both the seed beam and the signal beam copropagate in axial direction of the rod. This configuration is better for preservation of beam quality since thermal gradient created by both beams has azimuthal symmetry and creates equivalent of a thick convex lens and the highest gain is achieved on the axis of the gain medium [50,51]. End pumping is not suitable for long active media since the population inversion drops as the pump beam is absorbed and pump intensity becomes insufficient from some point.

In case of side pumping, the pump intensity is then limited only by geometrical considerations and brightness of pump diodes. A drawback of this setup is distribution of population inversion with maximum close to the surface of the rod and complex temperature profile leading to decrease of beam quality. Anyway, such a configuration is often used for high power and high energy applications (Figure 5.10) [52,53].

General disadvantage of the rod-type lasers is strong thermal lensing at high average power since it can be cooled only on the surface, where the cooling often collides with the pump source. Other geometries are therefore preferred for state-of-the-art high power lasers.

5.3.2 Fiber amplifiers

Fiber lasers can be considered as extremely thin and very long end-pumped rod-type lasers (Figure 5.11). Unlike other geometries, optical fibers use the advantage of waveguide effect, that is, laser beam under amplification can propagate a very long distance without diffraction and change of its parameters. When a core of the fiber is doped by active ions, the fiber can serve like a very long active medium. Although a host material for near-infrared emitting fibers is amorphous fused silica, cooling of fibers is efficient due to fiber thickness of only a few microns to a few tens of microns. Thus, ratio of the cooled surface area to volume of the active

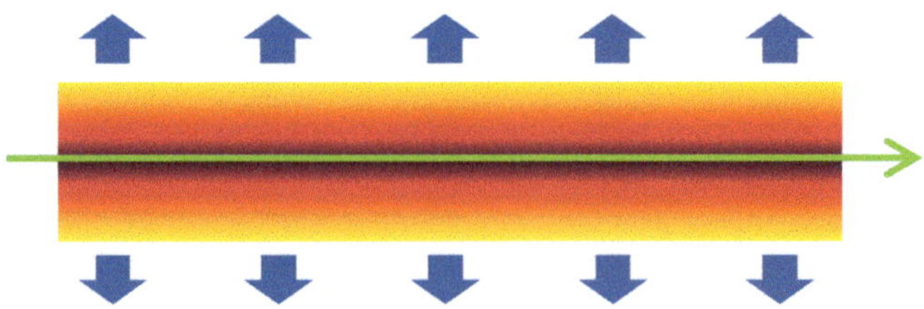

Figure 5.10 Rod with indicated beam path, heat extraction direction, and thermal gradients

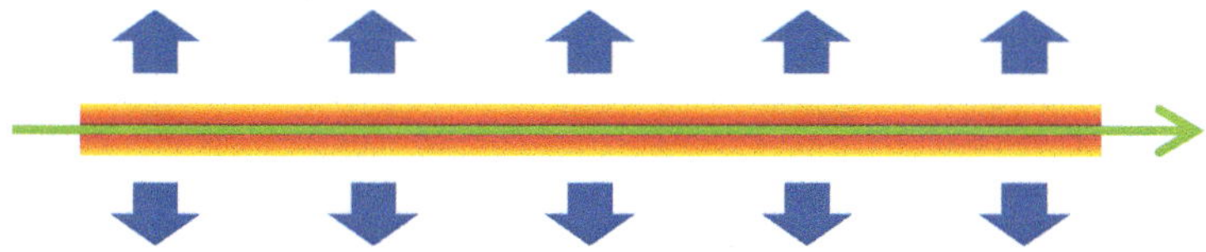

Figure 5.11 Fiber with indicated beam path, heat extraction direction, and thermal gradients

medium itself is extremely high. Because of the long active medium and high gain, ASE is the main limiting factor. Propagation of short pulses through the long gain medium induces Kerr effect (see Section 5.1.3), which can damage the fiber by self-focusing, and accumulates huge amount of material dispersion. Fiber amplifiers are therefore not suitable for amplification of high-energy (>mJ) ultrashort pulses.

Fiber active media are end pumped by fiber-coupled laser diodes. Drawback of complicated coupling of the pump radiation from poor beam quality laser diodes into a few micron thick fiber core was solved by so-called double clad optical fibers. Outer cladding with diameter ranging typically from tens to several hundred of microns guides the pump radiation until it is fully absorbed. Large, multimode-fiber-coupled laser diodes can reliably deliver pump power from few watts up to kW level.

Nowadays fiber lasers generate multi-kW beams, both in CW [54] and pulse regime [55–57]; however, pulse energy exceeding millijoule level even from a single large mode area (LMA) fiber is hardly feasible [58]. Even in case of coherent beam combining of multiple fibers and temporal pulse multiplexing [59,60] cannot by the state-of-the-art technology increase the energy of ultrashort pulses significantly over a few tens of millijoules (Figure 5.11).

5.3.3 Thin-disk and active mirror amplifiers (incl TRAM)

Thin-disk amplifier [61,62] is based on large size active medium (a few millimeters up to a few centimeters in diameter) with small thickness, which is typically a few hundred microns (Figure 5.12). The active medium is reflective coated from one side, antireflective coated from the other one, and attached to a heatsink or directly cooled by liquids. Such configuration behaves like an amplifying mirror and allows for very efficient heat extraction in axial direction over the whole area of the disk, which effectively suppresses radial thermal gradients and consequent thermal lensing. Also accumulated nonlinear effects and wavefront aberrations are negligible thanks to the low thickness of the gain medium. Disadvantage of this configuration is very low single-pass gain, therefore is this configuration ideal for highly doped materials and high power CW [63] or high repetition rate laser systems based on a short cavity or regenerative amplifiers [8,9,64–66], but rarely for multipass amplifiers [2,3]. High energy extraction is possible [1,4,5], but technologically complicated because of necessity to scale up the beam diameter for higher energy storage. Such conditions yield to evolution of ASE in transversal direction and depletion of gain [67,68]. Countermeasures like undoped caps of thin-disks [69]

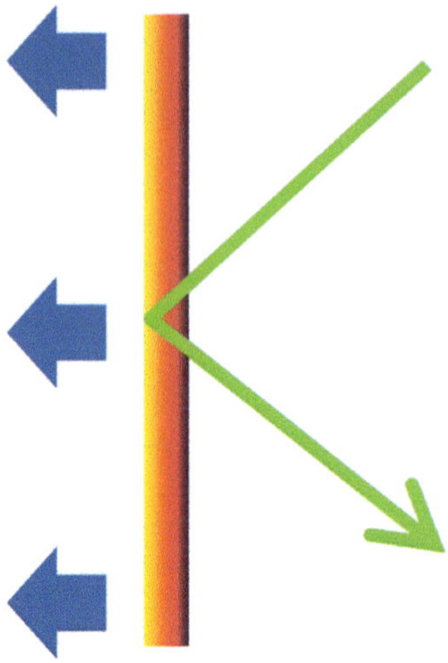

Figure 5.12 Thin-disk with indicated beam path, heat extraction direction, and thermal gradients

suppressing ASE by extraction of spontaneously emitted photons from the active medium have not provided satisfactory results yet [70], except for cryogenically cooled thick disks [71,72]. The highest energy extracted from a single disk is 300 mJ at 100 Hz [66] and 200 mJ at 5 kHz [4], from double disk regenerative amplifier than 0.6 J at 100 Hz repetition rate [5], and >1 J from a cryogenically cooled thick disk amplifier [72]. Interesting way of ultrashort pulse generation avoiding CPA are thin-disk femtosecond oscillators reaching average power of 250 W with microjoule energy level [73–75], or regenerative amplifiers for direct amplification of ultrashort pulses to energy >> a few tens of mJ [76,77]. Unlike fibers which can be considered like a long rod-type laser with a tiny aperture, thin-disks are opposite case limiting to a rod-type active medium with large aperture and extremely low thickness (Figure 5.12).

Thicker version of thin-disk amplifier is called active mirror (Figure 5.13). It doesn't have such good thermal properties, but enables higher energy storage and higher single pass gain, so it's suited for high energy amplifiers [78]. However, this concept is largely avoided today (Figure 5.13).

Special case of active mirror amplifier is total reflection active mirror (TRAM) [79] (Figure 5.14), it relies on total internal reflection of the beam in the active layer and therefore has better thermal contact with the heatsink that is not degraded by presence of high reflectivity layer from oxides with low thermal conductivity. TRAM is usually ceramics composite consisting of undoped bulk used for guiding the beam into the active layer and doped thin layer at one or several faces used to amplify the beam. The design was later upgraded to better resists ASE for large beams and 1 J output energy was achieved at 100 Hz repetition rate, but only for a short time due to

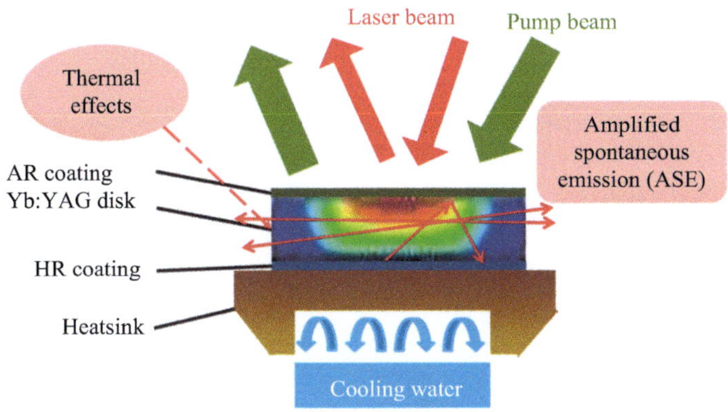

Figure 5.13 Active mirror/thin-disk with indicated beam path, heat extraction direction, and thermal gradients

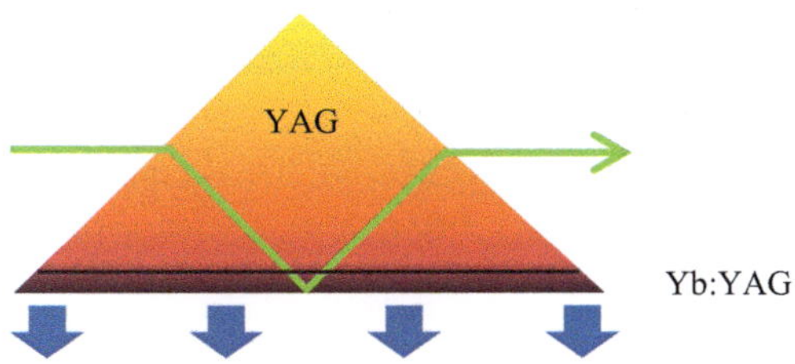

Figure 5.14 TRAM with indicated beam path, heat extraction direction, and thermal gradients

insufficient cooling [80]. The amount of material needed to guide the beam into the active layer is downside of this technology and makes scaling difficult (Figure 5.14).

5.3.4 Slab (zig-zag, multislab, innoslab)

Zig-zag [81] (Figure 5.15) slab uses both large faces for cooling and the beam is injected to the medium from the side. It is guided through the medium by total internal reflections and the beam propagates perpendicular to the gradient of temperature, thus minimizing acquired wavefront aberrations. Drawback of this design is nonsymmetrical beam in a form of rectangle with ratio between sides of around 10 (Figure 5.15).

Slab [82,83] (Figure 5.16) doesn't use any reflective surfaces, it's a simple plate from active medium whose front and back surface are cooled by contact with heatsink, by liquid coolants or by forced gas flow. Gas is usually Helium because of its high cooling capacity and low dependence of refractive index on its density. The slab can be either edge-pumped, which is for example case of the InnoSlab technology emitting >50 mJ picosecond pulses with 1.1 kW of average output power [84,85]. Other case is face-pumped slab together with gas flow cooling. Large cooling surface and propagation of the beam perpendicular to thermal gradients makes this geometry

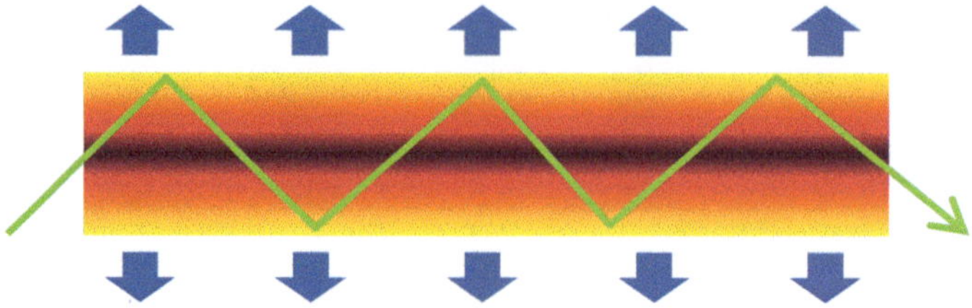

Figure 5.15 Zig-zag with indicated beam path, heat extraction direction, and thermal gradients

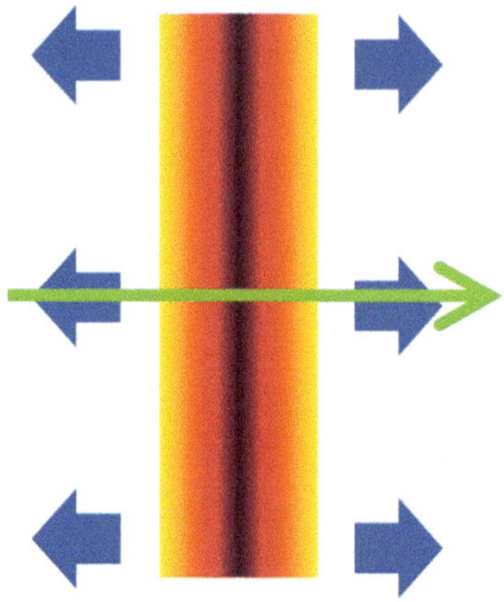

Figure 5.16 Slab with indicated beam path, heat extraction direction, and thermal gradients

ideal for high energy high average power amplifiers. This geometry is achieving highest energies in high power lasers (Figure 5.16) [7,8,32,33].

5.4 Thin-disk high power system

Thin-disk lasers are tightly connected with advent of affordable laser diodes and Yb:YAG active media (see Section 5.3.3), although many different kinds of thin-disks in near-infrared-based on LuAG [85,86], sesquioxides [87,88], calcium fluoride [19], or mixed garnets [89] were already demonstrated. Much of the effort has been also put to equivalent ceramic matrices [90]. Lately also thin-disk lasers emitting in mid-infrared spectral range based on Ho:YAG [91,92] have been constructed. However, active media different from Yb:YAG are still marginal.

5.4.1 Yb:YAG active medium

Yb:YAG is an active medium consisting of a popular yttrium-aluminum garnet, an isotropic gain medium showing very good thermo-mechanical properties and well established growing and polishing technology, and Yb^{3+}-active ions.

This combination offers very attractive mix of parameters like emission wavelength at 1,030 nm, close to in the past very popular Nd:YAG emitting at 1,064 nm. Unlike the Nd:YAG, broader emission bandwidth allows amplification of picosecond and subpicosecond pulses, and highly bright InGaAs laser diodes for efficient pumping at a broad absorption line close to 940 nm [21] are available. Long accumulation of pump energy due to upper laser level lifetime approaching 1 ms is possible.

Proximity of pump and signal photon energy caused by quasi-three-level system of energy levels in Yb ions (Figure 5.17) and a consequent low quantum defect of only 8.7% is advantageous for high power laser construction. Only small amount of delivered pump energy has to be extracted in a form of thermal energy and together with thin-disk geometry opens a way toward multi-kW laser systems [61,67,68]. Simplicity of Yb energy diagram consisting of only two levels leads to absence of parasitic processes like up-conversion or excited state absorption, which makes the laser action very efficient. On the other hand, the quasi-three-level behavior is sensitive to operational temperature. Properties of Yb:YAG gain media are summarized in [25,93]. Pros and cons of Yb:YAG disk laser are as follows:

- Efficient cooling thanks to large cooled area
- Axial heat flow
- Low thermal lensing
- Very low nonlinearities (*B*-integral), even in case many roundtrips
- Easy scalability by increasing the disk area
- Low single-pass gain.

Yb:YAG gain material then brings following properties:

- Very low quantum defect (8.7% at 940 nm, 5.9% at 969 nm)
- High doping concentration without quenching (15%) allows for construction of very thin-disk

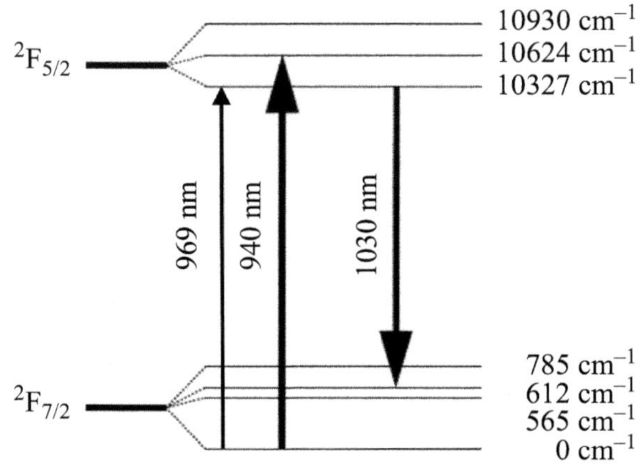

Figure 5.17 Energy levels of Yb:YAG gain medium

- Low fractional heating <12% (includes coating absorption and nonradiative transitions)
- No excited state absorption and up-conversion
- High thermal conductivity and tensile strength of YAG
- Long lifetime of upper laser level (approx. 1 ms) for effective pumping by reliable and affordable InGaAs pump diodes
- Broad emission bandwidth sufficient for picosecond pulse amplification
- Quasi-three-level behavior brings sensitivity to operational temperature.

5.4.2 Pump geometry for thin-disk lasers

Thin-disk lasers have a significant drawback in low single-pass gain for signal wavelength and low single-pass absorption for pump radiation. As already mentioned, the drawback can be solved by employment of regenerative amplifiers where the thin-disk serves like an active mirror in its resonator for signal wavelength of pulsed lasers. Situation for the pump beam is different and has to be solved by multipass configuration. A widely used configuration consisting of a parabolic mirror and two roof mirrors forming an angle of few degrees, which fold and shift the pump beam (Figure 5.18), was proposed by Giesen by the end of 1990s [62].

The configuration is very often optimized for 6 so-called V-passes through the active medium, that is, 12 passes through the medium back and forth. This number can be effectively doubled to 24 by installation of a backreflecting mirror. Yb doping concentration of optimized thin-disks is approximately 7 at.% and disk thickness of 220 µm. Under such conditions, more than 90 % of pump radiation is absorbed in the Yb:YAG active medium. Collimated pump beam can be either coupled to the laser head from an optical fiber, or directly from diode stacks and

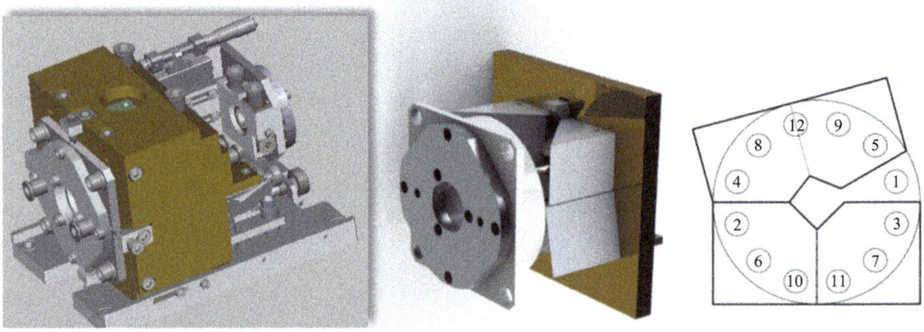

Figure 5.18 Three-dimensional model of a pump chamber for thin-disk lasers consisting of a parabolic mirror and roof reflectors (Hilase [8,9]). Right-hand figure shows order of pump beam passes on the pump optics

beam homogenizer. Pump spot on the thin-disk has then super-Gaussian spatial profile and total pump intensity approaches saturation intensity (20 kW/cm^2) for the pump wavelength. Generally, the pump spot diameter D on the thin-disk is given by the beam diameter of the pump beam and the focal length of the parabolic mirror (25–150 mm typically). For fiber coupled pump source the relation is:

$$D = \frac{f_2}{f_1} d$$

where d is the diameter of the pump fiber core, f_1 is the focal length of the collimating lens, and f_2 is the focal length of the parabolic mirror.

Besides the standard configurations, new configurations for thin-disk heads that double or triple the number of pump beam passes [94] through an active medium were proposed and allow for installation of thinner thin Yb:YAG disks or Ho:YAG disks, which are only 2–3 at.% doped because of up-conversion. For example, Yb:YAG thin-disk optimized for 48 pump beam passes has 10 % of Yb dopants and thickness of 100 μm only so the cooling is more efficient, OPD for the signal beam smaller, and achievable output average power higher.

5.4.3 Zero-phonon-line pumping

Even small reduction of generated heat can contribute to higher efficiency or improvement of laser behavior in high power laser. Besides the well-established pumping to a broad absorption line close to 940 nm, pumping at so-called zero phonon line at 968.8 nm has been employed for Yb:YAG active media recently (Figure 5.17) [95,96]. This approach reduces quantum defect from 8.7% down to only 5.9%. Zero-phonon-line means, in fact, pumping directly to upper laser level. However, unlike the absorption line close to 940 nm, zero phonon absorption line is extremely narrow and requires adoption of stabilized pump diodes with emission bandwidth <1 nm at 300 K.

Realizing of such a narrow bandwidth and locking the central wavelength is possible by adoption of narrow band volume Bragg gratings (VBG) reflecting a few percent of output radiation with the desired output characteristics back to the light source. This feedback stimulates the narrow band emission, although at the expense of output power. Crucial is also keeping the VBGs under stable conditions. Increase of their temperature leads to shifts in central wavelength and inefficient pump absorption in Yb:YAG.

In [94], we investigated benefits of the zero phonon line pumping for construction of high power thin-disk lasers, which brought a surprising fact. Temperature of Yb:YAG pumped at 940 nm rises much faster with rising pump power than expected upon taking into account quantum defect and fractional heating, and in addition, it follows nonlinear trend. We call it nonlinear phonon relaxation. In contrast, temperature rise of a zero-phonon-line-pumped disk was linear and corresponding to the calculated quantum defect (Figure 5.19). The data were acquired both for the fluorescence of the pumped thin-disk and for laser

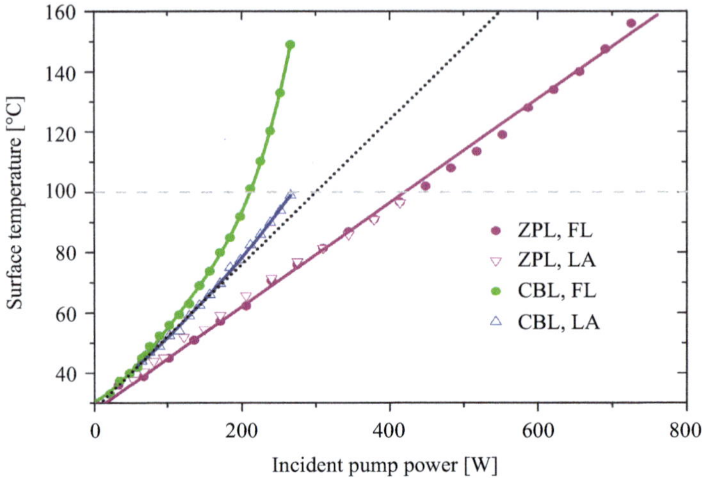

Figure 5.19 Temperature of a thin-disk surface measured by a thermal camera for pumping at 940 nm (CBL) and 968.8 nm (ZPL) both in the fluorescence regime (FL), and in laser operation (LA) [96]

operation as well. Mechanism of the heat generation is still unknown but impact on thin-disk laser construction is huge. It was proved that the disk temperature closely correlates with the thin-disk deformation and induced OPD which is experienced by the signal beam. Because thin-disk-based amplifiers are often based on regenerative amplifiers, induced OPD can shift stability point of the mirror resonator and event prevent lasing. Aspherical part of the OPD gives rise to diffraction losses in a spherical-mirror-based laser cavity. Spherical and aspherical (Figure 5.20) part of the OPD were analyzed by a Shack–Hartmann sensor [97] and this hypothesis was confirmed. Additionally, steep temperature rise in case of losing laser operation may destroy the disk module itself. Suppression of nonlinear phonon relaxation thanks to zero-phonon-line-pumped disks effectively eliminate this issue because of equality of disk temperature in both the fluorescence and the laser regimes.

5.4.4 Thin-disk module manufacturing

Critical for reliable operation of a thin-disk laser is quality of a thermal contact between the active medium and a heatsink. The very first Yb:YAG thin-disks used for CuW heatsinks, which had similar coefficient of thermal expansion to YAG, and the thermal contact was realized by an indium solder, Figure 5.21. The soldered disks usually have additional metallic coatings to improve adhesion and prevent diffusion of indium solder into the YAG. Thermal cycle up to the melting point of indium and back during the soldering procedure can generate permanent residual deformation of the thin-disk. Radius of curvature of such a thin-disk is not repeatable and causes problems during replacement of damaged disks in a laser resonator.

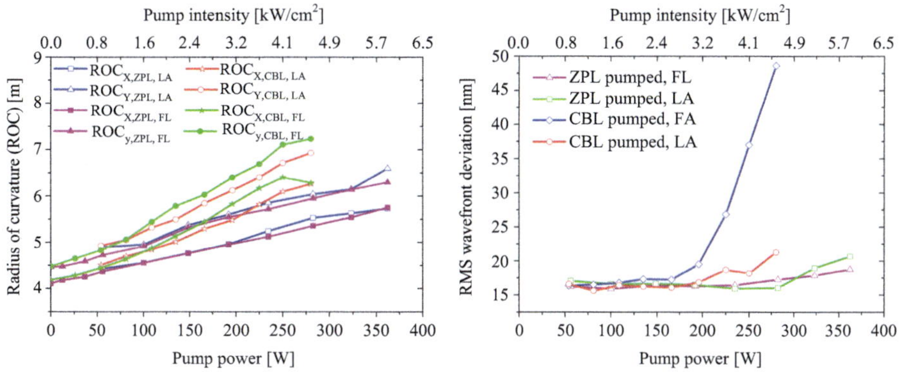

Figure 5.20 Thin-disk radius of curvature (ROC) calculated from an OPD measured in situ by a wavefront sensor (left) and aspherical part of the wavefront aberration (right) upon applying pump power at 940 nm (CBL) and 968.8 nm (ZPL). Difference between both pump wavelengths and fluorescence (FL) and laser operation regime (LA) are observable [96]

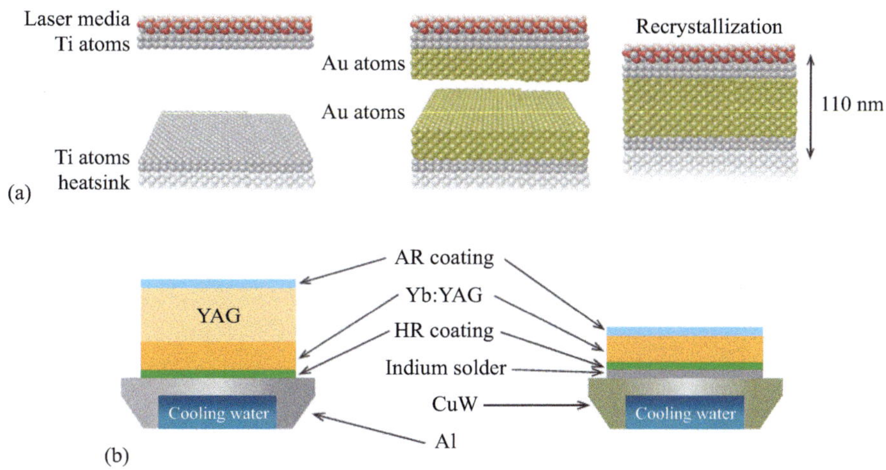

Figure 5.21 ADB procedure for bonding of Yb:YAG gain media to an aluminum heatsink (a). Structure of undoped-cap ADB bonded thin-disk (b) in comparison with indium-soldered Yb:YAG thin-disk, whose performance was experimentally compared [100]

Better alternative recently used for preparation of disk modules is epoxy bonding to a synthetic diamond substrate. The bonding procedure is carried out at room temperature and radius of curvature of the thin-disk is given by radius of curvature of the diamond substrate. Diamond also has excellent thermal properties, which implies better cooling and lower operational temperature of a thin-disk. Nowadays, such disks are used in majority of commercially available thin-disk lasers.

Other alternative is adoption of adhesive-free bonding. Absence of intermediate layer reduces thermal gradient over the contact layer. Lately bonding of a gain medium to an aluminum heatsink by atomic diffusion bonding (ADB) technique [98] was demonstrated. ADB of two surfaces with nanocrystalline Au films is a promising technique to achieve adhesive-free bonding at room temperature in air [99]. Nanocrystalline Au films (50 nm thickness) with fcc crystal structure were deposited on both Yb:YAG and Al heatsink surfaces, with Ti underlayers (5 nm thickness) with hcp crystal structure underneath, as shown in Figure 5.21. The Ti underlayers enhance the adhesion strength of the deposited films on the bonded surface and also control the crystallographic orientation of the deposited films. ADB with extremely thin Au films ensures better thermal management, mechanical tolerances, and prevention of out-gassing of materials in a vacuum or high temperature environment. Since the thickness of the nanocrystalline layer can be controlled within a certain percentage of variation, such a uniform bonding interface offers the fabrication of laser active medium with tight tolerances on thickness direction. Unlike thermal diffusion bonding or soldering, no annealing or curing at high temperatures is required and makes it possible to work with materials of different coefficients of thermal expansion. Performance of ADB-bonded disks and indium-soldered disks were compared [99]. Figure 5.22 shows disk surface temperature under comparable pump conditions.

5.4.5 High power regenerative amplifiers

0.5 kW picosecond regenerative amplifier Perla C operated at pulse repetition rate of 100 kHz recently developed at HiLASE Centre [8] [9,101,107] is a typical

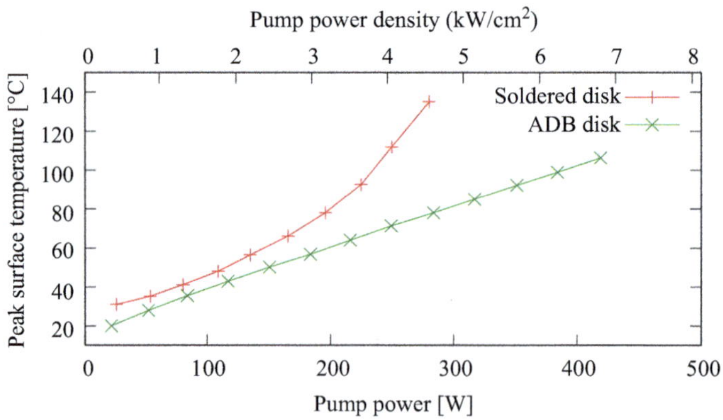

Figure 5.22 Performance of an ADB-bonded thin-disk (aluminum heatsink) and of a thin-disk soldered to a CuW heatsink described by achieved disk surface temperature. The soldered disk shows worse performance of cooling [99]

example of high average power thin-disk regenerative amplifier, one of the most powerful ones, with a single active medium in the main amplifier. The laser system is based on the CPA technology (see Section 5.2.3). For amplification from picojoule level is advantageous to combine several amplifier geometries, in this case fiber-based front-end and thin-disk-based power amplifiers.

Yb-doped, single-mode fiber oscillator and preamplifiers with enormous single-pass gain and excellent beam quality ($M^2 = 1.1$) shift pulse energy level in the front-end from picojoules to a fraction of microjoule. The front-end also contains pulse stretcher which stretches femtosecond pulses with 20 nm bandwidth from oscillator to pulse duration of approximately 0.5 ns. Pulse bandwidth is partially clipped in the stretcher down to 3 nm since gain bandwidth of Yb:YAG is narrower anyway. The stretcher is based on chirped fiber Bragg grating with chirp rate of 220 ps/nm. Its dispersion can be slightly tuned by applying thermal gradient along the grating which helps in later optimization of the system. Since oscillator repetition rate reaches 40 MHz, repetition rate is reduced by an acousto-optic pulse picker to 1 MHz. Lower repetition rate would boost parasitic effects in the fibers. The repetition rate is finally reduced to 100 kHz in the first regenerative amplifier. To compensate insertion losses of the stretcher and the pulse picker, the front-end contains two Yb-doped fiber preamplifiers for boosting the output pulse energy up to 1 μJ. Energy > 1 μJ is too high for the fibers since B-integral increases dramatically and nonlinear effects negatively influence pulse properties. Following amplification stage is therefore based on a small thin-disk-based regenerative amplifier.

The second-stage amplifier was seeded from the fiber preamplifier after passing a fiber polarization controller. This regenerative amplifier is a small footprint (900 × 1,200 mm) single Yb:YAG thin-disk amplifier with a standing wave cavity. Figure 5.23 shows optical scheme. More than 2 m long cavity is designed for a 2.7 mm pump spot on the Yb:YAG thin-disk. The amplification is switched on and off by a double BBO (beta barium borate) Pockels cell with 8 × 8 mm aperture. High repetition rate and high voltage switching of the Pockels cell can cause piezo ringing in the crystal and damage it if the crystal holder is wrongly designed. We developed two kinds of water cooled Pockels cell holders with spring loaded vibration absorbers that compensate these effects and we tested them for BBO crystals up to 12 × 12 mm aperture, 1 MHz repetition rate, and 10 kV operating voltage. The amplifier is pumped by continuous-wave fiber-coupled laser diodes at zero-phonon line wavelength (968.8 nm) as mentioned above (see Section 5.4.3). Amplified pulses from the amplifier reach 1.1 mJ of pulse energy in a nearly diffraction limited beam with M^2 parameter 1.3 at full power. Spectral bandwidth was 1.6 nm and bandwidth limit for sech^2 pulse shape is therefore 0.7 ps [8,9,100].

Fraction of the uncompressed output from the first stage (approx. 20 W, 200 μJ) was seeded into the main regenerative amplifier. Thin-disk was located in a 6 m long ring cavity with two V-passes through the disk per one roundtrip and footprint area of only 100 × 60 cm^2. Such a configuration compensates a main drawback of ring cavities comparing to standing wave cavities—only a single V-pass through an active medium per cavity roundtrip. On the other hand, scaling

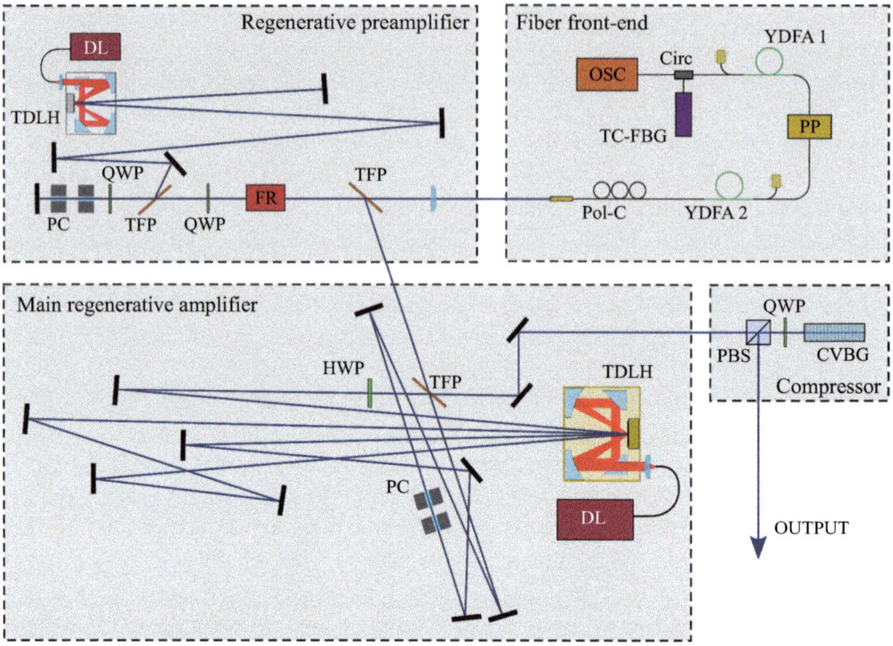

Figure 5.23 Optical scheme of the 500 W Perla C thin-disk laser system (OSC, oscillator; YDFA, Yb-doped fiber amplifier; PP, pulse picker; Pol-C, polarization controller; TFP, thin film polarizer; FR, Faraday rotator; QWP, quarter waveplate; PC, Pockels cell; TDLH, thin-disk laser head; DL, pump diodes; HWP, half waveplate; PBS, polarizing beam splitter; CVBG, chirped volume Bragg grating)

of the mode size is easier since ring cavities are shorter for given mode size than standing-wave cavities. Other pros of the ring cavity is absence of a Faraday rotator and isolator in the seed beam because of geometrically separated input and output [38]. But alignment is more complicated for the ring cavity. The thin-disk was pumped by a CW, 968.8 nm fiber-coupled diode laser on a pump spot with diameter of 5.2 mm. Dual Pockels cell with two BBO crystals in in-house developed holders was operated at 10-kV half-wave voltage and repetition rate of 50 or 100 kHz. Maximum CW output power of 565 W was obtained at 1.21 kW of pump power, and the optical-to-optical efficiency was 47% (Figure 5.24). In seeded operation with the input pulse energy of 0.2 mJ, obtained pulses had up to 5 mJ at 100 kHz (500 W average output power) with extraction efficiency of 43%. Pulses with 1.4 nm bandwidth (full width at half maximum, FWHM) had pulse duration of 1.8 ps (FWHM). In 50 kHz regime were generated 9 mJ pulses (450 W, uncompressed) with 1,130 W of pump power, that is, opt.-opt. efficiency reached an excellent value of almost 40%. In the 50-kHz regime, a several-hours-long stable operation was demonstrated with average output power of 320 W (pulse energy of 6.4 mJ) in a nearly diffraction-limited beam with M^2 parameter equal to

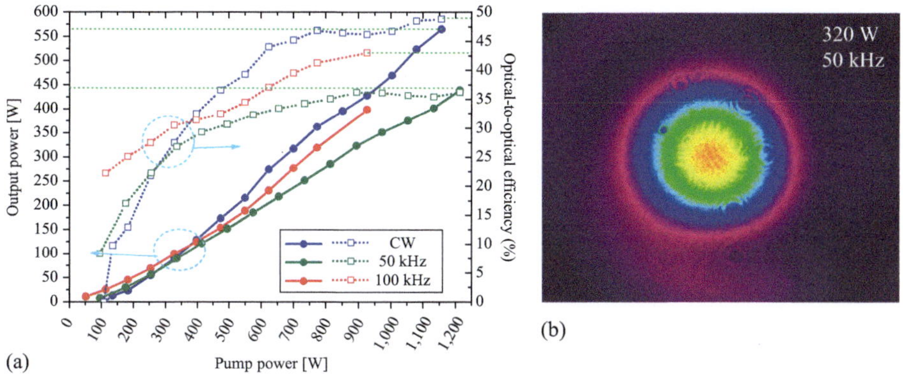

Figure 5.24 *(a) Output power and optical to optical efficiency of the Perla C ring cavity amplifier in CW operation, and pulsed operation at repetition rates of 50 and 100 kHz. (b) Near field beam profile of the Perla C at 320 W and 50 kHz pulse repetition rate*

1.4 before compression. The laser was operated with power fluctuation (rms) as low as 1.2% in a laboratory.

Since development of the laser system was pushed by requirements of industrial laser applications, we adopted a very compact, robust, and easy-to-align solution allowing efficient pulse compression with limited needs for space and low sensitivity to misalignment (Figure 5.25). The compressor was composed of a single chirped volume Bragg grating (CVBG) [100,101] with 8 × 8 mm aperture. CVBG is a kind of diffraction grating written into a block of photosensitive glass by an UV laser. Resulting modulation of refractive index can be used for near- and mid-infrared beams with no theoretical limit on average power. Practically, the grating has residual absorption causing thermal gradient and, as a consequence, depolarization of laser beam and beam degradation. For many applications the compact size and robustness prevails the drawbacks. This solution is unique in a high average power laser and to our knowledge, no similar high power system uses this solution. The compressor net efficiency measured at low output power reached 85%. Due to high average power of the output beam was the diffraction grating mounted in a water-cooled heatsink. However, the higher output power the more serious beam quality worsening is observable, especially for average power >200 W. Both the compressor grating and the pulse stretcher were optimally designed for 2.5 nm bandwidth (FWHM) and the stretcher was fitted to the CVBG dispersion. Compressed pulses measured by intensity autocorrelation have temporal length 1.6 ps ($sech^2$ expected pulse shape); however, wings are apparently visible in the intensity autocorrelation trace. Partial compensation of dispersion mismatch generated in the whole system is possible due to temperature control of the pulse stretcher (CFBG) [101]. To reach the bandwidth limit, higher order dispersion would have to be compensated fully in all dispersion orders.

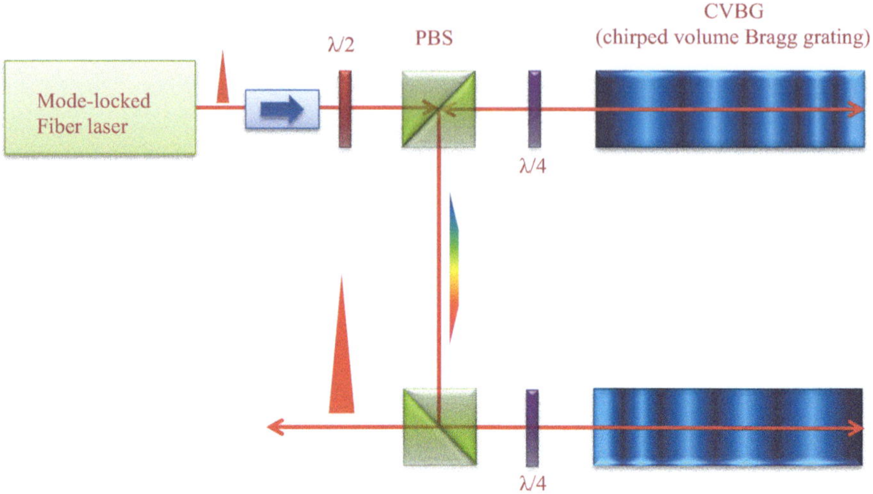

Figure 5.25 Principle of pulse stretching and pulse compression by a pair of chirped volume Bragg gratings (PBS, polarizing beam splitters; λ/4, quarter waveplate; λ/2, half waveplate)

5.4.6 Thin-disk-based multipass amplifier

Multipass amplifier in contrast to regenerative amplifier avoids the use of Pockels cell as the passes on the disk and therefore the gain is scaled by geometrically folding the beam so that each pass have dedicated beam path resulting in higher damage threshold at the cost of higher complexity of the propagation. On the other hand, they need to be seeded by a sufficiently powerful source as they exhibit lower total gain than regenerative amplifiers. The multipass amplifier reported uses collimated beam propagation scheme where collimated seed beam passes over thin-disk several times by individually adjusted mirrors. Similar setups have been demonstrated recently by other groups as well for amplification of high energy pulses. Room temperature operation of thin-disk multipass amplifiers requires quite complicated scheme realizing tens of passes through the active medium [2,3]. On the other hand, cryogenically cooled multipass thick disk amplifiers can be very compact [71,72].

Figure 5.26 shows the Perla B [9] multipass amplifier setup. The Yb:YAG thin-disk mounted on a diamond heatsink inside the laser head is pumped by 968.8 nm laser diodes stabilized by VBG. The setup consists of 12 individually adjustable plane mirrors arranged in an array for folding the beam path over disk and 5 folding mirrors arranged in two groups on either side of pump module to project onto a new mirror of array after every reflection from the disk. Folding mirrors can be a combination of plane and convex mirrors to compensate the thermal lensing effect of thin-disk and thus to maintain the nearly constant mode size on the thin-disk. The order in which the beam propagates through the

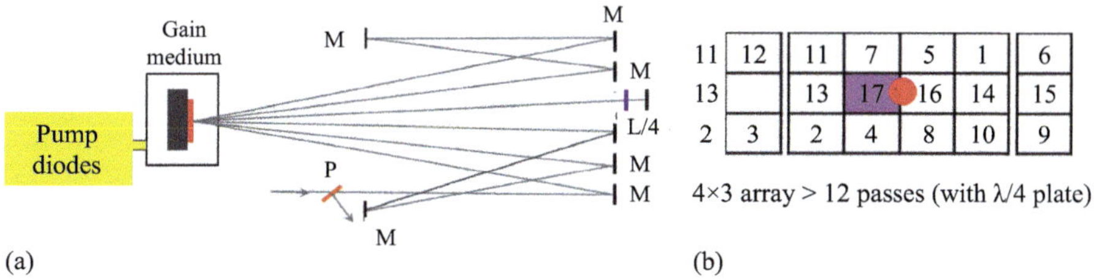

Figure 5.26 (a) Optical scheme of the thin-disk multipass amplifier Perla B [9] with a thin-disk gain medium. (b) Folding mirror array scheme showing order of signal beam passes (M, mirrors; P, polarizer; L/4, quarter waveplate)

individual folding and array mirrors is depicted in Figure 5.26. The linearly polarized collimated seed beam from regenerative amplifier similar to first thin-disk amplifier described in Section 5.4.6 passes mirror 1 on the array and reaches mirror 2 via reflection from the disk. It is then folded to mirror 3, which then folds the beam to mirror 4. After this the beam is reflected on to disk, mirror 5, mirror 6, mirror 7, disk and mirror 8. This beam propagation continues until mirror 17 is reached through a $\lambda/4$-waveplate. A total of six passes can be realized using this scheme and this number can be doubled by returning the beam back through the amplifier. The amplified beam is reflected off the thin-film polarizer. In general, N-passes requires N-mirror array (with quarter-wave plate and thin film polarizer combination) and (N/2)-1 folding mirrors. Such a system is able to generate multihundred mJ picosecond pulses at 100 Hz–1 kHz repetition rates.

5.5 Multislab high power system

5.5.1 Modeling [103,104]

Baseline parameters of the amplifier are usually given by required output energy. One usually tries to get the extraction fluence near the saturation fluence of the amplifier, when reasonable extraction efficiency is achieved without too high extraction fluence that could cause LIDT issues. The energy obtained from a high repetition rate amplifier is usually limited by thermal effects and ASE. Heat in the gain medium is caused mostly by the Stokes quantum defect, followed by absorption caused by nonradiative transitions, concentration quenching, and up-conversion.

The ASE process significantly limits the performance of the laser amplifier. When spontaneously emitted photons travel in a gain medium with population inversion, they are amplified and the population inversion is decreased and therefore the gain of the amplifier. The higher the gain and the larger the medium, the stronger the effect of ASE. Moreover, if the gain overcomes the reflection losses along closed paths of amplified photons, parasitic oscillations will occur, causing major depletion of the population inversion. To prevent parasitic oscillation and to keep ASE at a reasonably low level, the product of a transverse dimension of the slab and the small signal gain coefficient should be less than 3 [105].

For these reasons, the influence of ASE on the population inversion has to be precisely calculated. For early multislab design, it was done by a 3D ray-tracing code for ASE and population inversion calculations.

The model calculates pump energy absorbed by the gain medium, excited ion density, ASE, and heat sources in the Yb^{3+}:YAG laser slabs by solving rate equations in discrete time steps. Figure 5.27 shows a flow chart of the code. Absorption of the pump radiation and spontaneous emission are calculated independently during each step.

In the absorption phase, the energy from a polychromatic pump source is absorbed inside the medium proportionally to the number of nonexcited active ions and the wavelength-dependent absorption cross-section [106].

In the ASE phase, the spontaneously emitted photons with random polarization are generated according to the excited ion density inside the slab (Monte Carlo method is used to generate approximately 4 million rays in which photons are propagating through the slab). Their wavelength distribution is based on the probability density function which is derived from the emission cross section. Yb^{3+}: YAG is considered three-level system. Based on the temperature, the population of the ground state, the lower laser level, and the upper laser level are calculated from the Boltzmann statistics. Then reabsorption of emitted radiation is calculated.

Rays containing thousands of photons are traced through the medium and amplified in proportion to the population inversion and the path length in each cell. All slabs of the amplifier are included in the model and the generated rays propagate through all of them.

The amplification of the photons is expressed by the equation:

$$E_2 = E_1 \exp(lg)$$

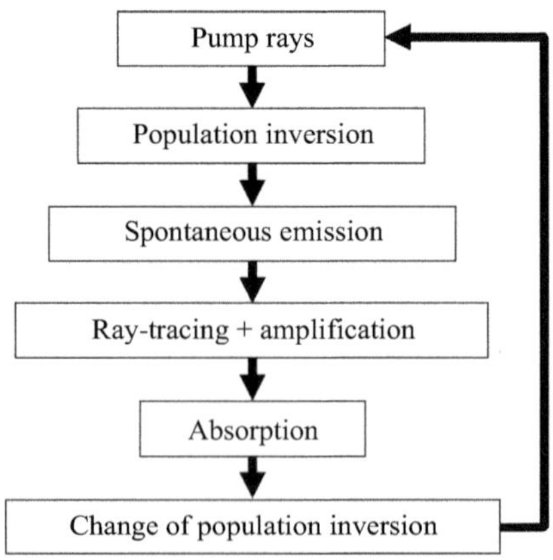

Figure 5.27 ASE model flow chart diagram [103]

where E_2 is the energy of the amplified ray after passing through a cell, E_1 is the energy carried by the ray at the entrance to the cell, l is the path length travelled by the ray in the cell, and g is the small signal gain coefficient.

At the back and front faces of the slabs, the rays are either reflected or transmitted depending on the angle of incidence. For angles $\alpha_i > \alpha_{TIR} = \arcsin(1/n_a)$, where n_a is the index of refraction of the amplifying medium, rays undergo total internal reflection and are trapped in the slab.

For $\alpha_i < \alpha_{TIR}$ the probability of transmission can be calculated according to the Fresnel reflection coefficients or can be described by anti-reflective (AR) coating parameters, that is, rays are not split but are transmitted or reflected based on the probability given by the reflection coefficient.

The side walls are surrounded by an absorptive layer of index-matched Cr^{4+}:YAG saturable absorber. The rays entering the saturable absorber are propagated and can be absorbed in any cell of the material in proportion to the nonexcited fraction of Cr^{4+} ions present. Photons that were not absorbed in the cladding propagate to its outer side. Photons reaching the outer side wall of Cr^{4+}:YAG can be transmitted out completely, scattered or reflected using Fresnel equations. The outer side is considered grinded, so incident rays will have different incidence angles upon impact. The angle of incidence of these rays is randomly generated. The probability of reflection is then calculated from the Fresnel equations. Transmitted photons are absorbed in surrounding material. Reflected photons pass through the absorptive layer again and, if they are not absorbed, they can propagate back to the gain medium and decrease the stored energy.

The modeled multislab amplifier consisted of eight slabs. The outer dimensions of each slab were 120 mm $\times$ 120 mm $\times$ 8 mm which included 60 mm $\times$ 60 mm Yb^{3+}:YAG crystal surrounded by a 30 mm thick absorptive layer of Cr^{4+}:YAG. Each slab had AR coatings on the front and back sides. The absorption coefficient of Cr^{4+}:YAG cladding was $\alpha = 0.34\ cm^{-1}$. Slabs were pumped from both sides by a polychromatic pump source, the laser diodes were characterized by their central wavelength and their bandwidth. Energy fluence of the pump was 5 J/cm^2 from each side with a super Gaussian (16th order) beam profile guaranteeing that 99% of energy from diodes goes into the amplifier. Pump pulse duration was 1 ms. Absorption and emission cross-sections were taken at temperature of 160 K.

The influence of diode parameters on the performance of a high energy laser amplifier was investigated. The stored energy in the gain medium was calculated for different central wavelengths ranging from 935 to 941 nm and for different bandwidths (FWHM from 3 to 7 nm). The results are summarized in Figure 5.28. The accuracy of our calculations is ± 0.5 J.

It was found that the commonly available laser diodes lasing at 938 nm with a bandwidth of 6 nm give the most stable pumping conditions and a stored energy of 192 J. Note that the energy fluctuation is less than 2% for a wavelength fluctuation of ± 3 nm and therefore there is little need to use more expensive diodes with tighter specifications.

Power scaling of the multislab system can be easily performed by increasing slab aperture due to one-dimensional temperature gradients and large cooling

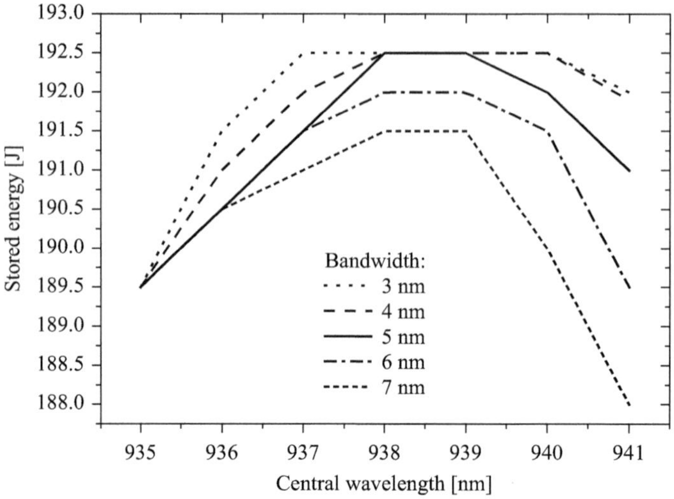

Figure 5.28 Stored energy in eight Yb^{3+}:YAG slabs with dimensions of 60 mm × 60 mm × 8 mm on diode parameters. Results were obtained for four-level system and complete absorption at Cr^{4+}: YAG side [103]

surfaces. However, while the slab size is increasing, the limitations resulting from ASE are getting more severe as the average path length of the photons in the slab increases as well. Depopulation of the gain medium is faster and the stored energy density is reduced. Figure 5.29 shows the calculated stored energy density as a function of the pumping fluence for different slabs dimensions. For larger slabs, one usually uses lower doping concentrations of active ions to keep ASE losses low.

5.5.2 System layout

The system DiPOLE/Bivoj [7,107] developed by STFC with participation of HiLASE incorporates a low-energy, fiber-based front-end oscillator (~nJ), followed by a regenerative Yb^{3+}:CaF_2 amplifier that increases the output energy to the mJ level and a rod Yb^{3+}:YAG multipass booster amplifier to raise the output to 50 mJ. Two diode pumped, helium gas-cooled large aperture power amplifiers then increase the output energy to 7 J (main preamplifier) and finally to 100 J (main power amplifier). The schematic of the system is in Figure 5.30.

5.5.2.1 Front-end

The front-end starts with a temperature-stabilized wavelength tunable CW fiber oscillator. The wavelength of the oscillator is matched to the peak of the gain curve of the cryogenically cooled amplifiers and the spectral bandwidth is less than 70 kHz. The CW output is then temporarily shaped in an A-O (acousto-optic) modulator to 200 ns long pulses with repetition rate of 10 kHz. The pulses are then amplified in a fiber amplifier and subsequently shaped by an E-O (electro-optic) modulator to

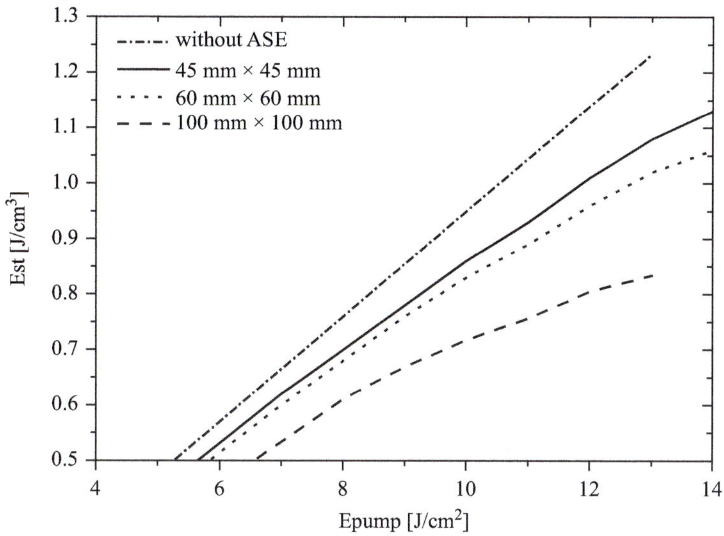

Figure 5.29 Stored energy density dependence on slab dimensions with constant thickness and varied pump fluence. Results were obtained for four-level system and complete absorption at Cr^{4+}:YAG side [103]

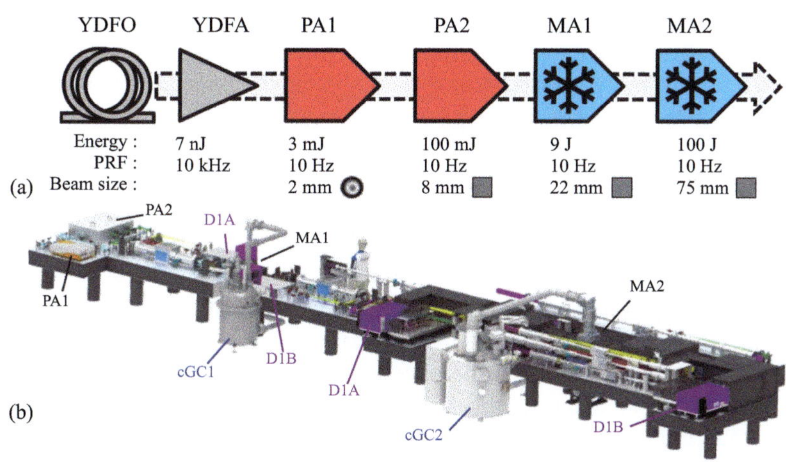

Figure 5.30 (a) Schematic of DiPOLE100/Bivoj amplifier chain showing maximum output performance after each amplifier stage, including free-space beam size and shape: YDFO = Yb-silica fiber oscillator; YDFA = Yb-silica fiber amplifier (inc. temporal pulse shaping); PA = room-temperature pre-amplifier (1 = Yb:CaF_2 regenerative, 2 = Yb:YAG multipass); MA, main cryogenic amplifier (ceramic Yb: YAG multislab). (b) 3D model of DiPOLE100 system: D, Diode pumps; cGC, cryogenic gas coolers [7]

produce 2–14 ns pulses with arbitrary shape. A semitriangular pulse shape is usually used to counter gain depletion in the power amplifiers. The temporal resolution of the shaper is below 200 ps. Before the fiber amplifier, the pulses can be phase modulated by 2 GHz and 4 GHz modulators to increase the bandwidth of the pulses to prevent SBS (Stimulated Brillouin scattering) and SRS (Stimulated Raman scattering) in the fiber amplifier, but this modulation is typically not needed for output energies below 10 nJ at which the amplifier operates. The pulses are then sent to regenerative rod amplifier (Amplitude Systemes) based on Yb^{3+}:CaF_2 that boosts the energy to ~4 mJ. A Pockels cell used to trap the pulses in the cavity also decreases repetition rate to 10 Hz. The Gaussian beam coming from the regenerative amplifier is then spatially shaped to a square super Gaussian profile in a beam shaper consisting of π-shaper and serrated aperture with spatial filter. After the shaper, the beam is relay-imaged through the whole amplifier chain. The pulse is further amplified to ~50 mJ in a multipass booster amplifier (Lastronics GmBH) based on Yb^{3+}:YAG. The booster amplifier preserves the square super Gaussian beam profile. The beam is spatially expanded after the booster amplifier and is injected into the 10 J main preamplifier.

5.5.2.2 10 J main preamplifier

The 10 J main preamplifier is based on a multislab design. It consists of four circular Yb^{3+}:YAG slabs. Each pair with different doping levels of Yb^{3+} (1.1, 2.0 at.%). The different doping levels are needed to uniformly divide the heat load among slabs. Each circular slab has diameter of 45 mm and thickness of 5 mm and the pumped area is a square of 23 mm $\times$ 23 mm. The pump beam is homogenized light from two diode pump modules operating at 939 nm and each producing 29 kW of peak power in 600 µs long laser pulses at a repetition rate of 10 Hz. The Yb^{3+}:YAG is clad with 5 mm Cr^{4+}:YAG absorber (absorption coefficient of 6 cm^{-1}) that prevents ASE and parasitic oscillations. The amplifier is cooled by forced Helium gas flow with pressure around 8 bar and operates at temperature of 150 K. Sapphire pressure windows separate He circuit from vacuum insulation and the vacuum is separated by fused silica windows from ambient environment.

The beam is injected into the amplifier through a dichroic mirror and then is image-relayed with spatial filtering ($f = 1$ m) to a back reflector and then relay-imaged back to the amplifier head. Each pass is propagated by a set of separate mirrors and lenses. A square deformable mirror with 49 actuators is placed in the amplifier after the third pass. This mirror together with lens arrays placed on linear stages is used to compensate thermal wavefront distortion acquired in the amplifier head. The sapphire pressure windows change the polarization of the passing beam and therefore sets of quarter and half waveplates are used to compensate this on pass 1 and 5. After seven passes, the beam is ejected from the amplifier with pulse energy of 7 J in the beam with a size of 22 mm $\times$ 22 mm. The beam is then sent to application areas for experiments or expanded and sent to 100 J amplifier.

5.5.2.3 100 J power amplifier

The 100 J power amplifier is also based on the multislab design. It consists of six square Yb^{3+}:YAG slabs with three doping levels of Yb^{3+} (0.4, 0.6, 1.0 at.%). The volume of each slab is 100 mm × 100 mm × 8.5 mm and the pumped area is square with dimensions of 78 mm × 78 mm. The parameters of the pump light are similar to the 10 J amplifier, but the peak power is higher and reaches 250 kW from each module and the pulse duration is longer, around 800 μs. The Yb^{3+}:YAG is clad with a 10 mm wide Cr^{4+}:YAG absorber (absorption coefficient of 3 cm^{-1}) that prevents ASE and parasitic oscillations.

The beam is injected into the amplifier under a range of angles around 5°, so any overlap with the pump beam outside the amplifier head is avoided and no dichroic mirrors are used. Each additional pass is image-relayed with spatial filtering by a dedicated telescope ($f = 3\ m$) back to the amplifier head. After the first pass, a deformable mirror is implemented to prevent degradation of the wavefront on subsequent passes. After four passes, the beam is ejected from the amplifier with pulse energy of up to 105 J with beam size around 75 mm × 75 mm. The average fluence of the amplified beam is 2 J/cm^2. The amplifier is cooled by forced Helium gas flow with pressure around 10 bar and at temperature of 150 K. Sapphire pressure windows separate He circuit from vacuum insulation and the vacuum is separated by fused silica windows from ambient environment.

5.5.3 Output parameters

Diode pumping enabled the amplifier to reach excellent energy stability, beam profile quality, and moderate optical-to-optical efficiency above 27% (Figure 5.31). At 9 J of output energy, pump diodes operate at only 60% of their capacity, so there is possibility to increase the energy further, but it is currently limited by laser-induced damage on optical elements. Combination of high average power and high fluence often decreases LIDT values specified by manufacturers up to ten times (Figure 5.31).

Comparison between model and experimentally acquired data is also very good (Figure 5.32). Data were obtained for 10 Hz repetition rate at temperature of 137 K.

High energy stability is required for acquiring similar results during repetitive experiments or for material treatment. Stability of 0.1% RMS was domain of low power diode-pumped fiber oscillator. With advances and price drop of laser diodes, such stability can be reached even in high energy systems. This was impossible for high energy flash lamp pumped systems. Example of energy stability obtained in Bivoj system is shown in Figure 5.33.

Beam quality offers P-V modulation in near field less than 15% with excellent wavefront quality and focusability (Figure 5.34).

For 100 J amplifier, the optical-to-optical efficiency drops to around 22% due to large losses through ASE. Nevertheless, the pump modules operate at 80% of

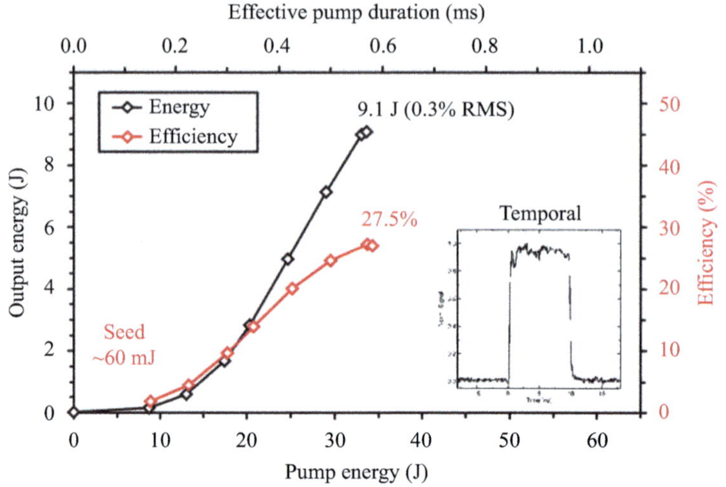

Figure 5.31 Output energy and laser optical-to-optical efficiency dependence on pump pulse duration. Inset shows pulse temporal profile at maximum energy [108]

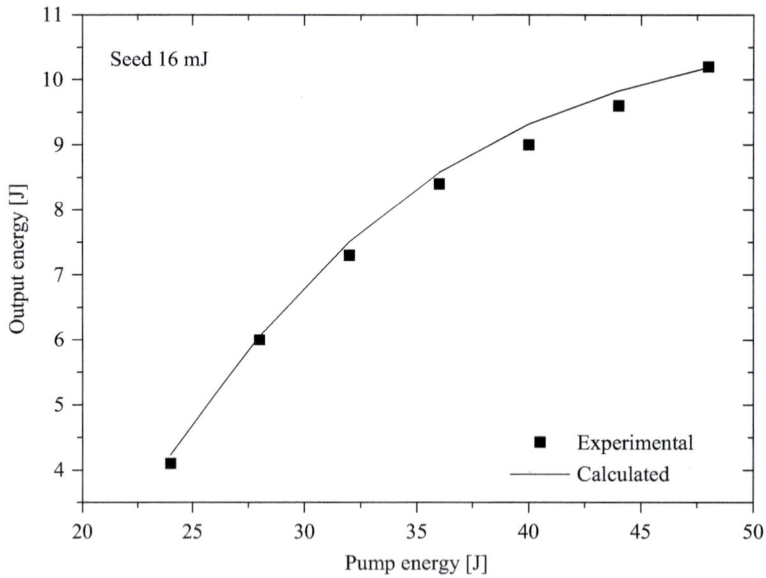

Figure 5.32 Energy output from 10 J amplifier measured versus calculated

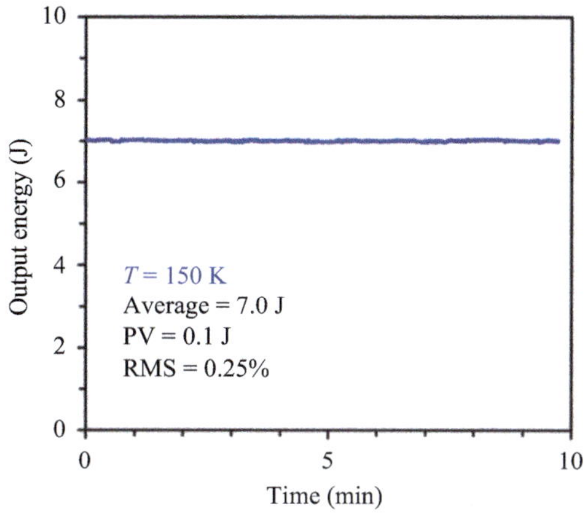

Figure 5.33 Energy stability at output energy of 7 J and repetition rate of 10 Hz [108]

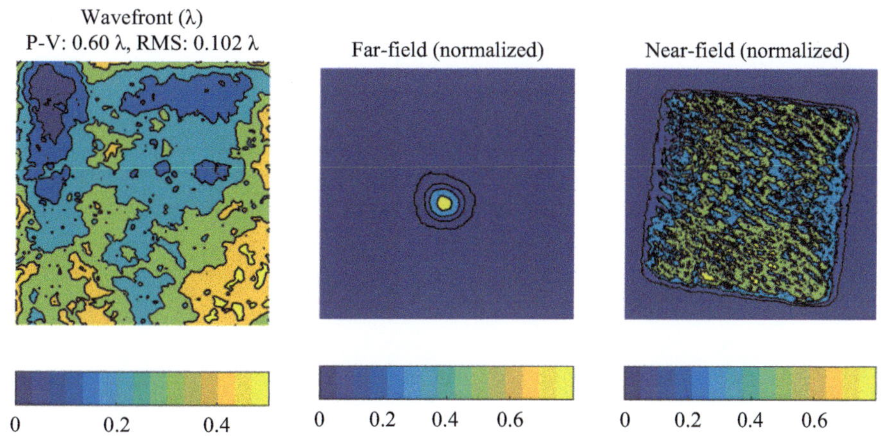

Figure 5.34 Output wavefront, far-field and near-field patterns at 7 J and 10 Hz [108]

their capacity and the output energy can be increased further, when LIDT issues are solved (Figure 5.35).

Energy stability (Figure 5.36) and near field beam profile (Figure 5.37) for 100 J amplifier are also excellent. The wavefront and thus far field beam profiles are slightly worse in comparison to 10 J amplifier, but the recorded images were obtained without closed loop operation of deformable mirror. Closed loop operation is unfortunately not supported on this mirror (Figures 5.36 and 5.37).

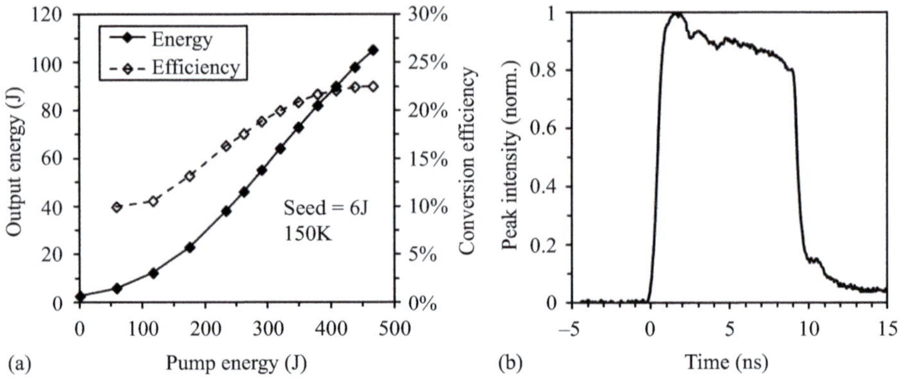

Figure 5.35 (a) Output energy from 100 J amplifier and its optical-to-optical efficiency in dependence on pump energy. (b) Pulse temporal profile for output energy of 100 J [7]

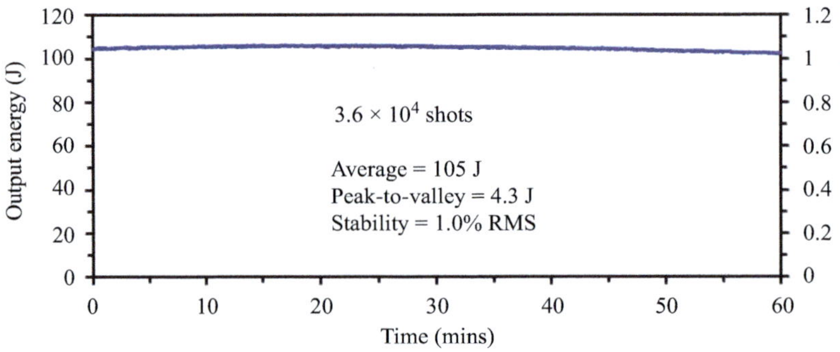

Figure 5.36 Energy stability at output energy of 105 J and repetition rate of 10 Hz [7]

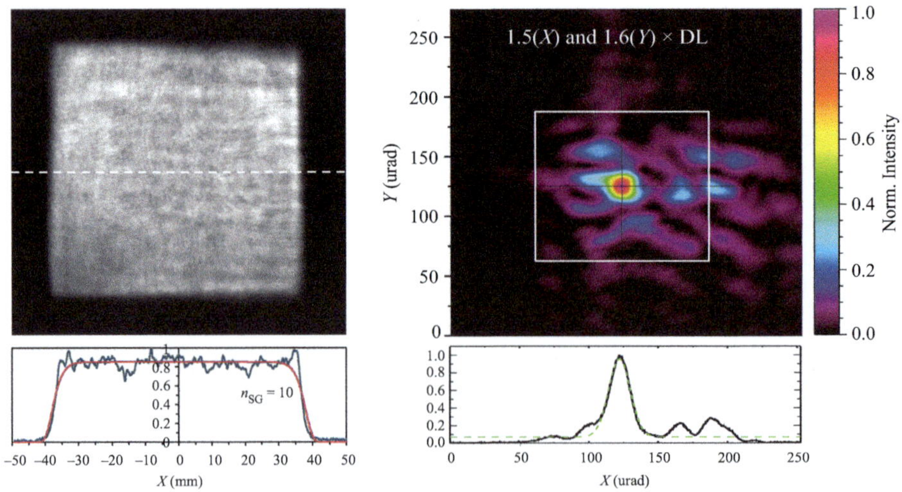

Figure 5.37 Near-field (left) and far-field (right) patterns and their horizontal cross-sections (below) at 105 J and 10 Hz [7]

References

[1] Schad S.S., Stolzenburg C., Michel K., and Sutter D. "Latest advances in high brightness disk lasers." *Laser Technik Journal.* 2014;11(2):49x–53

[2] Negel J.P., Loescher A., Voss A., *et al.* "Ultrafast thin-disk multipass laser amplifier delivering 1.4 kW (4.7 mJ, 1030 nm) average power converted to 820 W at 515 nm and 234 W at 343 nm." *Optics Express.* 2015;23(16): 21064–77

[3] Negel J., Loescher A., Bauer D., *et al.* "Second generation thin-disk multi-pass amplifier delivering picosecond pulses with 2 kW of average output power." *Proceedings of Advanced Solid State Lasers conference*, Boston, MA, USA, October–November 2016 (OSA Technical Digest, Optical Society of America, 2016), paper ATu4A.5

[4] Nubbemeyer T., Kaumanns M., Ueffing M., *et al.* "1 kW, 200 mJ picosecond thin-disk laser system." *Optics Letters.* 2017;42(7):1381–84

[5] Jung R., Tümmler J., Nubbemeyer T., and Will I. "Thin-disk ring amplifier for high pulse energy." *Optics Express.* 2016;24(5):4375–4381

[6] Brons J., Pervak V., Fedulova E., *et al.* "Energy scaling of Kerr-lens mode-locked thin-disk oscillators." *Optics Letters.* 2014;39(22):6442–45

[7] Mason P., Divoký M., Ertel K., *et al.*, "Kilowatt average power 100 J-level diode pumped solid state laser." *Optica.* 2017;4(4):438–439

[8] Novak O., Miura T., Smrž M., *et al.*, "Status of the high average power diode-pumped solid state laser development at HiLASE," *Applied Sciences.* 2015;5(4):637–665

[9] Smrž M., Novák O., Mužík J., *et al.* "Advances in high-power, ultrashort pulse DPSSL technologies at HiLASE." *Applied Sciences.* 2017;7(10):1016–1–1016-12

[10] Koechner W. *Solid-State Laser Engineering*. 6th edn. New York: Springer-Verlag; 2006.

[11] Chyla M., Miura T., Smrz M., Jelinkova H., Endo A., and Mocek T. "Optimization of beam quality and optical-to-optical efficiency of Yb:YAG thin-disk regenerative amplifier by pulsed pumping." *Optics Letters.* 2014;39(6):1441–44

[12] Boyd R. *Nonlinear Optics*. 3rd edn. Amsterdam: Academic Press; 2008.

[13] Perry M.D., Ditmire T., and Stuart B.C. "Self-phase modulation in chirped-pulse amplification." *Optics Letters.* 1994;19(24):2149–51

[14] Strickland D., and Mourou G. "Compression of amplified chirped optical pulses." *Optics Communications.* 1985;56(3):219–221

[15] Silfvast W. *Laser Fundamentals*. Cambridge: Cambridge University Press; 2004.

[16] Beil K., Saraceno C.J., Schriber C., *et al.* "Yb-doped mixed sesquioxides for ultrashort pulse generation in the thin disk laser setup." *Applied Physics B.* 2013;113(1):13–18

[17] Petermann K., Fagundes-Peters D., Johannsen J., *et al.* "Highly Yb-doped oxides for thin-disc lasers." *Journal of Crystal Growth.* 2005;275(1–2):135–140

[18] Erlandson A.C., Aceves S.M., Bayramian A.J., *et al.* "Comparison of Nd: phosphate glass, Yb:YAG and Yb:S-FAP laser beamlines for laser inertial fusion energy (LIFE)." *Optical Materials Express.* 2011;1(7):1341–52

[19] Wentsch K.S., Weichelt B., Günster S., *et al.* "Yb:CaF2 thin-disk laser." *Optics Express.* 2014;22(2):1524–32

[20] Krupke W.F., Shinn M.D., Marion J.E., Caird J.A., and Stokowski S.E. "Spectroscopic, optical, and thermomechanical properties of neodymium- and chromium-doped gadolinium scandium galium garnet." *Journal of Optical Society of America.* 1986;4(1):102–114

[21] Körner J., Hein J., Kahle M., *et al.* "Temperature dependent measurement of absorption and emission cross sections for various Yb3+ doped laser materials." *Proceedings of SPIE Optics+Optoelectronics conference,* Prague, Czech Republic, June 2011 (Proceedings SPIE 8080, 2011), pp. 8080–2

[22] Aggrawal R.L., Ripin D.J., Ochoa J.R., and Fan T.Y. "Measurement of thermo-optic properties of Y3Al5O12, Lu3Al5O12, YAlO3, LiYF4, LiLuF4, BaY2F8, KGd(WO4)2, and KY(WO4)2 laser crystals in the 80–300 K temperature range." *Journal of Applied Physics.* 2005;98(10):103514-1–103514-14

[23] Beil K., Friedrich-Thornton S., Tellkamp F., *et al.* "Thermal and laser properties of Yb:LuAG for kW thin disk lasers." *Optics Express.* 2010;18(20):20712–22

[24] Cardinali V., Marmois E., Le Garrec B., and Bourdet G. "Thermo-optical measurements of ytterbium doped sesquioxides ceramics." *Proceedings of SPIE Photonics Europe conference*, Brussels, Belgium, May 2010 (SPIE Proceedings 7721, 2010), pp. 77210U–1

[25] Fan T., Ripin D., Aggarwal R., *et al.* "Cryogenic Yb3+ doped solid-state lasers." *IEEE Journal of Selected Topics in Quantum Electronics.* 2007;13(3):448–459

[26] Kim W., Florea C.M., Gibson D.J., *et al.*, "Crystal fibers for high power lasers." *Proceedings of SPIE Security+Defence conference*, Edinburg, UK, Nov 2012 (Proceedings SPIE 8601, 2013), pp. 86012Z

[27] Chenais S., Druon F., Forget S., Balembois F., and Georges P. "On thermal effects in solid-state lasers: the case of ytterbium-doped materials." *Progress in Quantum Electronics.* 2006;30(4):89–153

[28] Bourdet LG., Casagrande O., D-Robin N., and Le Garrec B. "Performance of cryogenic cooled laser based on Ytterbium doped sesquioxide ceramics." *Journal of Physics: Conference Series.* 2006;112(3):032054

[29] Petermann K., Fagundes-Peters D., Johannsen J., *et al.* "Highly Yb-doped oxides for thin-disk lasers." *Journal of Crystal Growth.* 2005;275(1–2):135–140

[30] Jiang Ch., Liu H., Zeng Q., Tang X., and Gan F. "Yb:phosphate laser glass with high emission cross-section." *Journal of Physics and Chemistry of Solids.* 2000;61(8):1217–23

[31] Injeyan H., and Goodno G.D. *High-Power Laser Handbook.* New York, NY: McGraw Hill; 2011.

[32] Bayramian A., Armstrong J., Beer G., *et al.* "High-average-power femto-petawatt laser pumped by the mercury laser facility." *Journal of Optical Society of America B.* 2008;25(7):B57–B61

[33] Haefner C.L., Bayramian A., Betts S., *et al.*, "High average power, diode pumped petawatt laser systems: a new generation of lasers enabling precision science and commercial applications." *Proceedings of SPIE Optics + Optoelectronics conference*, Prague, Czech Republic, June 2017 (Proceedings SPIE 10241, 2017), pp. 1024102

[34] Frantz L.M., and Nodvik J.S. "Theory of pulse propagation in a laser amplifier." *Journal of Applied Physics*. 1963;34(8):2346–49

[35] Murray J.E., and Lowdermilk W.H. "Nd:YAG regenerative amplifier." *Journal of Applied Physics*. 1980;51(7):3548–55

[36] Lowdermilk W.H., and Murray J.E. "The multipass amplifier: theory and numerical analysis." *Journal of Applied Physics*. 1980;51(5):2436–44

[37] Backus S., Durfee III C.G., Murnane M.M., and Kapteyn H. "High power ultrafast lasers." *Review of Scientific Instruments*. 1998;69(3):1207–23

[38] Yanovsky, V., Felix C., and Mourou G. "Why ring regenerative amplification (regen)?". *Applied Physics B*. 2002;74(Suppl 1):181–83

[39] Coe J.S., Maine P., and Bado P. "Regenerative amplification of picosecond pulses in Nd:YLF: gain narrowing and gain saturation." *Journal of Optical Society of America B*. 1988;5(12):2560–63

[40] Dörring J., Killi A., Morgner U., Lang A., Lederer M., and Kopf D. "Period doubling and deterministic chaos in continuously pumped regenerative amplifiers." *Optics Express*. 2004;12(8):1759–68

[41] Grishin M., Gulbinas V., and Michailovas A. "Dynamics of high repetition rate regenerative amplifiers." *Optics Express*. 2007;15(15):9434–43

[42] Ristau D. *Laser-induced damage in optical materials*. CRC Press; 2015.

[43] Mero M., Liu J., Rudolph W., Ristau D., and Starke K. "Scaling laws of femtosecond laser pulse induced breakdown in oxide films." *Physical Review B*. 2015;71(11):115109-1–115109-7

[44] Diels J.-C., and Rudolph W. *Ultrashort laser pulse phenomena*. 2nd edn. Elsevier; 2006.

[45] Fork R.L., Martinez O.E., and Gordon J.P. "Negative dispersion using pairs of prisms." *Optics Letters*. 1984;9(5):150–2

[46] Kafka J.D., and Baer T. "Prism-pair dispersive delay lines in optical pulse compression." *Optics Letters*. 1987;12(6):401–3

[47] Treacy E.B. "Optical pulse compression with diffraction gratings." *IEEE Journal of Quantum Electronics*. 1969;5(9):454–58

[48] Martínez O.E., Gordon J.P., and Fork R.L. "Negative group-velocity dispersion using refraction." *Journal of Optical Society of America A*. 1984;1(10):1003–6

[49] Fan T.Y., Sanchez A., and DeFeo W.E. "Scalable, end-pumped, diode-laser-pumped laser." *Optics Letters*. 1989;14(19):1057–59

[50] Sutton S.B., and Albrecht G.F. "Optical distortion in end-pumped solid-state rod lasers." *Applied Optics*. 1993;32(27):5256–69

[51] Clarkson W.A. "Thermal effects and their mitigation in end-pumped solid-state lasers." *Journal of Physics D: Applied Physics*, 2001;34(16):2381–95

[52] Bruesselbach H., and Sumida D.S. "A 2.65-kW Yb:YAG single-rod laser." *IEEE Journal of Selected Topics in Quantum Electronics*. 2005;11(3):600–3

[53] Moon H.J., Yi J., Han J., Cha B., and Lee J. "Efficient diffusive reflector-type diode side-pumped Nd:YAG rod laser with an optical slope efficiency of 55%." *Applied Optics*. 1999;38(9):1772–76
[54] Fomin V., Gapontsev V., Shcherbakov E., Abramov A., Ferin A., and Mochalov D. "100 kW CW fiber laser for industrial application." Presented at 16th International Conference on Laser Optics; St. Petersburg, Russia, 2014
[55] Eidam T., Hanf S., Seise E., *et al.* "Femtosecond fiber CPA system emitting 830 W average output power." *Optics Letters*. 2010;35(2):94–96
[56] Otto H.J., Stutzki F., Modsching N., Jauregui C., Limpert J., and Tünnermann A. "2 kW average power from a pulsed Yb-doped rod-type fiber amplifier." *Optics Letters*. 2014;39(22):6446–49
[57] Zhao Z., and Kobayashi Y. "Ytterbium fiber-based, 270 fs, 100 W chirped pulse amplification laser system with 1 MHz repetition rate." *Applied Physics Express*. 2015;9(1):012701–1–012701-4
[58] Richardson D.J., Nilsson J., and Clarkson W.A. "High power fiber lasers: current status and future perspectives [Invited]." *Journal of Optical Society of America B*. 2010;27(11):B63–B92
[59] Limpert J., Klenke A., Kienel M., *et al.* "Performance scaling of ultrafast laser systems by coherent addition of femtosecond pulses." *IEEE Journal of Selected Topics in Quantum Electronics*. 2014;20(5):1–10
[60] Klenke A., Muller M., Stark H., *et al.* "Coherent beam combination of ultrafast fiber lasers." *IEEE Journal of Selected Topics in Quantum Electronics*. 2018;24(5):1–9
[61] Giesen A., Hügel H., Voss A., Wittig K., Brauch U., and Opower H. "Scalable concept for diode-pumped high-power solid-state lasers." *Applied Physics B*. 1994;58(5):365–72
[62] Contag K., Karszewski M., Stewen C., Giesen A., and Hugel H. "Theoretical modelling and experimental investigations of the diode-pumped thin disk Yb:YAG laser." *Quantum Electronics*. 1999;29(8):697–703
[63] Stewen C., Contag K., Larionov M., Giesen A., and Hugel H. "A 1-kW CW thin disc laser." *IEEE Journal of Selected Topics in Quantum Electronics*. 2000;6(4):650–57
[64] Beyertt A., Nickel D., and Giesen A. "Femtosecond thin-disk Yb:KYW regenerative amplifier." *Applied Physics B*. 2005;80(6):655–60
[65] Fattahi H., Alismail A., Wang H., Brons J., Pronin O., and Buberl T. "High-power, 1-ps, all-Yb:YAG thin-disk regenerative amplifier." *Optics Letters*. 2016;41(6):1126–29
[66] Jung R., Tümmler J., and Will I. "Regenerative thin-disk amplifier for 300 mJ pulse energy." *Optics Express*. 2016;24(2):883–87
[67] Giesen A., and Speiser J. "Fifteen years of work on thin-disk lasers: results and scaling laws." *IEEE Journal of Selected Topics in Quantum Electronics*. 2007;13(3):598–609
[68] Speiser J. "Scaling of thin-disk lasers—influence of amplified spontaneous emission." *Journal of Optical Society of America B*. 2009;26(1):26–35
[69] Kouznetsov D., and Bisson J.F. "Role of undoped cap in the scaling of thin-disk lasers." *Journal of Optical Society of America B*. 2008;25(3):338–45

[70] Aleknavičius A., Gabalis M., Michailovas A., and Girdauskas V. "Aberrations induced by anti-ASE cap on thin-disk active element." *Optics Express.* 2013;21(12):14530–38

[71] Zapata L.E., Lin H., Calendron A.L., *et al.* "Cryogenic Yb:YAG composite-thin-disk for high energy and average power amplifiers." *Optics Letters.* 2015;40(11):2610–13

[72] Baumgarten C., Pedicone M., Bravo H., *et al.* "1 J, 0.5 kHz repetition rate picosecond laser." *Optics Letters.* 2016;41(14):3339–42

[73] Pronin O., Brons J., Grasse C., *et al.* "High-power 200 fs Kerr-lens mode-locked Yb:YAG thin-disk oscillator." *Optics Letters.* 2011;36(24):4746–48

[74] Brons J., Pervak V., Fedulova E., *et al.* "Powerful 100-fs-scale Kerr-lens mode-locked thin-disk oscillator." *Optics Letters.* 2016;41(15):3567–70

[75] Baer C.R.E., Heckl O.H., Saraceno C.J., *et al.* "Frontiers in passively mode-locked high-power thin disk laser oscillators." *Optics Express.* 2012;20(7): 7054–65

[76] Ueffing M., Lange R., Pleyer T., *et al.* "Direct regenerative amplification of femtosecond pulses to the multimillijoule level." *Optics Letters.* 2016;41(16): 3840–43

[77] Neuhaus J., Fink F., and Larionov M. "Generation of high-energy femtosecond pulses by use of spectral broadening effects in Yb:YAG thin-disk regenerative amplifiers." *Journal of Optical Society of America B.* 2017;34(5):959–67

[78] Goncalves-Novo T., Albach D., Vincent B., Arzakantsyan M., and Chanteloup J.C. "14 J/2 Hz Yb3+:YAG diode pumped solid state laser chain." *Optics Express.* 2013;21(1):855–66

[79] Furuse H., Kawanaka J., Takeshita K., *et al.* "Total-reflection active-mirror laser with cryogenic Yb:YAG ceramics." *Optics Letters.* 2009;34(21):3439–41

[80] Divoky M., Tokita S., Hwang S., *et al.* "1-J operation of monolithic composite ceramics with Yb:YAG thin layers: multi-TRAM at 10-Hz repetition rate and prospects for 100-Hz operation." *Optics Letters.* 2015;40(6):855–58

[81] Yasuhara R., Kawashima T., Sekine T., *et al.* "213 W average power of 2.4 GW pulsed thermally controlled Nd:glass zigzag slab laser with a stimulated Brillouin scattering mirror." *Optics Letters.* 2008;33(15):1711–13

[82] Orth C.D., Beach R.J., Bibeau C.M, *et al.*, "Design modeling of the 100 J diode pumped solid state laser for project Mercury." *Proceedings of SPIE International Conference on High Power Laser and Application*, San Jose, California, Jan 1998 (SPIE proceedings 3265, 1998), pp. 114–129

[83] Ertel K., Hernandez-Gomez C., Mason P. D., Musgrave I. O., Ross I. N., and Collier J. L. "Concept for cryogenic kJ-class Yb:YAG amplifier." *Proceedings of Advanced Solid-State Photonics*, San Diego, CA, USA, 2010 (OSA Technical Digest Series, Optical Society of America, 2010), paper AWB20

[84] Russbueldt P., Mans T., Rotarius G., Weitenberg J., Hoffmann H.D., and Poprawe R. "400 W Yb:YAG Innoslab fs-amplifier." *Optics Express.* 2009;17(15):12230–45

[85] Russbueldt P., Mans T., Weitenberg J., Hoffmann H.D., and Poprawe R. "Compact diode-pumped 1.1 kW Yb:YAG Innoslab femtosecond amplifier." *Optics Letters.* 2010;35(24):4169–71

[86] Beil K., Fredrich-Thornton S.T., Tellkamp F., *et al.* "Thermal and laser properties of Yb:LuAG for kW thin disk lasers." *Optics Express.* 2010;18(20): 20712–22

[87] Peters R., Kränkel C., Petermann K., and Huber G. "Broadly tunable high-power Yb:Lu2O3 thin disk laser with 80% slope efficiency." *Optics Express.* 2007;15(11):7075–82

[88] Peters R., Kränkel C., Fredrich-Thornton S.T., and Beil K., Petermann K., Huber G. "Thermal analysis and efficient high power continuous-wave and mode-locked thin disk laser operation of Yb-doped sesquioxides." *Applied Physics B.* 2011;102(3):509–14

[89] Diebold A., Jia Z., Graumann I. J., *et al.* "High-power Yb:GGG thin-disk laser oscillator: first demonstration and power-scaling prospects." *Optics Express.* 2017;25(2):1452–62

[90] William P. Latham, Ahmed Lobad, Tim C. Newell, and Don Stalnaker 6.5 kW, Yb:YAG Ceramic Thin Disk Laser," AIP Conference Proceedings 1278, 758 (2010)

[91] Zhang J., Schulze F., Mak K.F., *et al.* "High-power, high-efficiency Tm:YAG and Ho:YAG thin-disk lasers." *Laser & Photonics Reviews.* 2018;12(3): 1700273

[92] Zhang J., Mak K.F., and Pronin O. "Kerr-Lens Mode-Locked 2-μm Thin-Disk Lasers." *IEEE Journal of Selected Topics in Quantum Electronics.* 2018;24(5): 1–11

[93] Petit J., Viana B., Goldner P., Roger J.-P., and Fournier D. "Thermomechanical properties of Yb3+ doped laser crystals: experiments and modeling." *Journal of Applied Physics.* 2010;108(12):123108–123108-11

[94] Schuhmann K., Hänsch T.W., Kirch K., *et al.* "Thin-disk laser pump schemes for large number of passes and moderate pump source quality." *Applied Optics.* 2015;54(32):9400–08

[95] Smrž M., Miura T., Chyla M., *et al.* "Suppression of nonlinear phonon relaxation in Yb:YAG thin disk via zero phonon line pumping." *Optics Letters.* 2014;39(16):4919–22

[96] Weichelt B., Voss A., Ahmed M.A., and Graf T. "Enhanced performance of thin-disk lasers by pumping into the zero-phonon line." *Optics Letters.* 2012;37 (15):3045–47

[97] Miura T., Chyla M., Smrz M., *et al.* "*In-situ* optical phase distortion measurement of Yb:YAG thin disk in high average power regenerative amplifier." *Proceedings of SPIE Photonics West conference*, San Francisco, CA, USA, Feb 2013 (Proceedings SPIE 8603, February 2013), pp. 860303-1– 860303-8

[98] Shimatsu T., and Uomoto M. "Atomic diffusion bonding of wafers with thin nanocrystalline metal films." *Journal of Vacuum Science & Technology B* 2010;28(4):706–14

[99] Nagisetty S.S., Severova P., Miura T., *et al.* "Lasing and thermal characteristics of Yb:YAG/YAG composite with atomic diffusion bonding." *Laser Physics Letters.* 2016;14(1):015001-1–015001-6

[100] Smrž M., Mužík J., Novák O., Chyla M., *et al.* "Progress in kW-class picosecond thin-disk lasers development at the HiLASE." *Proceedings of SPIE Photonics West conference*, San Francisco, CA, USA, Feb 2016 (Proceedings SPIE 9726, February 2016), pp. 972617

[101] Glebov L.B., Smirnov V., Rotari E., *et al.* "Volume-chirped Bragg gratings: monolithic components for stretching and compression of ultrashort laser pulses." *Optical Engineering.* 2014;53(5):051514

[102] Želudevičius J., Danilevičius R., and Regelskis K. "Optimization of pulse compression in a fiber chirped pulse amplification system by adjusting dispersion parameters of a temperature-tuned chirped fiber Bragg grating stretcher." *Journal of Optical Society of America B.* 2015;32(5):812–17

[103] Sawicka M., Divoky M., Novak J., Lucianetti A., Rus B., and Mocek T. "Modeling of amplified spontaneous emission, heat deposition, and energy extraction in cryogenically cooled multislab Yb3+:YAG laser amplifier for the HiLASE Project." *Journal of Optical Society of America B.* 2012;29(6): 1270–76

[104] Sawicka M., Divoky M., Lucianetti A., and Mocek T. "Effect of amplified spontaneous emission and parasitic oscillations on the performance of cryogenically-cooled slab amplifiers." *Laser and Particle Beams.* 2013;31(4): 553–60

[105] Trenholme J. B., *Fluorescence amplification and parasitic oscillation limitations in disc lasers*, Naval Research Laboratory Memorandum Rep. 2480, 1972

[106] Jambunathan V., Koerner J., Sikocinski P., *et al.* "Spectroscopic characterization of various Yb3+ doped laser materials at cryogenic temperatures for the development of high energy class diode pumped solid state lasers." *Proceedings of SPIE Optics + Optoelectronics*, Prague, Czech Republic, May 2013 (Proceedings SPIE 8780, 2013), pp. 8780–15

[107] Divoky M., Smrz M., Chyla M., *et al.* "Overview of the HiLASE project: high average power pulsed DPSSL systems for research and industry." *High Power Laser Science and Engineering.* 2014;2(14):1–10

[108] Mason P.D., and Divoky M. "Commissioning and characterisation of a kW average power nanosecond pulsed DPSSL operating at 100 J, 10 Hz." Presented at 10th workshop on HEC DPSSL; Noboribetsu, Japan, 2017

Chapter 6

High average power large mode area (LMA) fiber amplifiers

Manoj Kanskar[1]

High power continuous wave (CW) fiber lasers have rapidly penetrated industrial materials processing applications formerly dominated by legacy laser technologies such as CO_2 and lamp-pumped solid-state lasers (SSLs). They are also enabling new applications such as additive manufacturing of 3D-printed metal parts that are impossible to machine. For cutting thin sheet metal at high speed, nearly diffraction-limited beam with $M^2 \leq 1.05$ is often required but for most materials processing applications, such as thick metal cutting and welding, $M^2 \leq 2$ is adequate. For these applications, the spectral width is not very important. Cost of ownership, reliability, serviceability, and industrial ruggedness are some of the key desirable attributes for these lasers. However, there is an emerging application in Defense industry that requires high average power and narrow-linewidth fiber amplifiers. These lasers are being used in Directed Energy systems for defending high-value military assets. Typically, many channels of narrowband, kilowatt-class fiber amplifiers are combined using spectral beam combining (SBC) or coherent beam combining (CBC) techniques. These applications aspire to achieve $\gg$100 kW Directed Energy systems. But these fiber amplifiers are technologically very demanding. As they must have very narrow linewidth to support long coherent length needed for CBC or high-density channel spacing required for SBC. Furthermore, these applications are extremely power-hungry, requiring diffraction-limited beam quality and needing to operate with extremely high efficiency in compact and lightweight configuration. On one hand, the industrial applications are driving down the cost, increasing the reliability and proliferating new applications yet, the demand for high fidelity for beam quality, efficiency, low SWaP, and narrow bandwidth is driving thrusts in fiber design, new fiber amplifier architectures suitable for SBC and CBC as well as high brightness, high efficiency, and low SWaP diodes. This chapter is devoted to summarizing the current state of industrial fiber lasers and focusing on the key challenges to extending the power levels of combinable narrowband fiber amplifiers to many multikilowatts suitable for Directed Energy application.

[1]nLIGHT, Inc., Vancouver, WA, USA

6.1 A brief history of fiber lasers

The benefits of using rare-earth-doped single mode fiber as a laser to provide robust single spatial mode beam quality was realized in 1961 by Snitzer [1,2] who was at America Optical Company. A few years later, flash-lamp pumped, Nd-doped multimode fiber laser was demonstrated by Snitzer and Koester [3]. A diode-pumped Nd-doped multimode fiber laser was demonstrated in the early 1970s by Stone and Burrus at Bell Laboratories [4]. A flurry of work occurred in the 1980s to develop single mode fibers which led to development of rare-earth doping in silica fiber using vapor-phase deposition [5]. The convergence of two key technologies, diode lasers and erbium-doped fiber amplifiers (EDFAs), sparked great interest in fiber laser and amplifiers. Increasing demand for data traffic led to rapid maturation and deployment of diode lasers and EDFAs. Enormous resources from the telecommunication industry led to widespread availability of active and passive fibers, fiber components, relevant technologies, and tools needed for the development of fiber lasers and amplifiers. Early fiber lasers and amplifiers were pumped by gas lasers, lamp-pumped SSLs, or single mode sub-watt diode lasers but were impractical and too expensive for any type of commercial application. It took some key innovations before high power fiber laser and amplifiers became commercially viable. First was the realization of cladding-pumped fiber design to convert the poorer brightness of diode lasers to diffraction-limited beam [6] in the mid-1980s. The second was the dramatic increase in brightness of fiber-coupled multimode diodes in the 1990s which exponentially took off after 2000s. A double-clad fiber uses a small diameter and small NA rare-earth-doped (Yb, Er or Er-Yb) core that is embedded in a much larger diameter and larger NA multimode pump guide. In this configuration, the central active core performs brightness conversion of highly multimode diode laser beam introduced in the large cladding. One of the first commercial applications was a watt-level fiber laser for printing applications. The demand for diode laser power and brightness for direct-diode, SSL pumping, and Directed Energy played a pivotal role in exponential improvement in diode laser brightness. Along with these advances, fiber optic components such as high-power pump and pump-signal combiners as well as isolators also improved. With the convergence of all these technologies, kilowatt-class fiber lasers have emerged to displace many legacy laser technologies [7,8] and enable new applications not possible before. Today, numerous companies provide broadband kilowatt-class single-mode fiber lasers and many tens-of-kilowatt multimode fiber lasers for industrial applications. Single spatial mode and narrowband multikilowatt fiber amplifiers are becoming prevalent in combinable Directed Energy systems that use SBC or CBC. The demand for higher power will push fiber and diode laser technologies to their physical limits in the future.

6.2 Advantages of fiber lasers

One of the key disadvantages of bulk SSL is degradation in beam quality as the average output power goes up. Fiber laser and amplifiers can be designed to support only the fundamental mode, that is, fundamental HE_{11} or LP_{01} mode. Fiber

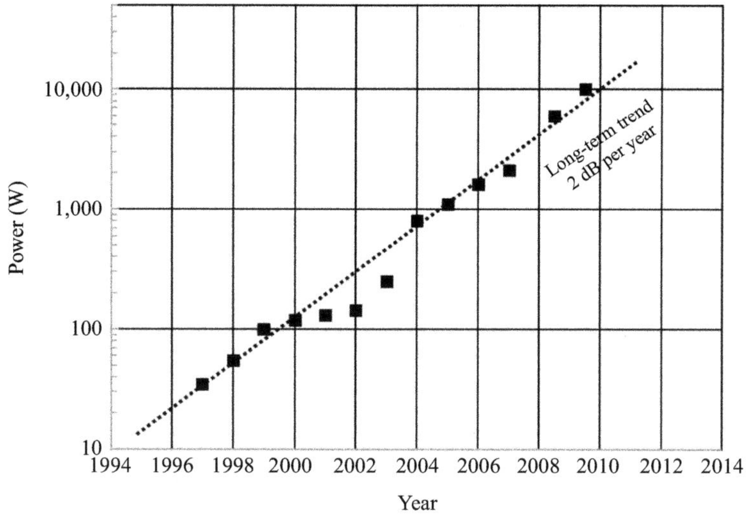

Figure 6.1 Progress in average output power from diffraction-limited and near-diffraction-limited fiber laser and amplifiers showing a trend of power improvement of 2 dB per annum

amplifiers with 10 kW of output power have been demonstrated with nearly diffraction-limited beam quality [9]. Bulk SSLs and disk lasers dissipate heat through three-dimensional or two-dimensional heat flow. This puts a severe limit on beam quality due to thermally induced mode distortions present in these SSL architectures. Rare-earth-doped fibers have a large gain bandwidth (up to ~20 THz) providing many channels for SBC or allowing generation of ultrashort pulses. Fiber amplifiers in a master oscillator power amplifier (MOPA) configuration can provide high gain (~20 dB). Pump absorption efficiency can be designed to be ~20 dB and optical-to-optical conversion efficiency can be very high (>85 percent) making the total electrical-to-optical power conversion efficiency unmatched by other diode-pumped SSLs. This technology has matured for a fully fiberized solution with no free-space optical components to align allowing for robust and compact systems. Progress in high average power fiber laser and amplifiers is summarized in Figure 6.1. For 20 years, output powers from single aperture fiber lasers grew at about 10 times every 3 years and most of these gains are directly attributable to improvements in diode laser power and brightness improvement. Fiber laser output power has been limited by physical and design attributes of the fiber itself since about 2010.

6.3 Rare-earth-doped fibers

6.3.1 Fundamentals of optical fibers

At the center of a typical single-clad optical fiber is a core of diameter $2r$ and refractive index n_{co}, surrounded by a cladding layer with refractive index n_{cl}. To make passive fibers, the core is typically doped with germanium which raises the index of refraction relative to the cladding and provides NA of less than 0.07. Phosphorous or aluminum doping can also be used to raise the refractive index. Alternatively, the cladding can be fluorine or boron doped to lower its refractive index providing NA ~0.22. The core and the innermost part of cladding comprises very high-purity glass deposited by using vapor phase deposition or direct nano-particle deposition (DND) processes. These processes yield very low transmission loss of just a few decibels per kilometer near 1 μm wavelength. A layer of high index acrylic coating is applied to protect the glass surface and strip away unwanted cladding light.

The relative index contrast of a step-index fiber is defined as:

$$\Delta = \frac{n_{\text{co}} - n_{\text{cl}}}{n_{\text{co}}} \tag{6.1}$$

and it is typically below 2 percent. In the weakly guiding regime, linearly polarized modes can be used to describe the modes of the fiber without invoking the vector nature of the fields. The NA is defined as:

$$NA = \sqrt{n_{\text{co}}^2 - n_{\text{cl}}^2} \tag{6.2}$$

an important parameter in an optical fiber is the normalized frequency which is defined as:

$$\text{V} = \frac{2\pi r \sqrt{n_{\text{co}}^2 - n_{\text{cl}}^2}}{\lambda} \tag{6.3}$$

where λ is the vacuum wavelength and r is the radius of the fiber core. The guided modes of a fiber can be derived by solving the Helmholtz eigenvalue equation using Maxwell's equations. A guided mode solution has a distinct spatial distribution and propagates with a propagation constant, β defined as:

$$\beta = \frac{2\pi n_{\text{eff}}}{\lambda} \tag{6.4}$$

Maxwell's equations are scale invariant. Once the normalized frequency is known, by proportionally scaling the core size and the wavelength, relative field distribution and effective index of the mode can be obtained by using (6.4). Typically, it falls between n_{co} and n_{cl}. Modes with $n_{\text{eff}} < n_{\text{cl}}$ have oscillating fields in the cladding and are no longer guided and they are referred to as radiation modes or leaky modes. The normalized propagation constant b is a good parameter that defines the strength of the confinement of the guided mode. For all guided modes $1 > b > 0$. It is no longer guided when $b \leq 0$:

$$b = \left(\frac{n_{\text{eff}}^2 - n_{\text{cl}}^2}{n_{\text{co}}^2 - n_{\text{cl}}^2}\right) \tag{6.5}$$

The total number of guided modes can be estimated from (6.6) and the fiber supports only a single mode for $V < 2.405$.

$$N = \frac{V^2}{2} = \left(\frac{2\pi r\sqrt{n_{co}^2 - n_{cl}^2}}{\lambda}\right)^2 \tag{6.6}$$

High power fiber amplifiers tend to use dual-cladding or even triple-cladding structures. In this case, the outermost cladding is a fluoro-acrylate which provides NA ~0.46.

Active or gain fibers comprise cores that are typically doped with rare earths such as Yb^{3+}, Er^{3+}, Tm^{3+}, and Nd^{3+}. High levels of Al^{3+} ions are added to silica glass host to form a homogeneously modified glass network which allows high quality incorporation of rare earths. Typically, much lower concentration of rare earths is possible in glass compared to crystalline host materials. This is not a problem because long lengths of low loss fibers can be used for efficient absorption of pump light. Chemical vapor deposition (CVD), solution doping of soot produced by CVD process and DND techniques are used to make active fibers. Of these three processes, DND appears to have the most control in RIP profile and RIP shaping capability due to very thin layers of materials that can be deposited at a time.

Yb-doped fiber amplifier (YDFA) has proven to be remarkably scalable for both industrial fiber lasers and combinable narrowband directed energy fiber amplifiers. Tm-doped fiber amplifiers are getting more attention due to their eye-safer emission wavelength around 2 μm and kilowatt-class [10] Tm-doped fiber laser has been reported, but they still cannot rival YDFAs in efficiency. The reason Yb-ions have become the dopant of choice for high power fiber lasers and amplifiers is due to the availability of low quantum-defect pumps in the 915 nm to 976 nm wavelength range. With 976 nm pumping, optical-to-optical efficiency of ~90 percent has been demonstrated. Furthermore, it is a quasi-two-level system with little parasitic excited state absorption and cooperative upconversion. For these reasons, to date, all multikilowatt fiber laser and amplifiers have been demonstrated in Yb-doped active fibers.

6.3.2 Design of fiber amplifiers

In a quasi-two-level system with upper-level population N_2 and lower level population N_1, absorption and emission cross-section for signal and pump, the gain per length can be obtained from rate equations. Ignoring absorption at the signal wavelength:

$$g = \frac{g_0}{\left(1 + \frac{I_s}{I_{sat}}\right)} \tag{6.7}$$

where g_0 is the small signal gain and I_s and I_{sat} are the signal intensity and saturation intensity, respectively. The relation in (6.7) indicates significant reduction of gain at higher signal intensities. For low amplified spontaneous emission (ASE), I_{sat} is given by:

$$I_{sat} = \frac{\nu_s}{\nu_p}\left(\frac{\sigma_a^p + \sigma_e^p}{\sigma_a^s + \sigma_e^s}\right) I_p \tag{6.8}$$

where I_p is the pump intensity and $\nu_{(s,p)}$ is frequency (s for signal and p for pump). For a known population inversion $\eta = N_2/(N_2 + N_1)$.

$$g_0 = N_2\sigma_e^s - N_1\sigma_a^s \tag{6.9}$$

Local pump and signal intensities in a doped fiber can be calculated by solving:

$$\frac{dI_p}{dz} = \left(N_2\sigma_e^p - N_1\sigma_a^p\right) I_p(z) \tag{6.10}$$

$$\frac{dI_s}{dz} = \left(N_2\sigma_e^s - N_1\sigma_a^s\right) I_s(z) \tag{6.11}$$

The evolution of signal power, pump power, and population inversion can be generated from (6.10) and (6.11) and they are illustrated for Yb-doped fiber amplifier for copumping and counter-pumping configurations in Figures 6.2 and 6.3, respectively. For the copumping configuration, high inversion is achieved at the front end of the gain fiber. This leads to a rapid growth of signal at the beginning of the fiber compared to the counter-pumped configuration. Hence, for any intensity-dependent nonlinear effects, the effective interaction length in the fiber is longer for the copumped case. A higher inversion can be maintained at the output end of the fiber for the counter-pumped configuration. This leads to higher output power in the same length fiber making it more efficient. Hence, the

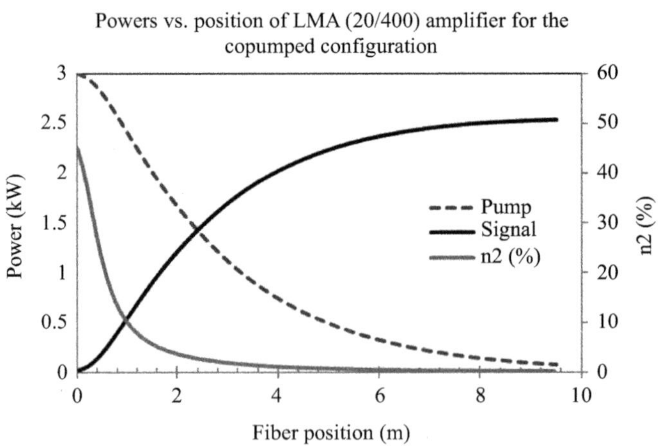

Figure 6.2 Plots showing signal power, pump power, and population inversion in an Yb-doped fiber amplifier for copumping configuration

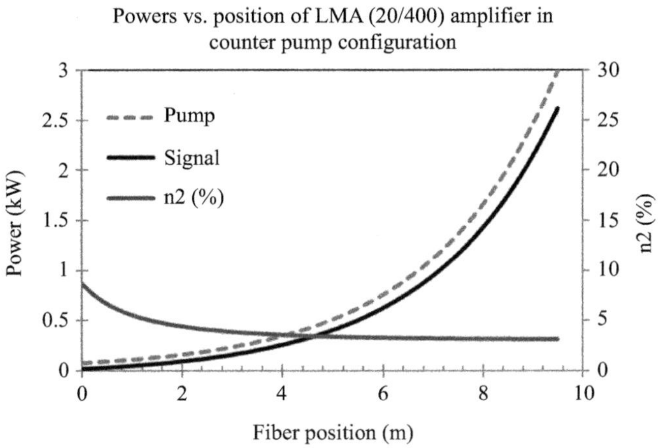

Figure 6.3 Plots showing signal power, pump power, and population inversion in an Yb-doped fiber amplifier for counter-pumping configuration

counter-pumping is more desirable for achieving higher efficiency and mitigating deleterious nonlinear effects. Currently, there are no reliable multikilowatt, signal-pump counter-combiners available to take advantage of this feature. Other than free-space counter-pumped experiments, most monolithic high-power fiber-amplifier results use the copumped configuration.

If spectral brightness is unimportant, then fiber laser output power is limited by the amount of pump power that can be coupled into the cladding and the efficiency, η_{laser}, of the laser which is ultimately constrained by the brightness of pump diode lasers, B_{pump}, as shown in (6.12) and (6.13).

$$P_{\text{laser}} = \eta_{\text{laser}} \times P_{\text{pump}} \tag{6.12}$$

$$P_{\text{laser}} = \eta_{\text{laser}} \times B_{\text{pump}} \times \left(\pi \cdot R^2\right) \times \left(\pi \cdot NA^2\right) \tag{6.13}$$

Fiber lasers are very efficient due to absorption of greater than 95 percent of input pump power. Consequently, absorption by the core is determined primarily by the ratio of the core to clad cross-sectional areas. Maximum achievable core absorption is limited by photo-darkening properties of the rare-earth doping methods but can be about 250 dB per meter at 976 nm or much higher. High-efficiency requirement parametrically links core and cladding radius given by (6.15):

$$P_{\text{unabsorbed}} = 10^{\frac{-\alpha_{\text{laser}}}{10}} \times P_{\text{pump}} \tag{6.14}$$

$$\alpha_{\text{laser}} = \alpha_{\text{core}} \cdot \frac{r^2}{R^2} \cdot L \Rightarrow R = r \cdot \sqrt{\frac{\alpha_{\text{core}} \cdot L}{\alpha_{\text{laser}}}} \tag{6.15}$$

6.4 Limitations of high average power Yb-doped fiber amplifiers

Brightness of diodes have not reached an asymptotic value yet. Current state-of-the-art diodes have demonstrated more than 300 W from 105 μm and 0.15 NA beam [11]. Using a tree-architecture and by combining seven of these using a commercial 7 × 1 pump combiner 2.1 kW of pump power can be generated from a 200 μm and 0.22 NA beam. Six of these can be used in conjunction with a signal-pump combiner to generate over 12 kW of pump power into a 400 μm and 0.46 NA DC-LMA fiber. One can envision a 10-kW diode-pumped fiber amplifier if some of the undesirable nonlinear effects could be mitigated. Diode laser brightness continues to advance with room for, at least, another one order of magnitude improvement. In the near term, fiber laser and amplifiers are unlikely to be limited by diode brightness.

There are several nonlinear effects that limit power scaling in diode-pumped large mode area (LMA) Yb-doped fiber lasers and amplifiers. Currently, stimulated Raman scattering (SRS), thermally induced multimode instability (TMI), and stimulated Brillouin scattering (SBS) are the key mechanisms that limit maximum power achievable in LMA fibers. Four-wave mixing (FWM) and self-phase modulation (SPM) can also introduce additional complexities. These deleterious effects lead to either back scattering of signal power, transfer of energy to unwanted spectral regions or coupling to higher order modes or severe degradation of spectral brightness. Typically, silica fiber doped with Yb^{3+} ions typically exhibits less than 10 dB per km loss in the core. For all practical purpose, waveguide losses do not impact optical efficiency since amplifiers have $\ll$50 m of gain fiber.

Theoretical maximum power from broadband fibers is then limited by SRS and thermal lens and it is given by [12]:

$$P_{\max} = 4\pi\sqrt{\left(\frac{\eta_{\text{laser}} k\lambda^2\,\Gamma^2\,\ln(G)}{2\,\eta_{\text{heat}}\frac{dn}{dT}g_R}\right)} \tag{6.16}$$

where η_{laser} ~ 0.85 is the laser efficiency, η_{heat} ~ 0.1 is the quantum defect generated heat, dn/dT ~ $12 \times 10^{-6}\ \text{K}^{-1}$, $G \geq 10$ dB is the gain, and g_R ~ $10^{-13}\ \text{m W}^{-1}$ is the Raman gain coefficient. This suggests that with conventional diode-pumped LMA fiber technology one can potentially achieve ~15 kW of power for broadband emission and SRS threshold is determined by:

$$P_{Cr}^{\text{SRS}} \approx 16\frac{A_{\text{eff}}}{g_R L_{\text{eff}}} \tag{6.17}$$

Most industrial fiber lasers and amplifiers do not require any spectral bandwidth requirement except for those used in higher harmonic generation. However, combinable fiber laser and amplifiers have a more stringent spectral bandwidth prerequisite. CBC and SBC techniques need narrow band amplifiers either for long coherence length or to fit largest number of channels for the available gain bandwidth. For these combinable fiber amplifiers, spectral bandwidth is necessary to be less than 25 GHz and narrower the better. This narrow bandwidth constraint on the spectrum poses a challenge from a SBS perspective. SBS has a lower threshold compared to SRS due to two orders of magnitude higher Brillouin gain. SBS is a phase-matched third-order nonlinear interaction that couples acoustic phonons to the photons of the laser field and the backscattered Stoke light. In SBS, a pump photon is annihilated to generate a Stokes photon and an acoustic phonon through electrostriction process. This process is further stimulated by the presence of Stokes photons and acoustic phonons in the fiber. Momentum conservation leads to propagation of Stokes photons opposite to the propagation direction. Energy and momentum conservation in SBS process leads to:

$$\nu_a = \nu_p - \nu_s, k_a = k_p - k_s \tag{6.18}$$

where ν_a, ν_p, and ν_s are the frequency for acoustic phonon, pump, and Stokes, respectively, and k_a, k_p, and k_s are the wave vectors for acoustic phonon, pump, and Stokes, respectively. In silica fibers, the acoustic frequency is ~11 GHz at 1.55 μm and the acoustic velocity is 5.994 km s^{-1}. The phonon lifetime is less than 10 ns which corresponds to a transit distance of less than 60 μm. SBS threshold depends very strongly on the spectral bandwidth of the laser:

$$P_{Cr}^{SBS} = 21 \frac{A_{\text{eff}}}{g_B L_{\text{eff}}} \left[1 + \frac{\Delta\nu_s}{\Delta\nu_a} \right] \tag{6.19}$$

where $g_B \sim 3\text{–}5 \times 10^{-11}$ m W^{-1} is the peak Brillouin gain coefficient and is mostly wavelength independent; $\Delta\nu_a \sim$ 10–1,000 MHz and $\Delta\nu_s \leq 25$ GHz are Brillouin gain bandwidth and laser spectral width respectively. Near 1 μm emission wavelength, measured Brillouin shift and bandwidth are in the range of 15 GHz to 16 GHz and 65 MHz to 85 MHz, respectively. A_{eff} is the area of the mode and L_{eff} is the effective Brillouin interaction length in the fiber where maximum output power P_0 and power distribution $P(z) = P_0 f(z)$ is:

$$L_{\text{eff}} = \int_0^L f(z) dz \tag{6.20}$$

In the counter-pumped fiber amplifier with length L, the signal grows exponentially as $f(z) = e^{gz}$ and the effective length can be estimated as:

$$\frac{L_{\text{eff}}}{L} = \frac{1 - e^{-gL}}{g} \tag{6.21}$$

From (6.17) and (6.19) it is evident that enlarging the effective area by increasing the mode field diameter is an effective means to suppress SRS and SBS. Early on this viewed to use of larger diameter fibers with 20 to 25 μm core sizes. Though these fibers are not purely single mode and can support numerous higher order modes, a technique such as tight coiling [13] (~10 cm diameter) was employed to suppress higher order modes. Unbeknownst to researchers for some time, the higher order mode suppression ratio was not sufficient enough using this method. As a result, a new nonlinear phenomenon was revealed, and it is referred to as TMI [14,15] or stimulated thermal Rayleigh scattering (STRS) [16]. The origins of TMI is a result of interplay between fundamental mode and one higher order mode inter-reference and thermal effects. For TMI to occur, at least, one higher order mode (typically LP_{11}) must persist in the fiber. Two fiber modes (e.g., LP_{01} and LP_{11}) with spatially varying amplitudes $A_i(r, \varphi, z, t)$ interfere. The Mach–Zender-like beating between the fundamental and the higher order mode results in a spatial longitudinal interference pattern given by (6.22). Local heating in the fiber due to quantum-defect heating (6.23) between the pump and signal is proportional to local inversion along the length of the fiber. This heat imprints an index-of-refraction, $n_{\mathrm{NL}}(r, \varphi, z, t)$, grating that mimics the spatial interference pattern through thermo-optic effect shown in (6.25). This refractive index is related to the nonlinearly induced polarization that is related to the change in temperature of the fiber and the electric field given by (6.26). Period of the beating, and therefore the grating period, is determined by the difference in the propagation constants between the two beating modes:

$$E_s(r,\varphi,z,t) = A_1(r,\varphi,z)e^{i(\beta_1 z - w_1 t)} + A_2(r,\varphi,z)e^{i(\beta_2 z - w_2 t)} + c.c. \tag{6.22}$$

$$I_s(r,\varphi,z,t) = E_s^2(r,\varphi,z,t) \tag{6.23}$$

$$Q_{QD}(r,\varphi,z,t) = \left(\frac{\lambda_s}{\lambda_p} - 1\right) g(r,\varphi,z,t) I_s(r,\varphi,z,t) \tag{6.24}$$

where, λ_s and λ_p are the signal and pump wavelengths, $g(r, \varphi, z, t)$ and $I_s(r, \varphi, z, t)$ are the bulk material gain and field intensity.

$$\rho C \frac{\partial \Delta T}{\partial t} = \kappa \nabla^2 \Delta T(r,\varphi,z,t) + Q_{QD}(r,\varphi,z,t) \tag{6.25}$$

where, ρ is the density, κ is the thermal conductivity of fiber, and C is the specific heat capacity.

$$P_{NL}(r,\varphi,z,t) = \varepsilon_0 \eta T(r,\varphi,z) E_s(r,\varphi,z,t) \tag{6.26}$$

where, ε_0 is the permittivity and η is the thermo-optic coefficient.

$$\nabla^2 E_s(z,t) = \frac{(n_0 + n_{NL}(z,t))^2}{c^2} \frac{\partial^2}{\partial t^2} E_s(z,t) \tag{6.27}$$

Solving for the time-independent wave (6.27) leads to coupled mode equation for the case of two-beam coupling which predicts TMI threshold. It is critical that the solution of the heat equation has both the spatial and temporal dependence to account for the energy transfer between the transverse mode [17–19]. Time-dependent equations must be solved to accurately describe TMI temporal dynamics above threshold [20]. Assuming the powers of two modes are represented by $P_1(z)$ and $P_2(z)$

and the amplifier gains are independent of propagation distance represented by g_1 and g_2, then power coupling between the two modes can be represented as follows for negligible background loss:

$$\frac{dP_1^N}{dz} = -g_1\chi e^{g_2 z}P_1^N(z)P_2^N(z) \tag{6.28}$$

$$\frac{dP_2^N}{dz} = -g_1\chi e^{g_1 z}P_1^N(z)P_2^N(z) \tag{6.29}$$

$$P_1^N(z) = P_1(z)e^{-g_1 z} \tag{6.30}$$

$$P_2^N(z) = P_2(z)e^{-g_2 z} \tag{6.31}$$

when the amplifier gain is large, that is, $g_1 \gg 1$ and power depletion in mode 1 due to mode coupling is very small, a simple relationship can be found which depends on the power of the first mode and the nonlinear coupling constant only.

$$P_2(z) = P_2(0)e^{g_2 L}e^{\chi P_1(L)} \tag{6.32}$$

The nonlinear coupling coefficient χ can be calculated using $\eta \sim 3.5 \times 10^{-5}\ \mathrm{K}^{-1}$, $\kappa \sim 1.4\ W\ (\mathrm{K\text{-}m})^{-1}$ and $\rho C = 1.67 \times 10^6\ \mathrm{J(K\text{-}m^3)}$ for a range of frequencies. The solution has a typical nonlinear Lorentzian gain curve shape. The solution to the equation for LP_{01} and LP_{11} shows peak gain at frequency of kilohertz and depends on the transverse thermal diffusion rate. Typically, LP_{01} loses power to LP_{11} in one propagation direction and then from LP_{11} back to LP_{01} mode in millisecond time scale. The nonlinear coupling coefficient also has dependence on the modal overlap function Γ_i and therefore on the fiber design. Temporal dynamics of TMI is different for different fiber amplifiers. The dependence of the spectral properties of χ on the core size and V-parameter of step-index fiber showed frequency and peak decreases with increasing core size and the maximum value of χ decreases with decreasing V-parameter. As a result, temporal dynamics of TMI is expected to be slower for large core fibers and threshold is expected to be higher for fibers with low V-parameter [21]. Higher order mode suppression ratio plays a key role in establishing TMI threshold as shown by (6.32). It also helps to have high power seed in the LP_{01} mode.

6.5 Overcoming the limitations of power scaling in combinable LMA fibers

6.5.1 Mitigating SBS

For broadband fiber lasers and amplifiers with 1 to 10 nm bandwidth, SRS is a relatively weak effect. Of the key deleterious nonlinear effects, SRS, SBS, and TMI that were previously discussed, SBS threshold is much lower compared to SRS due to greater than two orders of magnitude higher Brillouin gain compared to Raman gain. Since SBS signal counter propagates, it is very dangerous and causes catastrophic damage in the fiber. In the high-power regime, near single-frequency or very narrow linewidth fiber sources is typically configured as a MOPA. This configuration avoids the stability problem of a high single-frequency oscillator and takes advantage of high gain and high efficiency in the amplifier. When the spectral width is less than 10 MHz, that is, less than Brillouin gain bandwidth, SBS is nominally 500 times stronger than SRS. In this regime, state-of-the-art LMA fiber amplifiers with core size of 20 μm diameter reaches SBS threshold at 300 watts and 4-m effective fiber length and it degrades to 50 W at 10-m effective length as shown by the cutback experiment results displayed in Figure 6.4. Several methods have been devised to overcome this power scaling limit.

One of the most effective methods for increasing SBS threshold has been to increase the fiber amplifier bandwidth. A standard technique is to use a white noise source (WNS) with a pass-band filter and RF amplifier to phase modulate the seed to generate a desirable bandwidth. This method creates a continuous broadened spectral line-shape and may be unsuitable for SBS mitigation techniques that employ thermal gradient and acoustic tailoring. The spurious nature of the source may also lead to random pulsation in counter-propagation. An Even more effective method which avoids issues associated with WNS is to use a phase modulator that is driven by a pseudo random binary sequence (PRBS) to broaden the spectral bandwidth. While the source is no longer single frequency, even bandwidth of 25 GHz has been found to be sufficiently good for SBC and CBC techniques. When the PRBS modulation format is used, the resulting optical power spectral density contains discrete, periodic features that are a function of the modulation frequency and the PRBS pattern length. A PRBS pattern is denoted at 2^n-1 where n indicates the shift register length used to create the pattern. It contains all possible combination of bits of 0 and 1 except the null pattern. The 0 and 1 represent phase modulation depths of 0 and π, respectively. The spacing between each bit is determined by the PRBS modulation frequency. The modulated spectrum comprises a series of modes with a frequency spacing of $\Delta\nu = \nu \times (2^n - 1)^{-1}$,

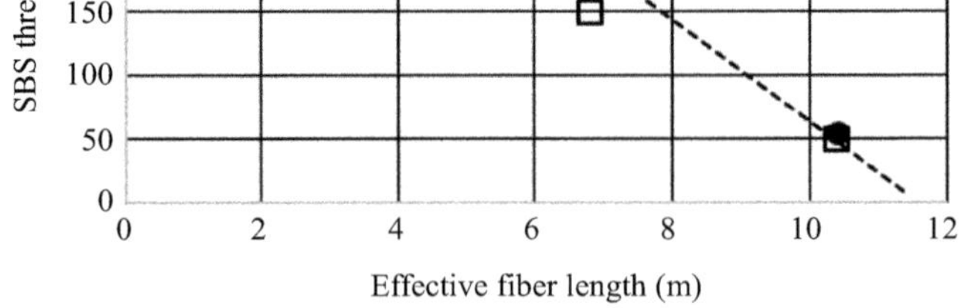

Figure 6.4 Plot showing SBS threshold power vs effective fiber length for narrow-band source (<10 MHz)

where ν is the PRBS modulation frequency. In addition, the spectral power density function envelope carries a sinc^2 dependence with nulls at the integer multiples of the modulation frequency. The key advantage of this technique over the random WNS is that the separation among the adjacent sidebands within the sinc^2 envelope can be chosen to be greater than the spontaneous Brillouin gain bandwidth by judicious choice of the modulation frequency for a given PRBS pattern. The spacing can be made to be deterministic. This implies that the sidebands generated by the PRBS act independently and the SBS threshold is primarily determined by the modes carrying the highest optical power [22]. This control is not possible with the WNS.

SBS threshold for copumped 10.5-m fiber amplifier using 2^7-1 PRBS pattern for various bandwidth less than 20 GHz is shown in Figure 6.5. Since the maximum SBS suppression depends on both the frequency spacing and the PRBS pattern, these measured values are not the best-case scenarios since only 2^7-1 PRBS pattern was used for all modulation frequencies. PRBS technique poses a challenge for CBC since this technique displays a linear drop in visibility resulting in a drop in combining efficiencies for small path-length deviations. Techniques have been developed to use a low-pass RF filter to mitigate high-frequency components thereby suppressing the sidelobes in the optical spectrum. This leads to a Gaussian visibility function and moderately improves coherence length. There is also a small enhancement in SBS suppression [23].

There are other methods that have also been used to further mitigate SBS in conventional LMA fibers. Multitone signal has been used to generate laser gain competition which results in large $\Delta P/\Delta z$ and $\Delta T/\Delta z$. Shorter effective length and larger thermal gradients help mitigate SBS and this technique is compatible with PRBS modulation as well as acoustically tailored fibers. As more techniques to mitigate SBS were uncovered, TMI has become the key bottleneck for power scaling combinable fiber amplifiers.

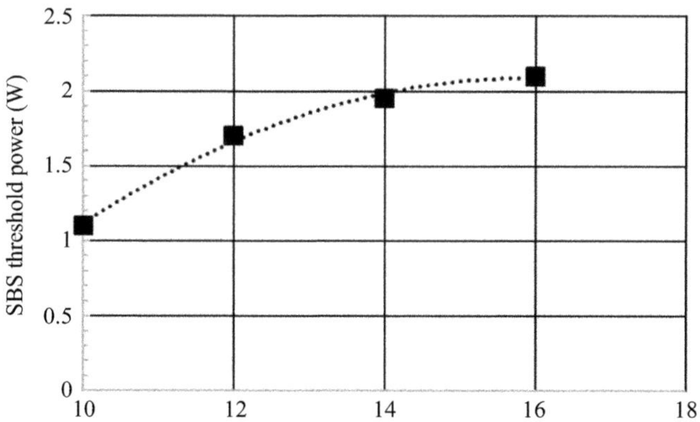

Figure 6.5 Plot showing onset of SBS as a function of input seed bandwidth and (2^7-1) PRBS pattern

6.5.2 Mitigating TMI

In Section 6.4, it was shown that TMI results from the existence of, at least, one higher order mode and quantum-defect heating with a modest dependence on the seed power. The most effective TMI mitigation techniques have relied on either suppressing higher order mode(s) or reducing quantum-defect heating. In principle, reducing quantum-defect heating is the most straightforward technique. Tandem-pumping has been used to achieve low quantum-defect heating. Diode lasers are not suitable for this application primarily because their brightness is not sufficiently high enough. Hence, fiber lasers are the only alternatives that can be used as tandem pumps and 10 kW of nearly single-mode fiber amplifier has been demonstrated. Such a technique poses a huge challenge from size, weight, and power consumption (SWaP) perspective for high energy lasers (HEL) used in Defense applications. As a result, much effort has been placed on suppressing higher order modes. The most straightforward method is to tightly coil a conventional LMA fiber to suppress higher order mode. TMI threshold improves as tighter coiling diameter is employed as shown in Figure 6.6. There is a limit to reducing the coiling diameter. Fundamental mode bend loss increases significantly below about 10-cm coiling diameter making it impractical when efficiency is critical. A more robust strategy is to implement higher order mode suppression techniques into the design of LMA fibers. One promising method that has been suggested is to use gain-tailoring or confined-doping profile. In such a fiber, the core of the fiber is only 80 to 90 percent doped. This provides preferential gain to the fundamental mode and suppresses higher order modes compared to fully doped cores [24]. In another design, cladded and linearly graded (CLING) refractive index profile has been suggested to reduce the optical modal overlap between the fundamental mode and the higher order modes. This helps to reduce the coupling between these modes

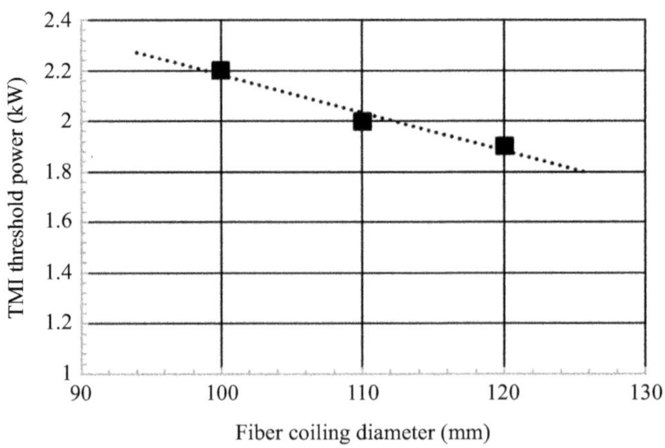

Figure 6.6 TMI threshold of 20/400/0.064 DC LMA fiber amplifier as a function of fixed fiber-coiling diameter with an input seed bandwidth at 20 GHz and (2^7-1) PRBS pattern

thus increasing TMI threshold power [25]. Another approach that has been experimentally successful in mitigating TMI is the use of chirally coupled core (3C) dual-clad LMA fiber.

6.5.3 Chirally coupled core fiber technology

3C fiber is a new type of LMA fiber with a built-in higher order mode suppression mechanism that does not require coiling. It comprises a rare-earth-doped central core surrounded by one or more passive side cores separated by a finite gap which are helically wound along the longitudinal direction of the fiber as shown in Figure 6.7. As a result, it has helical translational symmetry instead of linear-translational symmetry obeyed in conventional cylindrical fibers. Due to the helical nature of the fiber the interaction of 3C fiber modes involves both spin and angular momentum of the optical waves. This fiber comprises two types of wave-guiding cores within one glass cladding and the central core is active and straight. It is typically round when less than 35 μm in diameter and polygonal when larger. This central core is surrounded by a single or multiple guiding side cores that are in optical proximity with the core to form a weakly coupled system. There is no gain in the side cores. The central core runs along the axis of the fiber akin to conventional fibers and provides amplification. The side core(s) are passive and helically wound around the central core with a helix period of Λ (typically several millimeters long). The separation between the central and side core(s) are typically several micrometers. The primary purpose of the helical side core(s) is to control modal properties of the central core. The design consideration takes advantage of the quasi phase matching possibilities that exist in this fiber. Due to helical symmetry of the fiber, the linear propagation constants can be quasi phase matched (QPM) to both spin and orbital angular momentum of the helical modes. This can be described by

$$\beta_{l_1m_1} - \beta_{l_2m_2} \times \sqrt{1 + K^2R^2} - \Delta m \times K = 0 \tag{6.33}$$

where $\beta_{l(1)m(1)}$ and $\beta_{l(2)m(2)}$ are the propagation constants of $LP_{l(1)m(1)}$ and $LP_{l(2)m(2)}$ modes in the straight central and side cores, respectively, and $(1 + K^2R^2)^{1/2}$ is a

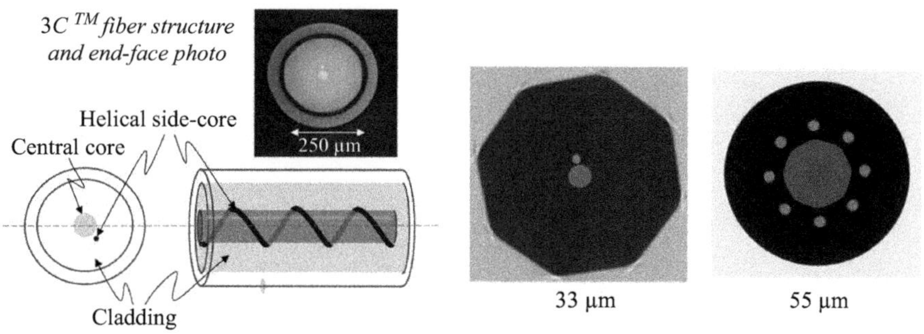

Figure 6.7 Schematic of 3C fiber and cross-sections of a 33 μm 3C fiber with a single side core and a 55 μm 3C fiber with eight side cores

helical correction factor for the side core and the $\Delta m = \Delta l + \Delta s$ where the value for Δl runs through four possible combinations of $\Delta l = \pm\, l_1 \pm l_2$ and Δs runs through five possible integers $\Delta s = -2, -1, 0, +1, +2$. For instance, interaction between central-core LP_{01} group and side core LP_{11} group is described by $\Delta m = -3, -2, -1, 0, +1, +2, +3$. The origin of multitude of QPM resonances stated by (6.33) is related to three types of perturbations that are causing three distinct types of coupling between the center and side core modes. The first type of coupling is the mutual perturbation of the center and side cores due to their proximity. This causes coupling between the modes with the same circular polarization (CP), that is, RCP to RCP and LCP to LCP. This corresponds to Δm values with $\Delta s = 0$. The second interaction arises from the linear birefringence created in the spacing between side core and center core due to mismatch in coefficient of thermal expansion of the two cores. This perturbation causes coupling between orthogonally polarized modes, that is, RCP to LCP and vice versa. This corresponds to Δm values with $\Delta s = -2$ and $+2$. The third type of perturbation arises from the shear birefringence created in the spacing between the two cores due to viscosity of the glass flow during the spinning of a heat-softened fiber perform during the 3C fiber draw process. This perturbation causes mixing between longitudinal and transverse modal field components and it is described by Δm values with $\Delta s = +1$ and -1. To take advantage of these resonances, one can judiciously choose central and side core sizes and their NAs, the spacing in-between and the helix period of the side cores to introduce >50 dB/m suppression for the next higher order mode while keeping the losses of the fundamental mode in the central core to a negligible value.

In Figure 6.8, a schematic of a copumped Yb-doped 22/400/0.059 DC 3C fiber amplifier is shown. This 3C amplifier was seeded with approximately 20 W and 20 GHz linewidth (phase modulated 2^7-1 PRBS pattern) 1,064 nm signal. Even though the 3C fiber was not tightly coiled (>30 cm coiling spiral diameter), onset of TMI was observed only near 2.9 kW (see Figure 6.10) of CW output power. Based on the extrapolation using Figure 6.6, at this coiling diameter, similar core size regular LMA fiber is expected to reach onset of TMI well below 500 W.

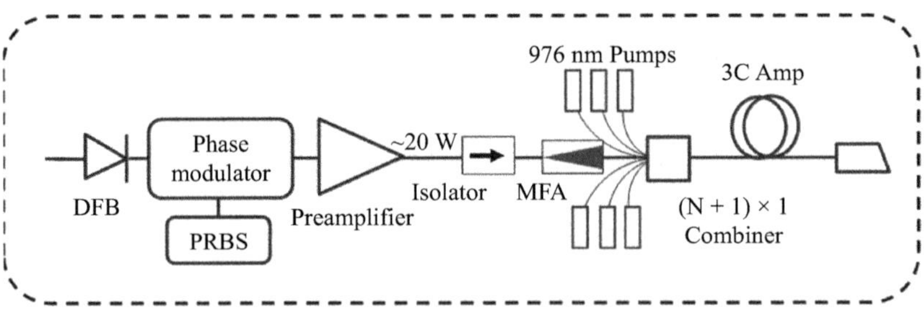

Figure 6.8 Schematic showing narrowband 3C amplifier coiled at >30 cm spiral diameter. It is comprised of DFB seed which is phase modulated with (2^7-1) PRBS pattern followed by several stages of preamplifier and power amplifier pumped by VBG-locked 976 nm diodes

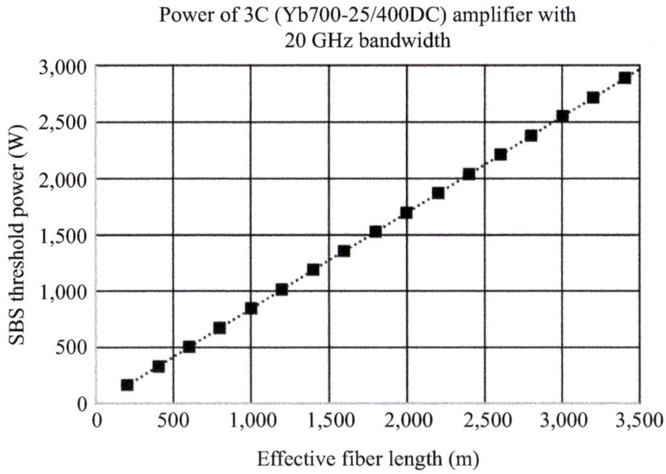

Figure 6.9 Plot showing output power of 3C amplifier versus absorbed 976 nm pump power. No SBS or TMI was observed up to maximum measured power of 2.9 kW

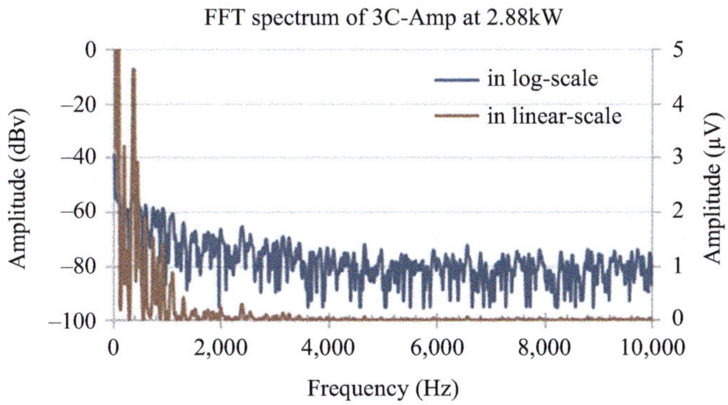

Figure 6.10 Plot showing onset of TMI near 2.9 kW

6.5.4 Conclusions

Is the current understanding that the state-of-the-art high average power single mode fiber laser and amplifier output power is primarily limited by TMI, SBS, and SRS. Broadband Yb-doped single-mode fiber lasers are approaching their fundamental limits. However, the combinable narrow band (<25 GHz) fiber amplifiers are still a factor of two to three under practical limits. Although the nonlinear challenges are nontrivial, there are numerous clever fiber designs such as gain-tailoring, CLING fiber, and 3C being considered to overcome power-scaling issues. Innovative architectures such as two-tone and low quantum-defect pumping have the potential to further mitigate power-scaling bottleneck. Opportunities exist to make significant in-roads in this direction and they will certainly be exploited in the coming decade.

Acknowledgments

I would like to thank nLIGHT's Management and Advanced Technology Group for their valuable input. All the figures are courtesy of nLIGHT.

References

[1] E. Snitzer, "Optical maser action of Nd3+in a barium crown glass," *Phys. Rev.* **7**(12), 444–446 (1961).
[2] E. Snitzer, "Proposed fiber cavities for optical maser," *J. Appl. Phys.* **32**(1), 36–39 (1961).
[3] C.J. Koester and E. Snitzer, "Amplification in a fiber laser," *Appl. Opt.* **3**(10), 1182–1186 (1964).
[4] J. Stone and C.A. Burrus, "Neodymium-doped fiber lasers: room temperature CW operation with an injection laser pump," *Applied Optics*, **13**, 1256–8 (1974).
[5] S.B. Poole, D.N. Payne and M.E. Fermann, "Fabrication of low loss optical fibers containing rare earth ions," *Electron. Lett*, **21**, 737–8 (1985).
[6] E. Snitzer, H. Po, F. Hakimi, R. Tuminelli and B.C. MaCollum, "Double-clad, offset core Nd fiber laser," (Post deadline paper PD5), in *Proceedings of Conference on Optical Fiber Sensors*, Optical Society of America (OSA) (1988).
[7] K.I. Ueda, H. Sekiguchi and H. Kan, "1 kW CW output from fiber-embedded disk lasers," in *Proceedings of the Conference on Lasers and Electro-Optics* (Post deadline paper CPDC4), Optical Society of America (OSA) (2002).
[8] Y. Jeong, J.K. Sahu, D.N. Payne and J. Nilsson, "Yb-doped large-core fibre laser with 1 kW of continuous-wave output power," *Electron. Lett.* **40**, 470–2 (2004).
[9] E. Stiles, "New developments in IPG fiber laser technology," in *Proceedings of the 5th International Workshop on Fiber Lasers*, SPIE (2009).
[10] T. Ehernreich, R. Leveille, I. Majid, T. Kanishka, G. Rines and P. Moulton, "1-kW, all-glass Tm: fiber laser," Post-Deadline Session, *Fiber Lasers VII: Technology, Systems and Applications, Photonics West Conference 7580*, SPIE (2010).
[11] M. Kanskar, L. Bao, Z. Chen, *et al.*, "Improvement in reduced-mode (REM) diodes enable 300 W from 105-µm 0.15-NA fiber-coupled modules," in *Proceedings of SPIE Conference: High-Power Diode Laser Technology XVI*, SPIE (2018).
[12] J.W. Dawson, M.J. Messerly, R.J. Beach, *et al.*, "Analysis of the scalability of diffraction limited fiber lasers and amplifiers to high average power," *Opt. Express* **16**, 13240–66 (2008).

[13] J.P. Koplow, D.A.V. Kliner and L. Goldberg, "Single-mode operation of a coiled multimode fiber amplifier," *Opt. Lett.* **25**(7), 442–444 (2000).
[14] C. Jauregui, T. Eidam, J. Limpert and A. Tünnermann, "Impact of modal interference on the beam quality of high-power fiber amplifiers," *Opt. Express* **19**, 3258–71 (2011).
[15] B. Ward, C. Robin and I. Dajani, "Origin of thermal modal instabilities in large mode area fiber amplifiers," *Opt. Express* **20**, 11407–11422 (2012).
[16] A.V. Smith and J.J. Smith, "Mode instability in high power fiber amplifiers," *Opt. Express* **19**, 10180–92 (2011).
[17] B. Ward, C. Robin and I. Dajani, "Origin of thermal modal instabilities in large mode area fiber amplifiers," *Opt. Express* **20**, 11407–11422 (2012).
[18] K.R. Hansen, T.T. Alkeskjold, J. Broeng and J. Lægsgaard, "Theoretical analysis of mode instability in high-power fiber amplifiers," *Opt. Express* **21**, 1944–1971 (2013).
[19] L. Dong, "Stimulated thermal Rayleigh scattering in optical fibers," *Opt. Express* **21**, 2642 (2013).
[20] S. Naderi, I. Dajani, T. Madden and C. Robin, "Investigations of modal instabilities in fiber amplifiers through detailed numerical simulations," *Opt. Express* **21**, 16111–16129 (2013).
[21] K.R. Hansen, T.T. Alkeskjold, J. Broeng and J. Lægsgaard, "Thermally induced mode coupling in rare-earth doped fiber amplifiers," *Opt. Lett.* **37**, 2382–2384 (2012).
[22] C. Zeringue, I. Dajani, S. Naderi, G.T. Moore and C. Robin, "A theoretical study of transient stimulated Brillouin scattering in iptical fibers seeded with phase-modulated light," *Opt. Lett.* **20**(19), 21196–213 (2012).
[23] B. Anderson, A. Flores and I. Dajani, "Filtered pseudo random modulated fiber amplifier with enhanced coherence and nonlinear suppression," *Opt. Express*, **25**(15), 17671–82 (2017).
[24] J.R. Marciante, R.G. Roides, V.V. Shkunov and D.A. Rockwell, "Near-diffraction-limited operation of step-index large-mode-area fiber lasers via gain filtering," *Opt. Lett.* **35**(11), 1828–30 (2010).
[25] J.P. Leidner and J.R. Marciante, "Three fiber designs for mitigating thermal mode instability in high power fiber amplifiers," in SPIE LASE, *Proceedings Volume 10512, Fiber Lasers XV: Technology and Systems; 105121Z*, SPIE (2018).

Chapter 7

Optical fibers for high-power operation

Cesar Jauregui[1] *and Jens Limpert*[1,2]

Optical fibers have demonstrated, against all early predictions, that their geometry is extremely well-suited for handling high optical powers. However, over the years, many refinements have been done to the basic structure of these waveguides to allow guiding and generating high-powers. As a result of this development the fibers employed in high-power laser systems differ significantly from those typically used in other fields such as telecommunications and/or sensors. Therefore, this chapter is devoted to describing and explaining the main intricacies of those optical fiber designs presently used for high-power operation. In this context, the main goal of this chapter is that the reader gets an overview of the different fiber designs and building blocks of high-power optical fibers.

The chapter is divided into three main parts: a brief historical overview of the evolution of this technology, a description of the main constituents of fibers for high-power operation (including the core, the pump cladding and materials), and, finally, an outlook in which novel trends in the design of these waveguides are briefly described.

7.1 A brief historical overview

Optical fibers were originally developed as passive transmission lines for communications [1] with the intention of developing long-haul, high-capacity telecommunication systems. This development was based on the promise of fiber optical cables being able to offer significantly lower losses than the usual copper wires [2]. In those times very few people were thinking of alternative uses for optical fibers [3]. However, the excellent beam quality ensured by the waveguide nature of fibers, together with the high optical gain they can provide, attracted the attention of researchers seeking for new laser active media [4,5]. Nevertheless, even though the first fiber lasers were demonstrated in the early 1960s [6], it still took nearly two decades for fiber technology to mature [7] to the point at which the first Nd-doped fiber laser with good beam quality could be demonstrated [8].

[1]Institute of Applied Physics, Friedrich-Schiller University Jena, Jena, Germany
[2]Fraunhofer IOF, Jena, Germany

These early demonstrations sparked the interest of the scientific community which saw, at first, the possibility of developing all-fiber transmission systems. Thus, the work of many scientists around the world led to the unprecedented rise of the average power of fiber laser systems [9–11] over the last 30 years shown in Figure 7.1.

The nearly exponential increase of the output average power of fiber laser systems shown in Figure 7.1 both in continuous wave (CW) (blue diamonds) and ultrashort pulse operation (red diamonds) remains unequaled by any other laser technology. Thus, in a time span of roughly 20 years, fiber lasers went from laboratory curiosities to being the laser concept emitting the highest output average power with diffraction-limited beam quality (in CW operation at least).

As indicated in Figure 7.1, the power evolution of CW fiber laser systems has been determined by the introduction of the double-clad concept around 1988 [38] based on an idea first proposed in 1974 [39]. The concept of double-clad, which will be explained in detail later in this chapter, allowed using low-brightness, high-power diodes to pump fiber lasers, thus freeing fiber technology from its main limitation for power scaling: the absence of high-power, high-brightness pump sources. In fact, it can be argued that it was the introduction of a double-clad which gave fibers their *raison d'être* as the highly efficient brightness converters they are today.

Thus, after the incorporation of a double-cladding in fiber designs the output power of CW systems (blue diamonds in Figure 7.1) grew at an almost exponential rate for roughly two decades profiting from the continuous advances of pump diodes. This exponential increase of the output power could only be sustained over such a long period of time due to the continuous refinements done to the active fibers, although none of them was as disruptive as the incorporation of the double-cladding.

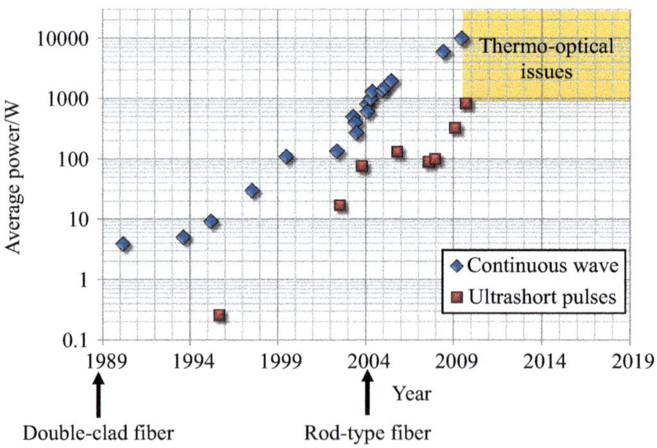

Figure 7.1 Evolution of the maximum average power emitted by fiber laser systems both in continuous-wave (blue diamonds) [12–29] and in ultrashort pulse (red squares) [30–37] regime over the last 30 years

The power scaling of fiber laser systems emitting ultrashort pulses (red squares in Figure 7.1) took off a few years after that of their CW counterparts. At that time these systems could benefit from the double-clad technology, so access to high-power pumps was not restraining them. The main limitation that fibers emitting ultrashort pulses had to face at the time (and still today) is the onset of nonlinear effects [40], which can significantly degrade the performance of these systems. This is because these systems, in order to be useful, strive at scaling the output average power through an increase of the emitted pulse energy, which results in extremely high peak powers in the fibers. There are two very successful strategies to mitigate the onset of nonlinear effects: the first one is the use of so-called chirped-pulse amplification (CPA) first proposed by Strickland and Moreau in 1985 and which was awarded with the Physics Nobel Prize in 2018 [41]; the second one is the scaling of the core area of the fiber, as it will be discussed later on. The incorporation of CPA schemes in ultrashort pulse fiber laser systems allowed for an exponential growth of the output average power. Nevertheless, in spite of CPA, larger core sizes were still required at the time (around the year 2002) to further mitigate nonlinear effects. Unfortunately, the maximum size of single-mode fiber cores in step-index fibers was technically limited by fabrication tolerances, which led to a stagnation of the output power emitted by these systems by around 2005. The adoption of photonic crystal fibers (PCF) and rod-type fiber [42] designs allowed for unprecedented core sizes in active fibers, which enabled resuming the power scaling of these systems.

Therefore, it can be stated that the power scaling of fiber laser systems has been mainly enabled by an impressive, sustained evolution of the design of active optical fibers. Moreover, according to theoretical studies the power scaling of these systems is far from over since they predict that average powers of several tens of kW per (Ytterbium-doped) fiber should be achievable [43,44]. However, in order to achieve this some present challenges have to be overcome, as it will be discussed in the following.

On a different account, it is worth noting that the thermal conductivity of the host material of fibers, that is, usually silica glass, is roughly ten times lower than that exhibited by the crystalline materials of which competing high-power solid-state laser technologies (such as thin-disk [45,46] and slab lasers) are made. In this context, it is truly impressive that fibers have achieved the highest output power with diffraction-limited beam quality demonstrated by any laser technology to date. This achievement speaks volumes of the qualities of the fiber geometry to evacuate heat. In fact, the large ratio of surface to active volume in a fiber has become one of its most important assets for the power scaling of this technology. In spite of this, the extremely high average powers already demonstrated in fiber laser systems have led to the onset of thermo-optical effects which, since 2010, have become the main limitation for the further power scaling of this technology, as illustrated in Figure 7.1. Judging from the past, however, we should rest assure that the scientific community will be able to find a solution (maybe in the form of a fiber design) to the thermo-optical issues that are currently affecting fiber laser systems.

7.2 High-power fiber laser systems: performance against the odds

As hinted above, the power scaling of fiber laser systems has been far from straightforward, since many challenges had to be overcome along the way. In fact, the thin and long fiber geometry that is so beneficial for evacuating heat from the fiber, and thus for successfully mitigating the thermal lensing problems that plagued the first solid-state lasers, is also the root of one of the main limitations of this technology: the onset of nonlinear effects. This is because the tight confinement of the light in the small cores, sustained over meter long lengths, makes fibers an ideal platform for the study of nonlinear phenomena since the power threshold of most nonlinear effects is significantly lower than in other optical elements [47]. Even though nonlinear phenomena are extremely interesting and can be rather useful, they are generally damaging or limiting in the context of power scaling, as will be discussed in the following.

An overview of the most relevant nonlinear and thermal effects in the context of high-power fiber laser systems will be given in this section. This is by no means intended to be an exhaustive list of nonlinear phenomena, just a quick overview of the most common effects and how they impact the performance of high-power fiber laser systems.

7.2.1 Stimulated Brillouin scattering

This effect describes the inelastic scattering that an optical wave undergoes when it interacts via electrostriction with acoustic phonons [40]. Hereby the optical field is reflected at an acoustic wave that propagates along the fiber locally changing its refractive index. This results in a back-reflected wave, the so-called Stokes wave, that is Doppler shifted to lower frequencies (by roughly ~11 GHz in Silica fibers). Since the lifetime of the acoustic phonons is of the order of several ns, the intrinsic gain bandwidth of this process for the reflected wave is ~30–100 MHz. This also means that stimulated Brillouin scattering (SBS) mostly affects narrowband signals and its onset threshold can be significantly increased when broadening the spectral bandwidth of the light to some GHz or beyond. SBS is a harming effect since it not only sets an upper limit to the power that can be transported by narrowband signals in a fiber, but also because it generates a backward propagating beam that can result in system instabilities or, even, in the destruction of optical elements at the front of the amplification chain.

7.2.2 Stimulated Raman scattering

Raman scattering is, in many ways, similar to Brillouin scattering only that the inelastic scattering is created by the interaction of the light with optical phonons (i.e., lattice vibrations) [40]. As in the case of SBS, this interaction results, mainly, in a frequency downshifted beam only that in this case the frequency shift of ~13 THz is significantly larger than that for SBS. The Raman scattered beam, also called Stokes wave, can copropagate with the signal along the fiber, which can result in a very strong amplification of this frequency downshifted radiation. Since the amplification of the Stokes wave happens at the expense of the signal energy, stimulated Raman scattering (SRS) limits the maximum (peak) power that can be transmitted through a fiber while preserving its spectral purity.

7.2.3 Self-phase modulation

This effect is a manifestation of the more general Kerr-effect, which is caused by the nonlinear dependence of the refractive index of materials on the intensity of the light [40]. Self-phase modulation (SPM) is particularly relevant in pulsed systems where the temporally changing envelope of the optical pulses results in a strong temporal modulation of the optical phase via nearly instantaneous changes of the refractive index of the fiber. Such a temporal change of the optical phase can lead to spectral broadening. Moreover, in CPA systems such optical phase changes are difficult to compensate for. This, upon compression of the stretched pulses, may result in the degradation of the temporal pulse quality emitted by such fiber laser systems.

7.2.4 Self-focusing

This effect is the spatial counterpart of SPM. Hereby the transverse intensity profile of the optical beam in the fiber (usually a Gauss function) creates, via the Kerr-effect, a focusing lens in (solid core) fibers. The focusing strength of this lens grows with the (peak) intensity of the optical pulse and it can become strong enough to cause a reduction of the beam size in the optical fiber. This may lead to a positive reinforcement because a smaller beam has a higher (peak) intensity, which will generate an even stronger lens. This situation can start a runaway process which may end up with the physical destruction of the optical medium. Remarkably, the threshold at which this process starts does not depend on the size of the beam, but only on its (peak) power, which for silica fibers is ~4 MW when using linearly polarized light (and ~6 MW when using circularly polarized light). This threshold is known as the critical power. Due to self-focusing leading to a catastrophic damage of the optical fiber, the critical power is the ultimate limit for the power that can be transported in a solid-core optical fiber.

All the limiting effects described above result from the nonlinear interaction of the light with the optical material fibers are made of. The onset threshold of nonlinear effects, with the only exception of self-focusing, depends on the intensity of the light in the fiber core. Therefore, they can be efficiently mitigated by using the CPA technique [41] or by scaling the size of the fiber core [9]. This is because the CPA technique leads to a reduction of the temporal peak power of the pulses, whereas a larger core directly reduces the intensity of the light beam by spreading the power over a larger area. Besides, reducing the fiber length also leads to higher onset-power thresholds in most nonlinear effects. Over two decades, the quest for higher powers in fiber laser systems was akin to a quest to mitigate nonlinear effects, which naturally led to the development of larger and shorter fibers. This evolution enabled a spectacular rise in the output power of fiber laser systems until 2010, as shown in Figure 7.1.

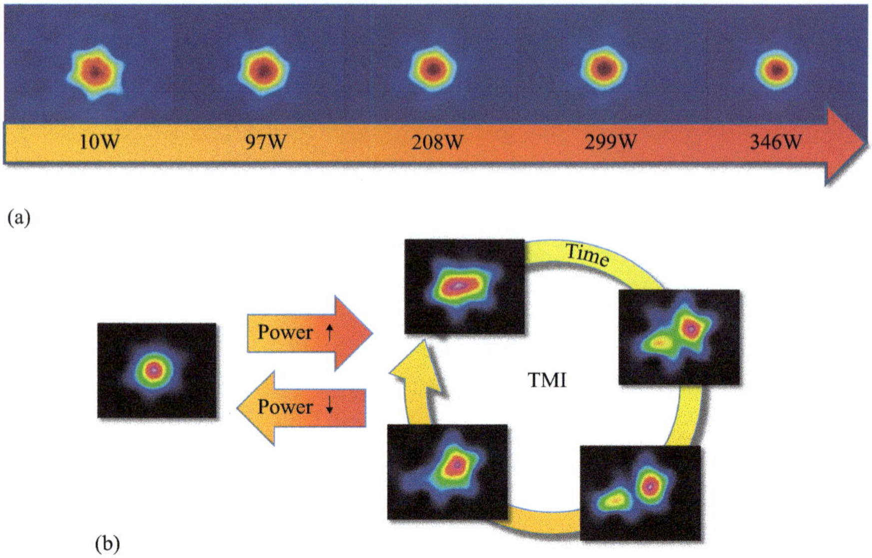

Figure 7.2 Thermal effects that can be observed in high-power fiber laser systems: (a) mode shrinking [48] and (b) transverse mode instabilities (TMI) [49]

The problem with this approach is that combining higher output average powers with shorter and shorter fibers leads to a strong increase of the heat-load of fiber laser systems. The heat-load in a fiber laser system has three main sources: the quantum defect (i.e., the energy difference between the pump and the signal photons), the quantum-efficiency (i.e., the average number of active ions that can be excited per pump photon) and absorption in the host material. Therefore, since the heat-load has been increasing over the years, it was unavoidable that, at some point, thermal effects would appear in high-power fiber laser systems, in spite of the thermal resilience of this concept. What was unexpected was the manifestation of these thermal effects which, since their first observations in 2010, continue to be the most challenging limitation for the average-power scaling faced by fiber laser systems. The two main manifestations of thermal effects are schematically shown in Figure 7.2 and they will be briefly described in the following.

7.2.5 Mode shrinking

This effect, as illustrated in Figure 7.2(a), sees the diameter of the beam in an optical fiber becoming smaller as the output average power increases [48]. The reason for this behavior is that, due to the heat-load originated during laser operation, a temperature gradient is created in the fiber, which mimics the intensity profile of the beam (i.e., the central part of the fiber core will exhibit the highest temperature). This temperature gradient is translated, through the thermo-optic effect, into a refractive index gradient which acts as a focusing lens that, in turn, results in a reduction of the beam size. This leads to a rise in the intensity of the light, which has the unwanted side effect of increasing the impact of nonlinear effects in the fiber. On the positive side, though, this modal shrinking is usually accompanied by a cleaning effect (i.e., by a smoothing of the edges of the beam) which may even improve the beam quality of the emitted light.

7.2.6 Transverse mode instabilities

Transverse mode instabilities (TMI) is a significantly more damaging and complex problem than the previous one. In this case what happens is that the beam emitted by the high-power fiber laser system breaks up and starts to fluctuate in time once that a certain average power threshold has been reached, as schematically illustrated in Figure 7.2(b). These beam fluctuations are the result of an unstable energy transfer that takes place between transverse modes in an optical fiber. Such an energy transfer between orthogonal modes only becomes possible because they create an interference pattern which, by getting imprinted in the inversion profile of the fiber, gives rise to a thermally induced index grating through the thermo-optic effect [9,50,51]. This grating, having been ultimately written by the modal interference pattern, has the right characteristics, in terms of period and symmetry, to couple energy between the interfering modes if the system is perturbed (e.g., by noise). The TMI effect was first reported in 2010 [49] and it quickly became the single most limiting effect for the average-power scaling of fiber laser systems. TMI is, to date, the only known representative of thermally induced nonlinear effects in optical fibers.

As can be seen from the descriptions provided above, the thermal effects are significantly different in nature that the nonlinear ones. This is because the onset of these effects is determined by a certain extracted average power which imposes a thermal gradient in the cross-section of the fiber, thus altering its guiding properties. Since this thermal gradient is not affected by the external cooling of the fiber, the mitigation of thermal effects is rather complex and challenging. This fact should not, however, give the impression that cooling high-power fiber lasers is not important. In fact, external cooling is mandatory to prevent thermal failures, for example, fiber fuse [52] or degradation of the protective coating materials of the fiber (which usually start degrading when they reach a temperature of ~70°C – 80°C) [53], and to increase the operational lifetime of the system.

Challenging as the thermal issues described above are, with all the insight gained on these effects over the last decade, it seems plausible to develop effective mitigation strategies, some of which have already been proposed and demonstrated [54–58].

7.3 Optical fibers for high-power operation

As discussed above, optical fibers face tremendous challenges to reach high-power operation. This, as already hinted before, conditions the fiber design to the point that many of the fibers used for high-power operation are significantly different in their morphology and internal structure from the well-known telecommunication fibers.

The internal structure of high-power fibers comprises, in general, the components schematically illustrated in the left-hand side of Figure 7.3.

- *Core*
  The core is where the signal light is guided. In active fibers this region is either completely or partially doped with laser-active particles. For high power operation these are usually Yb^{3+} (emitting at around 1 μm wavelength), Er^{3+} (emitting at around 1.5 μm wavelength), Tm^{3+} (emitting at around 2 μm wavelength), or Ho^{3+} (emitting at around 2.1 μm wavelength). Since the signal light is guided and amplified and the pump light is absorbed in this region, the core is an extremely important part of the fiber. In fact, the ability of the core to guide just one or, on the contrary, many different modes usually determines the quality of the beam emitted by the high-power fiber laser system. In fact, one of the main challenges in the design of fibers for high-power operation is to enlarge the core size while preserving the quality of the high-power emitted beam.
- *(Pump) Cladding*
  The cladding region is, in fact, responsible for the guidance of the light in the core. This cladding region either has a lower refractive index than the core (in step-index fibers) or it is structured in such a way that the light can be contained in the fiber core (in photonic-crystal and bandgap fibers). Attending to the optical function of this region it is possible to differentiate between *single-clad* and *double-clad fibers*. In *single-clad fibers*, akin to conventional telecommunication fibers, the cladding is only there to ensure guidance of the light in the core. This means that, when these fibers are active, the pump light has to be coupled into the core, which requires extremely high-brightness pump sources. In *double-clad fibers*, on the other hand, the cladding region not

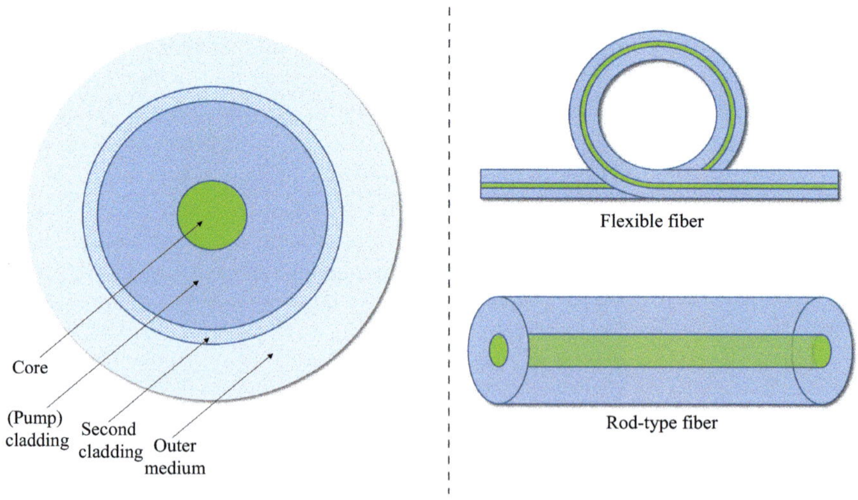

Figure 7.3 Morphology of fibers for high power operation attending to their inner structure (left) or to their geometry (right)

only ensures the guidance of light in the core but it doubles up as a waveguide itself which supports the propagation of low-brightness light. This allows coupling the light of high-power, low-brightness pump diodes into this part of the fiber, which significantly eases pumping fiber laser systems. In double-clad fibers this region is also known as pump-cladding or, even, pump-core. As already mentioned before, the presence of the pump-cladding makes possible that active fibers transform the low-brightness pump light is transformed into high-brightness signal light. In other words, *double-clad active fibers* are, in essence, highly efficient brightness converters.

- *Second cladding*
  In single-clad fibers this second cladding is usually a high-refractive index polymer used to provide protection and mechanical integrity. In double-clad fibers, on the other hand, this second cladding is responsible for the guidance of the light in the primary cladding. Therefore, in those cases, this region has a lower refractive index than the inner cladding. There are three main approaches to achieve this: the use of low-index polymers, the use of a ring of tightly packed air-holes (structure known as air-clad), or the use of a low-index glass [59].
- *Outer medium*
  The outer medium can still be part of the fiber or not. For example, in fibers in which the second cladding is a polymer, this outer medium is usually either air or water (for cooling purposes). On the other hand, in fibers which second cladding is an air-clad or a low-index glass ring this outer medium is usually pure silica glass. In any case the outer medium is always optically inert and it only serves the purpose of evacuating the heat and/or providing mechanical stability to the fiber.

Fibers for high-power operation can also be classified, attending to their geometry, into flexible and rod-type fibers, as seen in the right-hand side of Figure 7.3. Flexible fibers is the kind of geometry typically associated with optical fibers, that is, that of a long (several meters), extremely thin (hundreds of μm in diameter) and flexible cylinder. On the other hand, rod-type fibers are thick (millimeter diameters), short (usually around 1 m long) unbendable cylinders. Rod-type fibers were introduced in 2005 [42] to allow for an unprecedented scaling in the core size of fibers. The idea came from the recognition that the size of large beams collapses when a fiber is bent [60], which enhances the impact of nonlinear effects, as previously discussed. Therefore, the main idea to keep a large beam size during propagation through the fiber was to give up on one of the most popular characteristics of optical fibers, namely its flexible nature, and keep it straight. In order to achieve this, the fiber was sheathed with a thick glass outer medium, which makes the fiber stiff and prevents it from bending, giving them the appearance of a glass rod. Due to their relatively unpractical geometry, these fibers have to be kept short, usually in a length range between a few tens of cm and 2 m. Additionally, these fibers, due to their large sizes, are difficult to splice and are, therefore, employed in free-space setups, which further lessens their appeal for some users. In spite of their disadvantages, these fibers have become an extremely invaluable workhorse for high-power, high-energy, ultrafast fiber laser systems (i.e., those emitting ultrashort pulses). The reason is that rod-type fibers have allowed unleashing the full potential of the core-area scaling of optical fibers. Thus, these fibers have demonstrated record-breaking beam diameters of 100 μm in an active fiber [61] and more than 200 μm in a passive one [62]. This ability to scale the core size, together with their short lengths, make rod type fibers extremely effective in mitigating nonlinear effects. This has allowed reaching unprecedented performance levels with these fibers [9].

The introduction of rod-type fibers has sparked an interesting debate in the scientific community as to what makes a fiber. Even though there are some voices claiming that a fiber has to be flexible to be worth its name, actually the propagating light would see no difference if that same flexible fiber would be placed in a rigid package that keeps it straight. From this point of view it seems somewhat arbitrary and simplistic to reduce the main characteristic of optical fibers to their mechanical flexibility. A more physically relevant definition should attend to the wave-guiding characteristics of a fiber. Indeed, it can be argued that a fiber should only be called like that if it is able to guide the light (with low attenuation) from its input to its output end. This seems a rather academic discussion which, at first glance, does not really have more practical consequences than being able to brand a particular optical waveguide as a fiber. However, upon closer examination there is a genuinely interesting and physically relevant topic about the limits of wave-guiding that needs to be addressed. The reason is that, as the beams propagating in fibers grow in size, so does their Rayleigh length (i.e., half of the distance over which a beam is considered collimated in free-space). This means that, in principle, if the size of the (single-mode) core becomes large enough, it is possible to reach a situation in which the beam would remain collimated through the fiber even in the absence of a core. At this point no stable guidance can be formed in the fiber and, therefore, a great deal of the control offered by fibers on the propagation of the electromagnetic radiation will be lost. Thus, in order to ensure a stable guidance of the light, in our experience the minimum length of a fiber (or any other optical waveguide for that matter) should be at least four times the Rayleigh length of its fundamental mode (FM) beam.

In the reminder of this section the different components of a fiber will be described in detail. These include the fiber core, the cladding, and the materials fibers are made of.

7.3.1 The fiber core: guiding mechanisms

As already mentioned above, the fiber core is the region where the signal light is confined and where optical amplification takes place (in active fibers). However, there are several approaches to achieve the guidance of light that will be reviewed in this subsection.

7.3.1.1 Step-index fibers

Step-index is the classic form to obtain wave-guidance in fibers. It is schematically illustrated in the left-hand side of Figure 7.3. Hereby the core has an average refractive index that is higher than that of the surrounding cladding. Therefore, the light is guided by total internal reflection, which keeps the light confined in the core. When solving the wave equation in these fibers [63] it can be seen that, depending on the index difference between the core and the cladding, the light can propagate along the fiber by adopting different intensity profiles. These are referred to as transverse modes. A particular characteristic of fibers, which is not shared by waveguides in other electromagnetic bands, is that they always guide at least one mode which has a nearly Gaussian intensity profile. This is called the fundamental mode (FM) of the fiber, and it is usually the one that most fiber laser systems strive to excite, amplify, and deliver. The reason is that, of all the beam shapes, a Gaussian beam of a given size can be focused to the smallest spot, which is beneficial for most applications since it increases the effectiveness of light-matter interactions. The number of transverse modes that can be guided in a step-index fiber is governed by the so-called V-parameter, which is given by [64]:

$$V = \frac{2\pi}{\lambda} a \sqrt{n_1^2 - n_2^2}$$

where λ is the wavelength of the light in vacuum, a is the radius of the core, n_1 is the refractive index of the core and n_2 is the refractive index of the cladding. Moreover, the square-root term in the equation of the V-parameter is known as the numerical aperture of the fiber (NA).

If the V-parameter is smaller than 2.405, then the optical fiber will only support the propagation of the FM through its core [63]. Such fibers are referred to as single-mode. Since, as already mentioned, most applications demand that laser systems emit Gaussian beams, single-mode fibers are, by far, the most beloved type of fiber. They are also one of the reasons why fibers gained popularity as a power-scalable concept able to deliver excellent beam quality (i.e., Gaussian beams) regardless of the output power.

Whereas achieving single-mode operation with small core sizes is not a problem, keeping it while increasing the core diameter is technically challenging. The reason is that, with common fabrication methods, it is difficult to obtain an NA below 0.05 in a controllable way, which, for example, limits the maximum core diameter of single-mode step index fibers operating at ~1 μm wavelength to ~16 μm. Thus, fibers with cores larger than this, the so-called large-mode-area (LMA) fibers, will intrinsically support the propagation of more than one mode in their cores. When the number of modes supported by the fiber is low (below 10), these are referred as few-mode fibers.

At this point it is important to say that having fibers supporting more than one mode does not automatically mean that the quality of the emitted beam will be low, since there are several ways to discriminate the higher-order modes (HOMs), that is, all transverse modes that are not the fundamental one. Some of them are as easy and widely used as coiling the fiber with relatively small bending radii since the HOMs have higher bending losses than the FM [65–67]. This allows having intrinsically few mode fibers with core diameters of up to 20 μm or 25 μm operate regularly in a single-mode regime. However, the ability to discriminate the HOM from the FM becomes more and more difficult as the core increases in size.

For the reasons mentioned above it is very rare to find step-index fibers emitting a single-mode beam with core diameters larger than 30 μm. One way of achieving this, however, is by using additional elements that can provide higher losses for the HOMs. One example of this approach is the *chirally coupled-core* (CCC) fiber [68]. In this design one or more smaller satellite cores are winded around the main signal core. These satellite cores are designed and placed in such a way that they provide a resonant coupling for some of the HOMs of the main signal core. Thus, as the HOMs propagate along the main signal core, they are progressively coupled out of it and into the satellite cores, where, due to the tight winding, they experience high bending losses. Using this design, single-mode step-index fibers with cores around 55 μm have been demonstrated [69].

Another approach to obtain larger beam diameters with a step-index fiber is to fully embrace its multimode nature and coupled the energy into a HOM. The reason to do this is because both the diameter of the HOMs and their resilience against perturbations (e.g., bend-induced distortions, intermodal coupling, etc.) grow with their order. This is exactly the strategy followed when using the so-called higher-order-mode-fibers [70]. These fibers have two concentric step-index cores. The inner, smaller one only supports the propagation of one mode, whereas the outer, bigger one is multimode. The light here is injected into the smaller core and then, via a long-period grating (LPG) [71], it is coupled into one HOM of the outer core, where it propagates and, in the case of active fibers, it gets amplified. Later on, the HOM light is converted into a fundamental Gaussian beam either by means of an LPG or with axicons [72].

7.3.1.2 Photonic-crystal fibers

Photonic crystal fibers (PCF) are structurally and functionally different from step-index fibers [73,74]. Their main characteristic is that the cladding region contains an arrangement of inclusions (usually spatially periodic) with a different refractive index, as can be seen in Figure 7.4. In PCFs, the core is created in a region of the fiber that is free of inclusions. Apart from this, PCFs for high-power fiber laser

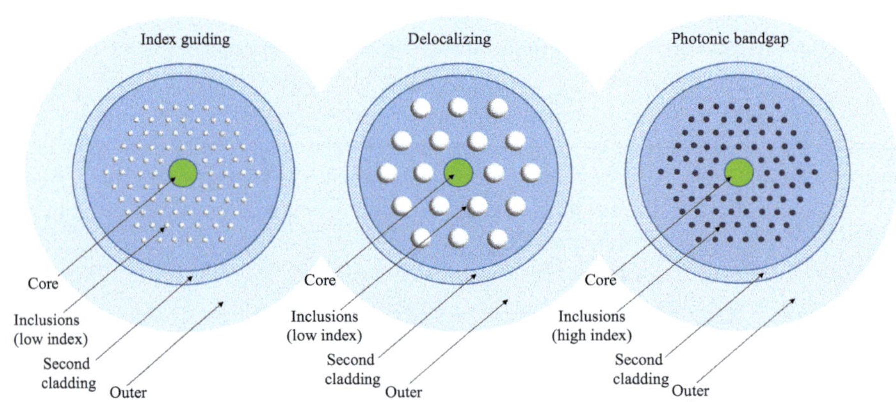

Figure 7.4 Morphology of photonic crystal fibers for high power operation. Attending to their inner structure they can be classified in: index-guiding (left), delocalizing (center), and photonic bandgap (right)

systems may also have a second cladding and an outer medium. The arrangement of the inclusions offers a high flexibility in the design of these fibers, which allows achieving performance features that are no attainable with step-index fibers. Thus, depending on the characteristics of this arrangement of inclusions, it is possible to distinguish three main operating principles of PCF: index-guidance, delocalization, and photonic-bandgap.

7.3.1.2.1 Index guiding PCF

These PCFs are characterized by an arrangement of a large number of small inclusions (sizes comparable to the wavelength of the signal light), which refractive index is lower than that of the host material of the fiber (usually Silica), as seen in the left-hand side of Figure 7.4. This arrangement of small inclusions reduces the effective refractive index of the cladding region surrounding the core, which implies that the light will be guided by a modified total internal reflection condition [73,74]. In other words, this kind of fibers operates in a similar way that step-index fibers, with the great advantage that the index step can be very accurately controlled by the geometric arrangement of the inclusions. This has allowed significantly increasing the core size of fibers beyond the limitations imposed by step-index technology. In fact, due to this reason, rod-type fibers based in this approach became the workhorse of ultrafast fiber laser systems between 2005 and 2010. These fibers, which were even commercially available, had core diameters spanning 40 to ~100 μm [42,75,76]. These unprecedentedly large fibers allowed reaching, for the first time with fiber laser systems, MW peak powers in the fiber [76], GW peak powers after compression in a fiber CPA system [75], and millijoule pulse energies [76–78]. In spite of these impressive results, index-guiding PCFs with cores larger than 50 μm were cumbersome to work with since they did support the propagation of at least two modes in the core, which implies that the coupling of the seed was of critical importance to ensure a high-quality beam at the output.

7.3.1.2.2 Delocalizing PCFs

These fibers are characterized by a structure of low index inclusions that acts as a low-pass spatial filter for the core region of the fiber. This way only the mode with the lowest spatial frequency (i.e., that with the smoothest features, which is the FM), can propagate in the core. The HOM, on the other hand, are deformed in such a way that a big portion of their intensity resides outside of the core, as schematically illustrated in Figure 7.5. This effect is called HOM-delocalization and it has been proven to be a very robust mechanism to obtain very large core area fibers [60]. Even though delocalizing fibers are, strictly speaking, multimode structures they still manage to emit their energy in a single mode at the end of the fiber. This ability allows speaking of an effective single-mode operation. The reason for this is that a HOM-delocalization such as that shown in Figure 7.5 has two benefits: on the one hand, most of the energy will be coupled into the FM if a Gaussian beam (of adequate dimensions and centered in the fiber) is launched into the fiber; on the other hand, in active fibers, the overlap of the FM with the amplifying region of the core (76% in the example shown in Figure 7.5) is significantly larger than that of the HOMs (26% in

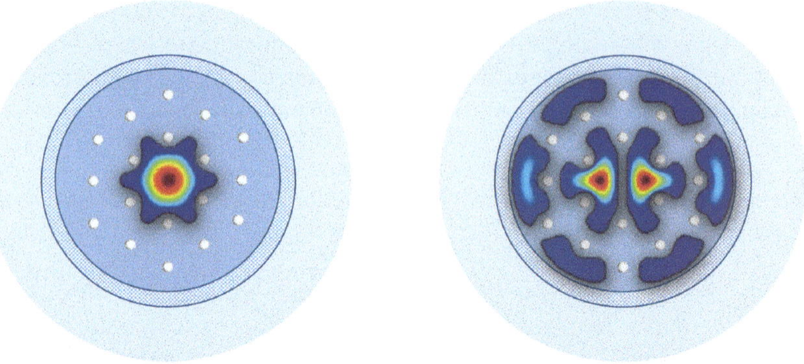

Figure 7.5 Principle of the delocalization of HOMs. In these fibers the FM has a high overlap with the core region (left), whereas a large portion of the intensity of the HOMs falls outside this region (right)

our example). This leads to a preferential amplification of the FM against the HOM. The combination of these two effects implies that, at the end of the fiber, most of the energy will be contained in the FM [79]. Strictly speaking, the first fiber design that exploited a precursory effect to HOM delocalization was the leaky-channel fiber, which working principle is based on introducing losses for the HOM by letting them leak out of the fiber core at a higher rate than the FM [80,81]. Unfortunately the operating principle of this structure, the so-called modal sieve concept [73], does not work for double-clad fibers, since the introduction of a closed optical barrier around the cladding makes all modes virtually lossless, especially in rod-type fibers (because they cannot discriminate the HOM using bending losses). An evolution of the leakage-channel fiber concept for high-power, double-clad fibers is the so-called Large-Pitch-Fiber (LPF) [60,82]. This design was also the first representative of a fiber fully basing its operation in HOM delocalization. The structure of a LPF, which is schematically shown in Figure 7.4 (center) and Figure 7.5, has a relatively low number of low-index inclusions with a diameter that is several times larger than the wavelength of the signal light. These inclusions have a center-to-center separation (i.e., Pitch) that is several tens of times the wavelength of the propagating light in the fiber. Such an open structure allows for a strong delocalization of the HOMs, as exemplarily shown in Figure 7.5 and, with it, for a robust effective single-mode operation. LPFs have been extremely successful at scaling the size of the core while still keeping effective single-mode operation. Thus, active and passive LPFs with core diameters up to 130 μm [61] and 230 μm, respectively, have already been demonstrated. The unprecedented ability of LPFs to scale the size of the core has made them the current workhorse of ultrafast, high-power fiber laser systems. Such systems have achieved record-breaking performance levels in recent years, thanks to this design. Thus, systems using LPFs have emitted femtosecond pulses with several millijoules of pulse energy and almost 4 GW of peak power [83], but they have also reached 26 mJ of pulse energy in Q-switched operation [84] and up to 2 kW of average power [85]. Over the years several refinements to the LPF design have been proposed in order to improve their delocalization of HOMs, for example, by breaking the symmetry of the arrangement of low-index inclusions [79,86,87]. Another design that has been recently proposed and that exploits HOM delocalization is the multi-trench fiber [88]. The core of these fibers is surrounded by several concentric, thin rings of low-index materials. If the spacing and thickness of the rings is carefully chosen, it is possible to achieve HOM delocalization. However, there has been to date no high-power fiber laser demonstration using this design.

7.3.1.2.3 Photonic Bandgap fibers

This kind of fibers is characterized by the presence of inclusions which refractive index is higher than that of the surrounding material (usually Silica). In these fibers the core is formed by a region void of inclusions, as schematically illustrated in Figure 7.4. If the arrangement of the inclusions (in terms of spacing, size, and periodicity) is properly chosen, it is possible to generate optical bandgaps, that is, spectral regions where no modes are allowed in the cladding region of the fiber. In other words, in the wavelength range where these bandgaps exist the light is confined to the core region. However, it is possible to design these fibers so that the FM is better confined than the HOMs, so that, after a certain propagation distance effective single-mode operation can be achieved [89]. These are, intrinsically speaking, resonant designs which only work in certain wavelength bands. Since the edges of these wavelength bands can be very steep, these fibers can be exploited to mitigate the onset of Raman scattering [90]. Up to now, an Yb-doped photonic-bandgap fiber with core diameter as large as 50 μm and emitting 75 W has been demonstrated [91]. In the case of passive bandgap fibers the largest core diameter demonstrated to date, while still operating in the effective single-mode regime, is ~85 μm [92].

An interesting design approach is to combine the bandgap and index guiding PCF designs into a hybrid concept. This is done by including an arrangement of both low- and high-index inclusions around the core area (which is again free of inclusions) [93–95]. Hereby the main guidance is provided by the low index inclusions, whereas the high-index ones filter out the HOMs from the core, thus ensuring effective single-mode operation in certain wavelength bands. This approach has been successfully implemented in a design known as "Distributed Mode-Filtering" (DMF) rod-type fiber. DMF fibers with core diameters of up to 85 μm have been able to deliver ~300 W of average power [94]. The main drawback of these fibers for high-power operation is that the wavelength range in which they work as intended is temperature-dependent, which means that, if not accounted for during the design phase, these fibers might cease to work at high powers [94].

7.3.1.3 General considerations

It is important to point out at this point that the temperature-gradient generated by the laser process in the fiber core can significantly alter its guiding properties. Thus,

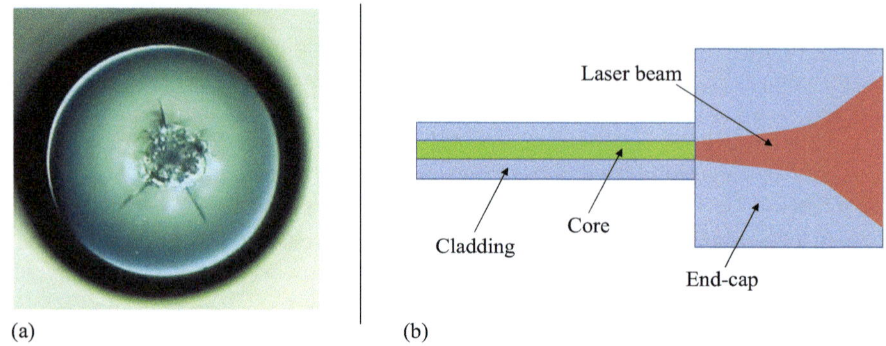

Figure 7.6 *(a) High fluence of the light in the core can lead to the destruction of the fiber end-facets (Photo: courtesy of IAP Jena). (b) This can be avoided with the use of end-caps, which let the beam expand before reaching the end-facet*

it has been shown that a single-mode step-index fiber can become multimode at high-powers due to the thermally-induced refractive index change [96]. Therefore, it is of the outmost importance to include the thermally induced changes of the refractive index profile of the fiber in the design process from the very beginning [79], especially for fibers with cores larger than 50 μm. In fact, these temperature-induced refractive index changes can be so dramatic that they may be exploited to obtain a purely thermally induced guidance in an optical fiber [97–99].

Another point to take into account is that in high-power fiber laser systems, especially in the pulsed regime, the fluence of the light in the core can be so high that it may destroy the end-facet of the fiber upon emission, as illustrated in Figure 7.6(a). The reason for this is the usually 2–5 times lower optical-damage threshold of the surface of a material with respect to the bulk [100]. A way to prevent this from happening is the use of so-called end-caps. As schematically illustrated in Figure 7.6(b), these are blocks of glass that are spliced at the ends of the fiber and serve the only purpose of letting the emitted laser beam expand in the bulk of the glass, which reduces its fluence before it reaches the end-facet. Thus, using this strategy a high-power laser beam can be emitted with a fluence that is below the damage threshold of the end-facets. End-caps are, therefore, widely used in high-power fiber laser systems and have become one of their more visually distinctive features.

7.3.2 The fiber cladding: added functionality

The cladding has two main functions in a fiber for high-power operation: it contains the structure required to ensure light guidance in the core region and, in most cases, it also functions as a waveguide for the pump light. As seen in Figure 7.7, attending to their cladding, fibers can be classified into: single-clad, double-clad, and triple-clad. The differences between these types of fibers will be briefly summarized in the following.

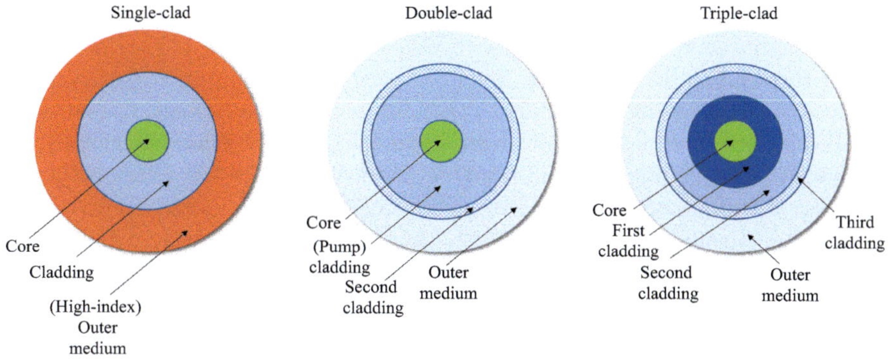

Figure 7.7 Different fibers attending to their cladding. From left to right: single-clad, double-clad, and triple-clad

7.3.2.1 Single-clad fibers

The morphology of these fibers can be schematically seen in the left hand-side of Figure 7.7. They comprise a core region, already described in Section 7.3.1, and a cladding region that is surrounded by an outer medium with a higher refractive index than that of the cladding. This outer medium is usually an acrylate coating mainly used to mechanically protect the fiber and chemically isolate it from the environment [59]. It has been shown that these coatings significantly improve the long term lifetime and reliability of flexible fibers [101]. Single-clad structures are usually the design of choice for passive fibers, which are mostly employed as transport channels between different subsystems or to deliver the light to the application. However, some active fibers also have adopted the single-clad configuration. In this case, since the cladding is not able to guide light (due to the higher refractive index of the outer medium) the pump light is coupled directly in the fiber core. This significantly restricts the choice of pump sources and usually means that single-clad fibers are mainly used in the low-power (<1 W) front-end of the system (i.e., in the laser oscillator or in the preamplifiers).

7.3.2.2 Double-clad fibers

These fibers have already been presented in Figure 7.3, since they are the usual configuration found in high-power fiber laser systems. The main difference between this and the single-clad configuration is that the medium surrounding the cladding has a refractive index that is lower than that of the fiber material. This implies that the cladding itself will be able to guide light, which is normally employed to pump the fiber. The reason why this is advantageous is that the cladding has a larger area than the core and, usually, also a higher numerical aperture[1], which makes it possible to

[1]In this context, for a fixed cladding area, the NA gives an idea of the maximum angle that a light ray can have and still be guided by the fiber. Higher NA values mean that rays with larger angles will be accepted (i.e., with lower brightness).

couple the light of low-brightness pump diodes in it. This fact widens the selection of available pump sources to now include very high-power diodes. The low index material surrounding the (pump) cladding can either be a polymer, a glass (usually fluorine doped), or a ring of tightly packed air inclusions (the so-called air-clad [102,103]). The first of these options is mostly employed for flexible fibers, whereas the other two can also be found in rod-type fibers. Nowadays low-index polymers allow for a NA of ~0.46, whereas a ring of fluorine-doped glass leads to a maximum NA ~0.22 and, finally, air-clads allow for NAs as high as 0.6. This has made air-clads the technology of-choice for many high-power fibers intended for ultrafast pulse operation since the high NA allows for a reduction of the cladding area, which leads to higher pump absorption levels and, therefore, to shorter device lengths. Furthermore, it has been demonstrated that air-clads have a negligible impact in the thermal conductivity of fibers [104]. Nevertheless, air-clads make all-fiber systems difficult, which is the main reason why all-glass solutions are usually preferred by some users, in spite of the lower NA they provide.

It has been shown that the circular geometry of the pump cladding schematically illustrated in Figure 7.7 leads to a low absorption of the pump power, since there are many skew rays that never cross the core region. Even though it has been shown that this situation can be improved by coiling and twisting the fiber [105], nowadays the preferred solution is to depart from such circular geometry [11,106,107], as schematically shown in Figure 7.8 (left and center). Thus, in double-clad fibers for high power operation it is common to find pump claddings with an octagonal shape (Figure 7.8 (left)) or with that of a broken circle (also known as D-shape, Figure 7.8 (center)). Obviously these are not the only possible shapes and many other geometries have been proposed over the years [11].

On a different account, it is worth mentioning that for many applications it is important that the fiber delivers a polarized beam (usually linearly polarized). In order to achieve this, it is possible to introduce linear birefringence in the fiber so that it shows two preferred polarization axes (the fast and the slow) [108]. Such fibers are known as polarization maintaining (PM). If the linear polarization of a

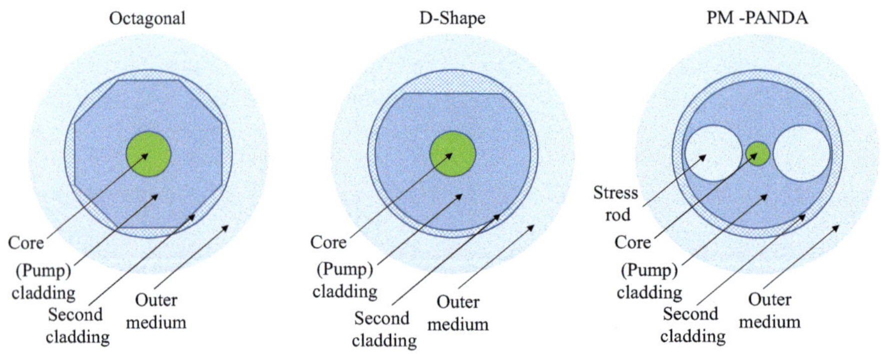

Figure 7.8 Different geometric configurations of the pump cladding

beam is aligned with one of these main axes in the PM fiber, then the state of polarization of the light will be maintained as it propagates along the fiber. There are several ways of inducing linear birefringence in a fiber but the two most widespread methods are: making the fiber core elliptical [109] or introducing stress-inducing elements in the fiber cladding [110]. This second method, which is schematically illustrated on the right-hand side of Figure 7.8, is the one most widely used in high-power fibers since it allows preserving the circular geometry of the core. The stress-inducing elements are rods with a glass composition that shows a different thermal expansion coefficient than that of the material the fiber is made of. Thus, after the fabrication process (in which the fiber is heated up to around 2,000 °C) some stress is frozen in the glass matrix, which leads to linear birefringence. Crucially, since these stress-inducing elements are placed in the pump cladding, they must have a lower refractive index than their surroundings. Otherwise they would act as a waveguide and trap part of the pump radiation, thus reducing the overall pump absorption of the fiber. These two constraints (a different expansion coefficient and a lower refractive index) significantly restrict the choice of glass compositions that can be used. In the praxis, these elements are typically made of Boron-doped silica glass.

Attending to the shape of these stress-inducing rods, the PM fibers receive different names. Thus, if they are circular (as in Figure 7.8) the fibers are called PANDA, and if they are segments of a disk the fibers are called Bow-Tie. Nowadays, not only step-index fibers, but also PCFs [111] and even rod-type fibers [112] can show a PM behavior. Furthermore, depending on the configuration of the stress-inducing elements, PM fibers may only guide one of the polarizations, thus becoming polarizing fibers (PZ) [113]. Even though this is not the only way to obtain PZ behavior, it is the most widespread in the context of high-power fiber laser systems.

Interestingly, double-clad PM fibers do have a round pump cladding. The reason is that it has been found out that the stress rods not only provide linear birefringence, but they also enhance the pump absorption [114].

7.3.2.3 Triple-clad fibers

As shown in the right-hand side of Figure 7.7, some fibers have an inner cladding surrounding the core region, a second cladding surrounding both, and a third cladding encompassing the complete structure. These are the so-called triple-clad fibers. In them the pump is guided in the second cladding, whereas the first cladding is used to lower the V-parameter of the core, that is, to reduce the number of modes supported by the fiber core. The advantage of this approach is that it allows using very high doping concentrations of the laser active ion (which increases the refractive index of the core) while keeping the fiber single or few-mode [87,115,116].

7.3.3 The fiber material: laser properties

As it has already been mentioned several times in this chapter, optical fibers for high-power operation are, in most cases, made of Silica glass. However, in order to

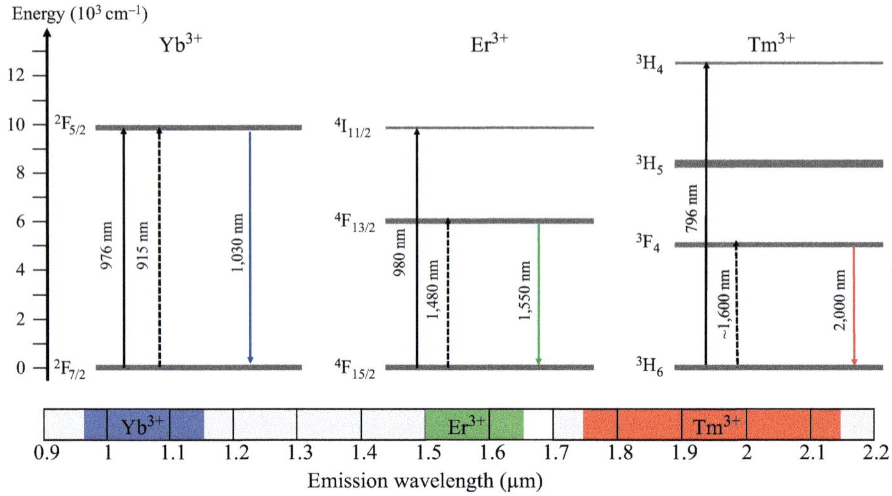

Figure 7.9 Main laser-active ions used in high-power fiber laser systems together with their simplified energy level diagram and their absorption (arrows pointing upwards, from which: solid—usual pump wavelength; dashed—alternative pump wavelength) and emission wavelengths (arrows pointing downwards). From left to right: trivalent Ytterbium (Yb^{3+}), trivalent Erbium (Er^{3+}), and trivalent Thulium (Tm^{3+})

modify its optical properties, for example, to obtain guidance through total internal reflection, the pure Silica glass is doped with other elements such as Aluminum, Phosphorous, Germanium, Boron, or Fluorine [59,117]. Some of these dopants, as is the case with Aluminum and Phosphorous, also increase the solubility of rare-earth dopants in the Silica glass matrix. The reason why this is important is because Silica glass is optically inert and, therefore, to obtain laser operation, it is necessary to incorporate laser-active particles in the glass host, which are usually rare-earth ions.

7.3.3.1 Laser active ions

As illustrated in Figure 7.9, in the context of high-power fiber laser systems the three main rare-earth ions used to obtain laser operation are: trivalent Ytterbium (Yb^{3+}), trivalent Erbium (Er^{3+}), and trivalent Thulium (Tm^{3+}). The properties of each one of these ions as a laser active particle will be briefly reviewed in the following.

7.3.3.1.1 Ytterbium

As can be seen in the left-hand side of Figure 7.9, Yb^{3+} is the ion with the simplest energy level diagram, since it only comprises two manifolds. This simplicity has many practical advantages. First and foremost, the pump and emission wavelengths are very close to one another, which allows for a very efficient operation with the generation of a low amount of waste heat. Besides, this simplicity implies that Ytterbium is devoid of other secondary effects (such as energy-transfer conversion processes) which may penalize its efficiency. Thus, it is not surprising Ytterbium-doped fiber laser systems have been demonstrated the highest average output-power: with 10 kW in single-mode CW operation [29], more than 4 kW with diffraction-limited beam quality [118,119] and nearly 1 kW in ultrafast pulse operation [37]. Yb^{3+} is usually pumped at 976 nm (vertical solid arrow), even though other wavelengths such as 915 nm (vertical dashed arrow), 940 nm, and 1,018 nm can also be used. In particular, the last wavelength is obtained from Yb-doped fiber lasers which are used to pump another Yb-doped fiber laser in a configuration known as tandem pumping. This configuration has been employed to obtain the highest CW single-mode output power demonstrated to date [29]. This is not a coincide since tandem pumping offers advantages for CW operation, such as a reduced quantum-defect heating and a lower nondepleted inversion [120]. However, tandem pumping is not very attractive for pulsed operation due to the limited amount of energy that can be stored in the fiber using this long pump wavelength.

The emission band from Yb^{3+} in silica fibers extends from ~980 nm [121] all the way up to 1,150 nm [122], but most CW systems operate in the range 1,040–1,080 nm, whereas most pulsed systems work around 1030 nm.

7.3.3.1.2 Erbium

The ion Er^{3+} is one of the most widely used in fiber laser technology since its emission band (going from ~1,500 to ~1650 nm), as shown in Figure 7.9, coincidentally overlaps with the minimum attenuation band of passive optical fibers (located around 1,550 nm). Thus, most of the laser and amplifier systems for telecommunications employ this ion. Er^{3+} can be pumped at ~980 nm (in which case, in order to improve the pump absorption, the fibers are co-doped with Ytterbium) or at 1,480 nm. Pumping at 980 nm has the drawback of a large quantum defect and, therefore, of the generation of a significant amount of waste heat and low efficiency. However, much higher pump powers are available at this pump wavelength than at 1,480 nm, which makes 980 nm usually the pump wavelength of choice for high-power laser systems. Thus, using 980 nm, an output power of almost 300 W has been demonstrated around 1,560 nm [123].

Interestingly, Er^{3+} is recently attracting attention because of different laser transitions that allow it emitting at ~2.75 μm and at ~3.45 μm [124]. Since these emission wavelengths are beyond the transparency band of Silica (which reaches up to ~2.2 μm), the fibers that exploit this transition are made of different materials such as fluorine or chalcogenide glasses [124].

7.3.3.1.3 Thulium

Thulium-doped fibers have attracted a significant amount of attention in recent years because they have the potential to outperform Yb-doped high-power fiber laser systems. The reason for this is that the onset power for most of the nonlinear effects limiting the performance of high-power fiber laser systems scales, at least, with the emission wavelength [40]. Thus, the longer emission wavelength of the

Tm^{3+} ion (which is centered at ~2 μm) is advantageous for the peak-power scaling of fiber laser systems. For example, in the particular case of self-focusing, the critical power scales with the square of the operating wavelength. Thus, it has been experimentally demonstrated that pulse peak-powers of 24 MW can be reached in a Tm-doped fiber laser system when using circularly polarized light [125]. Such a result would be unattainable in Yb-doped fibers. Furthermore, it has also been theoretically [126] and experimentally demonstrated [127–129] that Tm-doped fiber can withstand higher heat-loads than Yb-doped fibers before the onset of TMI.

The reasons above confer Tm-doped fibers a power-scaling potential that can see them outperforming Yb-doped fibers soon. However, harvesting the full potential of this technology is not an easy task, since there are several challenges that must be overcome. First of all, the emission wavelength, that is so beneficial for power-scaling, makes it challenging to propagate the light in air, especially for spectrally broadband systems. The reason is that the emission band of the Tm^{3+} ion in Silica (that goes from ~1.75 μm all the way up to ~2.15 μm, as illustrated in Figure 7.9) has a significant overlap with water-vapor absorption lines in atmospheric air [130]. Thus, as the light emitted by a high-power Tm-doped fiber laser system propagates through air it will, in most cases, get partially absorbed. This, in turn, results in a loss of power, deformation of the beam profile (due to the formation of a thermal lens) and in the degradation of the temporal contrast of the pulses [131]. Therefore, high-power Tm-doped fiber laser systems (especially ultrafast ones) must be operated in dry atmospheres or, even, in vacuum, which increases their complexity.

The second challenge that these systems have to overcome is the large amount of heat-load generated during laser operation. The reason is that, as illustrated in Figure 7.9, high-power Tm-doped fiber laser systems are usually pumped at 796 nm due to the availability of high-power pump sources at this wavelength. This results in a very high quantum defect that could potentially lead to the generation of a significant amount of waste heat. Fortunately, the Tm^{3+} is a very rich ion which shows energy transfer processes, such as, for example cross-relaxation [132]. Hereby, one ion can make the transition from the pump level ($^3H_4$) to the upper laser level ($^3F_4$) and the energy liberated when making this transition can be used to promote a neighboring ion from the ground state ($^3H_6$) to the upper laser level ($^3F_4$). Thus, it is potentially possible to get two ions excited per pump photon. Consequently, the maximum optical-to-optical efficiency of Tm-doped fiber laser systems is ~80%. In reality, though, this is extremely difficult to reach and the slope efficiency of these systems is typically around 50%–70% [133]. A way to get higher efficiencies would be to pump at longer wavelengths [133], for example, around 1.6 μm or even beyond. Unfortunately, there are no high-power pump sources at these long wavelengths readily available yet.

Despite these challenges, the performance of Tm-doped fiber laser systems has seen a dramatic development in the last few years. Thus, these fiber systems have broken the kW-average power barrier both in CW [134] and in ultrafast [127] operation and they can deliver up to 4 GW of peak power in few-cycle pulses [135].

This performance level puts ultrafast Tm-doped fiber laser systems currently on par with their Yb-doped counterparts but with better scaling prospects.

7.3.3.2 General considerations

Independently of the laser-active particle being used in the fiber, its geometrical distribution across the core can significantly impact the performance of the system. As seen in Figure 7.10, there are three main ways to distribute the laser-active ions: in a homogeneous way (left), confining them to an area not covering the whole core (center) or nanostructuring them (right).

The classic, and still most used, way of distributing the laser-active particles across the core is to do so in a homogeneous way (Figure 7.10 (left)). However, it has been shown both in theory [136] and in experiments [137] that confining the dopants to an area not covering the whole core (but still centered with it) or, alternatively, using a nonhomogeneous distribution of them across the complete core region may be beneficial. The reason is that this provides a preferential gain effect for the FM of the fiber, thus allowing effective single-mode operation in few-mode fibers. The problem of these two approaches is that it becomes increasingly technologically challenging to control the refractive index-step required to keep a low V-parameter as the core diameter increases. This can be significantly easier to do by nanostructuring the core [138]. In this technique rods of glass doped with laser-active particles (which increase the refractive index of the glass [139]) are tightly packaged with passive, low-index glass rods and fused together to create the core region, as illustrated in the right-hand side of Figure 7.10. Thus, nanostructured cores have island of doped-glass (with dimensions smaller than the wavelength of the light) in a matrix of low-index glass. This allows to geometrically control the NA of the fiber core with a degree of accuracy far higher than that achievable by classic fabrication methods.

Attractive as nanostructuring the core is, it has been pointed out that it can have drawbacks with respect to photodarkening in Yb-doped silica fibers [140]. This effect describes a photo-induced degradation [141–143] that leads to a higher

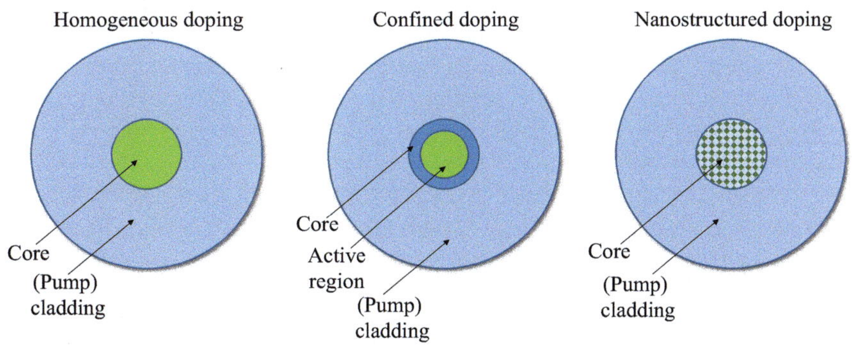

Figure 7.10 Usual distributions of the laser-active dopants in the fiber core: homogeneous (left), confined (center), and nanostructured (right)

absorption of the light in the fiber and, thus, to an increase of the heat generated during laser operation [144]. In this context, recently a link between photodarkening and TMI has been established and it has been shown, both theoretically [145] and experimentally [146], that even small amounts of photodarkening can drastically reduce the power threshold of TMI. Therefore, low-photodarkening fibers are required for the highest average-power operation.

One way to reduce the photodarkening in a fiber is to use a homogeneous distribution of the laser-active particles in the active region of the core. This is because this effect depends quadratically on the local concentration of Yb^{3+} ions [143]. Thus, homogeneously distributing the Yb^{3+} ions across the active region allows reducing the local ion concentration in comparison with the highly doped islands of Yb-doped material required for a nanostructured core. Another way of reducing photodarkening in a fiber is to co-dope the Silica glass with Phosphorous [147]. However, since the presence of this element in Silica glass reduces the absorption and emission cross-sections of the Yb^{3+} ions [139,147], it is necessary to optimize the concentration ratio of Yb^{3+} ions to Phosphorous to maximize the TMI threshold [148,149].

In any case, as can be inferred from the brief discussion above, the selection of the fiber material is not a trivial step in the design process since it has wide-reaching consequences for the high-power operation of fiber laser systems.

7.4 Outlook: multicore fibers

As it has been discussed in the previous pages, a careful design of a fiber can lead to a significantly better performance in high-power operation. In order to do this properly, the different aspects of the performance (e.g., efficiency in laser operation, threshold of the nonlinear effects, guidance and laser changes under high thermal load, TMI threshold, etc.) have to be considered for the get-go as an integral part of the design process [79]. However, in spite of this, the diffraction-limited output power that can be extracted from a single fiber core is ultimately limited [43,44].

One way to get around the limitations of an individual fiber (regardless of them being caused by nonlinear or by thermal effects) is to use several fibers in a parallel architecture and then coherently combine their outputs in a single output beam. This approach works very efficiently and it has been already implemented in an ultrafast fiber laser system delivering up to 3.5 kW of average power, 430 fs, ~40 μJ energy pulses with four main amplifier fibers [150]. Moreover, in a different ultrafast system, it was shown that it is possible to efficiently combine 16 fibers and get 2.3 mJ, 240 fs pulses in a beam with 1.8 kW average power [151]. These coherently combined systems are currently the maximum exponent of the level of performance that can be achieved with this strategy. However, the main drawback of this approach is that the systems become bulky and very complex. Both drawbacks can be addressed by using multicore fibers.

Multicore fibers have been used in telecommunications for years [152,153], but they have also been explored for sensors [154,155] and to build laser oscillators

[153,156]. However, even though the use of multicore fibers in high-power systems is relatively recent, their usefulness has already been emphatically proven in an experimental demonstration in which, using a four-core fiber, the TMI threshold was increased by a factor >3.5 [157]. This represents the largest increment of the TMI threshold ever obtained.

One crucial point that differentiates the new breed of multicore fibers used for high-power operation from the ones used in most of the other applications cited above is that the individual cores are both optically and thermally isolated from one another [157]. Thus, a typical design of high-power multicore fibers sees all the cores sharing the same pump cladding, but they are separated from each other by thermal and optical barriers, as schematically illustrated in Figure 7.11(a). Such barriers can be something as simple as leaving enough separation between the cores; an approach that usually works well for step-index multicore fibers (as that shown in Figure 7.11(b)). However, the barriers need to become more sophisticated for multicore PCF structures (or if a more compact arrangement of the cores is desired). Thus, for example as shown in Figure 7.11(c), some big air-holes can be placed between the individual cores to isolate them.

The development of high-power multicore fiber laser systems is still at the beginning and significant performance leaps are expected in the near future. Currently a 16-core step-index multicore fiber laser system delivering ps pulses with 70 W of average power has been demonstrated [158].

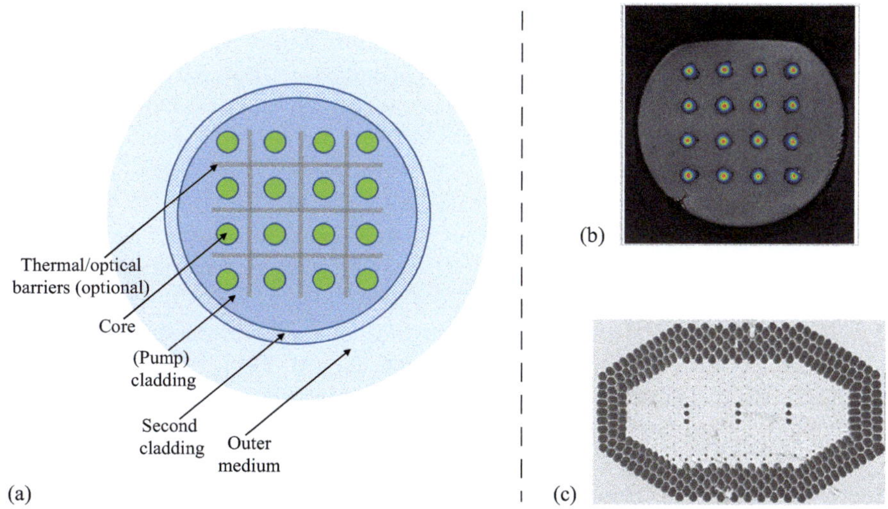

Figure 7.11 Multicore fibers used in high-power laser systems: (a) schematic illustration detailing the main constituents of such a fiber design; (b) 16-core step-index multicore fiber with overlaid output beams; (c) 4-core PCF in which the in-built thermal/optical barriers are clearly visible. Photos: courtesy of IAP Jena

Thus, it is fair to say that multicore fibers are the torch-bearers of the next generation of fibers for high-power operation. In fact, the development of better multicore fiber designs (borrowing from all the lessons learned from single-core fibers) presages a bright future for high-power fiber laser systems and their applications.

References

[1] D. Gloge, "Optical fibers for communication," Appl. Opt. **13**, 249–254 (1974).
[2] K. C. Kao, "Spectrophotometric studies of ultra low loss optical glasses I : single beam method," J. Sci. Instrum. **1**, 1063–1068 (1968).
[3] E. Snitzer, "Proposed fiber cavities for optical masers," J. Appl. Phys. **32**, 36–39 (1961).
[4] E. Snitzer, "Neodymium glass laser," *Proc. Int. Congress on Quantum Electronics*, Paris: Dunod, February 1964, 999–1019.
[5] E. Snitzer, "Glass lasers," Appl. Opt. **5**, 1487 (1966).
[6] C. J. Koester and E. Snitzer, "Amplification in a fiber laser," Appl. Opt. **3**, 1182 (1964).
[7] S. B. Poole, D. N. Payne, and M. E. Fermann, "Fabrication of low-loss optical fibres containing rare-earth ions," Electron. Lett. **21**, 737 (1985).
[8] R. J. Mears, L. Reekie, S. B. Poole, and D. N. Payne, "Neodymium-doped silica single-mode fibre lasers," Electron. Lett. **21**, 738 (1985).
[9] C. Jauregui, J. Limpert, and A. Tünnermann, "High-power fibre lasers," Nat. Photonics **7**, 861–867 (2013).
[10] D. J. Richardson, J. Nilsson, and W. A. Clarkson, "High power fiber lasers: current status and future perspectives [Invited]," J. Opt. Soc. Am. B **27**, B63 (2010).
[11] M. N. Zervas and C. A. Codemard, "High power fiber lasers: a review," IEEE J. Sel. Top. Quantum Electron. **20**, 219–241 (2014).
[12] V. P. Gapontsev and L. E. Samartsev, "High-power fiber laser," *Proceedings on Advanced Solid State Lasers* (Optical Society of America, OSA, Salt Lake City, Utah, USA, 1990), pp. 258–262.
[13] H. Po, J. D. Cao, B. M. Laliberte, *et al.*, "High power neodymium-doped single transverse mode fibre laser," Electron. Lett. **29**, 1500 (1993).
[14] H. Zellmer, S. Unger, P. Albers, *et al.*, "High-power cw neodymium-doped fiber laser operating at 9.2 W with high beam quality," Opt. Lett. **20**, 578 (1995).
[15] H. Zellmer, A. Tünnermann, H. Welling, and V. Reichel, "Double-clad fiber laser with 30 W output power," in *Optical Amplifiers and Their Applications* (Optical Society of America, OSA, Victoria, Canada 1997), p. FAW18.
[16] V. Dominic, S. MacCormack, R. Waarts, *et al.*, "110 W fibre laser," Electron. Lett. **35**, 1158 (1999).
[17] N. S. Platonov, D. V. Gapontsev, V. P. Gapontsev, and V. Shumilin, "135W CW fiber laser with perfect single mode output," in *Summaries of Papers*

Presented at the Lasers and Electro-Optics. CLEO'02. Technical Digest (Optical Society of America, OSA, Long Beach, CA, USA, 2002), pp. CPDC3–1–CPDC3–4.

[18] J. Limpert, A. Liem, H. Zellmer, and A. Tünnermann, "500 W continuous-wave fibre laser with excellent beam quality," Electron. Lett. **39**, 645 (2003).

[19] V. P. Gapontsev, N. S. Platonov, O. Shkurihin, and I. Zaitsev, "400 W low-noise single-mode cw ytterbium fiber laser with an integrated fiber delivery," in *Conference on Lasers and Electro-Optics, 2003* (Institute of Electrical and Electronics Engineers, IEEE, Baltimore, MD, USA). CLEO'03 (2003).

[20] Y. Jeong, J. K. Sahu, R. B. Williams, D. J. Richardson, K. Furusawa, and J. Nilsson, "Ytterbium-doped large-core fibre laser with 272 W output power," Electron. Lett. **39**, 977 (2003).

[21] C.-H. Liu, A. Galvanauskas, B. Ehlers, *et al.*, "810-W single transverse mode Yb-doped fiber laser," in *Advanced Solid-State Photonics* (Optical Society of America, OSA, Santa Fe, New Mexico, USA, 2004), p. PDP17.

[22] Y. Jeong, J. K. Sahu, S. Baek, *et al.*, "Cladding-pumped ytterbium-doped large-core fiber laser with 610 W of output power," Opt. Commun. **234**, 315–319 (2004).

[23] Y. Jeong, J. K. Sahu, D. N. Payne, and J. Nilsson, "Ytterbium-doped large-core fibre laser with 1 kW of continuous-wave output power," Electron. Lett. **40**, 470 (2004).

[24] A. Liem, J. Limpert, H. Zellmer, *et al.*, "1.3 kW Yb-doped fiber laser with excellent beam quality," in *Conference on Lasers and Electro-Optics, 2004* (Optical Society of America, OSA, San Francisco, CA, USA) (CLEO, 2004).

[25] Y. Jeong, J. K. Sahu, D. N. Payne, and J. Nilsson, "Ytterbium-doped large-core fiber laser with 1.36 kW continuous-wave output power," Opt. Express **12**, 6088 (2004).

[26] G. Bonati, H. Voelckel, T. Gabler, *et al.*, "1.53 kW from a single Yb-doped photonic crystal fi ber laser," in *Photonics West 2005* (International Society for Optics and Photonics, SPIE, San Jose, CA, USA, 2005), pp. 5709–2a.

[27] V. Gapontsev, D. Gapontsev, N. Platonov, *et al.*, "2 kW CW ytterbium fiber laser with record diffraction-limited brightness," in *CLEO/Europe. 2005 Conference on Lasers and Electro-Optics Europe* Institute of Electrical and Electronics Engineers, IEEE, Munich, Germany, 2005), pp. 508–508.

[28] D. Gapontsev, "6 kW CW single mode ytterbium fiber laser in all-fiber format," in *Solid State and Diode Laser Technology Review* (Directed Energy Professional Society, Albuquerque, New Mexico, 2008).

[29] V. Gapontsev, V. Fomin, A. Ferin, and M. Abramov, "Diffraction limited ultra-high-power fiber lasers," in *Advanced Solid-State Photonics*, OSA Technical Digest Series (CD) (Optical Society of America, San Diego, CA, USA, 2010), p. AWA1.

[30] J. D. Minelly, J. E. Caplen, Z. J. Chen, *et al.*, "Femtosecond pulse amplification in cladding-pumped fibers," Opt. Lett. **20**, 1797 (1995).

[31] J. Limpert, T. Schreiber, T. Clausnitzer, *et al.*, "High-power femtosecond Yb-doped fiber amplifier," Opt. Express **10**, 628–38 (2002).

[32] J. Limpert, T. Clausnitzer, A. Liem, *et al.*, "High-average-power femtosecond fiber chirped-pulse amplification system," Opt. Lett. **28**, 1984 (2003).
[33] A. Tünnermann, S. Höfer, A. Liem, *et al.*, "Power scaling of high-power fiber lasers and amplifiers," **15**, 107–117 (2005).
[34] F. Röser, D. Schimpf, O. Schmidt, *et al.*, "90 W average power 100 μJ energy femtosecond fiber chirped-pulse amplification system," Opt. Lett. **32**, 2230 (2007).
[35] F. Röser, T. Eidam, J. Rothhardt, *et al.*, "Millijoule pulse energy high repetition rate femtosecond fiber chirped-pulse amplification system," Opt. Lett. **32**, 3495 (2007).
[36] T. Eidam, S. Hadrich, F. Roser, *et al.*, "A 325-W-average-power fiber cpa system delivering sub-400 fs pulses," IEEE J. Sel. Top. Quantum Electron. **15**, 187–190 (2009).
[37] T. Eidam, S. Hanf, E. Seise, *et al.*, "Femtosecond fiber CPA system emitting 830 W average output power," Opt. Lett. **35**, 94–6 (2010).
[38] E. Snitzer, H. Po, F. Hakimi, R. Tumminelli, and B. C. McCollum, "Double clad, offset core Nd fiber laser," in *Optical Fiber Sensors* (Optical Society of America, New Orleans, LA, USA, 1988), p. PD5.
[39] R. D. Maurer, "Optical waveguide light source," U.S. Patent 3,808,549 (1974).
[40] G. P. Agrawal, *Nonlinear Fiber Optics*, 3rd ed. (Academic Press, San Diego, CA, USA 2006).
[41] D. Strickland and G. Mourou, "Compression of amplified chirped optical pulses," Opt. Commun. **56**, 219–221 (1985).
[42] J. Limpert, N. Deguil-Robin, I. Manek-Hönninger, *et al.*, "High-power rod-type photonic crystal fiber laser," Opt. Express **13**, 1055 (2005).
[43] J. W. Dawson, M. J. Messerly, R. J. Beach, *et al.*, "Analysis of the scalability of diffraction-limited fiber lasers and amplifiers to high average power," Opt. Express **16**, 13240 (2008).
[44] M. N. Zervas, "Power scaling limits in high power fiber amplifiers due to transverse mode instability, thermal lensing, and fiber mechanical reliability," in *Proceedings SPIE Volume 10512. Fiber Lasers XV: Technology and Systems*, A. L. Carter and I. Hartl, eds. (The International Society for Optics and Photonics, SPIE, San Francisco, CA, USA 2018), p. 4.
[45] A. Giesen and J. Speiser, "Fifteen years of work on thin-disk lasers: results and scaling laws," IEEE J. Sel. Top. Quantum Electron. **13**, 598–609 (2007).
[46] P. Russbueldt, T. Mans, G. Rotarius, J. Weitenberg, H. D. Hoffmann, and R. Poprawe, "400W Yb:YAG Innoslab fs-amplifier," Opt. Express **17**, 12230 (2009).
[47] G. P. Agrawal, "Nonlinear fiber optics: its history and recent progress [Invited]," J. Opt. Soc. Am. B **28**, A1 (2011).
[48] F. Jansen, F. Stutzki, H.-J. Otto, *et al.*, "Thermally induced waveguide changes in active fibers," Opt. Express **20**, 3997–4008 (2012).
[49] T. Eidam, C. Wirth, C. Jauregui, *et al.*, "Experimental observations of the threshold-like onset of mode instabilities in high power fiber amplifiers," Opt. Express **19**, 13218–13224 (2011).

[50] C. Jauregui, T. Eidam, J. Limpert, and A. Tünnermann, "The impact of modal interference on the beam quality of high-power fiber amplifiers," Opt. Express **19**, 3258–3271 (2011).

[51] A. V. Smith and J. J. Smith, "Mode instability in high power fiber amplifiers," Opt. Express **19**, 10180–92 (2011).

[52] R. Kashyap, "The fiber fuse–from a curious effect to a critical issue: a 25th year retrospective," Opt. Express **21**, 6422–41 (2013).

[53] K. Tankala, D. Guertin, J. Abramczyk, and N. Jacobson, "Reliability of low-index polymer coated double-clad fibers used in fiber lasers and amplifiers," Opt. Eng. **50**, 111607 (2011).

[54] C. Jauregui, H.-J. Otto, F. Stutzki, F. Jansen, J. Limpert, and A. Tünnermann, "Passive mitigation strategies for mode instabilities in high-power fiber laser systems," Opt. Express **21**, 19375–19386 (2013).

[55] A. V Smith and J. J. Smith, "Increasing mode instability thresholds of fiber amplifiers by gain saturation," Opt. Express **21**, 15168–82 (2013).

[56] C. Jauregui, C. Stihler, A. Tünnermann, and J. Limpert, "Pump-modulation-induced beam stabilization in high-power fiber laser systems above the mode instability threshold," Opt. Express **26**, 10691 (2018).

[57] H.-J. Otto, C. Jauregui, F. Stutzki, F. Jansen, J. Limpert, and A. Tünnermann, "Controlling mode instabilities by dynamic mode excitation with an acousto-optic deflector," Opt. Express **21**, 17285–98 (2013).

[58] J. Montoya, C. Hwang, D. Martz, C. Aleshire, T. Y. Fan, and D. J. Ripin, "Photonic lantern kW-class fiber amplifier," Opt. Express **25**, 27543 (2017).

[59] K. Schuster, S. Unger, C. Aichele, *et al.*, "Material and technology trends in fiber optics," Adv. Opt. Technol. **3**, 447–468 (2014).

[60] J. Limpert, F. Stutzki, F. Jansen, *et al.*, "Yb-doped large-pitch fibres: effective single-mode operation based on higher-order mode delocalisation," Light Sci. Appl. **1**, e8 (2012).

[61] F. Stutzki, F. Jansen, A. Liem, C. Jauregui, J. Limpert, and A. Tünnermann, "26 mJ, 130 W Q-switched fiber-laser system with near-diffraction-limited beam quality," Opt. Lett. **37**, 1073 (2012).

[62] A. Steinkopff, C. Jauregui, F. Stutzki, *et al.*, "Transverse single-mode operation in a passive large pitch fiber with more than 200 μm mode-field diameter," Opt. Lett. **44**, 650 (2019).

[63] J. M. Senior and M. Y. Jamro, *Optical Fiber Communications Principles and Practice* 3rd Edition (Prentice Hall Europe/Pearsons Education Limited, England, 2009).

[64] A. W. Snyder and J. D. Love, *Optical Waveguide Theory* (Springer US, New York, NY, USA, 1984).

[65] J. P. Koplow, D. a. V. Kliner, and L. Goldberg, "Single-mode operation of a coiled multimode fiber amplifier," Opt. Lett. **25**, 442 (2000).

[66] L. Leandro, L. Grüner-Nielsen, and K. Rottwitt, "Mode resolved bend-loss analysis in few-mode fibers using spatially and spectrally resolved imaging," Opt. Lett. **40**, 4583–4586 (2015).

[67] R. T. Schermer and J. H. Cole, "Improved bend loss formula verified for optical fiber by simulation and experiment," IEEE J. Quantum Electron. **43**, 899–909 (2007).

[68] H.-W. Chen, T. Sosnowski, C.-H. Liu, *et al.*, "Chirally-coupled-core Yb-fiber laser delivering 80-fs pulses with diffraction-limited beam quality warranted by a high-dispersion mirror based compressor," Opt. Express **18**, 24699–705 (2010).

[69] X. Ma, C. Zhu, I. Hu, A. Kaplan, and A. Galvanauskas, "Single-mode chirally-coupled-core fibers with larger than 50μm diameter cores," Opt. Express **22**, 9206 (2014).

[70] S. Ramachandran, J. M. Fini, M. Mermelstein, J. W. Nicholson, S. Ghalmi, and M. F. Yan, "Ultra-large effective-area, higher-order mode fibers: a new strategy for high-power lasers," Laser Photonics Rev. **2**, 429–448 (2008).

[71] T. Erdogan, "Cladding-mode resonances in short- and long-period fiber grating filters," J. Opt. Soc. Am. A **14**, 1760 (1997).

[72] J. W. Nicholson, A. DeSantolo, P. S. Westbrook, *et al.*, "Axicons for mode conversion in high peak power, higher-order mode, fiber amplifiers," Opt. Express **23**, 33849 (2015).

[73] P. Russell, "Photonic crystal fibers," Science **299**, 358–62 (2003).

[74] J. Broeng, D. Mogilevstev, S. E. Barkou, and A. Bjarklev, "Photonic crystal fibers: a new class of optical waveguides," Opt. Fiber Technol. **5**, 305–330 (1999).

[75] J. Limpert, R. Fabian, D. N. Schimpf, *et al.*, "High Repetition Rate Gigawatt Peak Power Fiber Laser-Systems: Challenges, Design, and Experiment," IEEE J. Sel. Top. Quantum Electron. **15**, 159–169 (2009).

[76] C. D. Brooks and F. Di Teodoro, "Multimegawatt peak-power, single-transverse-mode operation of a 100 μm core diameter, Yb-doped rodlike photonic crystal fiber amplifier," Appl. Phys. Lett. **89**, 111119 (2006).

[77] F. Röser, T. Eidam, J. Rothhardt, *et al.*, "Millijoule pulse energy high repetition rate femtosecond fiber chirped-pulse amplification system," Opt. Lett. **32**, 3495 (2007).

[78] O. Schmidt, J. Rothhardt, F. Röser, *et al.*, "Millijoule pulse energy Q-switched short-length fiber laser," **32**, 1551–1553 (2007).

[79] F. Stutzki, F. Jansen, H.-J. Otto, C. Jauregui, J. Limpert, and A. Tünnermann, "Designing advanced very-large-mode-area fibers for power scaling of fiber-laser systems," Optica **1**, 233 (2014).

[80] L. Dong, J. Li, and X. Peng, "Bend-resistant fundamental mode operation in ytterbium-doped leakage channel fibers with effective areas up to 3160 μm 2," 14, 11512–11519 (2006).

[81] L. Dong, X. Peng, and J. Li, "Leakage channel optical fibers with large effective area," J. Opt. Soc. Am. B **24**, 1689 (2007).

[82] F. Stutzki, F. Jansen, T. Eidam, *et al.*, "High average power large-pitch fiber amplifier with robust single-mode operation," Opt. Lett. **36**, 689–691 (2011).

[83] T. Eidam, J. Rothhardt, F. Stutzki, *et al.*, "Fiber chirped-pulse amplification system emitting 3.8 GW peak power," Opt. Express **19**, 255–60 (2011).

[84] F. Jansen, F. Stutzki, A. Liem, C. Jauregui, J. Limpert, and A. Tünnermann, "26-mJ pulse energy Q-switched large-pitch fiber laser system with excellent beam quality," in *Proc. SPIE 8237, Fiber Lasers IX: Technology, Systems, and Applications* (Optical Society of America, OSA, San Francisco, CA, USA 2012), p. 82371S–82371S.

[85] H.-J. Otto, F. Stutzki, N. Modsching, C. Jauregui, J. Limpert, and A. Tünnermann, "2 kW average power from a pulsed Yb-doped rod-type fiber amplifier," Opt. Lett. **39**, 6446–9 (2014).

[86] F. Stutzki, F. Jansen, C. Jauregui, J. Limpert, and A. Tünnermann, "Non-hexagonal Large-Pitch Fibers for enhanced mode discrimination," Opt. Express **19**, 12081–12086 (2011).

[87] R. Jamier, A. Benoît, K. Schuster, *et al.*, "Highly efficient higher-order modes filtering into aperiodic very large mode area fibers for single-mode propagation," Opt. Lett. **39**, 4561 (2014).

[88] D. Jain, C. Baskiotis, and J. K. Sahu, "Mode area scaling with multi-trench rod-type fibers," Opt. Express **21**, 1448–55 (2013).

[89] F. Kong, K. Saitoh, D. Mcclane, *et al.*, "Mode area scaling with all-solid photonic bandgap fibers," Opt. Express **20**, 26363–72 (2012).

[90] C. A. Codemard, N. T. Vukovic, J. S. Chan, *et al.*, "Resonant SRS Filtering Fiber for High Power Fiber Laser Applications," IEEE J. Sel. Top. Quantum Electron. **24**, 1–9 (2018).

[91] G. Gu, F. Kong, T. Hawkins, *et al.*, "Ytterbium-doped large-mode-area all-solid photonic bandgap fiber lasers," Opt. Express **22**, 13962–8 (2014).

[92] G. Gu, F. Kong, T. W. Hawkins, M. Jones, and L. Dong, "Extending mode areas of single-mode all-solid photonic bandgap fibers," Opt. Express **23**, 9147 (2015).

[93] T. T. Alkeskjold, M. Laurila, L. Scolari, and J. Broeng, "Single-mode ytterbium-doped large-mode-area photonic bandgap rod fiber amplifier," Opt. Express **19**, 7398 (2011).

[94] M. Laurila, M. M. Jørgensen, K. R. Hansen, T. T. Alkeskjold, J. Broeng, and J. Lægsgaard, "Distributed mode filtering rod fiber amplifier delivering 292W with improved mode stability," Opt. Express **20**, 5742–53 (2012).

[95] T. T. Alkeskjold, "Large-mode-area ytterbium-doped fiber amplifier with distributed narrow spectral filtering and reduced bend sensitivity," Opt. Express **17**, 16394 (2009).

[96] K. R. Hansen, T. T. Alkeskjold, J. Broeng, and J. Lægsgaard, "Thermo-optical effects in high-power ytterbium-doped fiber amplifiers," Opt. Express **19**, 23965–80 (2011).

[97] F. Jansen, F. Stutzki, H.-J. Otto, C. Jauregui, J. Limpert, and A. Tünnermann, "High-power thermally guiding index-antiguiding-core fibers," Opt. Lett. **38**, 510–2 (2013).

[98] C. R. Smith, N. Simakov, A. Hemming, and W. A. Clarkson, "Thermally guided Yb-doped fiber-rod amplifier and laser," Appl. Phys. B **125**, 32 (2019).

[99] W. Liu, J. Cao, and J. Chen, "Study on the adiabaticity criterion of the thermally-guided very-large-mode-area fiber," Opt. Express **26**, 7852 (2018).

[100] A. V. Smith and B. T. Do, "Bulk and surface laser damage of silica by picosecond and nanosecond pulses at 1064 nm," Appl. Opt. **47**, 4812 (2008).
[101] S. R. Schmid and A. F. Toussaint, "Optical fiber coatings," in *Specialty Optical Fibers Handbook* (Elsevier, Burlington, MA, USA, 2007), pp. 95–122.
[102] K. Furusawa, A. Malinowski, J. Price, *et al.*, "Cladding pumped Ytterbium-doped fiber laser with holey inner and outer cladding," Opt. Express **9**, 714 (2001).
[103] W. Wadsworth, R. Percival, G. Bouwmans, J. Knight, and P. Russell, "High power air-clad photonic crystal fibre laser," Opt. Express **11**, 48 (2003).
[104] J. Limpert, T. Schreiber, A. Liem, *et al.*, "Thermo-optical properties of air-clad photonic crystal fiber lasers in high power operation," Opt. Express **11**, 2982 (2003).
[105] P. Koška, P. Peterka, and V. Doya, "Numerical modeling of pump absorption in coiled and twisted double-clad fibers," IEEE J. Sel. Top. Quantum Electron. **22**, 55–62 (2016).
[106] D. Kouznetsov, J. V. Moloney, and E. M. Wright, "Efficiency of pump absorption in double-clad fiber amplifiers. I. Fiber with circular symmetry," J. Opt. Soc. Am. B **18**, 743 (2001).
[107] D. Kouznetsov and J. V. Moloney, "Efficiency of pump absorption in double-clad fiber amplifiers. II. Broken circular symmetry," J. Opt. Soc. Am. B **19**, 1259 (2002).
[108] J. Noda, K. Okamoto, and Y. Sasaki, "Polarization-maintaining fibers and their applications," J. Light. Technol. **4**, 1071–1089 (1986).
[109] A. Kumar, R. K. Varshney, and K. Thyagarajan, "Birefringence calculations in elliptical-core optical fibres," Electron. Lett. **20**, 112 (1984).
[110] R. Guan, F. Zhu, Z. Gan, D. Huang, and S. Liu, "Stress birefringence analysis of polarization maintaining optical fibers," Opt. Fiber Technol. **11**, 240–254 (2005).
[111] J. R. Folkenberg, M. D. Nielsen, N. A. Mortensen, C. Jakobsen, and H. R. Simonsen, "Polarization maintaining large mode area photonic crystal fiber," Opt. Express **12**, 956 (2004).
[112] O. Schmidt, J. Rothhardt, T. Eidam, *et al.*, "Single-polarization ultra-large-mode-area Yb-doped photonic crystal fiber," Opt. Express **16**, 3918 (2008).
[113] C.-H. Liu, A. Galvanauskas, V. Khitrov, *et al.*, "High-power single-polarization and single-transverse-mode fiber laser with an all-fiber cavity and fiber-grating stabilized spectrum," Opt. Lett. **31**, 17 (2006).
[114] M. Söderlund, J. Koponen, S. Tammela, V. Philippov, and H. Po, "Design considerations for large-mode-area polarization maintaining double clad fibers," in *Electro-Optical and Infrared Systems: Technology and Applications II*, R. G. Driggers and D. A. Huckridge, eds. (The International Society for Optics and Photonics, SPIE, Bellingham, Washington, USA, 2005), **Vol. 5987**, p. 59870A.
[115] C. Gaida, F. Stutzki, F. Jansen, *et al.*, "Triple-clad large-pitch fibers for compact high-power pulsed fiber laser systems," Opt. Lett. **39**, 209 (2014).
[116] J. Limpert, *High Power Ultrafast Fiber Laser Systems* (Mensch-und-Buch-Verlag, 2003).

[117] P. D. Dragic, M. Cavillon, and J. Ballato, "Materials for optical fiber lasers: A review," Appl. Phys. Rev. **5**, 041301 (2018).

[118] F. Beier, C. Hupel, S. Kuhn, *et al.*, "Single mode 4.3 kW output power from a diode-pumped Yb-doped fiber amplifier," Opt. Express **25**, 14892–14899 (2017).

[119] S. Ikoma, K. Uchiyama, Y. Takubo, M. Kashiwagi, K. Shima, and D. Tanaka, "5-kW single stage all-fiber Yb-doped single-mode fiber laser for materials processing," in *Fiber Lasers XV: Technology and Systems*, A. L. Carter and I. Hartl, eds. (SPIE, San Francisco, CA, USA, 2018), p. 11.

[120] C. Codemard, J. Sahu, and J. Nilsson, "Tandem cladding-pumping for control of excess gain in ytterbium-doped fiber amplifiers," Quantum Electron. IEEE ... 1–10 (2010).

[121] F. Röser, C. Jauregui, J. Limpert, and A. Tünnermann, "94 W 980 nm high brightness Yb-doped fiber laser," Opt. Express **16**, 17310–17318 (2008).

[122] M. Jacquemet, A. Mugnier, G. Le Corre, E. Goyat, and D. Pureur, "CW PM multiwatts Yb-doped fiber laser directly emitting at long wavelength," IEEE J. Sel. Top. Quantum Electron. **15**, 120–128 (2009).

[123] Y. Jeong, S. Yoo, C. A. Codemard, *et al.*, "Erbium:Ytterbium Codoped Large-Core Fiber Laser With 297-W Continuous-Wave Output Power," IEEE J. Sel. Top. Quantum Electron. **13**, 573–579 (2007).

[124] S. D. Jackson, "Towards high-power mid-infrared emission from a fibre laser," Nat. Photonics **6**, 423–431 (2012).

[125] C. Gaida, M. Gebhardt, F. Stutzki, C. Jauregui, and J. Limpert, "Self-compression in a solid fiber to 24 MW peak power with few-cycle pulses at 2 μ m wavelength," Opt. Lett. **40**, 5160–5163 (2015).

[126] J. J. S. Arlee V. Smith, "Mode instability thresholds for Tm-doped fiber amplifiers pumped at 790 nm," (2015).

[127] C. Gaida, F. Stutzki, C. Jauregui, M. Gebhardt, T. Heuermann, and J. Limpert, "Ultrafast thulium fiber laser system emitting more than 1 kW of average power," Opt. Lett. **43**, 5853 (2018).

[128] C. Gaida, M. Gebhardt, T. Heuermann, *et al.*, "Observation of transverse-mode instabilities in a thulium-doped fiber amplifier," in *Fiber Lasers XVI: Technology and Systems. Conference 10897* (SPIE, San Francisco, CA, USA, 2019), pp. 10897–1.

[129] G. D. Goodno, L. D. Book, J. E. Rothenberg, M. E. Weber, and S. B. Weiss, "Narrow linewidth power scaling and phase stabilization of 2μm thulium fiber lasers," Opt. Eng. **50**, 111608 (2011).

[130] L. S. Rothman, I. E. Gordon, Y. Babikov, *et al.*, "The HITRAN2012 molecular spectroscopic database," J. Quant. Spectrosc. Radiat. Transf. **130**, 4–50 (2013).

[131] M. Gebhardt, C. Gaida, F. Stutzki, *et al.*, "Impact of atmospheric molecular absorption on the temporal and spatial evolution of ultra-short optical pulses," Opt. Express **23**, 13776 (2015).

[132] S. D. Jackson, "Cross relaxation and energy transfer upconversion processes relevant to the functioning of 2 μm Tm3+-doped silica fibre lasers," Opt. Commun. **230**, 197–203 (2004).

[133] A. Sincore, J. D. Bradford, J. Cook, L. Shah, and M. C. Richardson, "High average power thulium-doped silica fiber lasers: review of systems and concepts," IEEE J. Sel. Top. Quantum Electron. **24**, 1–8 (2018).

[134] T. Ehrenreich, R. Leveille, I. Majid, K. Tankala, G. Rines, and P. Moulton, "1-kW, All-Glass Tm:fiber Laser," in *SPIE Photonics West 2010: LASE Fiber Lasers VII: Technology, Systems, and Applications. Conference 7580* (SPIE, San Francisco, CA, USA, 2010).

[135] M. Gebhardt, C. Gaida, F. Stutzki, *et al.*, "High average power nonlinear compression to 4 GW, sub-50 fs pulses at 2 μm wavelength," Opt. Lett. **42**, 747 (2017).

[136] M. Gong, Y. Yuan, C. Li, P. Yan, H. Zhang, and S. Liao, "Numerical modeling of transverse mode competition in strongly pumped multimode fiber lasers and amplifiers," Opt. Express **15**, 3236 (2007).

[137] T. Eidam, S. Hädrich, F. Jansen, *et al.*, "Preferential gain photonic-crystal fiber for mode stabilization at high average powers," Opt. Express **19**, 8656–8661 (2011).

[138] J. Limpert, O. Schmidt, J. Rothhardt, *et al.*, "Extended single-mode photonic crystal fiber lasers," Opt. Express **14**, 2715 (2006).

[139] H. Zimer, M. Kozak, A. Liem, *et al.*, "Fibers and fiber-optic components for high-power fiber lasers," in *Proceedings SPIE Volume 7914, Fiber Lasers VIII: Technology, Systems, and Applications*, J. W. Dawson, ed. (Optical Society of America, OSA, San Francisco, CA, USA 2011), **Vol. 7914**, p. 791414.

[140] C. Jauregui, H. Otto, S. Breitkopf, J. Limpert, and A. Tünnermann, "Optimizing high-power Yb-doped fiber amplifier systems in the presence of transverse mode instabilities," Opt. Express **24**, 7879 (2016).

[141] J. Koponen, M. Laurila, M. Söderlund, J. J. Montiel I Ponsoda, and A. Iho, "Benchmarking and measuring photodarkening in Yb doped fibers," in SPIE Proceedings Volume 7195, Fiber Lasers VI: Technology, Systems, and Applications; D. V. Gapontsev, D. A. Kliner, J. W. Dawson, and K. Tankala, eds. (The International Society for Optics and Photonics, SPIE, San Jose, CA, USA, 2009), **Vol. 7195**, p. 71950R–71950R–14.

[142] I. Manek-Hönninger, J. Boullet, T. Cardinal, *et al.*, "Photodarkening and photobleaching of an ytterbium-doped silica double-clad LMA fiber," Opt. Express **15**, 1606 (2007).

[143] S. Taccheo, H. Gebavi, A. Monteville, *et al.*, "Concentration dependence and self-similarity of photodarkening losses induced in Yb-doped fibers by comparable excitation," Opt. Express **19**, 19340–5 (2011).

[144] J. J. Montiel I Ponsoda, M. J. Söderlund, J. P. Koplow, J. J. Koponen, and S. Honkanen, "Photodarkening-induced increase of fiber temperature," Appl. Opt. **49**, 4139–43 (2010).

[145] C. Jauregui, H.-J. Otto, F. Stutzki, J. Limpert, and A. Tünnermann, "Simplified modelling the mode instability threshold of high power fiber amplifiers in the presence of photodarkening," Opt. Express **23**, 20203 (2015).

[146] H.-J. Otto, N. Modsching, C. Jauregui, J. Limpert, and A. Tünnermann, "Impact of photodarkening on the mode instability threshold," Opt. Express **23**, 15265 (2015).

[147] S. Jetschke, S. Unger, A. Schwuchow, M. Leich, and J. Kirchhof, "Efficient Yb laser fibers with low photodarkening by optimization of the core composition," Opt. Express **16**, 15540 (2008).

[148] J. Lægsgaard, "Optimizing Yb concentration of fiber amplifiers in the presence of transverse modal instabilities and photodarkening," Appl. Opt. **55**, 1966 (2016).

[149] C. Jauregui, F. Stutzki, A. Tünnermann, and J. Limpert, "Thermal analysis of Yb-doped high-power fiber amplifiers with Al:P co-doped cores," Opt. Express **26**, 7614 (2018).

[150] M. Müller, A. Klenke, A. Steinkopff, H. Stark, A. Tünnermann, and J. Limpert, "3.5 kW coherently combined ultrafast fiber laser," Opt. Lett. **43**, 6037 (2018).

[151] M. Müller, A. Klenke, H. Stark, *et al.*, "1.8-kW 16-channel ultrafast fiber laser system," in *Fiber Lasers XV: Technology and Systems*, A. L. Carter and I. Hartl, eds. (SPIE, San Francisco, CA, USA, 2018), p. 7.

[152] K. Nakajima, T. Matsui, K. Saito, T. Sakamoto, and N. Araki, "Multi-core fiber technology: next generation optical communication strategy," IEEE Commun. Stand. Mag. **1**, 38–45 (2017).

[153] A. M. Ortiz and R. L. Sáez, "Multi-core optical fibers: theory, applications and opportunities," in Selected Topics on Optical Fiber Technologies and Applications (InTech, 2018), **Vol. 2**, p. 64.

[154] A. Van Newkirk, J. E. Antonio-Lopez, A. Velazquez-Benitez, J. Albert, R. Amezcua-Correa, and A. Schülzgen, "Bending sensor combining multicore fiber with a mode-selective photonic lantern," Opt. Lett. **40**, 5188 (2015).

[155] A. Van Newkirk, E. Antonio-Lopez, G. Salceda-Delgado, R. Amezcua-Correa, and A. Schülzgen, "Optimization of multicore fiber for high-temperature sensing," Opt. Lett. **39**, 4812 (2014).

[156] C. Jollivet, A. Van Newkirk, J. Anderson, K. Schuster, S. Grimm, and A. Schülzgen, "Advances in multi-core fiber lasers," in Latin America Optics and Photonics Conference (Optical Society of America, 2014), p. LM1D.3.

[157] H.-J. Otto, A. Klenke, C. Jauregui, F. Stutzki, J. Limpert, and A. Tünnermann, "Scaling the mode instability threshold with multicore fibers," Opt. Lett. **39**, 2680–2683 (2014).

[158] A. Klenke, M. Müller, H. Stark, *et al.*, "Coherently combined 16-channel multicore fiber laser system," Opt. Lett. **43**, 1519 (2018).

Chapter 8
High power fiber lasers

Martin Richardson[1], Justin Cook[1], Patrick Roumayah[1], Alex Sincore[1] and Nathan Bodnar[1]

This review concentrates on the progress and developments that have led to explosive growth in high power continuous wave (CW) fiber lasers over the last decade. Here we arbitrarily defined high power fiber operation as being close to a single mode in beam output with powers in the kilowatt range. This growth has largely been led by Yb fiber lasers up to now. However, in this review, we also highlight the potential for high power, high brightness operation of two alternative approaches, thulium fiber lasers that operate at 2 μm wavelength, and CW Raman fiber lasers (RFLs), which have yet to reach kW power levels.

8.1 Introduction

Although the laser that was first demonstrated in 1960 was in a solid-state material, ruby [1], for over two decades following, high powers sufficient for meaningful applications in manufacturing and defense technologies such as cutting and welding [2] and directed energy (ref) could only be supplied with gas lasers. The CO_2 laser, first demonstrated with a power of ~1 mW [3], quickly improved in power to ~100 W by 1966, with commercial systems at the 1,000 W level appearing on the market in 1970 [2], and 10 kW becoming available in 1973 [4]. In parallel chemically energized lasers in the mid-IR were developed, largely for defense applications and ultimately reach powers so high that for many years following they were classified. Meanwhile despite massive investments in the development of high power solid-state lasers, using dozens of different media, their operation during the 1960s, 1970s, and 1980s was constrained to the largely pulsed mode by the sources available to pump the rare-earth ion or transition metal dopants used as the active medium. The primary method, the same optical pumping approach used in the first laser, was recognized as absurdly inefficient for CW operation. Millions of dollars were spent on the development of flashlamp light sources, all operating in the pulsed regime, enabling spectacular progress in pulsed laser systems for

[1]Laser Plasma Laboratory, CREOL, The College of Optics and Photonics, University of Central Florida, Orlando, FL, USA

applications ranging from metal surface processing to lidar, to laser fusion and X-ray lasers. All these applications were ultimately limited in repetition-rate by thermal effects occurring as a consequence of the absorption of out-of-band optical pump radiation in the laser medium and its surroundings, and by the constraints on the flashlamp and its driving circuits themselves. Even after 30 years of solid-state laser development, the highest average powers of high-repetition-rate-pulsed systems were only measured in 100s of watts [5]. Pumping efficiencies were dismally low (<1%), compared to values of 10% or higher for gas lasers with powers one or two orders of magnitude higher. Approaches other than flashlamps were explored, even to the point of considering nuclear reactors for pumping high power CW or quasi-CW (QCW) systems [6].

Thus by the end of the 1980s the availability of high power laser radiation across the spectrum was delineated by electrically pumped CW gas lasers in the long-wavelength region, CW or QCW chemical lasers in the mid-IR region—and pulsed solid-state lasers in the near-IR and pulsed (excimer) gas lasers in the visible-UV.

8.1.1 Diode-laser pumped solid-state laser media

It was long recognized during the time when flashlamps reigned supreme in the solid-state laser field, that population of the upper laser level through direct excitation to a nearby long-lived state would beneficially improve the overall efficiency. This is due to the fact that as the quantum defect defined as:

$$\mathrm{QD} = 1 - \frac{h\upsilon_l}{h\upsilon_p} \tag{8.1}$$

where $h\upsilon_l$ and $h\upsilon_p$ are the photon energies of the laser and pump wavelengths, approaches unity, thermal effects in the medium would be significantly decreased. But in the early days of the solid-state lasers, efficient pseudo-monochromatic light sources of sufficient power to directly populate the upper laser level did not exist. The first demonstration of such a source, pumping a Nd:YAG laser crystal, which has one of the highest stimulated emission cross-section coefficients, σ_{se} (cm^2) of all solid-state laser media in the near-IR, used diode laser radiation at 868 nm, occurred in 1974 [7].

The rapid evolution of diode-pumped solid-state (DPSS) CW and QCW lasers can be closely related to the development of diode source power at wavelengths that could effectively pump the upper laser levels of solid-state media Figure 8.1. Of the latter, Yb-doped YAG had by far the lowest QD (~8%).

8.2 High power Yb fiber lasers

The first fiber laser was demonstrated decades ago, not long after the invention of the ruby laser, by Elias Snitzer in 1961 [9]. Since it utilized flashlamp optical pumping it was of little practicality, though it is worth noting that Snitzer had demonstrated optical waveguiding in an optical fiber, both experimentally and theoretically before the first laser, in 1959 [10]. Snitzer and others went on

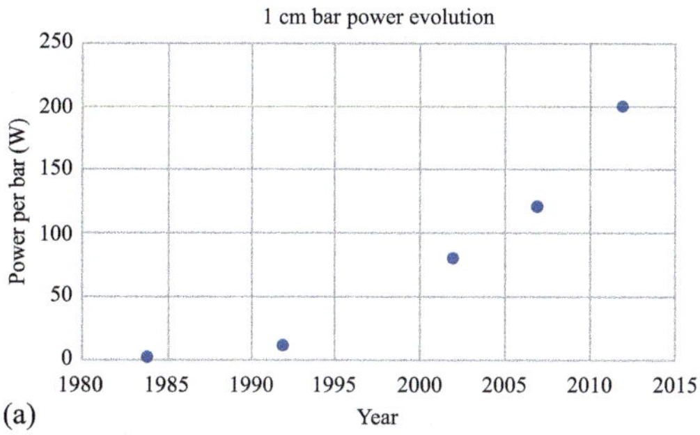

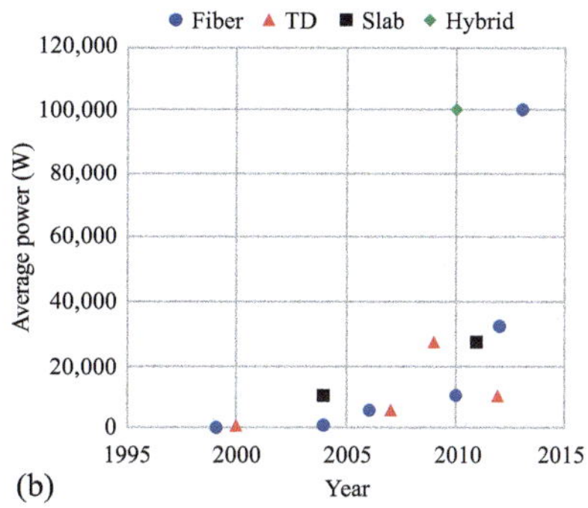

Figure 8.1 *Chronological development of (a) high power 970 nm diodes and (b) Yb solid-state lasers as either fiber lasers, thin disc lasers, slab, or hybrid lasers [8]*

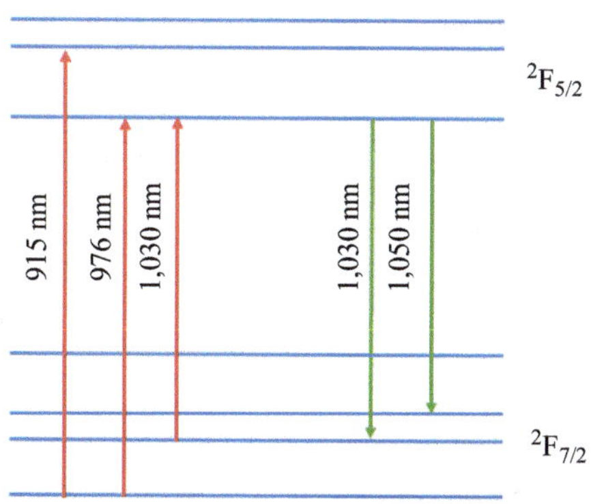

Figure 8.2 *Relevant energy levels in ytterbium. The two energy manifolds involved in lasing transitions are the $^2F_{5/2}$ and $^2F_{7/2}$. Pump wavelengths are in red while emission bands are in green*

developing various types of fiber lasers in the late 1960s and 1970s, but high power fiber lasers did not come onto the scene until the development of high-brightness diode pumps in the late 1970s and 1980s.

8.2.1 Yb:silica spectroscopy

Yb^{3+}-doped fused silica is the workhorse of modern high power fiber lasers in part due to its favorable energy level structure. Shown in Figure 8.2 is the energy level structure of Yb^{3+} with approximate absorption and emission wavelengths in fused silica. The ground state in fused silica is the $^2F_{7/2}$ level while the upper manifold is the $^2F_{5/2}$ level [11]. Ytterbium benefits from strong absorption bands at 915 and 976 nm with a broad emission band ranging from 976 to 1,150 nm, shown in Figure 8.3. Due to the amorphous nature of Yb:silica, the energy sublevels within

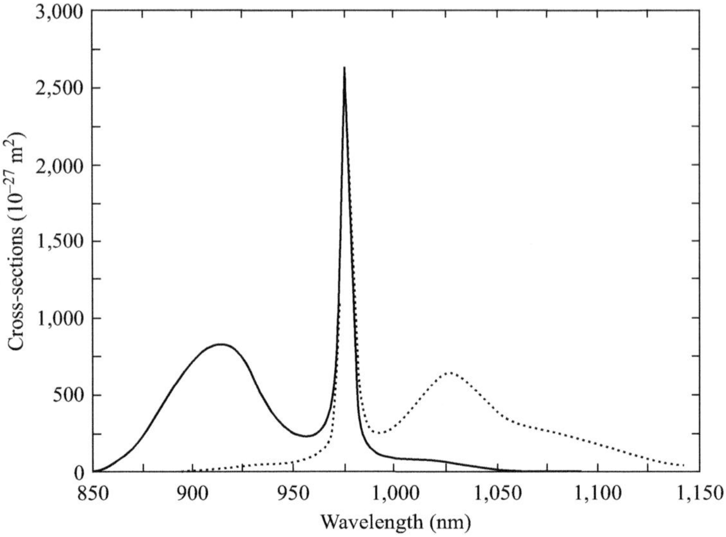

Figure 8.3 Typical absorption spectrum (solid line) and emission spectrum (dashed line) for ytterbium in germanosilicate glass [11]

the upper and lower manifolds are not fully resolved. Rather, homogeneous and inhomogeneous broadening mechanisms broaden these sublevels into continuous bands [12,13]. The energy level structure in Yb^{3+} is free from any parasitic energy transfer processes such as energy transfer upconversion (ETU) or excited state absorption which is a known issue in Er-doped fiber amplifiers [14].

The large absorption cross section at 976 nm allows for efficient pumping with modern high brightness fiber-coupled pump diodes. The broad emission band from ~976 nm to beyond 1,150 nm allows for quasi-three-level operation and is ideal for both wavelengths tunable and pulsed applications. By pumping at 976 nm and emitting around 1,030 nm, this allows for very efficient operation and a theoretical quantum defect as low at 6% making this an attractive gain medium for high average power generation.

In addition to direct diode pumping at 976 nm, in-band pumping has been shown to further reduce the quantum defect and allow for further power scaling. The idea behind in-band pumping involves, in the case of Yb:fibers, constructing a Yb:fiber laser around 1,018 nm to pump another Yb:fiber laser at $>$ 1,030 nm, giving a quantum defect $<$2% [15]. IPG has heavily invested in this method as the primary pumping architecture for their high power systems beyond 10 kW [16].

8.2.2 1 micron fiber lasers

The first single mode kW fiber laser was demonstrated by Southampton in 2004 [17]. Over the next decade, steady progress was made in further increasing the average power with 2 kW reported in 2005 [18] and 3 kW in 2014 [19]. The current output power record for single mode operation reported by an academic institution stands at 4.4 kW without the onset of transverse mode instability [20]. Record single mode power outputs from industry include 5 kW from Fujikura [21] and 20 kW from IPG [22]. Ultimately, the goal is to maximize the power scaling potential of single-mode fiber lasers and amplifiers.

Currently, most high-power fiber lasers in the 1 μm spectral region utilize low-NA fibers (NA $\leq$0.06) with large core diameters $\geq$20 μm. These fibers usually utilize long fibers in excess of 20 m with large pump claddings to lower the absorption per unit length and minimize the thermal load [23]. Traditional fiber laser architectures utilize low-index polymer claddings which have a maximum long-term operation temperature of $\approx$80°C. However, this imposes limitations on the laser or amplifier geometry and maximum output power that can be generated before thermal failure. Recent work has shown improved thermal stability of these low index polymers at temperatures >350 °C, enabling much higher temperature and thus average power operation [24].

Several alternatives exist to low-index polymer claddings. For example, Yu *et al.* have demonstrated 3 kW out of a 4 m amplifier by using a gold-coated fiber [25]. Other approaches rely on high-NA air claddings or all-glass claddings to confine pump light. Air claddings rely on an arrangement of air holes around the doped core region to confine pump light and can produce very high NA in excess of 0.8 [26,27]. All-glass claddings are usually implemented in a triple clad design and rely on a lower index glass composition surrounding the pump cladding to guide pump light via total internal reflection [27,28]. Glass claddings, due to the glass compositions used in high power optical fibers, typically have a lower NA $\geq$ 0.22 compared with polymer or air claddings [29–31].

8.2.3 Fiber laser architectures and fiber design

In 1988 Snitzer *et al.* introduced the concept of a double clad optical fiber [32]. This fiber had a step-index refractive profile and possessed two separate claddings around the fiber core (Figure 8.4). The inner cladding was much larger than the core diameter, allowing for the use of multimode pumping and reduced pump intensity on the facet. Snitzer proposed two different designs, one with a circular pump cladding and an offset core, and another with a noncircular pump cladding [32,33]. Both of these designs, particularly noncircular/nonsymmetric pump claddings, allow for scalability of the pump cladding dimensions and improved mode mixing down the length of the fiber. Since their invention, double clad fibers have been one of the most enabling technologies for increasing the average output power of fiber lasers and amplifiers, allowing for multi-kW single mode output from a single fiber.

Typical pump cladding geometries are shown in Figure 8.5(a). In each of these designs, the idea is to prevent the propagation of skew modes down the active fiber, as these modes have almost no overlap with the core. By breaking the circular symmetry of the cladding, this causes skew modes to scatter into meridional modes which have much higher overlap with the doped core [34]. In fact, by properly

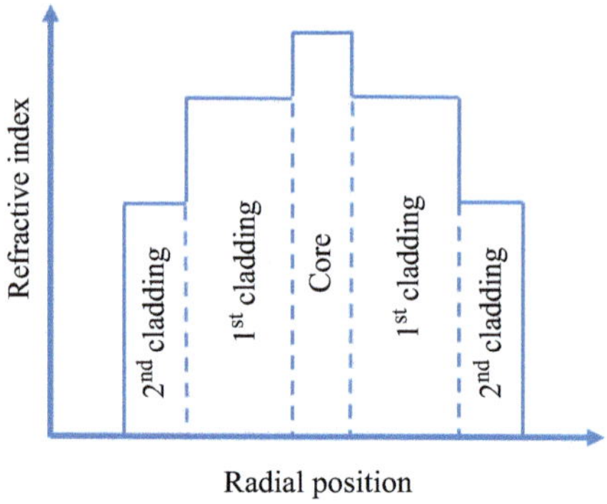

Figure 8.4 Refractive index profile of a double clad step index fiber

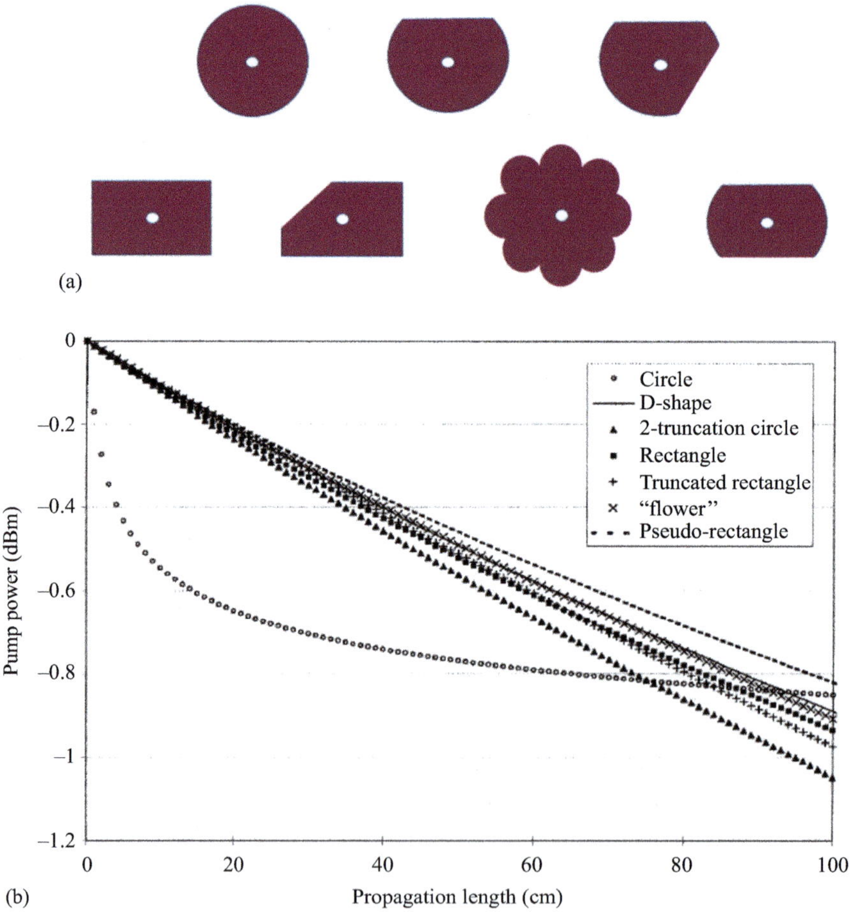

Figure 8.5 (a) Common pump cladding geometries. (b) The effective absorption induced by each pump cladding [34]. Note that all geometries perform better than circular claddings at longer lengths

designing the cladding geometry, the propagation of cladding modes can become chaotic. This situation is ideal for high power operation in a fiber amplifier since this produces the best mode mixing and absorption down the entire length of the fiber [27,34].

Figure 8.5(b) shows numerical simulations of the launched pump power versus fiber length for each of the cladding geometries in (a). The circular cladding has high absorption in the first few cm which quickly saturates at longer distances. This is due to the high absorption of meridional rays at the beginning of the fiber, after which get depleted. The remaining skew modes have very little overlap with the core, causing the absorbed pump power to saturate at some power level. In the other six cases, the pump power is absorbed in a more linear fashion down the fiber length indicating efficient mode mixing is occurring [34]. By making use of any noncircularly symmetric pump cladding, the absorption per unit length characteristics is drastically improved. In high power fiber amplifiers, constant absorption down the length of the fiber is key to distributing the thermal load and reducing the peak gain which can both lead to unwanted nonlinearities.

Besides the cladding geometry, the cladding diameter relative to the core dictates the pump absorption length scale. The pump absorption efficiency scales as $\eta \propto A_{\text{core}}/A_{\text{clad}}$ [27,34]. This parameter provides another degree of freedom when designing a fiber amplifier. Depending on the desired output power and cooling capabilities, the effective absorption can be tuned to maximize the performance of a given system simply by modifying the pump cladding diameter.

8.2.3.1 High power fiber designs

Step index designs are currently the industry standard for commercially available fiber intended for high average power applications. Step-index refractive index profiles [20,23] are the most predominantly used design and have been extended to include different dopant profiles in the fiber core, such as confined doping fibers [35], as well as multiple cores in the case of multicore fibers [36]. The step-index design has also been extended to include the use of satellite cores such as in chirally coupled core (3C) fibers [37,38].

In addition to step-index designs, large core diameter fibers such as large pitch fibers (LPF) [39,40], photonic bandgap fibers (PBG) [41–46] or photonic crystal fibers (PCF) [26,28,47], offer greatly reduced nonlinearities making them ideal for pulsed applications where intracavity peak powers tend to be very large. Due to the long interaction lengths typical of fiber lasers/amplifiers, nonlinearities such as stimulated Raman scattering and self-phase modulation become important. Photonic crystal-type fibers are thus ideal for pulsed applications owing to their large mode area doped cores, high absorption, and corresponding short length scales [48–50].

8.2.3.2 High power fiber laser architectures

Two different architectures for producing kW-class systems are commonly employed. Shown in Figure 8.6, the first is a monolithic design which utilizes fiber splicing to create an all-fiber architecture with no free-space components [51]. All-fiber architectures are made possible by past development of a variety of fiber components such as pump combiners, cladding pump strippers (CPS), high-reflective and partially reflective FBGs (HR- and PR-FBG), fiberized isolators (ISO) and endcaps. As an example, the seed source in Figure 8.6 is generated using single-mode active fiber spliced onto a pair of FBG's and pumped using a fiber coupled 976 nm laser diode. This seed signal is then amplified in a MOPA configuration through two cascaded amplifiers up to the ~2 kW level. Systems similar to this have demonstrated >5 kW with quasi-single mode output [21].

Monolithic designs such as this benefit from the lack of free-space components which removes the difficulties associated with free-space fiber coupling of seed and pump light into the fiber at high average powers. However, these systems are therefore susceptible to issues with fiber splicing such as scattering and heating at the splice point which can quickly lead to failure of the splice at high average power levels. Optimized splice routines which minimize scattering are critical to the successful implementation of all-fiber systems.

Free-space pumped kW-class fiber amplifiers have also been demonstrated with >4 kW output powers [20,23]. Shown in Figure 8.7 is a schematic of a free-space Yb:fiber amplifier. This system uses dichroic mirrors (DM) to discriminate between seed and pump light. Endcaps are used on either end of the fiber to reduce the pump and signal intensities on the glass/air interface. While these designs suffer from issues with free-space coupling stability at high powers, they offer much more flexibility in system layout since mode-matching between fibers at splice points is no longer required.

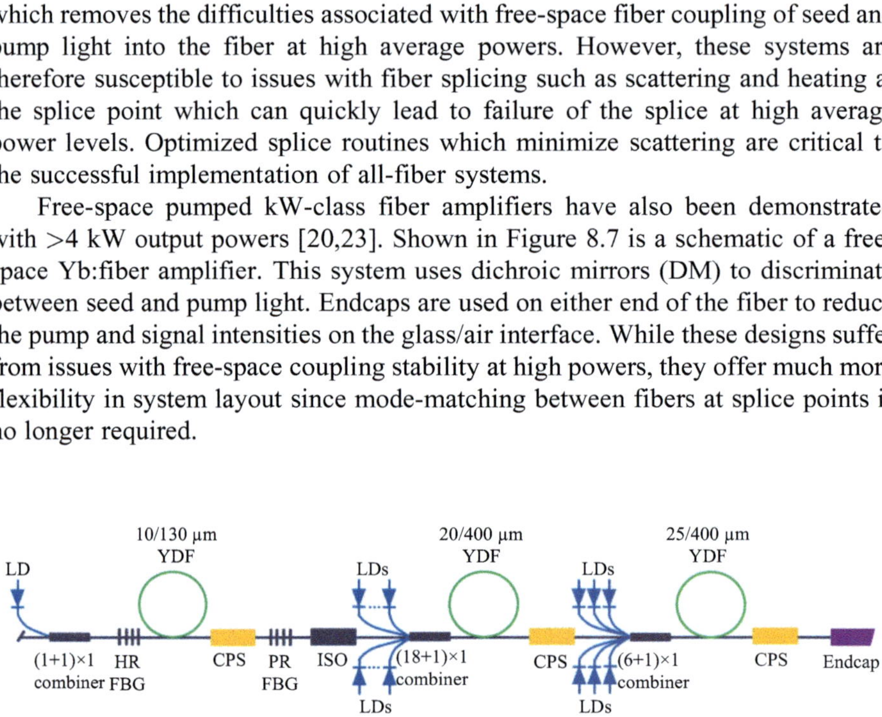

Figure 8.6 Layout of an all-fiber, monolithic fiber laser. CPS: cladding pump stripper. FBG: fiber Bragg grating. ISO: isolator. LD: laser diode. YDF: ytterbium-doped fiber [51]

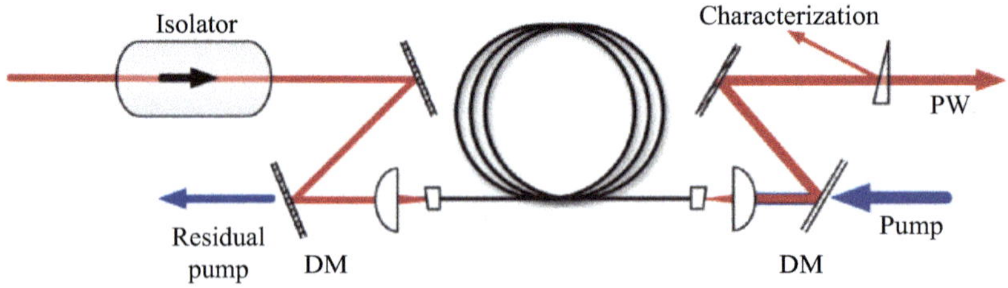

Figure 8.7 Schematic of a free-space pumped fiber amplifier [23]

8.2.4 Nonlinear optical loss mechanisms

8.2.4.1 Stimulated Raman scattering

First observed by Raman in 1928, stimulated Raman scattering (SRS) is a third-order nonlinear process which involves a pump beam propagating through an optical fiber to transfer energy to a secondary optical field [52]. This process is mediated by the phase matching of the pump beam with optical phonons present inside the medium. In the case of optical fibers, this is usually some composition of silica glass.

SRS can result in two different optical fields, one known as the Stokes field where pump photons are red-shifted in frequency and the anti-Stokes field where the pump photons are blue-shifted [52]. These processes are shown schematically in Figure 8.8. In both cases, the frequency difference between the pump and Stokes or anti-Stokes field is given by the Raman shift $\hbar\Omega$ of the medium. However, due to the phase-matching requirements inside the fiber waveguide, the probability of anti-Stokes Raman generation is much smaller than for Stokes light.

Like all third-order nonlinear effects, SRS is an intensity dependent process. Therefore, above a certain threshold pump power, SRS is expected to manifest itself and initiate the energy transfer process to the Stokes wave. This threshold was initially derived by Smith and is given by:

$$P_0^{\mathrm{crit}} = \frac{16 A_{\mathrm{eff}}}{g_R L_{\mathrm{eff}}} \tag{8.2}$$

where A_{eff} is the effective mode field area of the fiber and $L_{\mathrm{eff}} = \dfrac{1 - \mathrm{e}^{-\alpha_p \mathrm{L}}}{\alpha_p}$ is the effective fiber length determined by the background loss α_p at the pump wavelength [53]. For the anti-Stokes signal, the prefactor (2) is modified from 16 to 20. Due to this higher threshold power, the conversion to Stokes light usually dominates the SRS process. It is important to note that P_0^{crit} is proportional to A_{eff} and

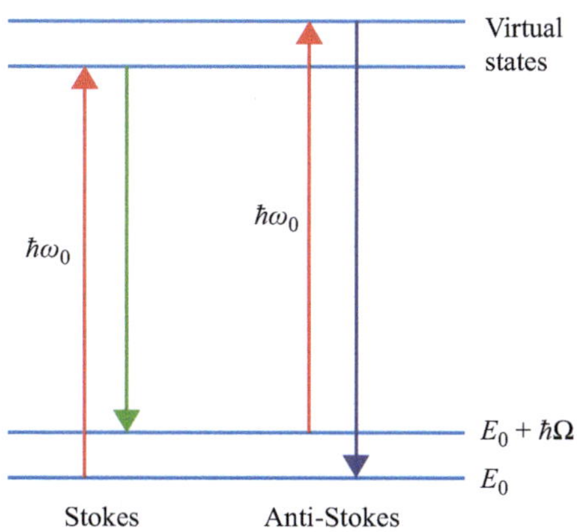

Figure 8.8 Energy level diagram for Stokes and anti-Stokes Raman scattering

inversely proportional to L_{eff}, indicating that larger mode field diameters and shorter fiber lengths are favorable to prevent the onset of SRS (Figure 8.9).

While SRS usually does not affect low power solid-state lasers, the effect becomes quickly apparent in the field of high power fiber lasers. Due to the high average powers obtainable from modern fiber lasers coupled with their potentially long length scales and small core diameters, SRS has been observed in a wide variety of fiber lasers in both CW and pulsed regimes [54–59]. In the CW regime, this phenomenon represents one of the major hurdles to overcome in terms of average power scaling [60,61]. As such, several mitigation strategies have been employed by various groups to hinder the onset of SRS, including engineering of the fiber refractive index profile [54], spectral shaping of fiber Bragg gratings used for output coupling [62], and the use of lumped spectral filters [63].

8.2.4.2 Stimulated Brillouin scattering

Stimulated Brillouin scattering (SBS) is very similar to SRS and in that the process is predicated upon the phase matching of a pump beam with phonons inside the medium. Rather than interacting with optical phonons, in the case of SBS the pump beam instead interacts with acoustic phonons. Due to the much smaller energy and momentum of the acoustic phonons compared with their optical counterparts, the phase matching bandwidth of SBS is severely reduced from 10 THz to around 10 MHz. However, due to random fluctuations in the refractive index profile and core diameter nonuniformities, the SBS bandwidth in silica has been experimentally observed to be on the order of 10 GHz [52,64,65].

Like the derivation for the Raman threshold, Smith also derived the following expression for the SBS threshold in silica fibers, given by:

$$P_0^{\text{crit}} = \frac{21\, A_{\text{eff}}}{g_B(\Omega_B) L_{\text{eff}}} \tag{8.3}$$

where $\hbar\Omega_B$ represents the Brillouin shift inside the medium [52,53]. This critical power follows the same scaling laws for length and mode area as the expression

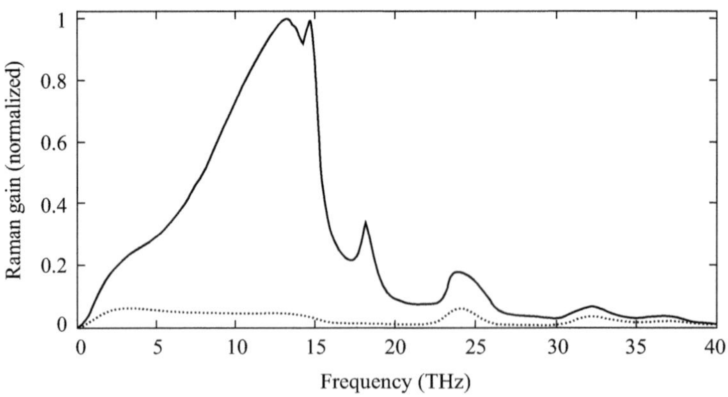

Figure 8.9 Raman gain spectrum in fused silica [52]

given for SRS. However, for SBS the gain coefficient is two to three orders of magnitude higher than for SRS at $g_B = 5 \cdot 10^{-11} \frac{m}{W}$, initiating the SBS process at much lower powers than required for SRS.

Due to the stringent phase matching requirements, SBS only becomes detrimental for fiber lasers with linewidths in the MHz range or less. While most applications do not require such narrow linewidths, several key technologies for power scaling such as spectral beam combining require this level of spectral control. If left unchecked, SBS can lead to a phenomenon known as modulational instability where the beam is modulated at Ω_B, creating sidebands at $\omega_0 \pm \Omega_B$ which can mix with the central tone through four-wave mixing (FWM) and cross-phase modulation (XPM) [52,66,67]. This eventually causes the beam to break-up into a chaotic or stochastic output in time and possibly initiate damage to the fiber [68]. Motivated by this, several methods have been developed to counteract the effects of SBS, especially at high powers.

Potentially the most widely applicable technique for SBS suppression is to use a phase modulator to broaden the spectral linewidth of the fiber laser system [69–71]. This allows for operation with linewidths in the 10s of GHz. Another promising technique is to seed the amplifier chain with two different tones within the gain bandwidth of the system [72–74]. Two-tone seeding relies on the laser gain competition between the two tones to suppress FWM and SBS, allowing for the efficient amplification of a narrow linewidth seed by suppressing parasitic lasing at shorter wavelengths near the peak of the gain bandwidth. While both techniques represent relevant methods to suppress SBS, their effectiveness must be assessed with respect to the experimentally achievable linewidth and the required linewidth for a certain application.

8.2.4.3 Transverse mode instability

Aside from SRS and SBS, the last main nonlinearity which impacts high average power fiber laser systems at 1 μm is thermal mode instability (TMI). This effect is unique in that it is dependent upon average power rather than peak power as is the case for third-order optical nonlinearities like SBS or SRS. TMI involves power transfer from the fundamental mode of a fiber waveguide into higher order modes and is observed in the form of degraded output beam quality of the optical system. TMI is observed to have a sharp threshold-like behavior beyond which the beam quality begins to degrade and has been reported in a wide variety of Yb:fiber systems [20,47,75–78].

The underlying physical mechanism of TMI involves two modes interfering in the doped fiber core. As the two modes propagate, due to their different effective indices, they form a modal interference pattern. This pattern then modulates the inversion within the fiber core. Through the thermo-optic properties of the fiber, this modulated inversion profile creates a refractive index grating in the fiber core. Due to its origin, this grating is exactly phase matched to transfer power between the fundamental mode and higher order mode that created the modal interference pattern [79–82].

8.3 High power TM fiber lasers

While high power 1 μm YDFLs have become the dominant fiber laser platform, fiber lasers operating at longer wavelengths are attracting attention as well. Thulium- and holmium-doped fiber lasers (TDFL and HDFL) produce output radiation in the 2 μm wavelength band [83,84]. This is the longest wavelength that can be efficiently produced in a fused silica host material because of the multi-phonon absorption edge near 2.4 μm. Other host materials enable emission beyond 2 μm, but with a large reduction in output power [85]. High power 2 μm sources are finding uses in materials processing [86–88], pump sources for nonlinear generation of mid-IR radiation [89,90], and directed energy/remote sensing utilizing high atmospheric transmission bands beyond 2.1 μm [91,92].

While CW HDFLs have been power scaled up to 407 W, these systems typically have low efficiencies <50% [93]. Issues with OH^- contamination and ion clustering have consistently restricted HDFLs from surpassing ~65 % laser efficiency. Fortunately, with improved fabrication techniques, recent reports have demonstrated HDFLs with >80 % efficiencies [94,95]. Nevertheless, high power HDFLs are pumped at 1.95 μm, which thus requires the development of high power TDFLs. As such, TDFLs still possess the highest potential for power scaling of all 2 μm fiber sources. This trend is expected to continue as improvements in fiber fabrication and high-brightness pump diodes are made.

8.3.1 Spectroscopic properties

While there are numerous electronic transitions for thulium-doped silica (Tm: silica), Figure 8.10(a) presents the radiative transitions that are most relevant as a 2 μm source. The most notable feature of Tm:silica is the extremely broad $^3F_4 \rightarrow {}^3H_6$ emission band. As seen in Figure 8.10(d), the emission extends from ~1,600 to ~2,200 nm and is one of the broadest emission bands for rare-earth doped silica. This has made Tm:silica an attractive platform for widely wavelength tunable 2 μm sources, as well as providing enough gain bandwidth to support amplification of ultrashort 2 μm pulses. Shown in Figure 8.10(c) and (d) are previously published absorption and emission cross-sections in the 2 μm band [83,94,96–101].

As seen in Figure 8.10(a), pumping at ~1,550–1,600 nm or 790 nm are the standard approaches for generating 2 μm emission. Pumping into the $^3H_5$ band has also been investigated using 1.06 μm sources, but it is less attractive because of the minuscule absorption cross-section. Pumping at 790 nm into the $^3H_4$ level is currently the standard excitation scheme for high power TDFLs. This is because commercially available, high power/brightness fiber-coupled diodes are now readily available from a variety of suppliers [105]. In addition, the $^3H_6 \rightarrow {}^3H_4$ transition has a large absorption cross-section, with published values ranging from 0.45 to 1.0 pm^2 [96,98,100]. This helps facilitate pump absorption from the highly multimode 790 nm pump diodes. However, the efficient absorption afforded by pumping at 790 nm generates a large ~60% quantum defect when extracting at 2,000 nm, which would be deposited as heat in the fiber. Such a large thermal load

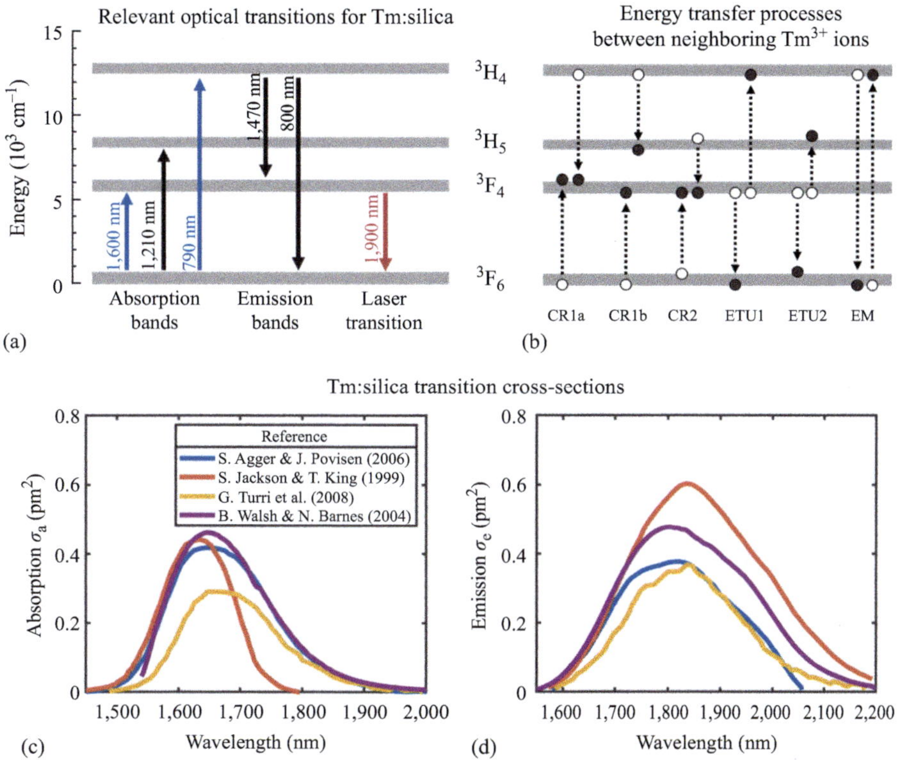

Figure 8.10 (a) Radiative transitions relevant for 2 μm TDFLs and (b) energy transfer processes that can occur between two neighboring thulium ions. Common pump bands are highlighted in blue and the laser transition is highlighted in red. (c–d) Selection of published $^3F_4 \leftrightarrow {}^3H_6$ absorption/emission cross-sections for Tm^{3+}-doped silica, reproduced from [97–99,102–104]

would prevent power scaling. Fortunately, the cross-relaxation process (CR) can generate two electrons in the $^3F_4$ state for each input pump photon [106]. The CR process is thus essential to improve overall efficiencies of TDFLs beyond 40%.

Several important ion–ion energy transfer processes which affect the lasing performance in Tm:silica are depicted in Figure 8.10(b). As mentioned, CR is extremely beneficial when operating in the 2,000 nm band since it assists in populating the upper laser level ($^3F_4$) and improves optical conversion efficiencies. The opposite process, ETU, is undesirable because it quenches the $^3F_4$ upper laser level. CR1a and CR1b are indistinguishable from the laser's perspective because the $^3H_5$ lifetime is exceedingly fast (~7 ns [103]). Therefore, these can be treated as just one process labeled CR1. CR1 constitutes the dominant energy transfer process in Tm:silica, while CR2 is very unlikely since it requires multiphonon assistance [106,107]. Multiphonon assistance is required for the ETU1 and ETU2 processes as well, therefore making them weaker than CR1. However, these ETU processes are still important to consider because they reduce CR efficiency and negatively affect laser performance [108]. Energy migration (EM) is a resonant process that redistributes neighboring ion populations between the $^3H_4$ and $^3H_6$ levels. This can be beneficial for CR because it effectively prolongs the excited $^3H_4$ level.

Due to the resonant nature of the energy transfer process, the efficiency of cross-relaxation is directly associated with the thulium ion–ion separation, and thus to the Tm^{3+} dopant concentration [109]. As the Tm^{3+} ion concentration increases, ion clustering becomes a major issue. Co-doping with Al^{3+} ions has been shown to increase the Tm^{3+} solubility in the silica host [110]. The Al^{3+} co-dopant modifies the local environment of Tm^{3+} ion by lowering the maximum phonon energy [111,112]. Al^{3+}:Tm^{3+} ratios >10:1 have been demonstrated experimentally to prevent concentration quenching [113]. Thus, highly doped fibers with $\geq$ 4.0 wt.% Tm^{3+} and large Al^{3+} co-doping levels are typically employed to maximize the effects of CR and improve TDFL efficiencies beyond 40% when pumped at 790 nm.

The slope efficiencies of several published Tm:silica lasers pumped at 790 nm are shown in Figure 8.11 as a function of doping concentration with data compiled from [96,106,113–130]. A clear trend is visible which indicates increasing slope efficiency with increasing dopant concentration. Efficient utilization of the CR process only occurs for doping concentrations >2.0 wt.%, which allows for >40% slope efficiencies. However, further increasing the Tm^{3+} concentration beyond

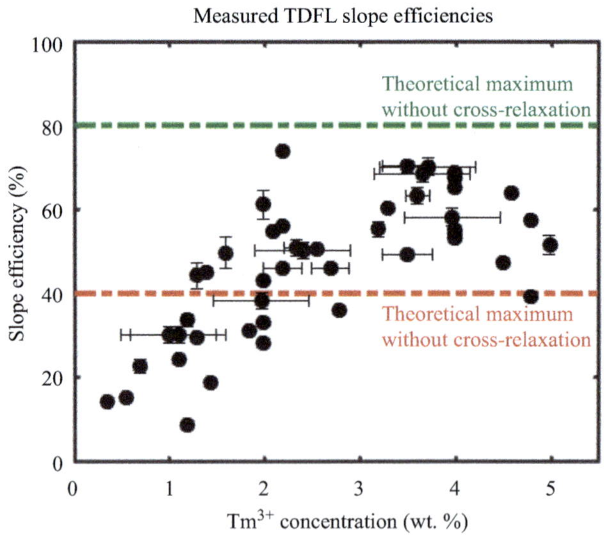

Figure 8.11 Slope efficiencies demonstrate improved CR with increasing Tm^{3+}-dopant concentration. Published slope efficiencies are reproduced from [96,104,106,113–130]

3.5 wt.% appears to have no further benefit in terms of increased efficiency. Previous studies have suggested that improving the fabrication of a flat-top dopant profile across the core region to ensure the entire core participates in the CR process may further improve the performance of TDFLs [129]. A recent breakthrough of fabricating a very highly doped (~8.5 wt.%) TDF has demonstrated 74% slope efficiencies, indicating near perfect CR efficiency [131]. These highly doped fibers may pave the way toward the next generation of high power, high-efficiency TDFLs.

Overall, there are several absorption bands which enable the use of various pumping schemes to excite the thulium ion. However, only those highlighted in blue in Figure 8.10(a) are of interest for high power systems. Pump wavelengths shorter than 790 nm are undesirable since they limit maximum conversion efficiencies to <35% due to quantum defect heating. While pumping at 1,210 nm into the $^3H_5$ level could likely reach 50%–60% efficiencies, suitable high power pump sources do not exist to implement this architecture. Current sources emitting at 1,210 nm rely on Raman-shifted YDFLs due to a lack of suitable rare-earth ion transitions at this wavelength [132]. Therefore, high power >100 W-class TDFLs are either pumped into the $^3H_4$ level at 790 nm or the $^3F_4$ level at 1,550–1,950 nm. Both of these pumping schemes benefit from access to high power/brightness sources and allow for optical-optical efficiencies exceeding 60%. Sections 8.3.2–8.3.3 will summarize notable advancements in high power TDFLs for these two pump bands.

8.3.2 Cross-relaxation pumping with 790 nm diodes

The first demonstration of a high power 790 nm pumped TDFL was a free-space pumped oscillator with 85 W output in 2005 [118]. TDFL output powers increased to the 300 W level by 2009, with 885 W available combined from the two output ports of an oscillator [124]. Also in 2009, a single frequency system emitting 608 W was demonstrated. This system was notable for its narrow linewidth (<5 MHz) and single mode beam quality with $M^2 < 1.06$ [125,133]. In 2010 the first kW-class TDFL was developed. This system utilized a two-stage, all fiber MOPA configuration with 53.2% slope efficiency [134]. For nearly a decade, this demonstration remained the highest CW power obtained from a 2 μm fiber laser until 2018 in which a thulium-doped PCF amplifier delivered 1,150 W with 61% efficiency and $M^2 < 1.1$ for an ultrafast laser [135]. This system utilized direct water-cooling of the TDF, which was double-end-pumped by eight high performance 790 nm pump diodes. Notable high power results have been compiled in Figure 8.12(a).

Since 2010, most work on high power TDFLs has emphasized improved system design and/or utilizing Tm^{3+}-doped silica's full emission bandwidth for ultrashort pulse generation, rather than CW power scaling beyond the 1 kW level. Thus far, TDFLs have exploited the broad $^3F_4 \rightarrow {}^3H_6$ transition to operate at the long wavelength edge near 2,200 nm as well as the short wavelength side down to 1,650 nm with 1,565 nm pumping [41,136,137]. Long fiber lengths and low upper state inversion are typical of 790 nm pumping, making long wavelength operation preferential.

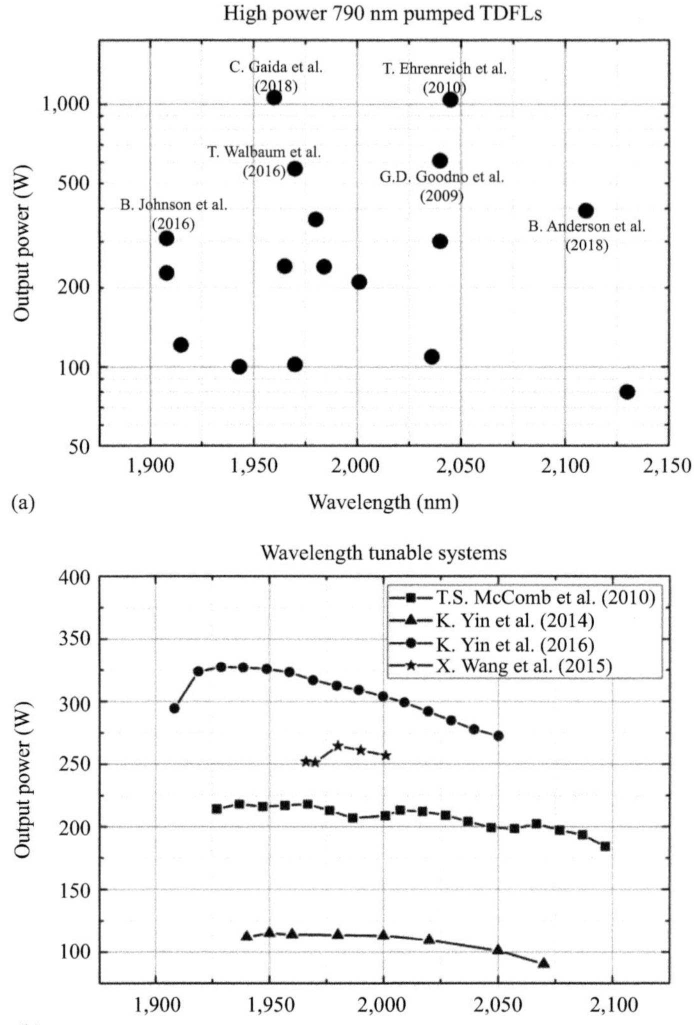

Figure 8.12 (a) Selection of notable high power TDFLs pumped by 790 nm diodes [104,124,125,133,134,138–148,152,153]. (b) High power, wavelength tunable TDFLs that utilize the broad emission of Tm:silica [104,127,149–151]

Therefore, 1,950–2,100 nm is the typical emission band for high power, 790 nm pumped TDFLs as depicted in Figure 8.12(b) [124,125,127,133,134,138–152].

T. McComb *et al.* demonstrated in 2010 the first high power, tunable TDFL with >185 W from 1,927 to 2,097 nm [127]. This system made use of a free-space coupled MOPA architecture with a diffraction grating in the oscillator for spectral control. Improvements in fiber component technology improved coupled with the increased commercial availability of 2 μm fiber components have since enabled more elegant, all-fiber architectures. An all-fiber architecture such as this was leveraged by Yin *et al.* to construct an all-fiber MOPA tunable from 1,910 to 2,050 nm and up to 327 W output power [151]. Another all-fiber MOPA using spectrally filtered ASE as the seed source generated >250 W with wavelength coverage from 1,966 to 2,001 nm [150]. Here it is worth noting that excellent temporal stability was provided by this spectrally filtered ASE seed source.

Due to the quasi-three-level nature of the Tm^{3+} ion, signal reabsorption becomes a major concern when operating <1,950 nm. Therefore, short wavelength operation below 1,950 nm requires short fiber lengths with high signal gain and pump absorption [123]. Using a single mode fiber with increased core/cladding overlap, 308 W output was demonstrated at 1,908 nm from a 2.3 m oscillator [139]. In contrast, long wavelength operation > 2,100 nm is dominated by ASE competition and large background losses. Published work by Anderson *et al.* has reported an output power of 393 W at 2,110 nm and 80 W at 2,130 nm [148,153]. To overcome the ASE and background loss issues, a technique known as two-tone seeding was employed where the main amplifier was simultaneously seeded at 1,950 and 2,110 nm [148,153]. The 1,950 nm light was reabsorbed in the long amplifier, further contributing to the 2,110 nm gain and providing 44% efficiency.

Most of the fiber laser systems discussed thus far rely on a MOPA architecture. By adopting a high power oscillator architecture as opposed to a MOPA, much of the complexity and cost of the system can be reduced to more manageable levels. To date, a 567 W all-fiber system at 1,970 nm remains the highest power TDF oscillator [146].

8.3.3 In-band pumping with 1,550—1,950 nm sources

Rather than relying on 790 nm diode pumps, an alternative approach to construct high power TDFLs involves optically pumping directly into the upper laser level $^3F_4$ via a process known as in-band pumping. In-band pumping provides access to much of the bandwidth available with 790 nm pumping but with the added benefits of higher efficiency and greatly reduced quantum defect heating. This approach also bypasses the need for efficient CR since the $^3H_4$ level is no longer involved in the laser dynamics. However, high power/brightness pump diodes in the 1,550–1,950 nm band do not currently exist [154]. As such, only two options currently exist which enable in-band pumping of a TDFL: pumping with erbium-doped fiber lasers (EDFLs) in the 1.5–1.6 μm band or pumping with another TDFL near 1.9 μm. Previous development of high power EDFLs has been limited to 300 W output powers with little to no potential for further power scaling [155,156]. This is due to issues from low absorption cross-sections, high quantum defect, and concentration quenching [157]. Recently in 2018, a record 656 W multimode EDFL was reported, but ~25% of the power was located in the cladding and the laser required nearly 2 kW of pump power [158]. Nevertheless, an in-band pumped TDFL with 415 W output and 60% slope efficiency was demonstrated in 2007 and relied on the incoherent combination of eighteen 40 W EDFLs to serve as the pump source [159]. To date, this remains the highest power for an in-band pumped TDFL.

As with YDFLs, the adoption of the in-band pumping technique benefits from the intrinsically higher conversion efficiencies and subsequently lower thermal loads compared with 790 nm pumping. However, excitation at ~1,550 nm with an EDFL and extracting at 2,000 nm caps the overall optical-optical efficiency of such systems at <80%. Since ~75% efficiencies are obtainable with 790 nm pumping, coupled with the availability and reliability of commercial 790 nm laser diodes, in-band pumping with EDFLs has no practical advantages over direct diode pumping.

A more promising alternative is to use a short wavelength TDFL, typically operating near 1,908 nm, to pump another longer wavelength TDFL [15]. Quantum defects <10% are easily obtainable with this approach [104,160]. This performance is much higher than systems pumped by EDFLs and are even similar to optimized YDFLs. Creeden *et al.* have previously demonstrated a TDFL pumped at 1,908 nm which produced 123 W emitting at 1,993 nm with 91.6% efficiency, demonstrating the feasibility of this approach for power scaling [161]. In addition, Figure 8.13 shows measured and modeled efficiencies for a low-power, in-band pumped TDF oscillator [104]. This system utilized 4.5 m of Tm-doped fiber and a 1908 nm pump source [104]. Measured efficiencies are >90% of the quantum defect limit with absolute efficiencies of >80% across over a ~100 nm tuning range. The drop in efficiency at longer wavelengths stems from increasing background loss as the 2.4 μm multiphonon edge is approached, stressing the need for extremely pure host materials with low $OH^{\mp}$ concentrations [162–164]. The low quantum defect from in-band pumping provides a stark reduction in thermal load as compared to 790 nm pumping. In fact, in-band pumping is a method used to generate multi-kW YDFLs by pumping with several shorter wavelength YDFLs [165,166]. Since this architecture has proven successful for high power YDFLs, it is anticipated that in-band pumping will produce both high power and high efficient TDFLs.

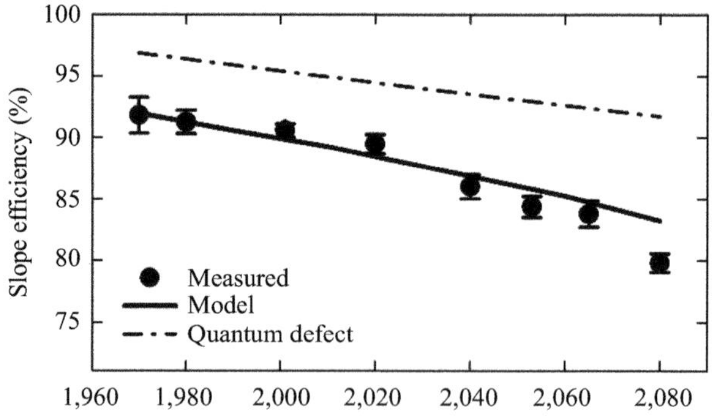

Figure 8.13 Slope efficiency performance of a single-mode TDFL in-band pumped at 1908 nm. Measured efficiencies are >80% across the 1,980–2,080 nm band [104]

8.4 Other fiber laser media

8.4.1 Er-fiber lasers

Erbium-doped fiber lasers (EDFLs) are attractive for their operating range (1.5–1.6 μm), which falls into the "eye safe" regime while maintaining high transparency in the atmosphere. Compared to ytterbium-doped fiber lasers (YDFLs), high power EDFLs have lagged behind significantly; with quasi-single-mode output power limited to 264 W [155] and overall maximum power limited to 656 W [158]. These low powers are largely due to difficulties in cladding pumping due to concentration quenching [167], low absorption cross-section at 980 nm, and excited state absorption.

8.4.1.1 Concentration quenching

Conceptually similar to CR pumping in Tm, concentration quenching, also known as pair induced quenching [168], is an energy transfer between two neighboring excited ions. As shown in Figure 8.14, the excited state ($I_{13/2}$) energy is transferred to another excited ion, where it is either decays nonradiatively or emits a 0.98 μm photon. In the former, more likely case, the excited state energy from the donor is lost to heat. In the latter case, both the donor and acceptor energies are lost to 0.98 μm radiation and heat. Various co-dopants are often used to reduce the effect of concentration quenching and achieve higher pump absorption.

8.4.1.2 Er:Yb codoped fibers

Erbium:ytterbium (Er:Yb) codoped fibers take advantage of the relatively high absorption of the Yb ions at the readily attainable 915–980 nm wavelength range. The Er ions are indirectly pumped by nonradiative energy transfer from the excited Yb ions. This method is subject to very low quantum efficiency and very high thermal load, but is still a popular method for achieving high powers in the 1.5 μm

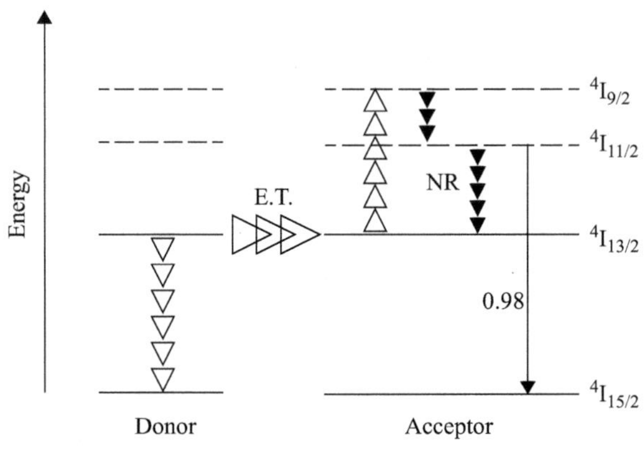

Figure 8.14 Illustration of the concentration quenching phenomenon [167]

range [155,169–179]. Unfortunately, it is subject to parasitic lasing at the Yb band, which can cause catastrophic damage to the laser system (Figure 8.15).

The current state of the art Er:Yb codoped fiber laser was produced 264 W in 2014, with a core diameter and NA of 17 μm and 0.08, respectively. It was pumped by 36 fiber lasers, each operating at 11 W at 1,535 nm to reduce the thermal load and avoid parasitic lasing in the Yb band [155].

8.4.1.3 Yb-free Er-doped fiber

Demonstrating high power at the expense of beam quality, researchers at the Optoelectronics Research Centre in Southampton have achieved 656 W at 1,601 nm with a measured $M^2 \sim 10.5$. In order to accommodate the high pump powers of the diodes used in this system while avoiding quenching and achieving moderate absorption, a custom 146 μm core, 700 μm inner cladding was developed (Figure 8.16). The 36 m fiber was double-end free space pumped with 977 nm diodes with DM used for feedback at the signal wavelength [158].

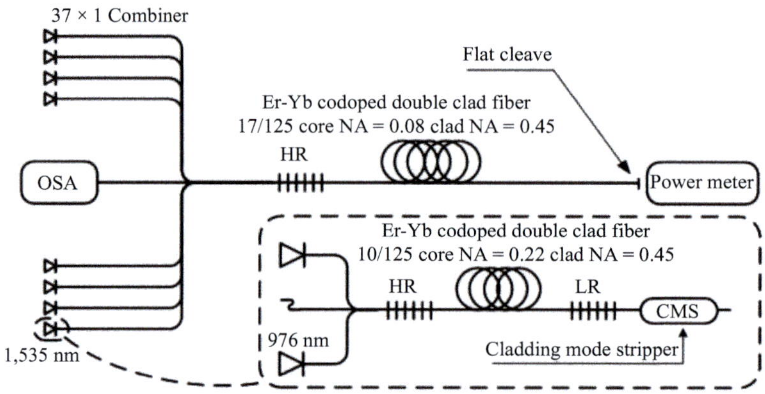

Figure 8.15 264 W Er:Yb fiber laser schematic [155]

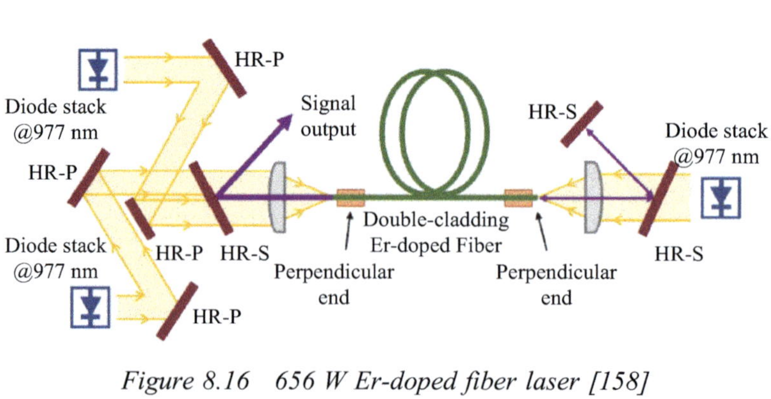

Figure 8.16 656 W Er-doped fiber laser [158]

8.5 High power Raman lasers

In Section 2.2.1, the stimulated Raman effect was described as a parasitic effect on high power fiber lasers. However, The Raman effect can be purposefully stimulated and used to generate a coherent light source. Though achievable in any medium, high efficiencies require high intensity over a long interaction length, so fibers or materials with very high Raman gain are desirable [180]. As it is associated with the vibrational effect of the medium, the materials do not have to be doped with a rare earth material in order to see gain, though germanium doping for higher non-linear refractive index is common in RFLs.

Though requiring brighter sources for pumping, RFLs offer several advantages compared to rare-earth fibers. First, Raman fibers are not limited by a specific pump band, as the Stokes signal due to the Raman Effect is a static frequency shift depending on the hose medium (maximum at 13.2 THz in silica fiber). In addition, the broad bandwidth of the Raman gain offers good tunability when coupled with a seed or wavelength selective optic such as a Bragg grating [180–183]. This allows full flexibility in output wavelength limited only by pump availability and the transparency window of the gain material [184]. Second, passive fibers are not susceptible to photo darkening: the gradual increase of absorption in a fiber associated with the formation of color centers. This increases the lifetime of the gain fiber and may delay the onset of transverse modal instability. TMI has not so far been demonstrated in a RFL, though numerical models have theorized that it would eventually occur [185,186]. Finally, the low gain per unit length of Raman fibers offers advantages on thermal load and eases cooling requirements of high powered systems. Raman lasers rarely have a linewidth small enough to promote SBS, so the generation of a parasitic second Stokes transition is the primary obstacle to high power.

Like conventional lasers, Raman lasers are typically separated into either wavelength conversion RFLs (core pumped) or brightness enhancement (cladding pumped or MM-GRIN), with different considerations for each [181,182,187–191].

8.5.1 Raman fiber lasers for wavelength conversion

In some cases, the Raman conversion takes place in the same fiber as the pump light generation. These configurations are sometimes known as hybrid Raman lasers and have reached power levels on the order of 4 kW [192]. The advantage of these systems is that additional control can be achieved by changing forward and backward pump intensities, granting some amount of control over output Raman power while suppressing second Stokes. In Figure 8.17, one such laser is shown to have achieved 3,890 W with an optical to optical efficiency of 70.9% from the 976 nm pump power of 5,487 W. The disadvantages to such a system compared to a passive RFL is the very high power propagating in a rare-earth-doped system, encouraging photo darkening in a relatively short lifetime. In addition, special couplers were required to prevent backwards lasing from damaging the pump laser diodes. In this case there is no brightness improvement from the ytterbium conversion to first Stokes.

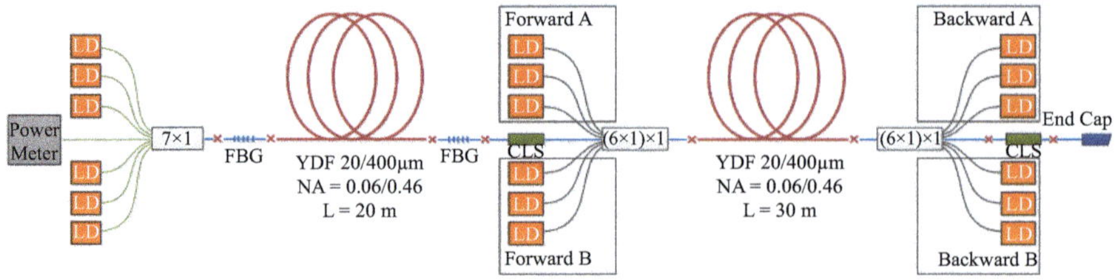

Figure 8.17 Hybrid Raman fiber laser. A 290 W 1,070 nm laser is amplified in 30 m of ytterbium-doped fiber. The conversion to first Stokes at 1,123 nm is seeded by spontaneous Raman noise in the oscillator [192]

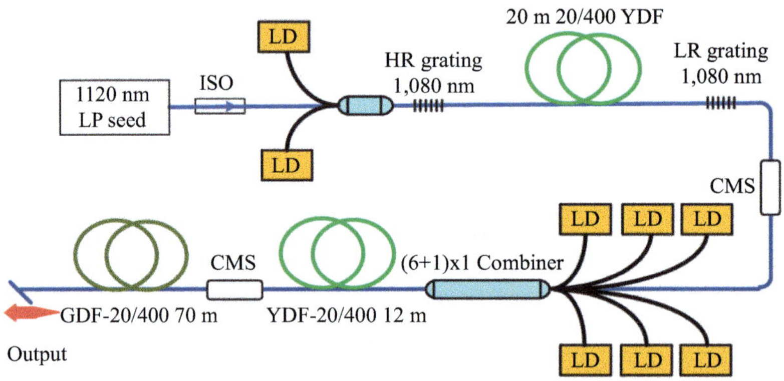

Figure 8.18 1.28 kW Raman fiber laser. The 1,120 nm seed is propagated through the entire MOPA and into the Raman conversion fiber [193]

More commonly, the Raman conversion fiber is a passive length that is either coupled to the high brightness pump laser or is spliced on after [193]. These can be thought of as Raman amplifiers, as there is no circulating power at the Stokes wavelength (Figure 8.18).

8.5.2 Raman fiber lasers for brightness enhancement

RFLs can produce light with a higher brightness than the pump beam. Here, brightness is defined as the quotient of the power and the beam parameter product (BPP), and is conserved in standard optics neglecting aberration. Raman lasers for brightness enhancement can be pumped by a variety of sources, including incoherently multiplexed fiber lasers [188,194], free space coupled or fiber coupled diode lasers [195,196], or other solid-state lasers.

It was found in [190] that for efficient first Stokes generation without parasitic loss to second Stokes in a cladding pumped double-clad fiber, the cladding to core area ratio must be <8, which puts a low limit on brightness enhancement in such fibers. This can be mitigated if there is sufficient loss mechanism for second Stokes inscribed in the fiber or inherent to the design of the laser system, such as a resonator configuration that does not allow feedback in the second Stokes wavelength.

It is well established that core pumped Raman lasers in graded index fibers (GIF) demonstrate brightness enhancement [189,191,195,197,198]. The more highly confined modes are in a medium with higher germanium content, which provides higher nonlinearity and higher Raman gain [189]. As the Raman signal will have higher confinement and lower NA than the pump, the brightness enhancement can be quite significant.

Of the RFLs incorporating brightness enhancement, the highest power reported is 250 W in 2018 [194], with a brightness enhancement of 3.8. An earlier design from the same group is shown in Figure 8.19 wherein two diode lasers are combined in free space and coupled into a GRIN fiber. The pump is double passed by way of mirror M3, but higher order Stokes are ejected at that point. The experimental setup for this laser is shown in Figure 8.20, and is an example of an all-fiber Raman laser setup.

As an alternative to diode or fiber lasers, Raman lasers can be pumped with solid-state lasers to improve brightness and offer some wavelength flexibility. Shown in Figure 8.21 is a high average power thin-disk pumped fiber laser that demonstrated with output peak power in the tens of kilowatt range, and average power of 70W, produced at the Laser Plasma Laboratory [199].

For systems such as this, a large area multimode graded index fiber (MM-GRIN) is required, as the pump does not have the high beam quality one would expect from a fiber laser.

The pump was pulsed, with a low power CW seed to have some control over efficiency at different powers. The very high peak power allows for much smaller

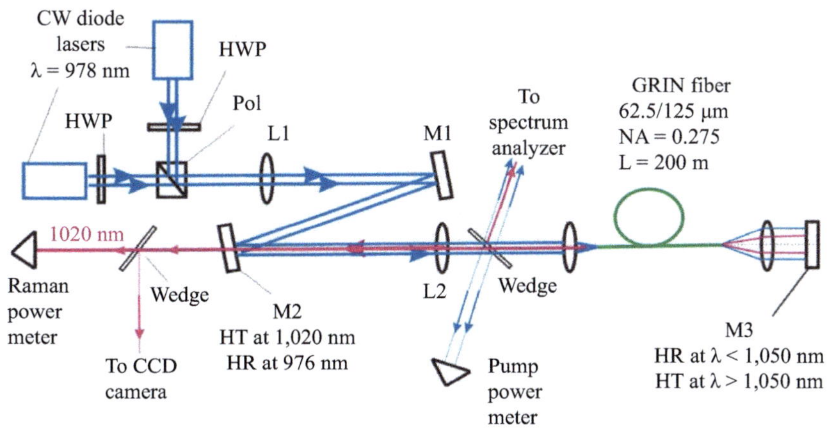

Figure 8.19 154 W output power fiber Raman laser, free spaced pumped by laser diodes. The pump is passed through the GRIN fiber twice, and there is a flat cleave on the left side of the fiber to act as a 4% feedback mechanism [195]

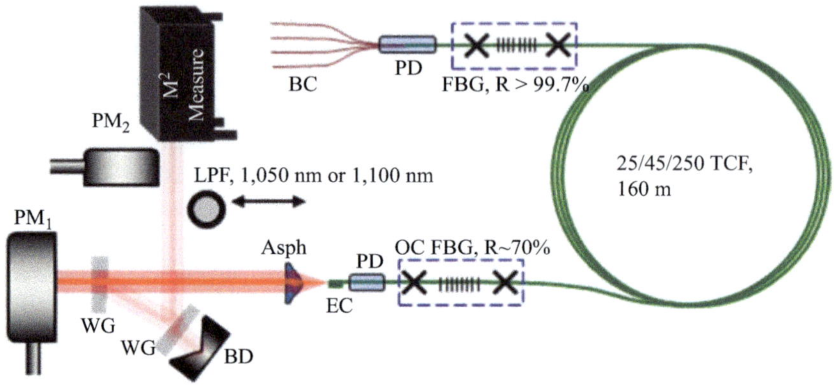

Figure 8.20 All fiber configuration with a high circulating power oscillator in triple-clad fiber, resulting in 250 W output power [194]

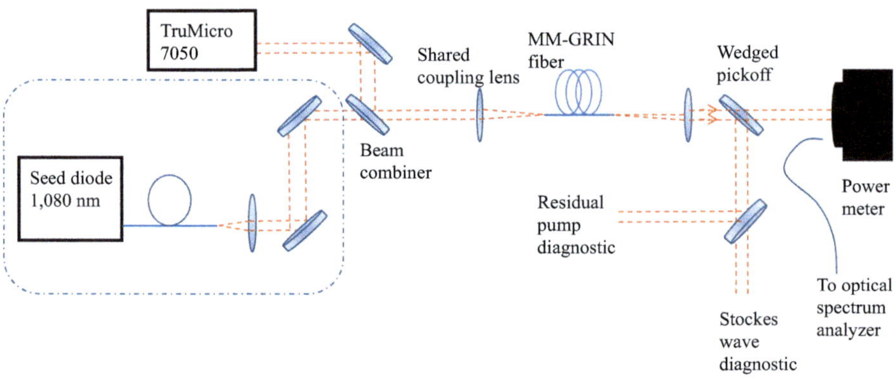

Figure 8.21 Solid-State pumped Raman fiber laser. A 30 ns pulsed disk laser is co-propagated with a low power seed in 200 μm MM GRIN fiber [199]

Raman gain fiber, and in this case the optimum fiber length for efficiency at ~5 mJ input power was 10–12 m, compared to 160 to 200 m in the aforementioned cases, and multiple kilometers in lower power configurations on the order of several watts. This is a good example of the potential for extreme scaling of brightness enhancing Raman lasers. In such configurations, length and seed optimization are key for good efficiency, as modes with higher confinement can be converted to second Stokes before modes with lower confinement are converted to first. Full depletion of the pump wavelength at 1,030 nm is never achieved, but at least four orders of Stokes are readily apparent in the spectral evolution by the time the input energy reaches 4 mJ.

Another common approach for Raman lasers with or without brightness enhancement is a resonator at the Raman wavelength. The resonator configuration offers several advantages: The circulating power is high, so the fiber length can be relatively small. The FBGs can offer high brightness mode selectivity, providing another path to brightness enhancement. Finally, and most importantly, the FBGs can selectively prevent the buildup of second Stokes, so the power in first Stokes can be scaled to otherwise unattainable levels while defeating the brightness limitations of step-index fibers [182,193,197,200–202].

Outside of the kW regime, the single mode wavelength conversion in Tellurite fiber or Chalcogenide fiber, or even gas-filled fibers [203–206]. These have been used to achieve tens of watts in the mid-IR, which is valuable for a number of medical surgery [207], spectroscopy, gas sensing, and defense [208]. The only limitation is the cumulative quantum defect of the Raman process, background loss, and the availability of pump sources within the transparency window.

8.6 Conclusions

In this chapter, we have introduced the concepts and technologies which have enabled the development of high power fiber lasers over the past two decades. Specifically, Yb-doped and Tm-doped fiber lasers currently possess the highest output powers of any rare-earth-doped fiber technology. RFLs which do not rely on active rare-earth dopants have made substantial progress in power scaling as well and can operate at wavelengths not attainable in Yb-doped or Tm-doped fibers. In addition, detrimental effects such as thermal failure and optical nonlinearities have been explored in the context of high power fiber lasers. As this is still an extremely active area of research, future progress in power scaling of fiber lasers at all wavelengths is expected.

References

[1] T. H. Maiman, "Stimulated optical radiation in ruby," Nature **187**, 493–94 (1960).
[2] S. W. H. J. Silvers Jr., "Perforating, welding and cutting plastic films with a continuous CO2 Laser," PA State University, Eng. Proc., 88–97 (1970).
[3] C. K. N. Patel, "Continuous-wave laser action on vibrational-rotational transitions of CO_2," Phys. Rev. **136**, A1187–A1193 (1964).
[4] D. A. Belforte, "Milestones in industrial laser materials processing," (www.industrial-lasers.com, 2010).
[5] W. F. K. J. L. Emmett and W. R. Sooy, "The potential of high-average-power solid state lasers," (UCRL-53571, 1984), pp. 1–48.
[6] G. H. Miley, "Review of nuclear pumped lasers," in *Laser Interaction and Related Plasma Phenomena*, G. H. Miley, ed. (Springer, Boston, MA, 1984), pp. 47–72.
[7] L. C. Conant and C. W. Reno, "GaAs laser diode pumped Nd:YAG laser," Appl. Opt. **13**, 2457–58 (1974).
[8] I. M. M. Richardson, S. Nolte, A. Tunnerman, I. Kelbassa, and R. Poprawe, "From femtosecond laser-materials processing to 3-D printing – laser-based

technologies that will transform manufacturing," in *15th Int. Symp. on Laser Precision Microfabrication* (Vilnius, Lithuania, 2014).
[9] E. Snitzer, "Optical maser action of Nd^{+3} in a Barium crown glass," Phys. Rev. Lett. **7**, 444–46 (1961).
[10] E. Snitzer and H. Osterberg, "Observed dielectric waveguide modes in the visible spectrum*," J. Opt. Soc. Am. **51**, 499–505 (1961).
[11] R. Paschotta, J. Nilsson, A. C. Tropper, and D. C. Hanna, "Ytterbium-doped fiber amplifiers," IEEE J. Quantum Electron. **33**, 1049–56 (1997).
[12] R. M. MacFarlane and R. M. Shelby, "Homogeneous line broadening of optical transitions of ions and molecules in glasses," J. Lumin. **36**, 179–207 (1987).
[13] J. T. Fournier and R. H. Bartram, "Inhomogeneous broadening of the optical spectra of Yb3+ in phosphate glass," J. Phys. Chem. Solids **31**, 2615–24 (1970).
[14] J. Nilsson, P. Blixt, B. Jaskorzynska, and J. Babonas, "Evaluation of parasitic upconversion mechanisms in Er^{3+}-doped silica-glass fibers by analysis of fluorescence at 980 nm," J. Light. Technol. **13**, 341–49 (1995).
[15] P. Zhou, H. Xiao, J. Leng, *et al.*, "High-power fiber lasers based on tandem pumping," J. Opt. Soc. Am. B **34**, A29-A36 (2017).
[16] E. Shcherbakov, V. Fomin, A. Abramov, *et al.*, "Industrial grade 100 kW power CW fiber laser," in *Advanced Solid-State Lasers Congress* (Optical Society of America, 2013), p. ATh4A.2.
[17] Y. Jeong, J. K. Sahu, D. N. Payne, and J. Nilsson, "Ytterbium-doped large-core fiber laser with 1.36 kW continuous-wave output power," Opt. Express **12**, 6088–092 (2004).
[18] V. Gapontsev, D. Gapontsev, N. Platonov, *et al.*, "2 kW CW ytterbium fiber laser with record diffraction-limited brightness," in *CLEO/Europe. 2005 Conference on Lasers and Electro-Optics Europe, 2005* (2005), p. 508.
[19] V. Khitrov, J. D. Minelly, R. Tumminelli, V. Petit, and E. S. Pooler, "3kW single-mode direct diode-pumped fiber laser," in *Proc. SPIE 8961, Fiber Lasers XI: Technology, Systems, and Applications* (SPIE, 2014), p. 89610V.
[20] F. Beier, F. Möller, B. Sattler, *et al.*, "Experimental investigations on the TMI thresholds of low-NA Yb-doped single-mode fibers," Opt. Lett. **43**, 1291–94 (2018).
[21] K. Shima, S. Ikoma, K. Uchiyama, Y. Takubo, M. Kashiwagi, and D. Tanaka, "5-kW single stage all-fiber Yb-doped single-mode fiber laser for materials processing," in *Proc. SPIE 10512, Fiber Lasers XV: Technology and System* (SPIE, 2018), p. 6.
[22] B. Shiner, "The impact of fiber laser technology on the world wide material processing market," in *CLEO: 2013* (Optical Society of America, San Jose, CA, 2013), p. AF2J.1.
[23] F. Beier, C. Hupel, S. Kuhn, *et al.*, "Single-mode 4.3 kW output power from a diode-pumped Yb-doped fiber amplifier," Opt. Express **25**, 14892–899 (2017).
[24] S. O. Selee Chang, Ilkwon Oh, Eunje Sung, Junghyun Oh, and Sukyoun Suh, "UV curable low refractive index clad coatings for high power fiber laser applications," Available from http://sscpusa.com/notice/UV%20Curable%20Low%20Refractive%20Index%20Clad%20Coatings%20for%20high%20power%20lasers.pdf (2016).

[25] C. X. Yu, O. Shatrovoy, T. Y. Fan, and T. F. Taunay, "Diode-pumped narrow linewidth multi-kilowatt metalized Yb fiber amplifier," Opt. Lett. **41**, 5202–05 (2016).

[26] J. Limpert, O. Schmidt, J. Rothhardt, *et al.*, "Extended single-mode photonic crystal fiber lasers," Opt. Express **14**, 2715–20 (2006).

[27] N. A. Mortensen, "Air-clad fibers: pump absorption assisted by chaotic wave dynamics?," Opt. Express **15**, 8988–96 (2007).

[28] J. Limpert, T. Schreiber, S. Nolte, *et al.*, "High-power air-clad large-mode-area photonic crystal fiber laser," Opt. Express **11**, 818–23 (2003).

[29] L. Dong, H. A. McKay, L. Fu, *et al.*, "Ytterbium-doped all glass leakage channel fibers with highly fluorine-doped silica pump cladding," Opt. Express **17**, 8962–69 (2009).

[30] T. Ehrenreich, R. Leveille, I. Majid, K. Tankala, G. Rines, and P. Moulton, "1-kW, all-glass Tm: fiber laser," in *Proc. SPIE 7580, Fiber Lasers VII: Technology, Systems, and Applications* (2010), p. 758001.

[31] C. X. Yu, O. Shatrovoy, and T. Y. Fan, "All-glass fiber amplifier pumped by ultrahigh brightness pump," in *Proc. SPIE 9728, Fiber Lasers XIII: Technology, Systems* (SPIE, 2016), p. 972806.

[32] E. Snitzer, H. Po, F. Hakimi, R. Tumminelli, and B. C. McCollum, "Double clad, offset core Nd fiber laser," in *Optical Fiber Sensors* (Optical Society of America, New Orleans, LA, 1988), p. PD5.

[33] H. Po, E. Snitzer, R. Tumminelli, *et al.*, "Double clad high brightness Nd fiber laser pumped by GaAlAs phased array," in *Optical Fiber Communication Conference* (Optical Society of America, Houston, TX, 1989), p. PD7.

[34] P. Leproux, S. Février, V. Doya, P. Roy, and D. Pagnoux, "Modeling and optimization of double-clad fiber amplifiers using chaotic propagation of the pump," Opt. Fiber Technol., **7**, 324–39 (2001).

[35] L. Liao, F. Zhang, X. He, *et al.*, "Confined-doped fiber for effective mode control fabricated by MCVD process," Appl. Opt. **57**, 3244–49 (2018).

[36] A. Klenke, M. Müller, H. Stark, *et al.*, "Coherent beam combination of pulses emitted by a 16-core ytterbium-doped fiber," in *Proc. SPIE 10897, Fiber Lasers XVI: Technology and Systems* (SPIE, 2019), p. 108971A.

[37] T. Sosnowski, A. Kuznetsov, R. Maynard, *et al.*, "3C Yb-doped fiber based high energy and power pulsed fiber lasers," in *Proc. SPIE 8601, Fiber Lasers X: Technology, Systems, and Applications* (SPIE, 2013), p. 86011M.

[38] X. Ma, C. Zhu, I. N. Hu, A. Kaplan, and A. Galvanauskas, "Single-mode chirally-coupled-core fibers with larger than 50μm diameter cores," Opt. Express **22**, 9206–19 (2014).

[39] A. Steinkopff, C. Jauregui, F. Stutzki, *et al.*, "Transverse single-mode operation in a passive large pitch fiber with more than 200 μm mode-field diameter," Opt. Lett. **44**, 650–53 (2019).

[40] F. Stutzki, F. Jansen, H.-J. Otto, C. Jauregui, J. Limpert, and A. Tünnermann, "Designing advanced very-large-mode-area fibers for power scaling of fiber-laser systems," Optica **1**, 233–42 (2014).

[41] G. Gu, F. Kong, T. Hawkins, *et al.*, "Ytterbium-doped large-mode-area all-solid photonic bandgap fiber lasers," Opt. Express **22**, 13962–68 (2014).

[42] F. Kong, G. Gu, T. W. Hawkins, *et al.*, "~1 kilowatt Ytterbium-doped all-solid photonic bandgap fiber laser," in *Proc. SPIE 10083, Fiber Lasers XIV: Technology and Systems* (SPIE 2017), p. 1008311.
[43] V. Pureur, L. Bigot, G. Bouwmans, Y. Quiquempois, M. Douay, and Y. Jaouen, "Ytterbium-doped solid core photonic bandgap fiber for laser operation around 980 nm," Appl. Phys. Lett. **92**, 061113 (2008).
[44] T. Matniyaz, M. T. Kalichevsky-Dong, T. W. Hawkins, *et al.*, "Single-mode Yb-doped double-clad all-solid photonic bandgap fiber laser generating 27.8W at 976nm," in *Laser Congress 2018 (ASSL)* (Optical Society of America, Boston, MA, 2018), p. AM6A.28.
[45] X. Fan, M. Chen, A. Shirakawa, K.-i. Ueda, C. B. Olausson, and J. Broeng, "A 98 W 1178 nm Yb-doped solid-core photonic bandgap fiber oscillator," Laser Phys. Lett. **10**, 105101 (2013).
[46] F. Kong, G. Gu, T. W. Hawkins, *et al.*, "Efficient 240W single-mode 1018nm laser from an Ytterbium-doped 50/400μm all-solid photonic bandgap fiber," Opt. Express **26**, 3138–44 (2018).
[47] C. Wirth, T. Schreiber, M. Rekas, *et al.*, "High-power linear-polarized narrow linewidth photonic crystal fiber amplifier," *Proc. SPIE 7580, Fiber Lasers VII: Technology, Systems, and Applications* (SPIE 2010), p. 75801H.
[48] S. Pierrot and F. Salin, "Amplification and compression of temporally shaped picosecond pulses in Yb-doped rod-type fibers," Opt. Express **21**, 20484–96 (2013).
[49] C. Gaida, F. Stutzki, F. Jansen, *et al.*, "Triple-clad large-pitch fibers for compact high-power pulsed fiber laser systems," Opt. Lett. **39**, 209–11 (2014).
[50] M. M. Jørgensen, S. R. Petersen, M. Laurila, J. Lægsgaard, and T. T. Alkeskjold, "Optimizing single mode robustness of the distributed modal filtering rod fiber amplifier," Opt. Express **20**, 7263–73 (2012).
[51] Y. Xu, Q. Fang, Y. Qin, X. Meng, and W. Shi, "2 kW narrow spectral width monolithic continuous wave in a near-diffraction-limited fiber laser," Appl. Opt. **54**, 9419–21 (2015).
[52] G. Agrawal, *Nonlinear Fiber Optics* (Academic Press, Cambridge, MA 2012).
[53] R. G. Smith, "Optical power handling capacity of low loss optical fibers as determined by stimulated Raman and Brillouin scattering," Appl. Opt. **11**, 2489–94 (1972).
[54] J. Kim, P. Dupriez, C. Codemard, J. Nilsson, and J. K. Sahu, "Suppression of stimulated Raman scattering in a high power Yb-doped fiber amplifier using a W-type core with fundamental mode cut-off," Opt. Express **14**, 5103–13 (2006).
[55] K. Hejaz, M. Shayganmanesh, R. Rezaei-Nasirabad, *et al.*, "Modal instability induced by stimulated Raman scattering in high-power Yb-doped fiber amplifiers," Opt. Lett. **42**, 5274–7 (2017).
[56] P. D. Rudych and N. V. Surovtsev, "Stimulated Raman scattering of light in optical silica fiber under subnanosecond pumping," Optoelectronics, Instrumentation and Data Processing **49**, 416–22 (2013).
[57] R. H. Stolen, E. P. Ippen, and A. R. Tynes, "Raman oscillation in glass optical waveguide," Appl. Phys. Lett. **20**, 62–64 (1972).

[58] R. H. Stolen, C. Lee, and R. K. Jain, "Development of the stimulated Raman spectrum in single-mode silica fibers," J. Opt. Soc. Am. B **1**, 652–7 (1984).
[59] C. Shi, R. T. Su, H. W. Zhang, *et al.*, "Experimental study of output characteristics of bi-directional pumping high power fiber amplifier in different pumping schemes," IEEE Photon. J. **9**, 1–10 (2017).
[60] M. N. Zervas and C. A. Codemard, "High power fiber lasers: a review," IEEE J. Sel. Top. Quantum Electron. **20**, 219–41 (2014).
[61] C. Jauregui, J. Limpert, and A. Tünnermann, "High-power fibre lasers," Nat. Photonics **7**, 861 (2013).
[62] H. Xu, M. Jiang, C. Shi, P. Zhou, G. Zhao, and X. Gu, "Spectral shaping for suppressing stimulated-Raman-scattering in a fiber laser," Appl. Opt. **56**, 3538–42 (2017).
[63] F. Jansen, D. Nodop, C. Jauregui, J. Limpert, and A. Tünnermann, "Modeling the inhibition of stimulated Raman scattering in passive and active fibers by lumped spectral filters in high power fiber laser systems," Opt. Express **17**, 16255–65 (2009).
[64] N. Shibata, R. G. Waarts, and R. P. Braun, "Brillouin-gain spectra for single-mode fibers having pure-silica, GeO2-doped, and P2O5-doped cores," Opt. Lett. **12**, 269–71 (1987).
[65] N. Shibata, K. Okamoto, and Y. Azuma, "Longitudinal acoustic modes and Brillouin-gain spectra for GeO2-doped-core single-mode fibers," J. Opt. Soc. Am. B **6**, 1167–74 (1989).
[66] K. Tai, A. Hasegawa, and A. Tomita, "Observation of modulational instability in optical fibers," Phys. Rev. Lett. **56**, 135–8 (1986).
[67] G. P. Agrawal, "Modulation instability in erbium-doped fiber amplifiers," IEEE Phot. Tech. Lett. **4**, 562–4 (1992).
[68] A. L. Gaeta, M. D. Skeldon, R. W. Boyd, and P. Narum, "Observation of instabilities of laser beams counterpropagating through a Brillouin medium," J. Opt. Soc. Am. B **6**, 1709–13 (1989).
[69] B. Anderson, A. Flores, R. Holten, and I. Dajani, "Comparison of phase modulation schemes for coherently combined fiber amplifiers," Opt. Express **23**, 27046–60 (2015).
[70] Y. Liu, Z. Lv, Y. Dong, and Q. Li, "Research on stimulated Brillouin scattering suppression based on multi-frequency phase modulation," Chin. Opt. Lett. **7**, 29–31 (2009).
[71] A. V. Harish and J. Nilsson, "Optimization of phase modulation formats for suppression of stimulated Brillouin Scattering in optical fibers," IEEE J. Sel. Top. Quantum Electron. **24**, 1–10 (2018).
[72] I. Dajani, C. Zeringue, C. Lu, C. Vergien, L. Henry, and C. Robin, "Stimulated Brillouin scattering suppression through laser gain competition: scalability to high power," Opt. Lett. **35**, 3114–6 (2010).
[73] I. Dajani, C. Zeringue, T. Justin Bronder, T. Shay, A. Gavrielides, and C. Robin, "A theoretical treatment of two approaches to SBS mitigation with two-tone amplification," Opt. Express **16**, 14233–47 (2008).
[74] C. Zeringue, I. Dajani, C. Lu, C. Vergien, and C. Robin, "Multi-tone approach to SBS mitigation in narrow linewidth high power fiber amplifiers,"

in *Proc. SPIE 7195, Fiber Lasers VI: Technology, Systems, and Applications* (SPIE, 2009), p. 71952I.

[75] D. Engin, W. Lu, H. Verdun, and S. Gupta, "High power modal instability measurements of very large mode area (VLMA) step index fibers," in *Proc. SPIE 8733, Laser Technology for Defense and Security IX* (SPIE 2013), p. 87330J.

[76] K. Brar, M. Savage-Leuchs, J. Henrie, *et al.*, "Threshold power and fiber degradation induced modal instabilities in high-power fiber amplifiers based on large mode area fibers," in *Proc. SPIE 8961, Fiber Lasers XI: Technology, Systems, and Applications* (SPIE, 2014), p. 89611R.

[77] T. Eidam, S. Hanf, E. Seise, *et al.*, "Femtosecond fiber CPA system emitting 830 W average output power," Opt. Lett. **35**, 94–96 (2010).

[78] T. Eidam, C. Wirth, C. Jauregui, *et al.*, "Experimental observations of the threshold-like onset of mode instabilities in high power fiber amplifiers," Opt. Express **19**, 13218–24 (2011).

[79] B. Ward, C. Robin, and I. Dajani, "Origin of thermal modal instabilities in large mode area fiber amplifiers," Opt. Express **20**, 11407–22 (2012).

[80] A. V. Smith and J. J. Smith, "Mode instability in high power fiber amplifiers," Opt. Express **19**, 10180–92 (2011).

[81] A. V. Smith and J. J. Smith, "Overview of a steady-periodic model of modal instability in fiber amplifiers," IEEE J. Sel. Top. Quantum Electron. **20**, 472–83 (2014).

[82] C. Jauregui, H.-J. Otto, F. Stutzki, J. Limpert, and A. Tünnermann, "Simplified modelling the mode instability threshold of high power fiber amplifiers in the presence of photodarkening," Opt. Express **23**, 20203–18 (2015).

[83] S. D. Jackson, "The spectroscopic and energy transfer characteristics of the rare earth ions used for silicate glass fibre lasers operating in the shortwave infrared," Laser & Photonics Reviews **3**, 466–82 (2009).

[84] A. Hemming, N. Simakov, J. Haub, and A. Carter, "A review of recent progress in holmium-doped silica fibre sources," Opt. Fiber Technol. **20**, 621–30 (2014).

[85] S. D. Jackson, "Towards high-power mid-infrared emission from a fibre laser," Nat. Photon. **6**, 423 (2012).

[86] I. Mingareev, N. Gehlich, T. Bonhoff, *et al.*, "Principles and applications of trans-wafer processing using a 2-μm thulium fiber laser," Int. J. Adv. Manuf. Technol. **84**, 2567–78 (2016).

[87] S. J. Xiaonong Zhu, "2μm lasers enable versatile processing of plastics," in *Industrial Laser Solutions for Manufacturing*. Available from http://www.industrial-lasers.com/articles/print/volume-31/issue-3/features/2-mu-m-lasers-enable-versatile-processing-of-plastics.html (2016).

[88] I. Mingareev, F. Weirauch, A. Olowinsky, L. Shah, P. Kadwani, and M. Richardson, "Welding of polymers using a 2μm thulium fiber laser," Opt. Laser Technol. **44**, 2095–9 (2012).

[89] I. Moskalev, S. Mirov, M. Mirov, *et al.*, "140 W Cr:ZnSe laser system," Opt. Express **24**, 21090–104 (2016).

[90] C. Kieleck, A. Berrou, B. Donelan, B. Cadier, T. Robin, and M. Eichhorn, "6.5 W ZnGeP2 OPO directly pumped by a Q-switched Tm-doped single-oscillator fiber laser," Opt. Lett. **40**, 1101–04 (2015).

[91] T. S. McComb, R. A. Sims, C. C. C. Willis, P. Kadwani, L. Shah, and M. Richardson, "Atmospheric transmission testing using a portable, tunable, high power thulium fiber laser system," in *CLEO/QELS: 2010 Laser Science to Photonic Applications* (2010), pp. 1–2.

[92] P. Sprangle, A. Ting, J. Penano, R. Fischer, and B. Hafizi, "Incoherent combining and atmospheric propagation of high-power fiber lasers for directed-energy applications," IEEE J. Quantum Electron. **45**, 138–48 (2009).

[93] A. Hemming, N. Simakov, A. Davidson, *et al.*, "A monolithic cladding pumped holmium-doped fibre laser," in *CLEO: 2013* (Optical Society of America, San Jose, CA, 2013), p. CW1M.1.

[94] A. Hemming, N. Simakov, M. Oermann, A. Carter, and J. Haub, "Record efficiency of a holmium-doped silica fibre laser," in *Conference on Lasers and Electro-Optics* (Optical Society of America, San Jose, CA, 2016), p. SM3Q.5.

[95] C. C. Baker, E. J. Friebele, A. A. Burdett, *et al.*, "Recent advances in holmium doped fibers for high-energy lasers (Conference Presentation)," in *Proc. SPIE 10637, Laser Technology for Defense and Security XIV* (SPIE, 2018), p. 1063704.

[96] B. M. Walsh and N. P. Barnes "Comparison of Tm : ZBLAN and Tm : silica fiber lasers; Spectroscopy and tunable pulsed laser operation around 1.9 μm," Appl. Phys. B **78**, 325–33 (2004).

[97] S. D. Agger and J. H. Povlsen, "Emission and absorption cross section of thulium doped silica fibers," Opt. Express **14**, 50–57 (2006).

[98] G. Turri, V. Sudesh, M. Richardson, M. Bass, A. Toncelli, and M. Tonelli, "Temperature-dependent spectroscopic properties of Tm3+ in germanate, silica, and phosphate glasses: A comparative study," J. Appl. Phys. **103**, 093104 (2008).

[99] P. Moulton, "High power Tm:silica fiber lasers: current status, prospects and challenges," in *CLEO/Europe and EQEC 2011 Conference Digest* (Optical Society of America, Munich, 2011), p. TF2_3.

[100] S. D. Jackson and T. A. King, "Theoretical modeling of Tm-doped silica fiber lasers," J. Lightwave Technol. **17**, 948 (1999).

[101] Thorlabs, "Online Datasheet for 'Thulium-doped single mode and large-mode-area optical fibers' (Webpage discontinued on July 05, 2017)," (Absolute cross-sections provided were referenced from [5]).

[102] B. M. Walsh and N. P. Barnes, "Comparison of Tm : ZBLAN and Tm : silica fiber lasers; Spectroscopy and tunable pulsed laser operation around 1.9 μm," Appl. Phys. B **78**, 325–33 (2004).

[103] S. D. Jackson and T. A. King, "Theoretical modeling of Tm-doped silica fiber lasers," Lightwave Technol. J. **17**, 948–56 (1999).

[104] A. Sincore, J. D. Bradford, J. Cook, L. Shah, and M. C. Richardson, "High average power thulium-doped silica fiber lasers: review of systems and concepts," IEEE J. Select. Topics Quantum Electron. **24**, 1–8 (2018).

[105] C. Ebert, T. Guiney, D. Irwin, and S. Patterson, "New advancements in 793 nm fiber-coupled modules for Th fiber laser pumping, including packages optimized for low SWaP applications," (2016), pp. 98340J–98349J.

[106] S. Jackson, "Cross relaxation and energy transfer upconversion processes relevant to the functioning of 2 μm Tm3+-doped silica fibre lasers," Opt. Commun. **230**, 197–203 (2004).

[107] J. Ganem, J. Crawford, P. Schmidt, N. W. Jenkins, and S. R. Bowman, "Thulium cross-relaxation in a low phonon energy crystalline host," Phys. Rev. B **66**, 245101 (2002).

[108] D. A. Simpson, G. W. Baxter, S. F. Collins, *et al.*, "Energy transfer up-conversion in Tm3+-doped silica fiber," J. Non-Crystal. Solids **352**, 136–41 (2006).

[109] D. L. Dexter, "A theory of sensitized luminescence in solids," J. Chem. Phys. **21**, 836–50 (1953).

[110] K. Arai, H. Namikawa, K. Kumata, T. Honda, Y. Ishii, and T. Handa, "Aluminum or phosphorus co-doping effects on the fluorescence and structural properties of neodymium-doped silica glass," J. Appl. Phys. **59**, 3430–36 (1986).

[111] B. Faure, W. Blanc, B. Dussardier, and G. Monnom, "Improvement of the Tm3+:3H4 level lifetime in silica optical fibers by lowering the local phonon energy," J. Non-Crystal. Solids **353**, 2767–73 (2007).

[112] Y. Qiao, N. Da, D. Chen, Q. Zhou, J. Qiu, and T. Akai, "Spectroscopic properties of neodymium doped high silica glass and aluminum codoping effects on the enhancement of fluorescence emission," Appl. Phys. B **87**, 717–22 (2007).

[113] S. D. Jackson and S. Mossman, "Efficiency dependence on the Tm3+ and Al3+ concentrations for Tm3+-doped silica double-clad fiber lasers," Appl. Opt. **42**, 2702–07 (2003).

[114] S. D. Jackson and T. A. King, "High-power diode-cladding-pumped Tm-doped silica fiber laser," Opt. Lett. **23**, 1462–4 (1998).

[115] R. A. Hayward, W. A. Clarkson, P. W. Turner, J. Nilsson, A. B. Grudinin, and D. C. Hanna, "Efficient cladding-pumped Tm-doped silica fibre laser with high power singlemode output at 2 /spl mu/m," Electron. Lett. **36**, 711–2 (2000).

[116] W. A. Clarkson, N. P. Barnes, P. W. Turner, J. Nilsson, and D. C. Hanna, "High-power cladding-pumped Tm-doped silica fiber laser with wavelength tuning from 1860 to 2090 nm," Opt. Lett. **27**, 1989–91 (2002).

[117] S. D. Jackson, T. Ryan, and S. Mossman, "High power Tm3+-doped silica fibre laser fabricated using chelate delivery deposition," Opt. Commun. **216**, 401–04 (2003).

[118] G. Frith, D. G. Lancaster, and S. D. Jackson, "85 W Tm^{3+}-doped silica fibre laser," Electron. Lett. **41**, 687–8 (2005).

[119] G. P. Frith and D. G. Lancaster, "Power scalable and efficient 790-nm pumped Tm3+-doped fiber lasers," in *Proc. SPIE, Fiber Lasers III: Technology, Systems, and Applications* (2006), p. 610208.

[120] Y.-J. Zhang, Y.-Z. Wang, Y.-L. Ju, and B.-Q. Yao, "Experimental investigation of Tm 3+ -doped silica double-cladding fibre laser," Chin. Phys. Lett. **23**, 2452 (2006).

[121] S. D. Jackson, A. Sabella, and D. G. Lancaster, "Application and development of high-power and highly efficient silica-based fiber lasers operating at 2 um," IEEE J. Select. Topics Quantum Electron. **13**, 567–72 (2007).

[122] M. Eichhorn and S. D. Jackson, "Comparative study of continuous wave Tm3+-doped silica and fluoride fiber lasers," Appl. Phys. B **90**, 35–41 (2008).

[123] G. Frith, A. Carter, B. Samson, and G. Town, "Design considerations for short-wavelength operation of 790-nm-pumped Tm-doped fibers," Appl. Opt. **48**, 5072–5 (2009).

[124] P. F. Moulton, G. A. Rines, E. V. Slobodtchikov, *et al.*, "Tm-doped fiber lasers: fundamentals and power scaling," IEEE J. Select. Topics Quantum Electron. **15**, 85–92 (2009).

[125] G. D. Goodno, L. D. Book, and J. E. Rothenberg, "Low-phase-noise, single-frequency, single-mode 608 W thulium fiber amplifier," Opt. Lett. **34**, 1204–1206 (2009).

[126] G. Frith, A. Carter, B. Samson, *et al.*, "Mitigation of photodegradation in 790nm-pumped Tm-doped fibers," (2010), pp. 75800A–75809A.

[127] T. S. McComb, R. A. Sims, C. C. C. Willis, *et al.*, "High-power widely tunable thulium fiber lasers," Appl. Opt. **49**, 6236–42 (2010).

[128] W. I. Ndebeka, A. Heidt, H. Schwoerer, and E. G. Rohwer, "Efficiency of Tm3+-doped silica triple clad fiber laser," in *International Summer Session: Lasers and Their Applications* (Optical Society of America, Changchun, 2011), p. Th4.

[129] P. C. Shardlow, D. Jain, R. Parker, J. Sahu, and W. A. Clarkson, "Optimising Tm-doped silica fibres for high lasing efficiency," in *2015 European Conference on Lasers and Electro-Optics – European Quantum Electronics Conference* (Optical Society of America, Munich, 2015), p. CJ_14_13.

[130] M. Tao, Q. Huang, T. Yu, P. Yang, W. Chen, and X. Ye, "Cross relaxation in Tm-doped fiber lasers," in *Proc. SPIE 8796, 2nd International Symposium on Laser Interaction with Matter* (2013), p. 87961W.

[131] R. Tumminelli, V. Petit, A. Carter, A. Hemming, N. Simakov, and J. Haub, "Highly doped and highly efficient Tm doped fiber laser (Conference Presentation)," in *Proc. SPIE 10512, Fiber Lasers XV: Technology and Systems* (SPIE, 2018), p. 105120M.

[132] C. Sang Won and T. Buma, "Injection-seeded Raman fiber amplifier for photoacoustic microscopy of lipids," in *2016 IEEE International Ultrasonics Symposium (IUS)* (2016), pp. 1–4.

[133] G. D. Goodno, L. D. Book, J. E. Rothenberg, M. E. Weber, and S. Benjamin Weiss, "Narrow linewidth power scaling and phase stabilization of 2-μm thulium fiber lasers," Opt. Eng. **50**, 111608–111610 (2011).

[134] T. Ehrenreich, R. Leveille, I. Majid, K. Tankala, G. Rines, and P. Moulton, "1-kW, all-glass Tm: fiber laser," in *Proc. SPIE* (2010), p. 758016.

[135] C. Gaida, M. Gebhardt, T. Heuermann, F. Stutzki, C. Jauregui, and J. Limpert, "Ultrafast thulium fiber laser system emitting more than 1 kW of average power," Opt. Lett. **43**, 5853–6 (2018).

[136] J. M. O. Daniel, N. Simakov, M. Tokurakawa, M. Ibsen, and W. A. Clarkson, "Ultra-short wavelength operation of a thulium fibre laser in the 1660-1750 nm wavelength band," Opt. Express **23**, 18269–76 (2015).

[137] Z. Li, Y. Jung, J. M. O. Daniel, *et al.*, "Exploiting the short wavelength gain of silica-based thulium-doped fiber amplifiers," Opt. Lett. **41**, 2197–200 (2016).

[138] Z.-Y. Hu, P. Yan, Q.-R. Xiao, Q. Liu, and M.-L. Gong, "227-W output all-fiberized Tm-doped fiber laser at 1908 nm," Chin. Phys. B **23**, 104206 (2014).

[139] B. R. Johnson, D. Creeden, J. Limongelli, H. Pretorius, J. Blanchard, and S. D. Setzler, "Comparison of high power large mode area and single mode 1908nm Tm-doped fiber lasers," in *Proc. SPIE Fiber Lasers XIII: Technology, Systems and Applications* (2016), p. 972810.

[140] X. Wang, P. Zhou, X. Wang, H. Xiao, and L. Si, "102 W monolithic single frequency Tm-doped fiber MOPA," Opt. Express **21**, 32386–92 (2013).

[141] J. Liu, K. Liu, F. Tan, and P. Wang, "High-Power Thulium-Doped All-Fiber Superfluorescent Sources," IEEE J. Select. Topics Quantum Electron. **20**, 497–502 (2014).

[142] F. Stutzki, C. Gaida, M. Gebhardt, *et al.*, "152 W average power Tm-doped fiber CPA system," Opt. Lett. **39**, 4671–4 (2014).

[143] J. Liu, H. Shi, C. Liu, and P. Wang, "Widely-tunable high-power narrow-linewidth thulium-doped all-fiber superfluorescent source," in *2015 Conference on Lasers and Electro-Optics (CLEO)* (2015), pp. 1–2.

[144] Y.-B. Xing, L. Liao, F. Bu, *et al.*, "Fabrication of Tm-doped fibers for high power and 121 W output all-fiber Tm-doped fiber laser," Chin. Phys. Lett. **32**, 034204 (2015).

[145] J. Liu, C. Liu, H. Shi, and P. Wang, "High-power linearly-polarized pico-second thulium-doped all-fiber master-oscillator power-amplifier," Opt. Express **24**, 15005–011 (2016).

[146] T. Walbaum, M. Heinzig, T. Schreiber, R. Eberhardt, and A. Tünnermann, "Monolithic thulium fiber laser with 567 W output power at 1970 nm," Opt. Lett. **41**, 2632–5 (2016).

[147] L. Shah, R. A. Sims, P. Kadwani, *et al.*, "Integrated Tm:fiber MOPA with polarized output and narrow linewidth with 100 W average power," Opt. Express **20**, 20558–63 (2012).

[148] B. Anderson, A. Flores, J. Grosek, and I. Dajani, "High power Tm-doped all-fiber amplifier at 2130 nm," in *Conference on Lasers and Electro-Optics* (Optical Society of America, San Jose, CA, 2017), p. SM1L.3.

[149] K. Yin, B. Zhang, G. Xue, L. Li, and J. Hou, "High-power all-fiber wavelength-tunable thulium doped fiber laser at 2 μm," Opt. Express **22**, 19947–52 (2014).

[150] X. Wang, X. Jin, P. Zhou, X. Wang, H. Xiao, and Z. Liu, "High power, widely tunable, narrowband superfluorescent source at 2 μm based on a monolithic Tm-doped fiber amplifier," Opt. Express **23**, 3382–9 (2015).

[151] K. Yin, R. Zhu, B. Zhang, G. Liu, P. Zhou, and J. Hou, "300 W-level, wavelength-widely-tunable, all-fiber integrated thulium-doped fiber laser," Opt. Express **24**, 11085–90 (2016).

[152] L. Pearson, J. W. Kim, Z. Zhang, M. Ibsen, J. K. Sahu, and W. A. Clarkson, "High-power linearly-polarized single-frequency thulium-doped fiber master-oscillator power-amplifier," Opt. Express **18**, 1607–12 (2010).

[153] B. M. Anderson, A. Flores, J. Grosek, and I. Dajani, "Thulium doped all-fiber amplifiers beyond 2.1 μm," presented at the SPIE Photonics West, San Francisco, 2018.

[154] T. Tanbun-Ek, R. Pathak, Z. Xu, *et al.*, "Improved long wavelength 14xx and 19xx nm InGaAsp/InP lasers," in *Proc. SPIE 9733, High-Power Diode Laser Technology and Applications XIV* (2016), p. 973307.

[155] M. A. Jebali, J. N. Maran, and S. LaRochelle, "264 W output power at 1585 nm in Er-Yb codoped fiber laser using in-band pumping," Opt. Lett. **39**, 3974–7 (2014).

[156] Y. Jeong, S. Yoo, C. A. Codemard, *et al.*, "Erbium:Ytterbium codoped large-core fiber laser with 297-W continuous-wave output power," IEEE J. Select. Topics Quantum Electron. **13**, 573–9 (2007).

[157] P. Myslinski, D. Nguyen, and J. Chrostowski, "Effects of concentration on the performance of erbium-doped fiber amplifiers," J. Lightwave Technol. **15**, 112–20 (1997).

[158] H. Lin, Y. Feng, Y. Feng, P. Barua, J. K. Sahu, and J. Nilsson, "656 W Er-doped, Yb-free large-core fiber laser," Opt. Lett. **43**, 3080–3 (2018).

[159] M. Meleshkevich, N. Platonov, D. Gapontsev, A. Drozhzhin, V. Sergeev, and V. Gapontsev, "415W single-mode CW thulium fiber laser in all-fiber format," in *CLEO/Europe and IQEC 2007 Conference Digest* (Optical Society of America, Munich, 2007), p. CP2_3.

[160] D. Creeden, B. R. Johnson, S. D. Setzler, and E. P. Chicklis, "Resonantly pumped Tm-doped fiber laser with >90% slope efficiency," Opt. Lett. **39**, 470–3 (2014).

[161] D. Creeden, B. R. Johnson, G. A. Rines, and S. D. Setzler, "High power resonant pumping of Tm-doped fiber amplifiers in core- and cladding-pumped configurations," Opt. Express **22**, 29067–80 (2014).

[162] T. Izawa, N. Shibata, and A. Takeda, "Optical attenuation in pure and doped fused silica in the ir wavelength region," Appl. Phys. Lett. **31**, 33–35 (1977).

[163] O. Humbach, H. Fabian, U. Grzesik, U. Haken, and W. Heitmann, "Analysis of OH absorption bands in synthetic silica," J. Non-Crystal. Solids **203**, 19–26 (1996).

[164] N. Simakov, A. Hemming, W. A. Clarkson, J. Haub, and A. Carter, "A cladding-pumped, tunable holmium doped fiber laser," Opt. Express **21**, 28415–22 (2013).

[165] H. Xiao, J. Leng, H. Zhang, L. Huang, J. Xu, and P. Zhou, "High-power 1018 nm ytterbium-doped fiber laser and its application in tandem pump," Appl. Opt. **54**, 8166–9 (2015).

[166] E. Stiles, "New developments in IPG fiber laser technology," in *5th International Workshop on Fiber Lasers* (2009), pp. 4–9.

[167] E. Delevaque, T. Georges, M. Monerie, P. Lamouler, and J. Bayon, "Modeling of pair-induced quenching in erbium-doped silicate fibers," IEEE Photon. Technol. Lett. **5**, 73–75 (1993).

[168] T. Georges, E. Delevaque, M. Monerie, P. Lamouler, and J. F. Bayon, "pair induced quenching in erbium doped silicate fibers," in *Optical Amplifiers and Their Applications* (Optical Society of America, Santa Fe, New Mexico, 1992), p. WE4.

[169] S. Alam, P. W. Turner, A. B. Grudinin, J. Nilsson, and J. A. Alvarez-Chavez, "High-power cladding pumped erbium-ytterbium co-doped fiber laser," in *Optical Fiber Communication Conference and International Conference on Quantum Information* (Optical Society of America, Anaheim, CA, 2001), p. TuI4.

[170] S. U. Alam, J. Nilsson, P. W. Turner, W. A. Clarkson, and A. B. E. D. M. C. Grudinin, "Erbium-ytterbium co-doped cladding-pumped fiber laser tunable from 1533 to 1600 nm with up to 6.7 W of output power," in *Advanced Solid-State Lasers* (Optical Society of America, Seattle, Washington, 2001), p. TuA5.

[171] J. K. Sahu, Y. Jeong, C. Codemard, *et al.*, "Tunable narrow linewidth high power erbium:ytterbium co-doped fiber laser," in *Conference on Lasers and Electro-Optics/International Quantum Electronics Conference and Photonic Applications Systems Technologies* (Optical Society of America, San Francisco, CA, 2004), p. CMK1.

[172] J. K. Sahu, Y. Jeong, D. J. Richardson, and J. Nilsson, "Highly efficient high-power erbium-ytterbium co-doped large core fiber laser," in *Advanced Solid-State Photonics* (Optical Society of America, Vienna, 2005), p. MB33.

[173] E.-L. Lim, S.-u. Alam, and D. J. Richardson, "Highly efficient, high power, inband-pumped Erbium/Ytterbium-codoped fiber laser," in *CLEO:2011 - Laser Applications to Photonic Applications*(Optical Society of America, Baltimore, MD, 2011), p. CTuI1.

[174] S. Yang, Z. Wu, Y. Yang, Y. Li, H. Chen, and M. Chen, "Terahertz repetition rate pulse generation in Erbium-Ytterbium co-doped fiber," in *Conference on Lasers and Electro-Optics* (Optical Society of America, San Jose, CA, 2017), p. JW2A.39.

[175] D. Y. Shen, P. J. Jander, J. K. Sahu, W. A. E. D. D. C. Clarkson, and I. Sorokina, "High-Power and Ultra-Efficient Operation of 1645 nm Er: YAG laser pumped by a 100W tunable Er,Yb fiber laser," in *Advanced Solid-State Photonics (TOPS)* (Optical Society of America, Vienna, 2005), p. 618.

[176] D. Y. Shen, J. K. Sahu, and W. A. Clarkson, "High-power widely tunable Tm:fibre lasers pumped by an Er,Yb co-doped fibre laser at 1.6μm," Opt. Express **14**, 6084–90 (2006).

[177] B. Ibarra-Escamilla, E. A. Kuzin, O. Pottiez, J. W. Haus, Q. Zhan, and P. E. Powers, "Tunable high power Er:Yb fiber laser," in *Frontiers in Optics 2011/Laser Science XXVII* (Optical Society of America, San Jose, CA, 2011), p. FWE3.

[178] L. Wang, X. Dong, P. P. Shum, X. Liu, and H. Su, "Random laser with multiphase-shifted Bragg grating in Er/Yb-codoped fiber," J. Lightwave Technol. **33**, 95–99 (2015).

[179] E. Fujita, Y. Mashiko, S. Asaya, M. Musha, and M. Tokurakawa, "High power narrow-linewidth linearly-polarized 1610 nm Er:Yb all-fiber MOPA," Opt. Express **24**, 26255–60 (2016).

[180] R. H. Stolen, E. P. Ippen, and A. R. Tynes, "Raman oscillation in glass optical waveguide," Appl. Phys. Lett. **20**, 62–64 (1972).

[181] E. Bélanger, M. Bernier, D. Faucher, D. Côté, and R. Vallée, "High-power and widely tunable all-fiber Raman laser," J. Lightwave Technol. **26**, 1696–1701 (2008).

[182] R. Vallée, E. Bélanger, B. Déry, M. Bernier, and D. Faucher, "Highly efficient and high-power Raman fiber laser based on broadband chirped fiber Bragg gratings," J. Lightwave Technol. **24**, 5039–43 (2006).

[183] C. Lin, R. H. Stolen, W. G. French, and T. G. Malone, "A cw tunable near-infrared (1.085–1.175-μm) Raman oscillator," Opt. Lett. **1**, 96–97 (1977).

[184] S. G. Grubb, T. Erdogan, V. Mizrahi, *et al.*, "1.3 μm Cascaded Raman Amplifier in Germanosilicate Fibers," in *Optical Amplifiers and Their Applications* (Optical Society of America, Breckenridge, CO, 1994), p. PD3.

[185] S. A. Naderi, I. Dajani, J. Grosek, and T. Madden, "Analysis of modal instability in Raman fiber amplifiers," in *Frontiers in Optics 2015* (Optical Society of America, San Jose, CA, 2015), p. FTh2F.3.

[186] S. Naderi, I. Dajani, J. Grosek, and T. Madden, "Theoretical and numerical treatment of modal instability in high-power core and cladding-pumped Raman fiber amplifiers," Opt. Express **24**, 16550–65 (2016).

[187] D. Hollenbeck and C. D. Cantrell, "Multiple-vibrational-mode model for fiber-optic Raman gain spectrum and response function," J. Opt. Soc. Am. B **19**, 2886–92 (2002).

[188] C. A. Codemard, P. Dupriez, Y. Jeong, J. K. Sahu, M. Ibsen, and J. Nilsson, "High-power continuous-wave cladding-pumped Raman fiber laser," Opt. Lett. **31**, 2290–2 (2006).

[189] N. B. Terry, K. T. Engel, T. G. Alley, and T. H. Russell, "Use of a continuous wave Raman fiber laser in graded-index multimode fiber for SRS beam combination," Opt. Express **15**, 602–7 (2007).

[190] J. Ji, C. A. Codemard, M. Ibsen, J. K. Sahu, and J. Nilsson, "Analysis of the Conversion to the First Stokes in Cladding-Pumped Fiber Raman Amplifiers," IEEE J. Select. Topics Quantum Electron. **15**, 129–39 (2009).

[191] A. K. Sridharan, J. E. Heebner, M. J. Messerly, J. W. Dawson, R. J. Beach, and C. P. J. Barty, "Brightness enhancement in a high-peak-power cladding-pumped Raman fiber amplifier," Opt. Lett. **34**, 2234–6 (2009).

[192] Q. Xiao, P. Yan, D. Li, *et al.*, "Bidirectional pumped high power Raman fiber laser," Opt. Express **24**, 6758–68 (2016).

[193] L. Zhang, C. Liu, H. Jiang, *et al.*, "Kilowatt Ytterbium-Raman fiber laser," Opt. Express **22**, 18483–9 (2014).

[194] Y. Shamir, Y. Glick, M. Aviel, A. Attias, and S. Pearl, "250 W clad pumped Raman all-fiber laser with brightness enhancement," Opt. Lett. **43**, 711–4 (2018).

[195] Y. Glick, V. Fromzel, J. Zhang, N. Ter-Gabrielyan, and M. A. Dubinskii, "High Efficiency, 154 W CW, Diode-pumped Raman fiber laser with brightness enhancement," in *Lasers Congress 2016 (ASSL, LSC, LAC)* (Optical Society of America, Boston, MA, 2016), p. ATu4A.6.

[196] E. A. Zlobina, S. I. Kablukov, A. A. Wolf, A. V. Dostovalov, and S. A. Babin, "Nearly single-mode Raman lasing at 954 nm in a graded-index fiber directly pumped by a multimode laser diode," Opt. Lett. **42**, 9–12 (2017).

[197] M. N. Zervas and C. A. Codemard, "High power fiber lasers: a review," IEEE J. Select. Topics Quantum Electron. **20**, 219–41 (2014).

[198] G. Lopez-Galmiche, Z. Sanjabi Eznaveh, M. A. Eftekhar, *et al.*, "Visible supercontinuum generation in a graded index multimode fiber pumped at 1064nm," Opt. Lett. **41**, 2553–6 (2016).

[199] P. Roumayah, J. Bradford, J. Cook, J. E. Antonio-Lopez, R. A. Correa, and M. Richardson, "High average power pulsed multi-mode Raman fiber laser in graded index fiber," in *Proc. SPIE 10512, Fiber Lasers XV: Technology and Systems* (SPIE, 2018), p. 105121X.

[200] L. Zhang, H. Jiang, S. Cui, and Y. Feng, "Integrated ytterbium-Raman fiber amplifier," Opt. Lett. **39**, 1933–6 (2014).

[201] V. R. Supradeepa, F. Yan, and W. N. Jeffrey, "Raman fiber lasers," J. Optics **19**, 023001 (2017).

[202] J. Dong, L. Zhang, H. Jiang, *et al.*, "High order cascaded Raman random fiber laser with high spectral purity," Opt. Express **26**, 5275–80 (2018).

[203] X. Zhu and N. Peyghambarian, "High-power ZBLAN glass fiber lasers: review and prospect," Adv. OptoElectron. **2010** (2010).

[204] S. Aparanji, V. Balaswamy, S. Arun, and V. R. Supradeepa, "Simultaneous Raman based power combining and wavelength conversion of high-power fiber lasers," Opt. Express **26**, 4954–60 (2018).

[205] V. Balaswamy, S. Arun, G. Chayran, and V. R. Supradeepa, "All passive architecture for high efficiency cascaded Raman conversion," Opt. Express **26**, 3046–53 (2018).

[206] X. Wang, D. Chen, H. Li, L. She, and Q. Wu, "Random fiber laser based on artificially controlled backscattering fibers," Appl. Opt. **57**, 258–62 (2018).

[207] J. T. Walsh, T. J. Flotte, and T. F. Deutsch, *Er:YAG Laser Ablation of Tissue: Effect of Pulse Duration and Tissue Type on Thermal Damage*, Lasers Surg. Med. **9**, 314–326 (1989).

[208] C. Kumar and N. Patel, "Quantum cascade lasers and applications in defense and security," in *2011 IEEE Photonics Society Summer Topical Meeting Series* (2011), pp. 49–50.

Chapter 9

Beam combinable, kilowatt all-fiber amplifiers for directed energy

Angel Flores[1], Iyad Dajani[1], Brian Anderson[1] and Nader Naderi[1]

9.1 Introduction

Fiber laser systems with a broad range of industrial [1], medical [2], and directed energy (DE) [3] applications have evolved rapidly over the past decade. Generally, ytterbium (Yb)-doped fiber lasers offer several advantages over conventional bulk solid-state and gas lasers including compactness, high output efficiencies, and reduced size, weight, and power consumption (SWaP). In addition, monolithic or all-fiber architectures with minimal free-space and alignment-sensitive coupling optics further enable robust and reliable packaging for industrial and DE systems. In terms of industrial applications, multimode fiber lasers with low beam quality, broad output bandwidths and powers as high as 100 kW are now commercially available [4]. While such low-brightness fiber lasers are adequate for material processing, excellent or near diffraction limited beam quality is critical for DE applications. These high-brightness lasers at relevant power levels are more challenging to build and as a result, to scale overall power and brightness beam combining techniques where multiple fiber lasers are efficiently combined into one output beam are being actively pursued.

Several beam combining architectures suitable for DE applications have been reported, and are divided into two distinct categories: incoherent beam combination (IBC) and coherent beam combination (CBC). The purpose of each method is to increase the far field intensity when brightness and power of a single fiber amplifier is limited. In IBC, an array of fiber lasers is superimposed in the far field without control of the relative spectra or phases of the different fibers. No brightness enhancement is obtained from this method, limiting the maximum intensity in the far field. Although such beam combining has been successfully demonstrated by the Navy [5], this approach is limited to propagation ranges on the order of a few kilometers. In spectral beam combining (SBC), a separate class of IBC, incoherent

[1]Air Force Research Laboratory (AFRL), Directed Energy Directorate, Kirtland Air Force Base, Albuquerque, NM, USA

beams of different wavelengths are spatially overlapped to create a single output beam of multiple colors. The brightness of the propagating beam can be increased at the cost of increased spectral bandwidth. Similarly, SBC has the advantage of not requiring active phase control or mutual temporal coherence of the individual beams. Alternatively, coherent beam combining (CBC) schemes require proper phase, frequency, and polarization relationships for efficient beam combination. Overall, due to their brightness enhancements, for longer range and higher power/intensity DE applications, SBC and CBC are potentially more appealing.

For both SBC and CBC, it is highly desired that the output power from the individual fiber amplifiers be as high as possible. Furthermore, each beam combining architecture places strict constraints on the beam quality and spectral bandwidth of the fiber amplifiers in order to be combined with high efficiency. However, due to high optical intensities, high-power, single-mode fiber lasers are limited by the onset of detrimental nonlinear effects such as stimulated Raman scattering (SRS) and stimulated Brillouin scattering (SBS). The latter phenomenon is the lowest threshold nonlinear effect in single-frequency and narrow linewidth continuous-wave (CW) fiber amplifiers. Single frequency, in this context, refers to an optical wave whose spectral linewidth is less than the intrinsic gain bandwidth of the spontaneous Brillouin scattering process. Typically in Yb-doped fibers operating at ~1 μm near-infrared wavelengths, the spontaneous Brillouin bandwidth is ~50 MHz (0.2 pm at ~1 μm). Along those lines, narrow linewidth refers to larger spectral bandwidths on the order of ~10 GHz (0.04 nm) that are still compatible with fiber beam combining systems. Nevertheless, the maximum output power that can be extracted from single-frequency all-fiber amplifiers has been limited to 414 W [6], due primarily to the challenge of suppressing SBS.

SBS is a phase-matched third-order nonlinear interaction that couples acoustic phonons to the photons of the optical field and the backscattered Stokes light. Consequently, optical power is transferred from the laser light and into the nonlinear backward Stokes light; thus degrading amplification of the signal light and possibly damaging the fiber amplifier through pulsation. To maintain high power operation, numerous SBS suppression techniques have been reported. As this nonlinear effect is driven by the intensity of the optical field and the overall material interaction length, the most obvious means of suppression include increasing the mode area or decreasing the effective fiber length. In an amplifier, shorter fiber lengths reduce the amplifier pump-light absorption and decrease the overall optical efficiency. Fibers with higher pump absorption achieved through high active-ion dopant concentrations may be able to attain reduced fiber lengths for nonlinear suppression. Yet, solubility and dopant concentration limits of ytterbium as well as associated photodarkening effects limit these approaches.

In contrast, counter pumping fiber amplifier architectures preserve the physical fiber length while reducing the nonlinear interaction length. For counter pumping systems the pump wave travels in the opposite direction of the signal wave, resulting in a slow initial signal rise at the input end of the fiber and a steep rapid rise at the output end. Nevertheless, counter-pumped combiner devices have to simultaneously manage high-signal light combined with high-intensity pump light at the fiber output end. In addition, the added pump combiner delivery fiber at the output end can create a longer effective length, partially minimizing the enhancement from a counter-pumped architecture [7]. As a result, most monolithic all-fiber amplifiers are based on co-pumped configurations.

Similarly, increasing the fiber core area reduces the optical intensity and suppresses nonlinear effects. Here we utilize large mode area (LMA) fibers with core diameters on the order of 20 or 25 μm, much larger than typical single-mode (~5 μm) fiber dimensions. While LMA fibers are no longer strictly single-mode and guide a few modes, the higher-order modes (HOM) can be filtered out by coiling/bending the fiber [8]. Generally, effective HOM filtering is difficult to achieve for fiber diameters beyond 25 μm resulting in diminished beam quality for larger core fibers. It is important to note that LMA fibers are also susceptible to thermal effects such as transverse mode instability (TMI) [9]. TMI is a power dependent effect related to the quantum defect heating between the pump (nominally 976 nm) and signal wavelengths (typically 1,064 nm) and modal interference between the few modes in LMA fibers. In TMI energy is transferred from the fundamental mode to the higher order modes, resulting in a sharp and sudden degradation in beam quality. Analogous to SBS, mitigation of TMI and thermal effects in high-power fiber amplifiers is an active research subject [10,11], but beyond the scope of this chapter. Other novel fiber designs such as higher-order mode [12] and gain guided fibers [13] have been fabricated to further increase the effective area, but none have attained the desired kW class power levels.

The generation of Brillouin frequency shifts due to temperature or strain variation in the fiber has also been explored [7,14]. Such frequency shifts can be used to produce multiple Brillouin frequency sidebands and hence reduce the overall peak Brillouin gain. For example, SBS suppression can be achieved through utilization of the steep temperature gradient at the fiber output end or application of an external thermal gradient [14]. Likewise, another promising SBS mitigation technique involves tailoring of the transverse acoustic profile in the fiber to introduce multiple transverse Brillouin frequencies [15]. Notably, the combination of external thermal gradients and acoustic tailoring resulted in a record single-frequency amplifier output power of 811 W [16], albeit, in a free-space pump architecture (not all-fiber).

Another fundamental SBS mitigation technique revolves around spectral linewidth broadening [17]. In the 1980s, SBS suppression through phase modulation was investigated in the telecommunication industry for long haul applications [18]. In this case, the broadening is commonly achieved through radio frequency (RF) modulation and suppression occurs when the phase variation is on a time scale shorter than the phonon lifetime hence, inhibiting the rise and amplification of Brillouin scattering. Along those lines, narrow linewidth, kW class Yb-doped all-fiber amplifiers have been demonstrated through external phase modulation [19]. Although the effective linewidth of the laser would no longer be single-frequency, beam combining of high-power fiber amplifiers modulated at the GHz level has been demonstrated for both spectral and CBC architectures [20,21]; further driving interest in kW class narrow-linewidth fiber amplifiers. Table 9.1 summarizes

Table 9.1 Brief summary of typical nonlinear SBS mitigation techniques for fiber amplifiers

Nonlinear SBS mitigation	SBS threshold increase	Notes
Counter-pumping (shorter effective fiber length)	2	Typically implemented with free-space pumping; high-power all-fiber counter-pump combiners not as prevalent as co-pumped combiners [15,16]
Thermal or strain gradient	2–4	Thermal zones and high temperatures constrained by polymer outer clad material; total strain application may be limited by fiber failure [10,14–16]
Acoustic tailoring (dopants)	2–7	Challenging to maintain optical core uniformity for near diffraction limited beam quality [15,16]
Large mode area (LMA) fiber	~16–20	Fiber no longer single mode; transverse modal instability observed at high powers (beam quality degradation)
Linewidth broadening	>20	External RF phase modulation devices readily available; efficient beam combinability sets upper limit [17–21]

prominent SBS mitigation techniques and corresponding threshold increases for fiber amplifiers as shown below.

In addition, another effective SBS mitigation technique developed at the Air Force Research Laboratory (AFRL) titled laser gain competition (LGC) or two-tone is based on seeding the amplifier with two distinct sources. By seeding with the appropriate wavelengths and seed ratios, SBS suppression is realized due to a reduction in the effective nonlinear length as the desired laser signal near the output end of the active fiber rises sharply. This is similar to the sharp rise in the signal power encountered in a counter-pumped configuration; however, LGC preserves the favorable co-pumped all-fiber pump architecture. Accordingly, LGC fiber amplifiers have been shown to result in two times SBS threshold enhancement over conventional single-tone amplifiers [14].

Based on the commercial advent of LMA fibers and fiber-coupled lithium niobate ($LiNbO_3$) electro-optic modulators with low insertion loss, large frequency bandwidth, and low π-phase shift voltages; linewidth broadening via phase modulation with LMA fibers have become the preeminent SBS suppression techniques for high-power all-fiber amplifiers. Leveraging ubiquitous electronic RF modulation equipment, implementation of spectral linewidth broadening is less intricate than other corresponding nonlinear mitigation schemes. Despite promising results with phase modulation, scaling to multi-kW class levels has proven difficult, with broader linewidths (>20 GHz) required at 2 kW.

Such linewidths may hinder efficient beam combining, which is critical for further scaling to powers of DE interest. This is due to added path length matching complexities in CBC where the linewidth is inversely proportional to the temporal coherence of the laser (or coherence length). Thus, beam combining at linewidths >20 GHz would require sub-millimeter coherence lengths and path length matching, considerably adding to the system complexity. In SBC, broader linewidths reduce the number of channels that can be accommodated within the typical Yb gain spectrum (~1,045–1,075 nm). Moreover, the multilayer dielectric (MLD) gratings normally used for SBC are particularly linewidth sensitive. Here the combined output beam quality will degrade with broader linewidths, due to the strong angular dispersion of the MLD gratings.

Toward that end, we have extensively studied nonlinear suppression in high power Yb-doped all-fiber amplifiers. In this chapter, we describe and detail our theoretical and experimental research of SBS in fibers seeded with phase modulated light [22]. Various phase modulation schemes are analyzed in terms of nonlinear suppression, optical linewidth, and beam combinability [23]. Particularly, a novel filtered pseudo-random phase modulation approach for high-power fiber amplifiers that simultaneously improves both the coherence length and SBS threshold is presented [24]. In addition, we discuss efficient coherent combining of five individual kilowatt-class fiber amplifiers seeded with pseudo-random phase modulated light, using a diffractive optical element (DOE) [21]. Here a combined output of ~5 kW was attained with 82% combining efficiency and near diffraction-limited beam quality ($M^2 < 1.1$). This represented the first multi-kW class beam combining demonstration of fiber amplifiers seeded with pseudo-random phase modulated light.

Lastly, to further suppress nonlinear effects LGC is described. An additional advantage of LGC is its compatibility with other SBS mitigation schemes such as phase modulation. Therefore, we detail power scaling results of a phase modulated LGC all-fiber amplifier with 1 kW of output power at a modulation frequency of 2.5 GHz and excellent optical pump efficiency (85%) [25]. Overall, the power scaling results represent a significant reduction in spectral linewidth compared to that of commercially available narrow-linewidth Yb-doped all-fiber amplifiers.

9.2 Time-dependent nonlinear SBS theory and model

Phase modulation has become the preferred method of suppressing SBS in high-power Yb-doped all-fiber amplifiers. Subsequently, standard RF phase modulation techniques include driving the phase modulator with a sine wave [23], a white noise source (WNS) [26], or pseudo-random bit sequence (PRBS) [19]. In sinusoidal phase modulation, the spectral power exhibits a series of discrete sidebands at integer multiples of the modulation frequency. Yet, once the modulation frequency increases appreciably beyond the spontaneous Brillouin gain bandwidth, there is no further increase in the SBS threshold [23]. As a result, more spectrally efficient linewidth broadening schemes such as WNS or PRBS are preferred and have been utilized to attain kilowatt class, monolithic all-fiber amplifiers.

Accordingly, theoretical analysis of Brillouin scattering in optical fibers seeded with phase modulated light has been developed based on the transient coupled wave equations [22]. SBS suppression is examined in the time domain using a triply coupled set of nonlinear partial differential equations. These equations describe a three-wave interaction of two optical fields (laser and nonlinear Stokes) and an acoustic (phonon) field. We note that a wavelength of 1,064 nm, the SBS Stokes frequency shift is approximately 16 GHz. Therefore, spectral overlap between the GHz level phase modulated signal and the nonlinear Stokes wave can lead to seeding and enhancement of the SBS process [19]. To include the seeding effects, we derive the coupled wave equations using the total electric field in the slowly varying envelope approximation as shown in (9.1). Here we include the forward propagating electric field (A_{LF}), the backwards propagating electric field from Rayleigh scattered light and reflections (A_{LB}), and the backward propagating nonlinear Stokes field (A_S). To simplify the phase matching conditions, we have rewritten the backwards propagating laser field such that the optical frequency (k_L, ω_L) is written in terms of the Stokes shifted frequency (k_S, ω_S):

$$\begin{aligned}\tilde{E} &= A_{LF}(z,t)e^{i(k_L z-\omega_L t)} \\ &\quad + A_{LB}(z,t)e^{i(-k_S z-\omega_S t)}e^{-i\Omega_B(zn/c+t)} \\ &\quad + A_S(z,t)e^{i(-k_S z-\omega_S t)} + c.c.\end{aligned} \tag{9.1}$$

$$\nabla^2\tilde{E} - \frac{n^2}{c^2}\frac{\partial^2\tilde{E}}{\partial t^2} = \frac{1}{\varepsilon_0 c^2}\frac{\partial^2\tilde{P}}{\partial t^2} \tag{9.2}$$

$$\tilde{P} = \varepsilon_0\gamma_e\tilde{\rho}\tilde{E} \tag{9.3}$$

The interaction of the electric field and the density wave is given by the wave equation (9.2), where n is the index of refraction, c is the speed of light, ε_0 is the vacuum permittivity, and $\tilde{P}$ is the nonlinear polarization given by (9.3). In (9.3), γ_e is the electrostrictive constant and $\tilde{\rho}$ is the density wave given by $\tilde{\rho} = \exp(i(qz - \Omega_B t)) + c.c.$, where q is the phonon propagation constant and Ω_B is the Stokes shift. Inserting the electric field (9.1) into Maxwell's equations (9.2) and (9.3), and dropping the fast oscillating and nonphased matched terms yields the following equations:

$$\frac{c}{n}\frac{\partial A_{LF}}{\partial z} + \frac{\partial A_{LF}}{\partial t} = \frac{\omega\gamma_e}{2n^2\rho_0}\rho\left(A_S + A_{LB}e^{-i\Omega_B(zn/c+t)}\right) \tag{9.4}$$

$$-\frac{c}{n}\frac{\partial A_{LB}}{\partial z} + \frac{\partial A_{LB}}{\partial t} = \frac{\omega\gamma_e}{2n^2\rho_0}\rho^* A_{LF}e^{i\Omega_B(zn/c+t)} \tag{9.5}$$

$$-\frac{c}{n}\frac{\partial A_S}{\partial z} + \frac{\partial A_S}{\partial t} = \frac{\omega\gamma_e}{2n^2\rho_0}\rho^* A_{LF} \tag{9.6}$$

Likewise, the electric field (9.1) is inserted into the density equation below (9.7) to give the following relation in (9.8). Here, Γ_B is the Brillouin bandwidth, v_s is the speed of sound in the medium, and $\tilde{f}$ is the stochastic noise driving the phonon relation. In (9.8), the second derivative in the density field is kept.

However, in practice, this term has been found to add only a small (<3%) correction to the final solution for phase modulation frequencies of less than 10 GHz.

$$\frac{\partial^2 \tilde{\rho}}{\partial t^2} - \frac{\Gamma_B}{q^2} \nabla^2 \frac{\partial \tilde{\rho}}{\partial t} - v_S^2 \nabla^2 \tilde{\rho} = -\frac{1}{2} \varepsilon_0 \gamma_e \nabla^2 \langle \tilde{E}^2 \rangle + \tilde{f} \tag{9.7}$$

$$\begin{aligned} &\frac{\partial^2 \rho}{\partial t^2} + (\Gamma_B - 2i\Omega_B) \frac{\partial \rho}{\partial t} - i\Omega_B \Gamma_B \rho \\ &= \varepsilon_0 \gamma_e q^2 A_{LF} \left(A_S^* + A_{LB}^* e^{i\Omega_B (zn/c+t)} \right) - 2i\Omega_B f \end{aligned} \tag{9.8}$$

Due to our initial assumptions, the backwards oscillating laser field (A_{LB}) in (9.4)–(9.6) and (9.8) includes a fast oscillating phase term. For a single frequency source, the amplitude of this fast oscillating term will quickly average to "0" as the fiber length and integration time are significantly longer than the oscillating period. However, for narrow linewidth sources at GHz bandwidths, many frequency components will oscillate at the correct frequency to cancel this term and give the correct phase matching conditions. As a result, the time-dependent model was expanded to include the fast oscillating term.

The coupled differential equations shown in (9.4)–(9.8) are solved using the method of characteristics as described in [22]. Phase modulation effects are included through the initial and boundary conditions for the electric field at the input end of the fiber: $A_{LF}(0,t) = A_{LF}^0 e^{i\varphi(t)}$, $A_{LF}(z > 0, 0) = 0$, where $\varphi(t)$ is the phase modulation function and A_{LF}^0 describes the forward input electric field amplitude of the laser. The Stokes field is counter-propagating and is zero everywhere at $t = 0$. It is subject to the following conditions: $A_s(z,0) = 0, A_s(L,0) = 0$, where L is the length of the fiber. Furthermore, to investigate the effect of the SBS seeding process, we added the boundary condition shown below:

$$A_{LB}(z = L, t) = \sqrt{R}\, A_{LF}(z = L, t) \tag{9.9}$$

The slowly varying component of the backwards laser field is fixed at the end of the fiber by the reflected component of the forward laser field. Without the presence of laser gain, the reflectivity, R, was been estimated to be $1*10^{-5}$. In practice, this reflectivity is affected by a variety of properties including end cap length, output fiber cleave angle, AR coating on the end cap, and any partial reflections from free space elements. Correspondingly, the enhanced boundary conditions allow us to theoretically analyze the SBS seeding process when driven with phase modulated light. This is a critical component when designing high power Yb-doped fiber amplifiers with linewidths that are on the order of the Brillouin shift frequency (~16 GHz at 1,064 nm).

Through our time-dependent model, different modulation schemes were investigated in an effort to achieve the maximum SBS suppression at the smallest possible effective spectral linewidth. Experimental validation of the time-dependent model has been conducted for sinusoidal phase modulation, with the results agreeing with the theory and simulations [23]. Similarly, we performed an experimental study of SBS suppression in passive fibers via WNS and PRBS [23].

Through a cutback experiment we investigated SBS suppression as a function of spectral linewidth and fiber length. Notably, the WNS and PRBS phase modulation experiments further validated the time-dependent SBS model.

9.3 Phase modulation in kW class all-fiber amplifiers

Linewidth broadening via a filtered WNS was the standard approach toward SBS mitigation in kW class fiber amplifiers. Specifically, WNS modulation creates a continuum spectrum around the carrier with lineshape depending on the voltage amplification and electronic filtering setup. Optical linewidth can be controlled by varying the RF filter range. Particularly, WNS modulation has been employed to attain kilowatt class, monolithic fiber amplifiers at 10–20 GHz [27,28]. Along those lines, to further reduce SBS in high power fiber amplifiers we have investigated PRBS phase modulation. In contrast to WNS, PRBS generates a periodic, well-defined discrete optical power spectral density with features that are a function of the modulation frequency and pattern length. Moreover, phase shift keying (PSK) with pseudo-random patterns has been shown to provide significant SBS suppression in long-distance optical communication systems [29]. Likewise, our time-dependent models and simulations demonstrated considerable linewidth reduction in kilowatt class fiber lasers through PRBS modulation.

In digital communications, PRBS is an ideal test signal as it simulates the random characteristics of a digital signal and can be easily generated through linear shift registers. Specifically, it allows one to generate repeatable noise of known spectrum and amplitude, with adjustable linewidth, using prevalent and reliable digital circuitry. As such, robust, high-speed PRBS generators are widely available. In contrast to the continuum spectra generated by a WNS, PRBS modulation exhibits a discrete optical frequency comb. The spectral frequency comb is due to the periodically repeating pseudo-random pattern lengths. A nominal spectrum of an optical field modulated with a PRBS pattern is presented in Figure 9.1.

A PRBS pattern is typically denoted as $2^n - 1$. The power (n) indicates the shift register length used to create the pattern. The $2^n - 1$ patterns contain every possible

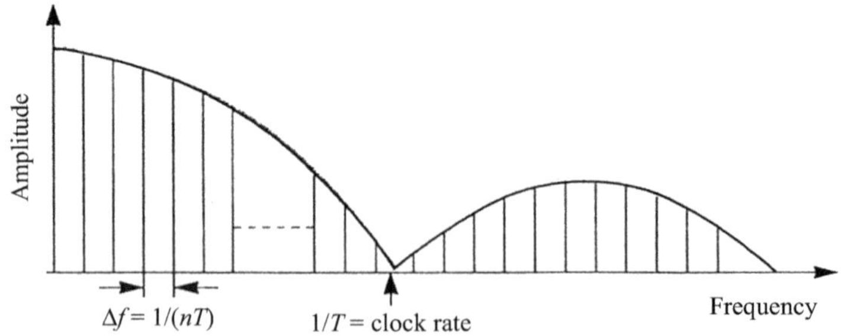

Figure 9.1 Nominal frequency spectrum of a PRBS pattern exhibiting a discrete frequency comb within a sinc² envelope

combination of n number of binary bits ("0" and "1"), except the null pattern. For PRBS phase modulation, the binary "0" and "1" data bits represent phase shifts of "0" and "π," respectively. Subsequently, a sinc^2 spectral lineshape is generated (Figure 9.1), with the spectral nulls occurring at the bit or clock rate ($1/T$; where T is the bit period). Here the frequency comb separation is given as $\Delta f = 1/nT$ and the full width at half maximum (FWHM) linewidth is approximately equal to the PRBS clock rate (~89%). Thus, using PRBS modulation the SBS spectrum can be preferentially altered through appropriate bit/clock rate and pattern length adjustments.

9.3.1 WNS and PRBS SBS suppression comparison

A monolithic, all-fiber, Yb-doped fiber amplifier at 1,034 nm was built to experimentally investigate laser power scaling via PRBS and WNS modulated signals. Here we investigated operation at shorter wavelengths to reduce quantum defect heating, which may be beneficial in suppressing thermal effects such as TMI [8]. Efficient operation at shorter wavelengths may also expand the wavelength range for SBC (current waveband: 1,045–1,075 nm), providing an appreciable increase in the number of fiber lasers that can be spectrally combined. However, significantly higher levels of amplified spontaneous emission (ASE) are expected in amplifiers utilizing conventional step-index silica fibers, thus rendering operation at <1,040 nm in these types of amplifiers challenging [30]. To that end, we developed a high efficiency, kilowatt-class narrow linewidth 1,034 nm all-fiber amplifier [31]. Through proper selection of amplifier fiber lengths, seed power and spectral content, high optical pump efficiency (81%) was attained. More importantly, utilizing the kilowatt fiber amplifier, SBS suppression via both WNS and PRBS phase modulation was directly compared and analyzed.

A schematic diagram of the 1,034 nm all-fiber amplifier is shown in Figure 9.2. The master oscillator (seed) was a single-frequency distributed feedback (DFB) diode laser module centered at 1,033.8 nm. An external fiber-coupled

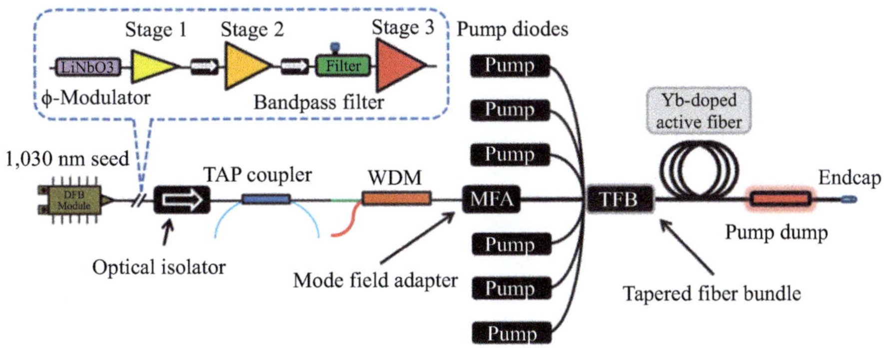

Figure 9.2 Experimental setup for the kilowatt 1,034 nm monolithic fiber amplifier system. A fiber-coupled optical bandpass filter and WDM were used to suppress ASE

$LiNbO_3$ electro-optic phase modulator driven by either a WNS or PRBS pattern generator was used to broaden the linewidth and suppress SBS. The spectrally broadened signal was then preamplified in three stages to provide ~20 W of seed power to the final stage. A cutback experiment was performed on each pre-amplification stage to suppress the ASE content from building up along the amplifier chain. To further minimize the ASE content, a fiber-coupled 1-nm wide band-pass optical filter was inserted between the second and third stages. Furthermore, a 1,035/1,064 nm wavelength division multiplexer (WDM) was inserted between the third and the final gain stages. The WDM had a small insertion loss and was effective in further filtering out ASE lying in the 1,045–1,075 nm range where the peak of the gain is typically expected.

A 0.01% tap coupler was spliced onto the output of the third amplifier stage to monitor the forward and backward propagating light. In order to characterize the SBS response, the backward tap was spliced onto a 50/50 splitter. This allowed us to simultaneously measure the backward power on a photodiode and capture the backward spectral content on a 0.01 nm (~3 GHz) resolution bandwidth optical spectrum analyzer (OSA). Backward power measurements are a standard metric for SBS characterization in fiber amplifiers. Several methods of characterizing the amplifier backward content are typically used to identify the SBS threshold. This includes a reflectivity in the range 0.05%–0.1%, appreciable departure of the reflectivity from linearity with pump power, relative power spectral intensity of the backward Rayleigh and Stokes light (with the latter being higher by 5–10 dB as captured by the OSA), and observation of backward pulses.

The three-stage preamplifier output was fusion spliced onto a mode field adapter (MFA) followed by a $(6 + 1) \times 1$ tapered fiber bundle (TFB) pump combiner with nonpolarization maintaining (non-PM) 20 (core)/400 (clad) μm signal feed through fiber and 200 (core)/220 (clad) μm pump input legs. The high-power stage was driven by six ~240 W pump diodes centered at 976 nm. A LMA non-PM 25/400 μm Yb-doped fiber from Coherent was spliced onto the TFB in a co-pumping configuration. The length of the LMA gain fiber in our experiment was 6.4 m. The output of the amplifier was terminated with a cladding mode stripper to remove residual pump light and a 400 μm diameter endcap to reduce intensity at the glass/air interface. Power scaling comparisons of the WNS and PRBS modulated amplifiers with identical fiber lengths and comparable effective spectral linewidths are shown in Figure 9.3. Initially, we measured an SBS threshold of 44 W for the single-frequency (i.e., unmodulated) case (blue curve). The nonlinear departure in backward power at 44 W signifies the SBS threshold.

Subsequently, we applied WNS phase modulation to the 1,034 nm all-fiber amplifier. The WNS scheme utilizes an RF thermal noise to produce the Gaussian broadened signal. For this comparison, a white noise FWHM linewidth of 3.5 GHz or 11 pm was used. We note from the backward power (black curve) that the SBS threshold has been reached at ~470 W in the WNS modulated amplifier. Pulsing of the backward power was also observed on a fast photodetector at this power level. Thus, an SBS enhancement factor of 10 dB was attained, where the SBS enhancement factor is defined as the increase in SBS threshold over the

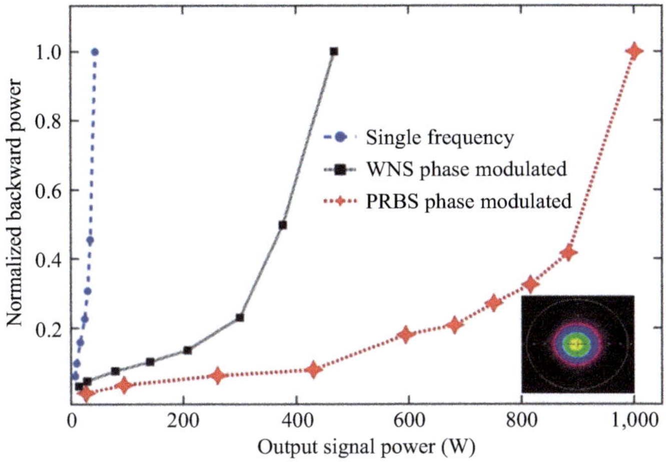

Figure 9.3 *Plots of normalized backward powers as functions of output signal power compared for single frequency (blue-dashed curve), WNS modulated (black-solid curve), and PRBS modulated (red-dotted curve) cases at approximately 3.5 GHz. Inset shows beam profile at 1 kW [31]*

corresponding single-frequency amplifier. A slope efficiency of 74% was obtained with near diffraction-limited beam quality ($M^2 < 1.2$). Here the lower slope efficiency is due to our nonwavelength stabilized diodes, which are configured to reach the optimal 976 nm pump wavelength at >1 kW.

We then applied PRBS phase modulation for the same amplifier configuration at a comparable 3.5 GHz clock rate and optimal PRBS pattern of 2^8-1. Particularly, with PRBS modulation we were able to achieve 1 kW of output power with 81% optical slope efficiency. Although no pulsing was observed here, further power scaling was once again limited by SBS as indicated from the backward power (red curve) and spectral content. In this case, the Stokes response was ~5 dB above the Rayleigh signal. An SBS enhancement factor of 14 dB was obtained and a greater than 2x enhancement in SBS threshold power was achieved with PRBS phase modulation over the corresponding WNS modulation case.

Previously, fiber length dependent behavior of various phase modulation schemes were theoretically and experimentally analyzed [22,23]. For WNS modulation, it was shown that the normalized SBS suppression decreases with fiber length. As discussed by Supradeepa [32], for short fibers the effective optical signal spectrum as encountered by the Stokes light appears sparse and random due to the small number of samples. Hence, the desired Gaussian lineshape of WNS is effectively not realized. Alternatively, the work presented in [23] shows that PRBS phase modulation can correct for this strong length dependence through proper selection of the pattern length. In addition, the periodic digital feature of PRBS modulation results in less random backward pulsing. These attributes, in part, result in the improved nonlinear SBS mitigation of PRBS phase modulation.

9.3.2 WNS and PRBS coherent beam combining analysis

In addition to their SBS mitigation properties, applicability of phase modulation schemes toward efficient beam combining is essential for high-power applications. Thus, to test the path length matching requirements and visibilities of WNS and PRBS modulation schemes, an actively stabilized CBC system was constructed. The beam combining system, similar to a Mach–Zehnder interferometer is shown in Figure 9.4. It consists of a phase modulated seed (WNS or PRBS) which was sampled by a 50/50 fiber coupled splitter. The lower arm then contains a variable delay line (VDL) for optical path length sweeping and an electro-optic phase modulator for active phase stabilization via the LOCSET phase locking technique developed at AFRL [33]. LOCSET applies a small RF phase dither or "tag" to each channel through sinusoidal phase modulation. The corresponding phase dithers are then measured with a photodetector as an interference beat note that contains the phase information needed for phase locking. Subsequently, error correction signals are applied to the LOCSET phase modulator to ensure optimal phase matching. The lower arm output is then sent to the input leg of a 2x2 fiber splitter. The top arm of the interferometer contains a high-power fiber amplifier to generate kilowatt class power levels. After the amplifier, an AR coated wedge is used to pick off a small portion of the beam and was coupled into a fiber collimator. The sampled output signal was then sent to the other input leg of the 2x2 splitter.

The visibility or coherence function can be derived from the interference between the sampled kilowatt class amplifier signal and the variable delay line

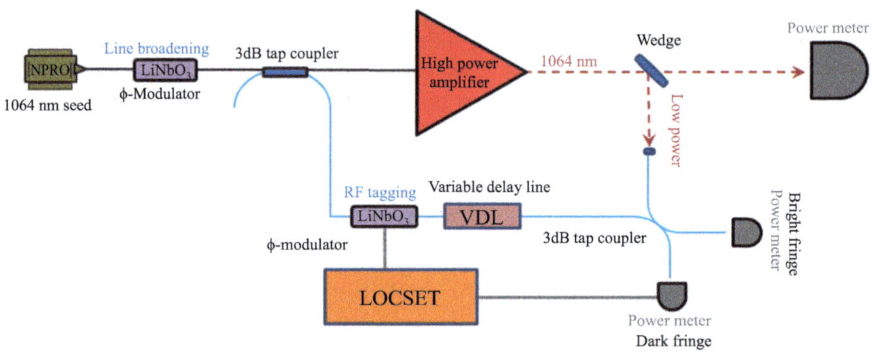

Figure 9.4 Experimental diagram for measuring the visibility of the phase modulated amplifier. The 1,064 nm seed is linewidth broadened using a phase modulator before being split into two paths using a 3dB (2x2) coupler. One portion of the beam is amplified to high power and an AR coated wedge is used to sample the beam. The other portion of the beam is RF tagged using a phase modulator controlled by LOCSET and path length matched to the high power beam. The two beams are interfered, and the bright and dark fringes are measured using a photodiode

controlled reference signal. At the combining port, the transmitted intensity is determined by the square of the sum of the electric fields. Here based on the coherence of the lasers and the optical length difference between the two paths, the intensity fluctuates between a maximum, $I_{\max}$, and minimum, $I_{\min}$, intensity. The combined intensity of the two channels is therefore given by:

$$I(\tau) = I_1 + I_2 + 2\sqrt{I_1 I_2}|V(\tau)|\cos(\alpha\tau - \delta) \tag{9.10}$$

where τ is the optical delay, α is the angular frequency, V is the visibility function, and δ is the phase offset. For a narrow-linewidth laser, the visibility function is defined by the Wiener–Khinchin theorem, imparting a Fourier transform relationship between the visibility and power spectral density (P_{SD}) [34]:

$$|V(\tau)| = \left|\int_{-\infty}^{+\infty} P_{\mathrm{SD}}(\nu)e^{-i\nu\tau}d\nu\right| = \frac{I_{\max} - I_{\min}}{I_{\max} + I_{\min}} \tag{9.11}$$

For a WNS phase modulated system, the broadened optical signal follows a Gaussian function due to the random distribution of voltage amplitudes applied to the modulator. In a PRBS system, the binary data pattern generates a sinc^2 spectral envelope with nulls located at the digital clock rate, ν_c. Accordingly, the power spectral density of a white noise and PRBS modulated laser can be represented, respectively, as follows:

$$P_{\mathrm{SD,WNS}}(\nu) \cong \frac{2\sqrt{\ln 2}}{\sqrt{\pi}\Delta\nu}\exp\left[-\left(2\sqrt{\ln 2}\frac{\nu}{\Delta\nu}\right)^2\right] \tag{9.12}$$

$$P_{\mathrm{SD,PRBS}}(\nu) = \frac{1}{\nu_c}\mathrm{Sinc}^2(\pi\nu/\nu_c) \tag{9.13}$$

where $\Delta\nu$ is the FWHM bandwidth of the optical spectrum. Based on (11–13), we can characterize the WNS and PRBS visibility functions as:

$$V_{\mathrm{WNS}}(x) = e^{\frac{-\pi(\Delta\nu)^2 x^2}{4\ \ln(2)c^2}} \tag{9.14}$$

$$V_{\mathrm{PRBS}}(x) = \mathrm{Tri}\left(\frac{x}{c}\nu_C\right) \tag{9.15}$$

where x is the optical path difference. In this work, we report the coherence length, L_c, as the delay where the visibility drops to 90% of its maximum; which corresponds to a combining efficiency of 95% in a two-channel beam combination system (Efficiency = (1 + V)/2). For an N-channel CBC array, the combining efficiency is equal to V + (1 − V)/N, indicating that the combining efficiency approaches the visibility function for a large number of channels.

Subsequently, we investigated the coherence length properties of PRBS and WNS modulated signals. For WNS modulation, a Gaussian broadened optical signal with 3.06 GHz FWHM was implemented. Similarly, a 2.88 GHz, 2^9-1 PRBS pattern was applied resulting in a sinc^2 spectrum. Next, through modification of the variable delay line, the visibility was measured as a function of optical path difference. These measurements were compared to the expected WNS and PRBS visibilities using the time-averaged spectral distributions as provided by (12–13). While both phase modulated signals have comparable spectral linewidths, the different spectral shapes of the modulation formats (WNS-Gaussian; PRBS-sinc^2) lead to different visibility curves and coherence lengths. The subsequent normalized visibility versus optical path difference for the 3.06 GHz linewidth WNS is plotted in Figure 9.5(a). The measured signal has strong correspondence with the theoretical Gaussian PSD, and a ~18 mm coherence length (visibility reduction of 10%) was determined. For PRBS modulation, the corresponding visibility versus OPD exhibits a linear dependence as captured by (15), with a ~13.5 mm coherence length as shown in Figure 9.5(b). As expected, such a spectrum reduces the coherence length. In general, for similar spectral FWHM linewidths, small path length mismatches lead to more severe drop in visibility for PRBS than WNS. In terms of coherence length WNS provides a favorable 33% improvement in coherence length over PRBS.

9.3.3 Filtered PRBS phase modulation

9.3.3.1 Filtered PRBS: coherent combining

Consequently, we proposed a novel approach to improve the coherence length and coherent combining performance of PRBS phase-modulated fiber amplifiers. Specifically, a filtered PRBS phase modulation scheme was developed where a low pass RF filter is utilized to mitigate the PRBS high-frequency components, thereby suppressing the sidelobes in the optical spectrum. To further investigate the modulated signal spectra, a heterodyne technique based on the interference between the phase modulated seed input and a second unmodulated frequency

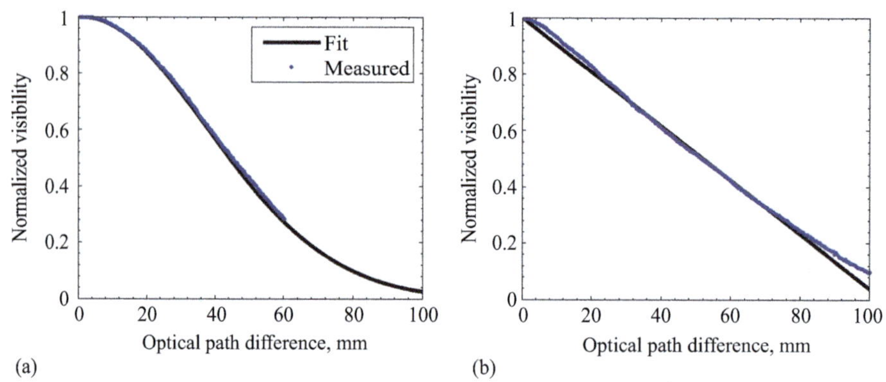

Figure 9.5 (a) Normalized visibility versus optical path difference for a 3.06 GHz WNS modulated system; experimental result (blue) and expected visibility (black). (b) Visibility measurements for the $2^9 - 1$ PRBS pattern with 2.88 GHz clock rate; experimental result (blue) and expected visibility (black)

detuned seed was implemented. The forward heterodyne signals for various WNS, PRBS, and filtered PRBS cases are presented in Figure 9.6. Through standard PRBS modulation with no RF filter (blue curve), the sidelobes generated by the 2.88 GHz PRBS pattern extend indefinitely well beyond 13 GHz, meaning only ~93% of the optical power resides within the central peak. With a corresponding linear visibility function, such a spectrum reduces the coherence length.

Next, we tested various RF low pass filters (LPF) to determine an optimal filter cutoff frequency. As the cutoff frequency of the low pass filter is reduced, the resulting sidelobes generated by the PRBS are suppressed. Using a filter equal to approximately half the modulation frequency (red curve) completely suppresses all of the generated sidelobes, producing an approximate Gaussian spectral distribution. For comparison, the forward spectrum of a Gaussian broadened WNS with 3.06 GHz linewidth was also measured (black curve). We note that the Gaussian spectrum generates a more advantageous Gaussian visibility function (based on Fourier transform relationship), where the gradual degradation in visibility leads to longer coherence lengths and greater tolerance to small path length deviations.

After, the visibility function of the filtered PRBS modulation scheme was measured and compared to the corresponding WNS and standard (unfiltered) PRBS modulation techniques. Based on the experimental setup depicted in Figure 9.4, the measured visibility versus optical path length differences are plotted in Figure 9.7. As expected, the reduction in the spectral sidelobes results in a longer coherence length. Based on the desired beam combining coherence length metric (90% visibility), the WNS modulated amplifier (black curve) has a coherence length of 17.8 mm, with a Gaussian visibility curve. However, the unfiltered PRBS (blue curve) has a coherence length of 13.4 mm with a linear drop in visibility. Filtering

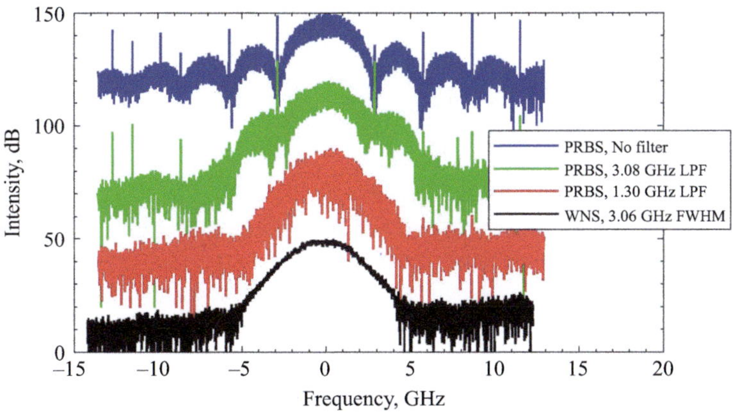

Figure 9.6 Heterodyne measurement of the optical spectrum for a variety of phase modulations schemes: (1) A 2.88 GHz PRBS signal with 2^9-1 pattern and no filter (blue), (2) PRBS signal with 3.08 GHz low pass RF filter (green), (3) PRBS signal with a 1.3 GHz low pass RF filter (red), and (4) a WNS with 3.06 GHz FWHM (black) [24]

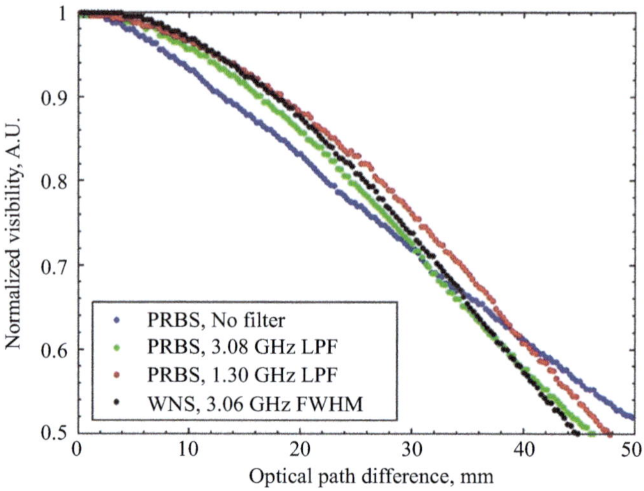

Figure 9.7 Measurements of the visibility comparing the unfiltered PRBS (blue), to the filtered PRBS (green, red) and the reference WNS (black) [24]

the RF signal with a 3.08 GHz LPF increases the coherence length to 16.7 mm (green curve), and further filtering using a 1.30 GHz LPF increases the coherence length to 18.6 mm (red curve).

After investigating several different types of filters, it was determined that filters with cutoff frequencies at approximately half the modulation frequency provide the best coherence length improvement. This is due to the complete side-lobe suppression at these cutoff frequencies, as shown in Figure 9.6. Significantly, by filtering the higher frequency components of the RF signal, the coherence length can be enhanced by 38.8% over standard unfiltered PRBS. The enhanced coherence length is even greater than the coherence length attained with a 3.06 GHz WNS (Filtered PRBS: 18.6 mm versus WNS: 17.8 mm).

9.3.3.2 Filtered PRBS: SBS suppression

In addition to the coherence length and beam combining properties, SBS suppression is equally critical for power scaling of kilowatt class Yb-doped all-fiber amplifiers. Thus, we analyzed the SBS threshold in filtered PRBS modulated fiber amplifiers. We note that the optimal PRBS modulation amplitude changes between unfiltered and filtered PRBS. Nevertheless, similar to the unfiltered PRBS scheme, the voltage of the filtered PRBS RF signal was tuned such that backwards reflected SBS signal was minimized. In Figure 9.8, the backward reflectivity versus output power was measured for the corresponding modulation schemes presented in Figures 9.6 and 9.7. Remarkably, both improved coherence length and SBS suppression were concurrently observed via filtered PRBS modulation.

For example, although WNS had the second longest coherence length, it has the lowest SBS threshold at 360 W (black curve). The unfiltered 2.88 GHz PRBS has a

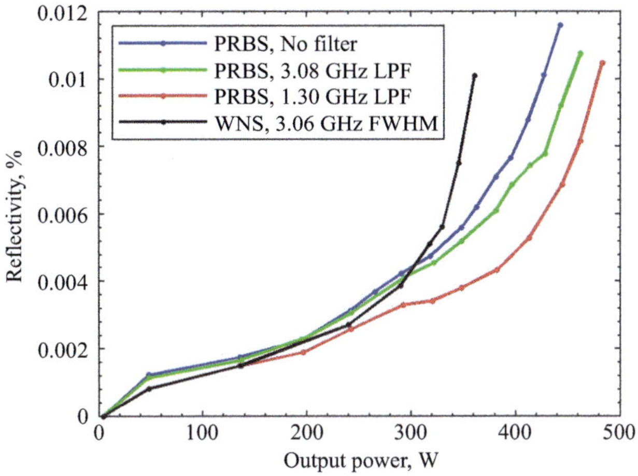

Figure 9.8 SBS threshold measured for the 2.88 GHz PRBS signal (blue) and compared to the filtered PRBS (green, red) and WNS (black) signals. Filtering the PRBS increases the SBS threshold due to the suppressed sidelobes [24]

larger SBS threshold of 426 W (blue curve). Alternatively, using a 3.08 GHz filter increases the SBS threshold to 454 W, and a 1.30 GHz filter increases the SBS threshold to 479 W (red curve) for a maximum improvement of 12.4% over unfiltered PRBS (and 33% SBS threshold improvement over WNS). The apparent power scaling improvement of filtered PRBS modulation is due to suppression of the SBS seeding phenomena observed in phase modulated narrow linewidth amplifiers.

Previously, we have shown that phase modulated signals at linewidths approaching the Brillouin frequency shift (~16 GHz) can result in seeding of the SBS process. Seeding effects due to spectral overlap of the Stokes and Rayleigh signals were shown to reduce the SBS threshold by up to 25%–30% [19]. Thus, due to the suppression of these sidelobes via RF filtering, we can assume that the increased SBS threshold of filtered PRBS is due to a seeding reduction. As discussed in Section 9.2, we have expanded our time-dependent model to account for backward reflectivity and the SBS seeding phenomena. Consequently, we simulated the seeding effects with unfiltered and filtered PRBS phase modulated light.

The SBS (9.4)–(9.9) were initially solved assuming no backward reflectivity ($R = 0$). Similar to Ref [22], a core diameter of 10 μm and fiber length of 9 m was simulated. We have shown that the SBS enhancement factors between 10 μm and LMA fibers (20–25 μm) do not vary when normalized to their corresponding single-frequency SBS thresholds [19]. Therefore, although we do not know the exact core dimensions of the commercial fiber amplifier, the SBS enhancement factors modeled should be applicable. The results are plotted in Figure 9.9(a) and (b) for the unfiltered and filtered PRBS cases, respectively. In Figure 9.9(a), the unfiltered PRBS results in the standard SBS threshold dependence on the modulation

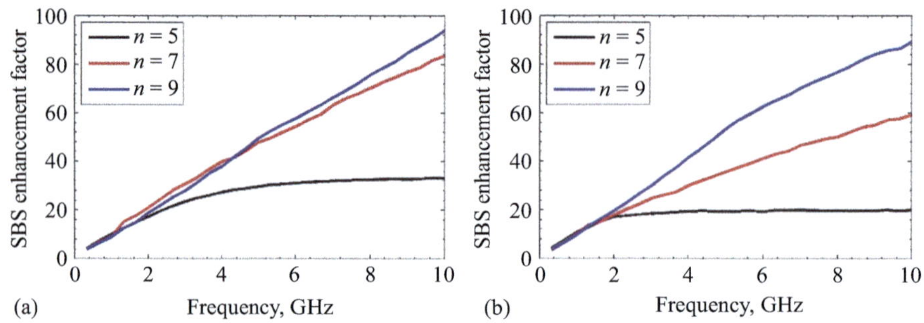

Figure 9.9 Theoretical modeling of the SBS enhancement factor without seeding. (a) No RF filter is used showing the expected SBS dependence on PRBS modulation frequency. (b) A 5th order Butterworth low pass filter is used to filter the PRBS reducing the SBS enhancement factor for low PRBS patterns [24]

frequency and pattern length. In Figure 9.9(b), the PRBS is filtered using a 5th order Butterworth low pass filter with a cutoff frequency set to half the modulation frequency. While filtering reduces the SBS enhancement for all cases, for the smaller patterns with lower number of sidebands (i.e., 2^5-1 or $n = 5$ and below), the reduction is more pronounced due to decreased number of sidebands in the main central lobe.

Alternatively, for higher PRBS patterns with larger number of sidebands, filtering has less effect on the overall SBS threshold. For the 2^9-1 ($n = 9$) pattern, which has higher SBS threshold enhancement and is more applicable for high power fiber amplifiers, there is little enhancement difference between the filtered and unfiltered PRBS cases. Here a small 10% increase is seen for modulation frequencies less than 8 GHz for the case without seeding. Beyond these frequencies the threshold is reduced by 5%. In practice, fiber amplifiers with some level of backward reflectivity represent a more realistic scenario. Among other things, this is due to minor reflections from end caps, splices, or free-space optics. Accordingly, the seeding effect was introduced by adding a reflectivity term, R to (9.9).

The SBS threshold enhancements versus modulation frequency for $R = 10^{-5}$ are shown in Figure 9.10. In Figure 9.10(a), the unfiltered PRBS is used, showing an approximately 20% reduction in the SBS threshold for all patterns caused by the backward reflectivity or seeding. Comparing Figure 9.9(a) to Figure 9.10(a), we observe similar enhancement factors at lower frequencies. However, as frequencies increase beyond 3–4 GHz, more pronounced reductions in SBS enhancement is seen. This is expected since seeding becomes more prominent at higher frequencies that start to approach the Brillouin frequency shift (~16 GHz). Additionally, the reduction is most prominent for the 2^9-1 pattern, where the small sideband separation leads to more spectral overlap between the Rayleigh and Stokes broadened signals.

On the other hand, filtering the PRBS signal does not cause a drastic reduction in SBS enhancement factors as observed by Figures 9.9(b) and 9.10(b). The SBS

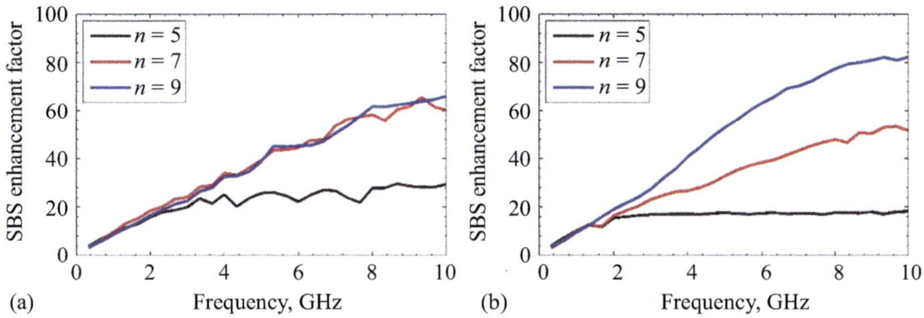

Figure 9.10 Theoretical modeling of the SBS enhancement factor with seeding. (a) No RF filter is used, the SBS thresholds are reduced due to spectral overlap. Fluctuations seen are due to the change in the overlaps of the Stokes and reflected spectrums. (b) An RF filter is used with the PRBS modulation, increasing the SBS enhancement factor to levels approaching the enhancement factor without seeding [24]

response for all patterns using filtered PRBS is nearly equal with backward reflectivity included in the model. This partly indicates that the filtered PRBS can mitigate the SBS seeding caused by the backward reflectivity. When compared to the unfiltered PRBS modulation with seeding (Figure 9.9(a)) for the optimal 2^9-1 pattern, we note that the filtered PRBS provides a 15%–20% improvement in SBS enhancement factor. For the smaller patterns, the SBS enhancement factors of the filtered PRBS are almost equal or slightly less than the unfiltered case. As the pattern number and hence number of sidebands increase, the filtering has less effect on the central lobe reduction. Significantly, by including the SBS seeding effect in the model we have shown that filtered PRBS can improve the SBS threshold of high power fiber amplifiers for optimal patterns (e.g., 2^9-1) over conventional unfiltered PRBS. Advantageously, filtered PRBS can improve both the SBS suppression and coherence properties of high power fiber amplifiers.

9.3.4 PRBS re-coherence

The periodic nature of the pseudo-random pattern allows for a re-coherence in the visibility, a phenomenon not seen with the continuous spectral distributions of WNS modulation or typical broadband sources. By increasing the path length difference to an integer multiple of the spacing between spectral lines $\{c(2^n - 1)/\nu_c\}$, coherence is restored. This re-coherence is illustrated in Figure 9.11 for both a 5 GHz (left) and 12 GHz (right) PRBS modulated 2^3-1 pattern. For the 5 GHz signal, the re-coherence distance is ~420 mm, agreeing well with the expected distance based on the pattern/sideband separation frequency of 0.714 GHz. More importantly, there is no drop in visibility or combining efficiency as the lasers periodically come back into phase.

By increasing the clock rate to 12 GHz, the pattern/sideband repetition frequency increases to 1.71 GHz, corresponding to the measured re-coherence distance

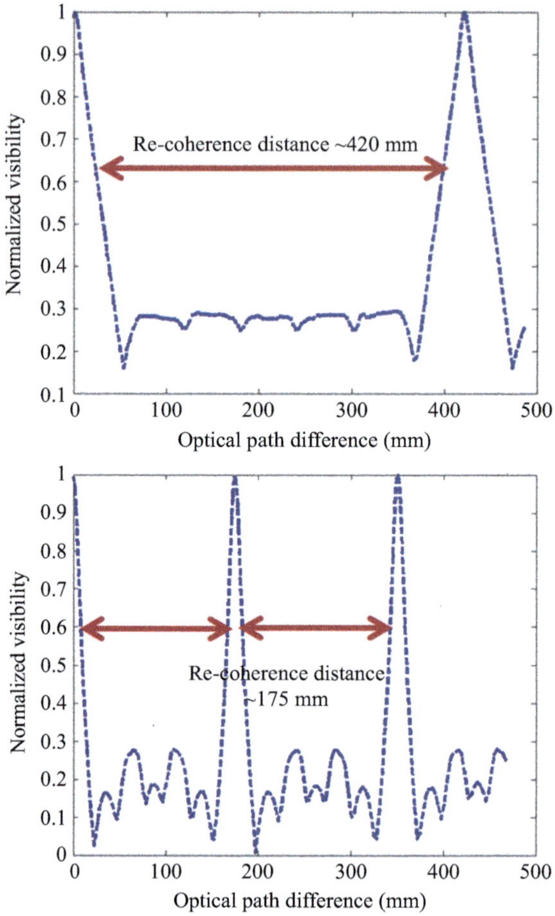

Figure 9.11 *Re-coherence measurements where re-coherence distance can be controlled by modifying the pattern repetition rate. (Left)* 2^3-1 *PRBS pattern at 5 GHz with re-coherence distance of 420 mm. (Right)* 2^3-1 *PRBS pattern length at 12 GHz with re-coherence distance of 175 mm [23]*

of 175 mm. Visibility measurements were also repeated using a 4.5 GHz WNS phase modulated signal for a path length difference of up to 500 mm and no re-coherence was observed as expected. In contrast to WNS modulation, which is based on random RF signals (no periodicity), re-coherence is an intrinsic property of periodic signals such as sinusoid and PRBS modulation. As a result, instead of absolute path length matching to one beam combining maximum, path length matching to any one of the periodically repeating maxima is sufficient. Significantly, this property may greatly simplify path length matching complexities in fiber laser CBC systems.

9.4 Multi-kW coherent beam combining of PRBS modulated fiber amplifiers

For phase modulated linewidth broadening, applicability toward efficient high power beam combining is another critical metric. As such, high power multi-kW fiber laser beam combining experiments based on traditional WNS phase modulation have been reported [27,28,35,36]. Nevertheless, in a direct comparison conducted at the kW level, PRBS phase modulation was shown to provide superior SBS suppression to that provided by WNS. In addition to superior nonlinear performance, we endeavored to demonstrate high power multi-kW beam combining of PRBS modulated fiber amplifiers. Here we established coherent combining of five commercial kilowatt-class fiber amplifiers using a DOE. It is well-known that DOE based beam splitters are convenient optical components for dividing an input beam into *N* output beams with precise angular spacing; at angles that represent the diffractive orders of the structure. Similarly, if mutually coherent beams are incident on the DOE at the proper angles and with appropriate phase matching/locking, the DOE can function in reverse as a beam combiner.

Five commercial kilowatt-class NuKW fiber amplifiers [37] from Nufern/Coherent were utilized for the combining experiments. These amplifiers are non-polarization maintaining and exhibit electrical-to-optical efficiencies of 35%–36%. Based on a master oscillator power amplifier (MOPA) configuration, as shown in Figure 9.12, a single-frequency non-planar ring oscillator (NPRO) operating at 1,064 nm from JDSU was used to seed the five channels. A PRBS generator and a fiber-coupled LiNbO3 electro-optic modulator (from EOSPACE) were then used to broaden the effective linewidth of the seed. Here a modulation frequency of 10 GHz and 2^9-1 PRBS pattern were applied. After spectral expansion, a small in-line fiber amplifier was constructed to amplify the seed signal before insertion into a 2 × 8 fused fiber splitter. The splitter provided identical narrow-linewidth signals for each of the five fiber amplifiers. Figure 9.13 displays a plot of the backward power as a function of output power for one of the amplifiers. As shown, the dependence is linear, which is indicative of operation below the SBS threshold.

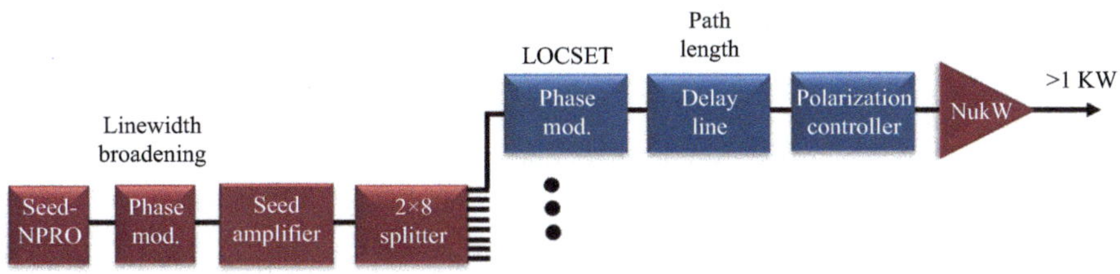

Figure 9.12 MOPA block diagram for the five channel DOE combining. LOCSET phase modulator, variable delay line, and polarization controller were used for phase stabilization, path length matching, and polarization locking, respectively

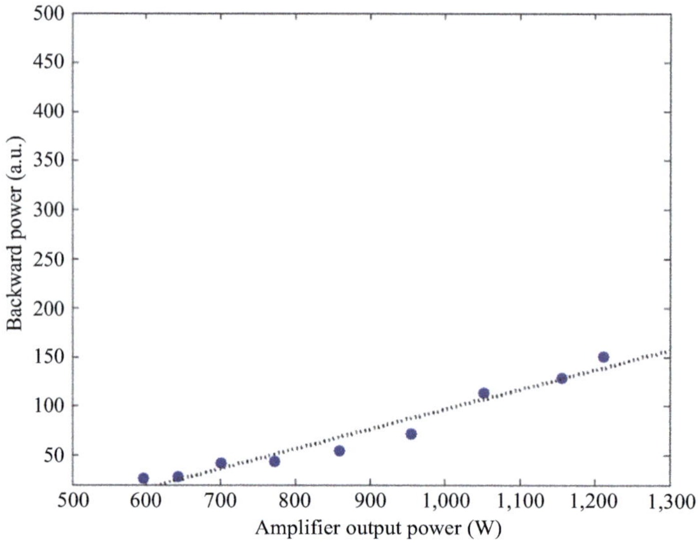

Figure 9.13 Backward power versus output power for one of the NuKW amplifiers driven with a 2^9-1 PRBS pattern at a frequency of 10 GHz [21]

Fiber-coupled electro-optic phase modulators were inserted into each of the channels to apply both small amplitude RF phase dithers and perform piston phase corrections in our LOCSET phase locking control board. Subsequently, tunable variable delay lines were inserted that can match the optical path lengths between fiber channels to submillimeter tolerances. A polarization tracker module with integrated controller from General Photonics was utilized for linear polarization locking. The polarization controller outputs were then used as inputs for the NukW fiber amplifiers. Particularly, accurate phase stabilization, path length matching and linear polarization control are critical for efficient coherent combining.

A schematic of the DOE combining experimental setup is shown in Figure 9.14. Initially, the beams were individually collimated and a low-power sample of each amplifier was utilized for active polarization locking using the aforementioned polarization controllers. After polarization locking, polarization extinction ratios (PER) of 15–16 dB were measured for each fiber amplifier. Next the five beams were directed onto the reflective DOE by a pair of turning mirrors at the appropriate diffraction angles. At 1,064 nm, the angle between diffracted orders was 8.87 mrad. The high-reflection (HR) coated DOE had dimensions of 15 mm × 15 mm × 6 mm on a fused silica substrate. The continuous surface relief profile DOE was uncooled similar to other DOE combining experiments [28,38]. Considerably, similar HR DOEs have been shown to exhibit ~17 ppm absorption [38] and high-power handling capabilities comparable to HR-coated mirrors [28].

The DOE combines most of the power into the $m = 0$ diffractive order, with the remaining power diffracted into the higher orders ($m > 0$). Hence, a copper beam block was inserted after the DOE to capture the stray light. The DOE was

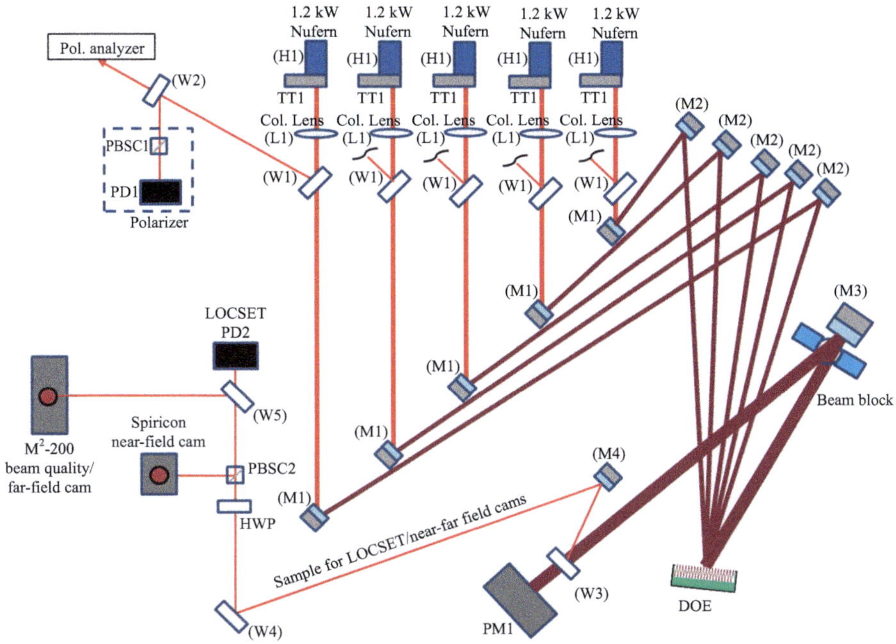

Figure 9.14 Schematic of the 1 × 5 DOE beam combining setup. After collimation, samples of each beam were used for polarization locking and a sample of the combined output beam was used for LOCSET phase locking and beam quality analysis. M1-4: reflective dielectric mirrors; W1-5: beam sampling wedges; HWP: half wave-plate; PBSC: polarizing beam splitter cube; PD1-2: photodetector; PM1: power meter

tilted slightly so the combined output beam is separated from the input beams and reflected out of the plane of diffraction. Subsequently, a low power sample of the combined output beam after the DOE was utilized for active phase stabilization and beam quality analysis. Active phase control was performed via LOCSET where a small RF phase dither or "tag" was applied to each channel through sinusoidal phase modulation. These corresponding phase dithers are then measured at the LOCSET photodetector (PD2) as an interference beat note that contains the phase information needed for phase locking. Consequently, error correction signals are applied to each phase modulator to ensure optimal phase matching between all of the amplifiers in the system.

After polarization and phase locking, the beams were coherently combined via the 1 × 5 DOE. Ultimately, a total output power of 4.9 kW was achieved with 82% combining efficiency. A plot of the input power versus combined output power is shown in Figure 9.15. The beam quality of the combined output beam was also measured as shown in Figure 9.16. The differences between the x and y curves are due in part to the ~4° angled end caps added to each fiber output to minimize

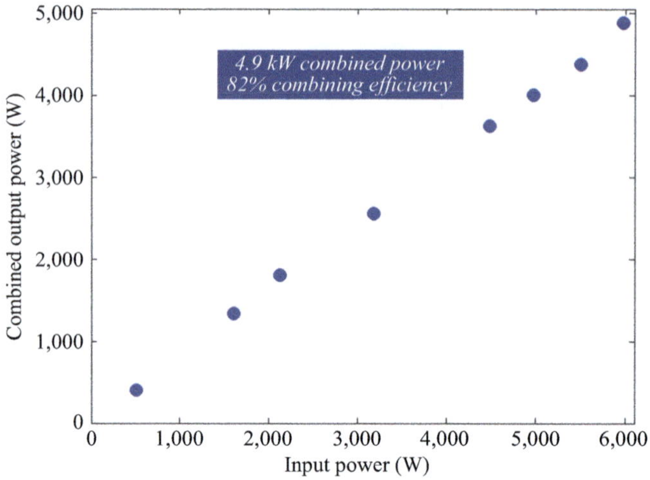

Figure 9.15 Input power (or power incident on DOE) versus combined output power for the 5 channel beam combining. An output power of 4.9 kW with 82% combining efficiency was demonstrated [21]

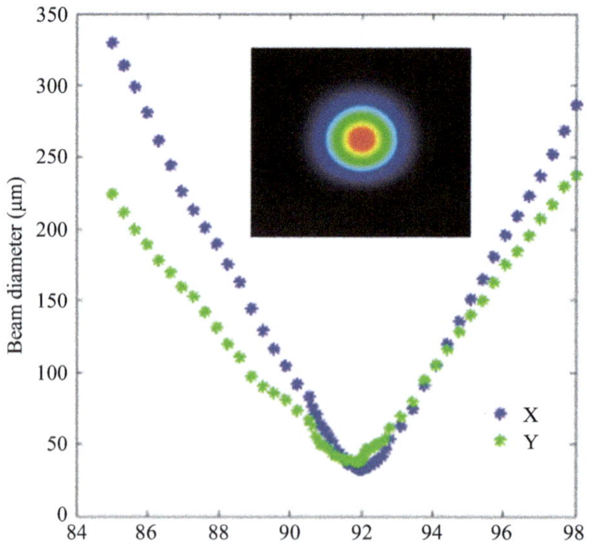

Figure 9.16 Near-diffraction-limited beam quality was measured with an M^2 of <1.1. Inset shows beam profile of combined output beam [21]

backward, reflectivity. Specifically, near-diffraction-limited beam quality was attained with a measured M^2 value of less than 1.1. It is important to note that the beam quality and combining efficiency were nearly constant at all power levels; thus inferring the absence of major thermal effects in the system.

We note that the overall combining efficiency is constrained by the cumulative effect of various aberrations and mismatches between the fiber amplifiers [39]. For example, the 1×5 DOE has an intrinsic splitter loss of ~5% when used as a combiner, with the remaining power diffracted into the higher orders. This is due to the low DOE channel count where there are less degrees of freedom to optimize the phase profile. Typically, higher efficiencies of ~97%–99% have been demonstrated for larger DOE channel counts from $N = 9 \ldots 81$ [40]. Similarly, losses due to nonideal polarization contributed to a reduction in efficiency. As a result, the measured 15–16 dB PERs contribute approximately 2%–2.5% loss. Residual phase errors from LOCSET, previously measured to be approximately $\lambda/60$ waves also impart an approximate 1%–1.5% loss in combining efficiency. The combining loss due to amplifier ASE content was roughly 1.6%. Moreover, other losses were believed to be due in part to slight optical misalignments/beam displacements and uncorrelated wavefront errors.

Particularly, higher combined powers and higher efficiencies can be attained through utilization of DOEs with higher channel counts. In addition, improved polarization controllers with PERs of >20 dB and incorporation of polished end facets should mitigate some of the uncommon wavefront errors. These attainable improvements may yield combining efficiencies approaching 90% for future experiments. In general, coherent combining has no established channel limit, though constraints due to cost, complexity and packaging will bound channel counts. Nevertheless, to our knowledge, the resulting 4.9 kW output with 82% efficiency represents the highest reported coherently combined power for a pseudo-random modulated fiber amplifier system.

9.5 Laser gain competition of all-fiber amplifiers

In addition to phase modulation, various other approaches for SBS suppression in fiber amplifiers have been reported, with some typical mitigation schemes summarized in Table 9.1. Generally, simultaneous utilization of multiple SBS suppression schemes which modify the Brillouin gain spectrum such as phase modulation, thermal/stress gradient, and acoustic tailoring may be not compatible due to overlapping spectra of the corresponding frequency shifts. In this regard, shortening of the effective fiber length is an advantageous nonlinear mitigation technique that can be used in conjunction with other schemes such as phase modulation and LMA fibers. As previously mentioned, counter-pumped combiners have been successfully used to reduce the nonlinear interaction length while preserving the long physical fiber length needed for adequate optical pump absorption and efficiency. Nevertheless, counter-pumped combiners have to simultaneously manage high-signal light combined with high-intensity pump light at the gain fiber output. In addition, the added pump combiner delivery fiber at the output end creates a longer fiber length, partially minimizing the enhancement from a counter-pumped architecture [41].

Similarly, an SBS mitigation technique developed at AFRL called laser gain competition (LGC) or two-tone can effectively reduce the nonlinear interaction length [14,42]. Here the fiber amplifier is seeded with two distinct sources. The shorter wavelength seed typically possesses broader spectral linewidth whereby SBS is not a consideration, while the longer wavelength seed is spectrally narrow. SBS suppression is realized due to a reduction in the effective nonlinear length as the narrow-linewidth signal, near the output end of the active fiber, rises sharply. This is similar to the aforementioned counter-pumped configuration, yet we are able to maintain the favorable co-pumped all-fiber architecture. Based on optimal selection of the seed power ratios and wavelength separation, this type of amplifier has been shown, both theoretically and experimentally, to result in 2 times SBS threshold enhancement over single-tone amplifiers. An additional advantage of the LGC or two-tone technique is its compatibility with other SBS mitigation techniques such as application of phase modulation and thermal gradients. As a result, we sought to demonstrate compatibility of LGC with PRBS phase modulation, extending upon previous work where LGC was successfully demonstrated in single-frequency fiber amplifiers.

9.5.1 Laser gain competition (two-tone): power scaling

The experimental setup, including the beam combining subsection 9.5.2, of our kilowatt two-tone all-fiber amplifier system is shown in Figure 9.17. One oscillator consisted of a single-frequency (<10 kHz linewidth) 1,064 nm NPRO seed laser externally modulated at an effective linewidth of a few GHz using a fiber-coupled

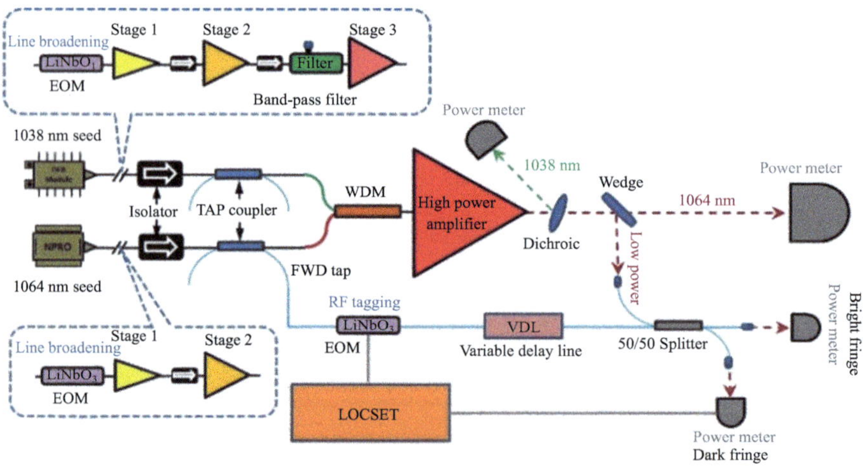

Figure 9.17 Power scaling and beam combining experimental setup for the kW two-tone amplifier

electro-optic modulator driven by either PRBS or WNS phase modulation. For the shorter or secondary wavelength seed, we used a single-frequency external cavity laser centered at 1,038 nm and we externally broadened the signal linewidth to 18 GHz using WNS phase modulation to mitigate nonlinear effects. As shown in the figure, each seed consisted of its own preamplification subsystem. The 1,064 nm signal was preamplified in two stages to 2 W before insertion into the final gain stage. Similarly, the 1,038 nm signal was preamplified in three stages to >20 W which provided sufficient seed signal power for kilowatt class amplification.

A fiber-coupled optical bandpass filter was inserted between the second and third stages to remove unwanted ASE content. Subsequently, a 1,035/1,064 nm WDM was used to combine the two signals before insertion into the high-power amplification stage. This section was comprised of a 10 m long non-PM 25/400 Yb-doped fiber from Nufern/Coherent that was powered by a ~1,400 W pump system centered at 976 nm. Here a $(6 + 1) \times 1$ TFB pump combiner was used to couple the pump light into the fiber. The amplifier output end was terminated with a cladding mode stripper to remove unabsorbed cladding/pump light, and an endcap possessing a 400 μm diameter. A 1,035 nm/1,064 nm dichroic mirror was used to separate the two signals.

To characterize SBS, tap couplers were inserted after both the 1,038 nm and 1,064 nm final preamplifier stages to monitor the backward light. The tap couplers were spliced onto 50/50 splitters for concurrent monitoring on a photodiode and an optical spectrum analyzer. Based on prior work, for effective SBS suppression through LGC, the seed power in the shorter wavelength signal (103x nm) is required to be more than ten times greater than that of the longer wavelength (106x nm) [14]. However, an important consideration is to ensure that most of the signal output power is at the longer primary wavelength with minimal ASE generated. Thus highly skewed ratios (toward shorter wavelength), while beneficial for the suppression of SBS, may not be conducive to achieving high efficiency at the desired longer primary wavelengths.

For our study, we compared seed ratios of 20:1 (1,038 nm:1,064 nm) and 40:1. Initially, we used the PRBS format in order to investigate SBS suppression when LGC is used in conjunction with phase modulation. Figure 9.18 shows a comparison of the backward power versus signal power among various amplifier seed scenarios: (1) 1,064 nm single-tone single frequency (i.e., no phase modulation; black curve), (2) PRBS-modulated (2.5 GHz) 1,064 nm single tone (gray curve), and (3 and 4) 1,064 nm two-tone amplifier, PRBS modulated at a clock rate of 2.5 GHz with seed ratios of 20:1 (blue curve) and 40:1 (red curve). As shown, the single-frequency amplifier reached an SBS threshold power of 33 W. Power scaling of the single-tone phase modulated amplifier, set at a clock rate of 2.5 GHz, yielded a maximum power of 470 W; which corresponded to an SBS threshold enhancement of ~11 dB. It should be noted, that in order to maximize SBS suppression, an optimal 2^8-1 PRBS pattern was applied in this case.

Next, through integration of LGC with a 20:1 seed ratio and PRBS phase modulation set at the same clock rate, an output power of 905 W was obtained.

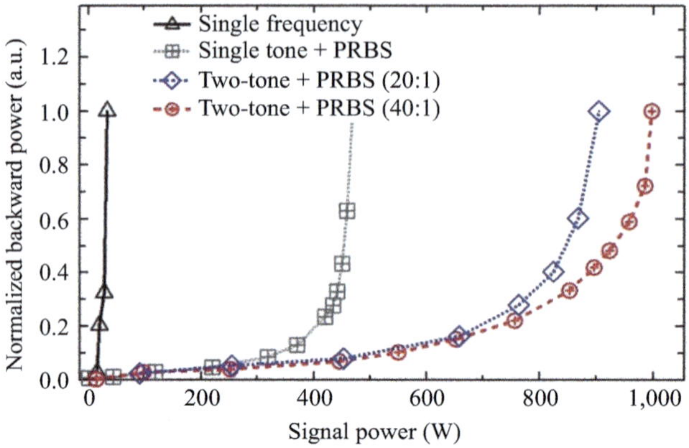

Figure 9.18 Measured normalized backward power versus output signal. Plots compare power scaling result for single-frequency (black curve), single-tone PRBS modulated (gray curve), and two-tone PRBS modulated cases with different power ratios (blue and red curves) [31], each PRBS amplifier was modulated at 2.5 GHz

The optical efficiency in the desired wavelength (1,064 nm) was 86% with <20 W left in the 1,038 nm signal. Since the effective nonlinear length in a two-tone amplifier is shorter than a single-tone one, the optimal PRBS pattern was found to be 2^6-1 for this case. Notably, the dependence of the normalized SBS threshold on fiber length in a phase modulated configuration was previously discussed [22,23,31], and the shorter PRBS pattern required for optimized two-tone operation was in accordance with the conclusions of those investigations.

By seeding the amplifier with a 40:1 ratio, we were able to suppress SBS further, leading to 1 kW of output power as shown in Figure 9.18 (red curve). The optical efficiency of 85% was slightly lower than that of the 20:1 case, due to more residual unconverted light from the secondary wavelength. The signal power versus launched pump power plots, for the two seed power ratio scenarios, are shown in Figure 9.19 and illustrate the slight difference between the optical efficiencies. Shown in the inset of the figure is the normalized backward spectral content in the vicinity of 1,064 nm for the two cases at maximum power as captured by the high-resolution OSA. For both cases, the amplifier output was scaled until the Brillouin-shifted Stokes light was ~5 dB higher than the Rayleigh light. Overall, optimization of the seed power ratio yielded a factor of 2.1x improvement in SBS threshold compared to the single-tone PRBS modulated case, while still maintaining high optical efficiency. However, as we will show further below, a lower relative SBS threshold enhancement was obtained when WNS phase modulation was applied.

Measured forward spectral content for the two-tone fiber amplifier is shown in Figure 9.20; indicating little ASE with ASE suppression levels of 52 dB (20:1 ratio) and 49 dB (40:1), respectively. Significantly, this corresponded to <1% ASE

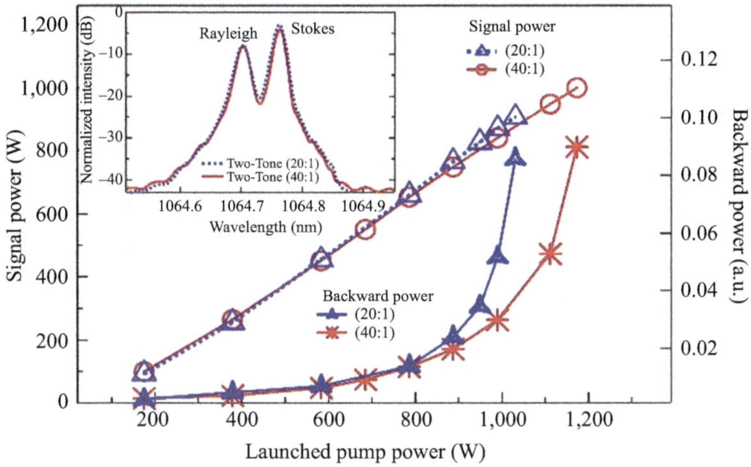

Figure 9.19 Signal power versus launched pump power for the two-tone amplifier at 2.5 GHz clock rate with 20:1 and 40:1 seed ratios. Inset shows the normalized backward light spectral content at maximum power for the two cases [31]

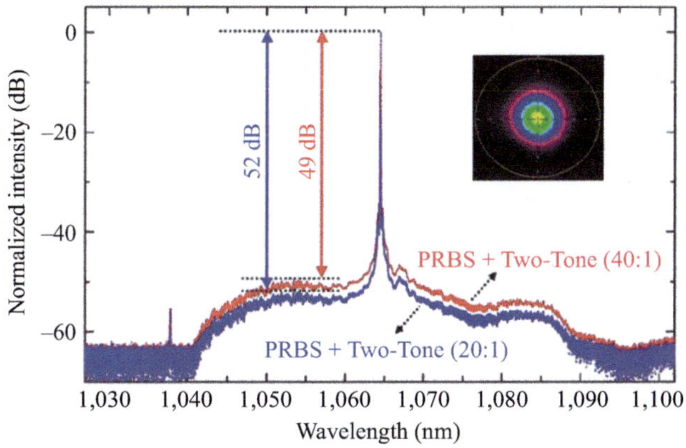

Figure 9.20 Normalized forward spectral content of the PRBS modulated two-tone amplifier for 20:1 and 40:1 seeding ratios captured at their respective maximum output power. Inset shows beam profile at 1 kW signal power indicating excellent beam quality [31]

content at the 1 kW power level. From a beam combining perspective, high ASE levels in amplifiers lead to a drop in combining efficiency regardless of the beam combining approach (CBC or SBC) utilized. As a result, the combination of narrow-linewidth performance and low ASE levels in this 1 kW amplifier, is beneficial to both beam combining approaches. The inset in Figure 9.20 illustrates the laser beam profile at 1 kW. Beam quality results indicated near diffraction-limited operation with M^2 values between 1.1 and 1.2 depending on the measurement axis.

Subsequently, the spectral linewidth of the two-tone amplifier output signal at full power was measured using a Fabry-Perot (FP) interferometer with a free spectral range (FSR) of 30 GHz. Figure 9.21 shows the spectral linewidth of the amplifier output signal at maximum signal power (1 kW). Note that the visible side-lobes in measured FP data are due to the PRBS sinc^2 spectra. At this power level, a FWHM linewidth of ~2.3 GHz (corresponding to the PRBS clock rate of 2.5 GHz) was measured, indicating no linewidth broadening. Similarly, for the PRBS modulated single-tone fiber amplifier, no spectral broadening was observed at its maximum SBS threshold power.

In an effort to compare, in the same two-tone configuration, the performance of WNS phase modulation, the 1,064 nm seed was spectrally broadened to 4 GHz using a WNS. As shown in Section 9.3.1, broader linewidths for SBS suppression are typically required for WNS in order to obtain similar power levels to those obtained by optimal PRBS patterns. Figure 9.22 shows the power scaling results of our WNS phase modulated amplifier. The SBS threshold for the single-tone amplifier and the two-tone amplifier with seed ratio of 20:1 were reached at ~530 W and ~850 W, respectively. In contrast, power scaling at the higher seed ratio (i.e., 40:1) was found to be difficult due to the onset of backward pulsation at only a few hundred watts which was captured on a fast photodetector.

Notably, in the WNS phase-modulated amplifier with a seed ratio of 20:1, LGC provided only 1.6x additional improvement in SBS threshold over the single-tone case. Yet (as shown above), at the same seed ratio in the PRBS phase-modulated amplifier, LGC provided 1.9x additional improvement in threshold. The drop in threshold enhancement for the former case is attributed to the fiber length dependence of SBS suppression via WNS modulation as discussed earlier. Especially when considering the reduced nonlinear effective length of the 1,064 nm

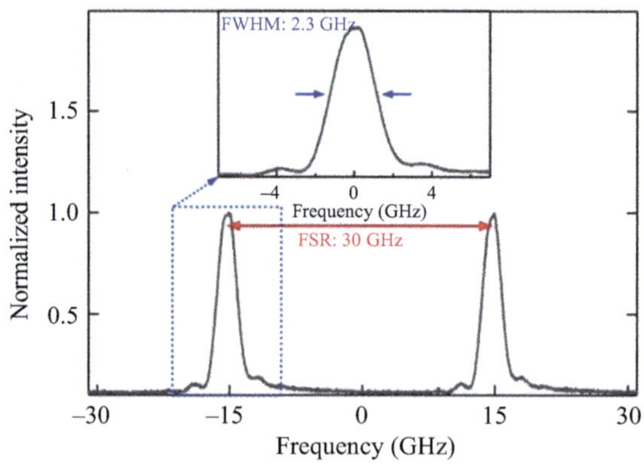

Figure 9.21 Spectral linewidth measurement of PRBS-modulated two-tone amplifier at 1 kW using a Fabry-Perot interferometer [31]

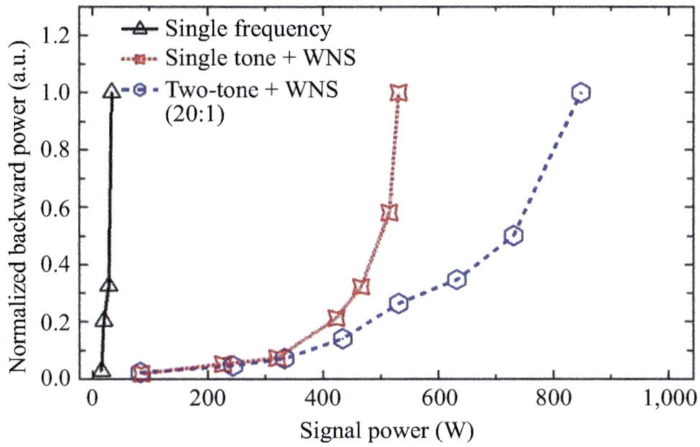

Figure 9.22 Measured normalized backward power versus output signal for the two-tone amplifier modulated with WNS at 4 GHz. Due to reduced SBS suppression in WNS a lower maximum output power of 850 W was reached [31]

light due to LGC. Alternatively, it was shown that PRBS phase modulation can partly correct for its own length dependence through the proper selection of pattern length. Therefore, by adjusting the PRBS pattern from 2^8-1 to 2^6-1, we were able to recover almost the full benefit of SBS suppression via LGC. Overall, the power scaling results represent a significant reduction in spectral linewidth compared to that of commercially available narrow-linewidth Yb-doped fiber amplifiers.

9.5.2 Laser gain competition (two-tone): beam combining

To examine the CBC performance of amplifiers utilizing LGC, coherence length and fringe visibility properties of the amplifier at 1 kW output power were compared to the single-tone amplifier. The CBC setup is shown in Figure 9.17. Prior to insertion into the high-power stage, a low power output sample was taken from the 1,064 nm main forward tap and sent to our LOCSET phase stabilization sub-system. Accordingly, a fiber-coupled variable delay line was inserted to adjust the optical path length difference for the fringe visibility measurements. The high-power stage was separately amplified to the maximum power with the single-tone and two-tone configurations. At the final stage output end, a small portion of the amplified 1,064 nm signal was then picked off using a wedge and free-space coupled to a PM fiber patch cable. The picked off signal from the high-power channel was then interfered with the low-power forward signal using a 50/50 fiber coupled splitter. The splitter outputs were coupled onto two detectors acting as bright and dark ports. To measure the fringe visibility, the variable delay line was set to sweep the optical path difference between the two channels.

The corresponding normalized visibility curves as functions of path length difference are shown in Figure 9.23, representing the single-tone amplifier and the

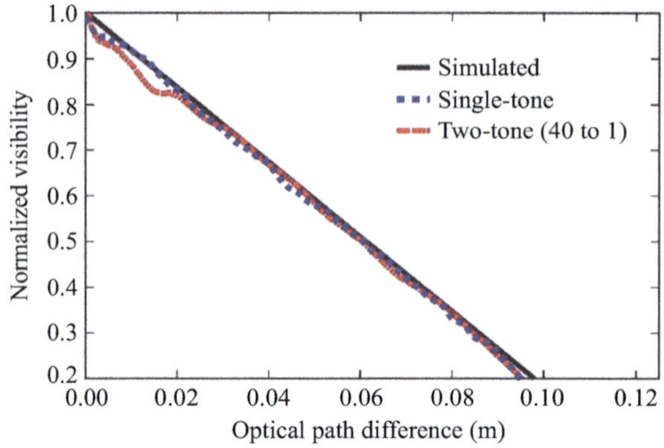

Figure 9.23 Coherence length comparison between the single-tone and two-tone amplifiers. Dashed lines are experimental data and solid line is the expected visibility plot

two-tone amplifier with 40:1 seed power ratio. In both cases, PRBS phase modulation at a clock rate of 2.5 GHz was applied and the visibility measurements were conducted at the respective maximum power of each amplifier. As shown by the simulated plot, a linear visibility function with coherence length of ~10 mm is expected. This corresponds to the Fourier transform of the sinc^2 spectral power density of the PRBS modulated 1,064 nm signal. Specifically, the measured visibilities are in good agreement with expected coherence lengths for both the two-tone and single-tone cases. Small discrepancies in visibility for the two-tone amplifier are due to the uneven coupling efficiency of the forward output light as the collimating optics alignment tended to drift due to heating at the 1 kW level. Overall, high fringe contrast was obtained for both amplifiers with ~90% combining efficiency. Here the small drop in efficiency is likely due to the power difference between the coupled output beam and the fixed low-power forward tap signal. Nevertheless, due to the comparable combining efficiencies, visibility functions and measured coherence lengths, no deleterious linewidth broadening was encountered with the phase modulated two-tone amplifier at high-power.

9.6 Conclusion

Compact, efficient, and robust all-fiber amplifiers have made tremendous strides in the last decade for an extensive number of industrial and DE applications. In particular, fiber amplifiers suitable for beam combination have stricter requirements than those in industrial environments, leading to a difficult trade space between the desired output power and the nonlinear limitations resulting from the linewidths needed for efficient beam combination. Bandwidth efficient techniques to suppress SBS, while maintaining suitably narrow linewidths for efficient beam combination have been explored. Toward that end, the improved nonlinear SBS suppression and narrow linewidths achieved through pseudo-random phase modulation can have a significant impact on the beam combining of kW class fiber lasers; where broader linewidths may hinder efficient coherent and SBC. In a direct comparison at the kW level, PRBS phase modulation was shown to provide superior SBS suppression (up to 2x) to that provided by the standard WNS modulation. In terms of coherent combining, however, pseudo random signals display a linear drop in visibility, leading to pronounced drops in combining efficiencies for small path deviations. As such, we have developed a novel filtered PRBS modulation approach for enhanced combining efficiency and coherence length performance. Through this approach, improved coherence lengths of ~40% and SBS threshold enhancements of ~10%–20% were observed. The simultaneous SBS suppression and coherence length improvements of the filtered PRBS approach can offer significant advantages for power scaling and coherent combining of many lasers.

Accordingly, high-power multi-kW beam combining of PRBS modulated fiber amplifiers was demonstrated for the first time. Five-kilowatt-class fiber amplifiers seeded with PRBS modulated light were coherently combined using a 1x5 DOE, demonstrating combined output power of ~5 kW with 82% combining efficiency and near diffraction-limited beam quality ($M^2 < 1.1$). Another effective SBS mitigation technique developed at AFRL called laser gain competition, based on seeding the amplifier with two distinct seed sources was described. SBS suppression is realized due to a reduction in the effective nonlinear length. More importantly, LGC can be used in conjunction with phase modulation for even further SBS mitigation in high-power fiber amplifiers. Notably, integration of LGC with PRBS modulation at a clock rate of 2.5 GHz, yielded 1 kW of output power. A factor of 2 improvement in SBS threshold power over the corresponding single-tone case was achieved. The beam quality was measured to be near-diffraction limit and high optical efficiency of 85% was attained. Overall, the power scaling results represent a significant reduction in spectral linewidth compared to that of commercially available narrow-linewidth Yb-doped fiber amplifiers.

Acknowledgments

The authors would like to thank the Air Force Office of Scientific Research (AFOSR) and the Directed Energy-Joint Transition Office (DE-JTO) for supporting our high-power fiber laser research through numerous research grants. The authors acknowledge Northrop Grumman and DE-JTO for providing the DOE used for the beam combining experiments. We thank Dr. Ben Pulford for fruitful discussions and constructive comments for this chapter. Similarly, the authors would like to thank other current and former colleagues who have contributed greatly to the power scaling and beam combining fiber research at AFRL.

References

[1] L. Quintino, A. Costa, R. Miranda, *et al.*, "Welding with high power fiber lasers – A preliminary study," Materials and Design **28**, 1231–7 (2007).

[2] S. D. Jackson and A. Lauto, "Diode-pumped fiber lasers: a new clinical tool?," Lasers in Surgery and Medicine **30**, 184–90 (2002).

[3] P. Sprangle, J. Peñano, B. Hafizi, and A. Ting, "Incoherent combining of high-power fiber lasers for long-range directed energy applications," Journal of Directed Energy **2**, 273–84 (2007).

[4] V. Fomin, V. Gapontsev, E. Shcherbakov, A. Abramov, A. Ferin, and D. Mochalov, "100 kW CW fiber laser for industrial applications," 2014 International Conference Laser Optics, IEEE, St. Petersburg, pp. 1–1 (2014).

[5] P. Sprangle, B. Hafizi, A. Ting, and R. Fischer, "High-power lasers for directed-energy applications," Applied Optics **54**, F201–F209 (2015).

[6] L. Huang, H. Wu, R. Li, *et al.*, "414 W near-diffraction-limited all-fiberized single-frequency polarization-maintained fiber amplifier," Optics Letters **42**, 1–4 (2017).

[7] Anping Liu, "Stimulated Brillouin scattering in single-frequency fiber amplifiers with delivery fibers," Optics Express **17**, 15201–09 (2009).

[8] J. P. Koplaw, D. Kliner, and L. Goldberg, "Single-mode operation of a coiled multimode fiber amplifier," Optics Letters **25**, 442–4 (2000).

[9] T. Eidam, C. Wirth, C. Jauregui, *et al.*, "Experimental observations of the threshold-like onset of mode instabilities in high power fiber amplifiers," Optics Express **19**, 13218–24 (2011).

[10] B. Ward, C. Robin, and I. Dajani. "Origin of thermal modal instabilities in large mode area fiber amplifiers," Optics Express **20**, 11407–22 (2012).

[11] A. V. Smith and J. J. Smith, "Mode instability in high power fiber amplifiers," Optics Express **19**, 10180–92 (2011).

[12] M. D. Mermelstein, S. Ramachandran, J. M. Fini, and S. Ghalmi, "SBS gain efficiency measurements and modeling in a 1714 μm^2 effective area LP08 higher-order mode optical fiber," Optics Letters **30**, 3225–7 (2005).

[13] T. Kokki, J. Koponena, M. Laurila, Y. Changgeng, "Fiber amplifier utilizing an Yb-doped large-mode-area fiber with confined doping and tailored refractive index profile," SPIE Proceedings **7580**, 758016 (2010).

[14] C. Zeringue, C. Vergien, and I. Dajani, "Pump-limited 203 W single-frequency monolithic fiber amplifier based on laser gain competition," Optics Letters **36**, 618–20 (2011).

[15] C. Robin and I. Dajani, "Acoustically segmented photonic crystal fiber for single-frequency high-power laser applications," Optics Letters **36**, 2641–3 (2011).

[16] C. Robin, I. Dajani, and B. Pulford, "Modal instability-suppressing, single-frequency photonic crystal fiber amplifier with 811 W output power," Optics Letters **39**, 666–9 (2014).

[17] F. W. Willems and J. S. Leong, "Simultaneous suppression of stimulated Brillouin scattering and interferometric noise in externally modulated lightwave AM-SCM systems," IEEE Photonics Technology Letters **6**, 1476–8 (1994).

[18] A. Hadjifotou and G. A. Hill, "Suppression of stimulated Brillouin back-scattering by PSK modulation for high-power optical transmission," IEE Proceedings: Optoelectronics **133**, 256–8 (1986).
[19] A. Flores, C. Robin, A. Lanari, and I. Dajani, "Pseudo-random binary sequence phase modulation for narrow linewidth, kilowatt, monolithic fiber amplifiers," Optics Express **22**, 17735–44 (2014).
[20] E. C. Honea, R. S. Afzal, M. Savage-Leuchs, *et al.*, "Advances in fiber laser spectral beam combining for power scaling," SPIE Proceedings **9730**, 97300Y (2016).
[21] A. Flores, I. Dajani, R. Holten, T. Ehrenreich, and B. Anderson, "Multi-kilowatt diffractive coherent combining of pseudorandom-modulated fiber amplifiers," Optical Engineering **55**, 096101 (2016).
[22] C. Zeringue, I. Dajani, S. Naderi, G. T. Moore, and C. Robin. "A theoretical study of transient stimulated Brillouin scattering in optical fibers seeded with phase-modulated light," Optics Express **20**, 21196–213 (2012).
[23] B. Anderson, A. Flores, R. Holten, and I. Dajani, "Comparison of phase modulation schemes for coherently combined fiber amplifiers," Optics Express **23**, 27046–060 (2015).
[24] B. Anderson, A. Flores, and I. Dajani, "Filtered pseudo random modulated fiber amplifier with enhanced coherence and nonlinear suppression," Optics Express **25**, 17671–82 (2017).
[25] N. Naderi, A. Flores, B. Anderson, and I. Dajani, "Beam combinable, kilo-watt, all-fiber amplifier based on phase-modulated laser gain competition," Optics Letters **41**, 3964–7 (2016).
[26] R.S. Williamson III, "Laser coherence control using homogeneous linewidth broadening," U.S. Patent No. 7,280,568, October 9 (2007).
[27] G. D. Goodno, S. J. McNaught, J. E. Rothenberg, *et al.*, "Active phase and polarization locking of a 1.4 kW fiber amplifier," Optics Letters **35**, 1542–4 (2010).
[28] S. J. McNaught, P. A. Thielen, L. N. Adams, *et al.*, "Scalable coherent combining of kilowatt fiber amplifiers into a 2.4-kW beam," IEEE J. of Sel. Top. In Quantum Electronics **20**, 174–81 (2014).
[29] Y. Aoki, K. Tajima, and I. Mito, "Input power limits of single-mode optical fibers due to stimulated Brillouin scattering in optical communication systems," IEEE J. Lightwave Technology **6**, 710–9 (1988).
[30] M. M. Hildebrandt, M. Frede, and D. Kracht, "Narrow-linewidth ytterbium-doped fiber amplifier system with 45 nm tuning range and 133 W of output power," Optics Letters **32**, 2345–7 (2007).
[31] N. Naderi, I. Dajani, and A. Flores, "High-efficiency, kilowatt 1034 nm all-fiber amplifier operating at 11 pm linewidth," Optics Letters **41**, 1018–21 (2018).
[32] V. R. Supradeepa, "Stimulated Brillouin scattering thresholds in optical fibers for lasers linewidth broadened with noise," Optics Express **21**, 4677–87 (2013).
[33] A. Flores, B. Pulford, C. Robin, C. A. Lu, and T. M. Shay, "Coherent beam combining of fiber amplifiers via LOCSET," *Coherent Laser Beam Combining*: Chapter 2, 45–73, Wiley-VCH, Weinheim (2013).

[34] J. W. Goodman, *Statistical Optics*, John Wiley & Sons, New York (1985).
[35] C. X. Yu, S. J. Augst, S. M. Redmond, *et al.*, "Coherent combining of a 4 kW, eight-element fiber amplifier array," Optics Letters **36**, 2686–8 (2011).
[36] S. M. Redmond, D. J. Ripin, C. X. Yu, *et al.*, "Diffractive coherent combining of a 2.5 kW fiber laser array into a 1.9 kW Gaussian beam," Optics Letters **37**, 2832–4 (2012).
[37] Nufern Product Brief, NukW: Kilowatt laser amplifier platform.
[38] E. C. Cheung, J. G. Ho, G. D. Goodno, *et al.*, "Diffractive-optics based beam combination of a phase-locked fiber laser array," Optics Letters **33**, 354–6 (2008).
[39] G. D. Goodno, C. Shih, and J. E. Rothenberg, "Perturbative analysis of coherent combining efficiency with mismatched lasers," Optics Express **18**, 25403–14 (2010).
[40] J. E. Rothenberg and G. D. Goodno, "Advances and limitations in beam combination of kilowatt fiber amplifiers," SPIE Proceedings **7686**, pp. 768613-1–768613-8 (2010).
[41] Anping Liu, "Stimulated Brillouin scattering in single-frequency fiber amplifiers with delivery fibers," Optics Express **17**, 15201–09 (2009).
[42] I. Dajani, C. Zeringue, and T. M. Shay, "Investigation of nonlinear effects in multitone-driven narrow-linewidth high-power amplifiers," IEEE Journal of Selected Topics in Quantum Electronics **15**, 406–14 (2009).

Chapter 10

Applications of high-power 2 μm thulium fiber lasers in materials processing

Ilya Mingareev[1,2], Fabian Weirauch[2], Tobias Bonhoff[2], Nils Gehlich[2], Myungkoo Kang[2], Kathleen Richardson[2] and Martin Richardson[2]

Recent advances in the development of high-power laser sources emitting at the wavelength around 2 μm have enabled a novel material processing regime that is currently finding many new industrial and scientific applications, from automotive manufacturing and biomedical engineering to microelectronics and infrared photonics. High output powers and system reliability enabled by a robust fiber laser architecture bring many new opportunities to laser materials processing community, but also many scientific and technical challenges. On the one hand, a large number of new and unique processes have been introduced, from absorber-free welding of transparent polymers and volume-selective micro-processing of semiconductors to local refractive index modifications in infrared optical materials. On the other hand, with a steadily growing selection of commercially available continuous-wave (CW) and pulsed thulium-doped fiber lasers, processing limits for many conventional laser processing techniques such as cutting, drilling, and welding, will need to be reassessed. In addition, the knowledge of materials' response to irradiation at 2 μm is still incomplete for many important classes of materials, and a limited availability of beam delivery equipment optimized for this wavelength makes industrial implementation of thulium fiber lasers challenging. This chapter is aiming at improved understanding of fundamental principles of light-matter interaction at 2 μm, discusses various new applications, and contributes to implementation of thulium fibers lasers as a promising tool for many industrial areas.

10.1 Introduction

Laser materials processing has been a very intense area of research and development for several decades and is now one of the backbones of modern

[1]College of Engineering and Science, Florida Institute of Technology, Melbourne, FL, USA
[2]College of Optics and Photonics, CREOL, University of Central Florida, Orlando, FL, USA

manufacturing industry and the economy. While for many classes of materials laser processes like cutting, welding, marking, and drilling processes have reached maturity with wide industrial acceptance, steady progress in the development of new materials demands for new processing technologies. At present day, a large selection of cost-efficient high-power lasers emitting at wavelengths around 1 micron and based on fiber, disk, and diode technologies are available. However, for processing of certain classes materials like semiconductors and ceramics with high spatial selectivity, laser sources emitting in the infrared are required.

For semiconductors [1] and infrared optical materials [2], processing at wavelengths larger 1 μm allows utilizing materials' transparency and enables 3D processing capability. For example, intrinsic silicon's energy bandgap is around 1.1 μm at room temperature, and the material is semitransparent at larger wavelengths. With precisely tailored energy distributions in the infrared silicon wafers could be processed not only at the surface, but also in the volume of the wafer as well as at the wafer's back surface, without affecting the front surface. The trade-off between linear and nonlinear absorption defines the laser-matter interaction regime and is a matter of intense research interest currently. While some relevant applications are already accepted in the industry, for example, stealth-dicing, many other emerging applications are currently at relatively low technology readiness levels (TRL).

Similarly, optical materials suitable for operation in the infrared, specifically, mid-wave and long-wave infrared spectral ranges, are opaque at wavelengths around 1 μm. High-power, high-brightness sources such as the thulium-fiber lasers are essential to fabrication of graded-index (GRIN) optics based on chalcogenides and other infrared materials. A combination of linear and nonlinear absorption is responsible for the local refractive index in these materials, allowing custom optical devices to be fabricated.

In polymers, differences in linear absorption between 1 μm and at larger wavelengths enable new processing regimes such as additive-free welding [3]. For some materials, efficient welding using laser radiation at the commonly used wavelengths around 1 μm is only possible with additives that increase absorption of laser light. Using additives may considerably increase the number of required process steps and therefore the overall manufacturing costs [4–6]. Furthermore, their utilization is sometimes prohibited in biomedical applications due to the toxicity of the additives. In contrast, laser radiation produced by Tm-doped fiber lasers offers high CW output powers and good beam quality that can be directly absorbed in the material.

In addition, the wavelength of 2 μm is considered to be in the so-called "eye safe" spectral region, which supports the general acceptance of Tm-fiber lasers in industrial environment. This is due to the high absorption of laser light in the eye's cornea in this spectral region. Thus the laser light cannot threaten the high sensitive retina.

10.2 Interaction of 2-μm laser light with materials

Interaction of 2 μm laser light with thermoplastic polymers, intrinsic and doped semiconductors, as well as infrared glass-ceramics, has been extensively studied in several experimental studies and modeling efforts. In experiments shown here, laser radiation was produced by a thulium-doped fiber laser system that could operate in q-switch and CW modes. The laser system was designed using the master-oscillator-power-amplifier (MOPA) architecture, with several design options available both for the oscillator stage (e.g., using a single-mode fiber) and for the amplifier stage (e.g., using a photonic crystal fiber). In all studies, the wavelength was set to approx. 1,985 nm, the pulse durations were varied between 6.9 and 100 ns, the repetition rates were varied between 1 and 100 kHz, and the beam quality was $M^2 < 1.2$. Laser beam polarization was set either linear or random, depending on the application. Subsections 10.2.1–10.2.3 provide a basic overview of laser-matter interaction phenomena, and summarizes experimental observations for polymers, semiconductors, and infrared glass-ceramics processed with laser light at 2 mm.

10.2.1 Polymers

The spectral properties of selected polymer materials were studied using a UV-VIS-NIR spectrometer (PerkinElmer Lambda 1050). All polymers under investigation were natural, nonpigmented materials, including low- and high-density polyethylene (PE-LD, PE-HD), polymethylmethacrylate (PMMA), polypropylene (PP), polyoxymethylene (POM), and glycol-modified polyethylene terephthalate (PETG). The sample thickness was $d = 1.6$ mm for all materials.

As a prerequisite for efficient polymer processing including processes like marking, cutting, and joining, the materials must allow a sufficient optical penetration depth. In particular, for transmission welding of similar and dissimilar polymers, the energy deposited at the weld interface must be large enough to generate a joint melt pool between the joint partners. A basic study of the optical properties was carried out to evaluate the new process windows enabled by the use of Tm-fiber lasers, as well as to identify any differences in optical absorption compared to laser processing at more commonly utilized, shorter wavelengths. Transmission and reflection spectra of the polymer samples obtained in the spectral range $\lambda = 0.5 - 2.5$ μm were used for the calculation of laser light absorption (Figure 10.1).

As a general trend for all materials, higher absorption around $\lambda = 2$ μm was observed compared to the wavelength range $\lambda = 0.8 - 1$ μm, denoting the typical emission wavelength range of the most common diode and fiber laser systems used for polymer processing. For instance, the materials PMMA and POM revealed about ten times higher absorption at $\lambda = 2$ μm than at $\lambda \approx 1$ μm, the emission wavelength of the Nd- and Yb-fiber lasers widely used for polymer marking and welding. Consequently, a more efficient energy deposition into the material can be enabled at 2 μm. It should be noted, however, that higher absorption leads to a smaller optical penetration depth in the long-wavelength range.

In addition to investigation of spectral properties, polymer melting behavior was studied for exposure of polymers to CW laser light at various powers. This allows the determination of modification thresholds, provides useful data on laser-induced surface morphology changes and volume expansion, and simulates materials' response during various laser processes relevant to industry. In the initial

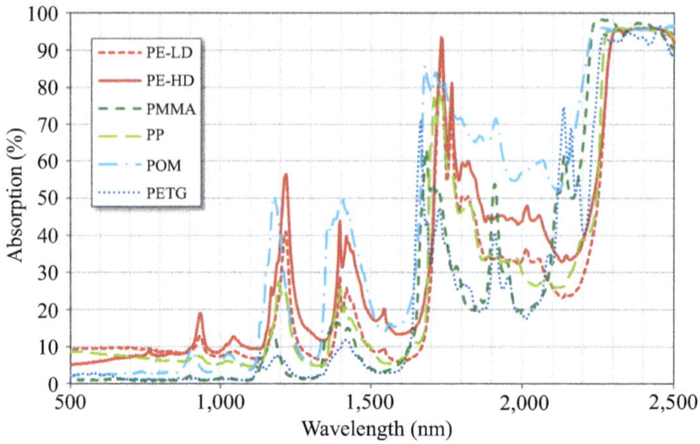

Figure 10.1 Spectral absorption properties of selected polymers

set of experiments, polymer samples with dimensions 80 × 8 × 1.6 mm^3 were processed by focusing laser light using a biconvex lens with the focal length of f = 150 mm, with the focus located approx. 0.8 mm below the surface and translating the target at a constant speed v = 5 mm/s perpendicular to the laser beam. Single melt tracks were produced that are separated by a distance of 2 mm from each other, and sufficient time was given for the samples to cool down before applying a new set of process parameters. In order to simulate a butt-welding process, a linear spring-loaded sample holder was designed and built that applies a constant force the samples perpendicularly to the melt tracks. This ensures a pressure of $\sigma = 0.6$ N/mm^2 at the virtual weld interface and simulates the case of greatest possible proximity of the two joint partners for butt-welding.

The laser power was changed in 1-Watt steps in the range $P = 1$–15 W, and an appropriate range of applicable laser powers was found for each material. The modification power threshold (P_{min}) was identified as the point at which surface and/or bulk modification was first observed with optical microscopy, while the beginning material decomposition, for example, burning or charring, combined with generation of soot indicated the maximum power level (P_{max}) applicable for a particular material.

Evaluation of experimental results (Figure 10.2) revealed an increase of the melt track depth and width with increasing laser power. This behavior was observed until a maximum depth corresponding to the full sample thickness was achieved. Thereafter due to the lower viscosity of the melt and the force applied to the sample, the track width decreased as the excessive material was pushed out of the laser-matter interaction zone. In most materials, the melted material was transferred to the surfaces, where a bead was formed after solidification. The shape of the bead appeared to be material-dependent defined by the solidification behavior and viscosity of the material.

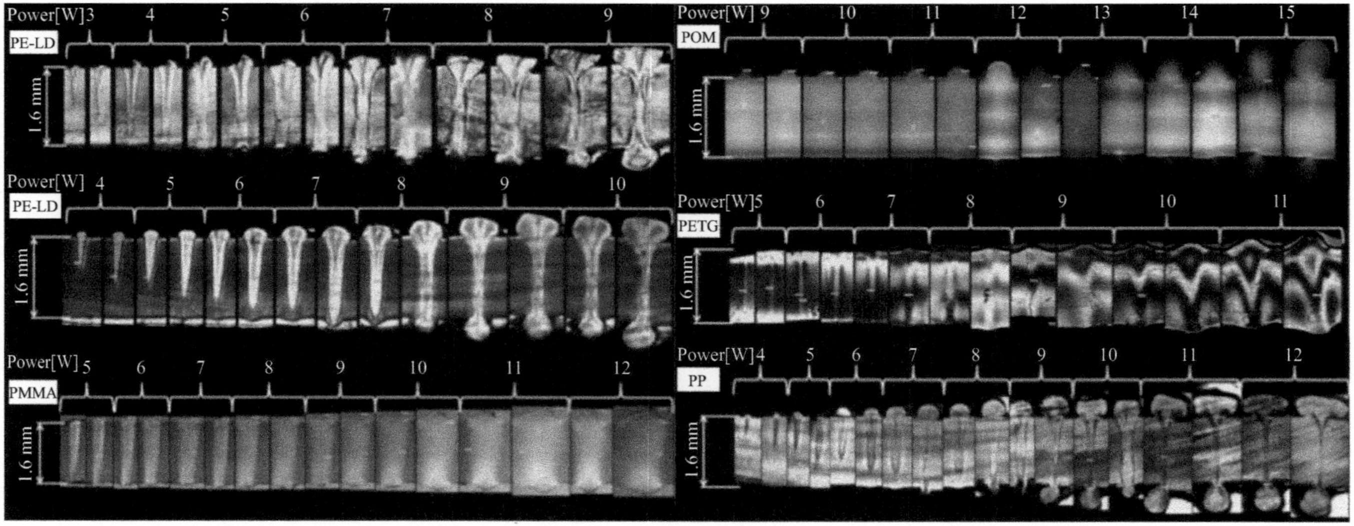

Figure 10.2 Cross-sections of melted tracks produced at various laser powers at the wavelength 2 μm (continuous-mode regime) in different polymers

10.2.2 Semiconductors

Extensive modeling and experimental efforts were performed to characterize interaction of 2-μm laser light with intrinsic and doped semiconductors. In general, for the semiconductors' "transparency" regime, the mechanisms of light absorption and energy transfer leading to internal material modifications in semiconductors are not well studied. It is essential to review the underlying physical phenomena accompanying semiconductor processing, in order to obtain a conclusive evaluation of the experimental results obtained at the wavelength 2 μm. Similarities and differences to processing in the nontransparent regime, that is, with photon energies exceeding the material's bandgap E_g need to be studied. For example, while the front surface modification thresholds for Si and GaAs are usually very similar to the values obtained for processing at shorter wavelengths when linear absorption dominates, modification morphology observed in the bulk of semiconductors indicates that nonlinear absorption processes may play an important role. Consequently, irradiation of pure semiconductors with photon energies below the band gap may enable a new, different processing regime which allows a spatially selective modification of the wafers' bulk and processing of the wafer's back surface without damaging the front surface.

10.2.2.1 Absorption in semiconductors in the infrared

For laser light with photon energies greater than the band gap, that is, in UV and visible, three primary absorption mechanisms in semiconductors can be identified: intrinsic, free-carrier, and defect-based absorption. Intrinsic absorption is governed by interband electronic transitions, while free carrier absorption is based on photo-induced intraband transitions and resulting in the energy transfer to the lattice. The third absorption mechanism is inherent to impure and contaminated materials, is initiated by defects and impurities leading to a lowered energy band gap. Multiphoton absorption is usually not considered an important absorption mechanism for semiconductor processing with photon energies $\hbar\omega > E_g$, because material modifications such as melting and ablation occur at much lower intensities than needed for multiphoton absorption to engage.

In contrast, for photon energies below the bandgap, such as produced by the 2 μm thulium fiber lasers, and for pure semiconductors, intrinsic absorption vanishes, and free-carrier absorption becomes negligible. Therefore, multiphoton absorption starts to play an important role in laser-semiconductor interaction, and requires further consideration involving a thorough analysis of the energy budget for different experimental conditions. In addition, electron avalanche ionization is an important, secondary nonlinear carrier excitation mechanism, which requires background (seed) electrons generated by another (primary) absorption process. Precise knowledge of the origin and the density of photo-excited seed electrons is essential for the evaluation of the laser-semiconductor interaction mechanisms for different photon energies.

Using a simple, first-principles model describing laser heating of semiconductors [7], the temperature increase of a material irradiated at fluence F can be estimated. Assuming the optical energy is added to the material uniformly within the heating depth L_H, the temperature rise ΔT can be estimated as:

$$\Delta T = \frac{F(1-R)}{\rho c L_H} \tag{10.1}$$

where R is the reflectivity, c is the specific heat of the material, and ρ is the mass density of the material. The heating depth L_H can be estimated using the following approximation:

$$L_H \approx \frac{1}{\alpha} + L_D + L_T \tag{10.2}$$

where L_D is the carrier diffusion length, and L_T is the thermal diffusion length. The total absorption coefficient α can be expressed as a sum of α_{SP}, α_{MP}, and α_{FC} representing single-photon, multiphoton, and free-carrier absorption coefficients, respectively. At the wavelength 2 μm, intrinsic absorption disappears, and the free carrier absorption can be considered negligible, $\alpha_{FC} << 1\ \mathrm{cm}^{-1}$ for pure semiconductors like Si and GaAs. Consequently, multiphoton absorption can be considered the only primary absorption process that supplies seed electrons for avalanche ionization followed by the material modification phenomena such as heating and melting. The parameter L_D depends on the carrier lifetime and the material temperature, and is usually considered to be on the order of 1 μm or smaller for Si and GaAs irradiated with nanosecond pulses. The thermal diffusion depth $L_T \sim t_p^{1/2}$ is negligibly small as well, for irradiation with nanosecond laser pulses. As a consequence, assuming no other primary photoionization processes are involved in semiconductor laser processing at $\lambda = 2$ μm, (10.2) can be simplified to $L_H \approx \alpha^{-1} \approx \alpha_{MP}^{-1}$. In silicon, a possible multiphoton absorption mechanism is the two-photon absorption with the 2PA coefficient $\beta_{Si} \approx 0.25$ cm/GW [8]. In GaAs, two-photon absorption is not possible at $\lambda = 2$ μm, and at least three photons are needed to exceed the band gap. The corresponding 3PA coefficient is $\gamma_{GaAs} \approx 0.18\ \mathrm{cm^3/GW^2}$ [9]. By calculating α_{MP} for both materials, and using the reported, experimentally obtained modification threshold values in (10.1), the corresponding temperature rise induced by a single laser pulse would be on the order of only 1 K. In fact, in order to increase the surface temperature to the melting point, a total absorption coefficient on the order of $\alpha \sim 10^3$–$10^4\ \mathrm{cm}^{-1}$ would be required. Therefore, the contribution of the multiphoton absorption processes to the material modification is very small, but not negligible as it could still provide seed electrons for avalanche ionization.

10.2.2.2 Nonlinear material response in semiconductors

A qualitative validation of these generalized considerations for Si and GaAs can be obtained in an open-aperture Z-scan experiment, a simple and powerful technique that allows studying nonlinear optical properties of transparent materials. For this experiment, a triplet lens ($f = 25$ mm) was used for beam focusing. The repetition rate was set to $f_{rep} = 10$ kHz, and the pulse duration was set to $t_p = 100$ ns, to reduce the risks of wafer damage and to provide sufficient laser power for reliable measurements. For a Z-scan experiment, transmission through the wafers was determined as a function of the position z and the pulse energy E_p. Measurements were repeated five times, and the wafers were translated in the xy-plane to a new spot to avoid possible interferences with previously exposed areas.

The results of the Z-scan measurements (Figure 10.3) revealed a transmission decrease for both materials as the focused beam passed through the wafer. In Si, the transmission dropped from 53.4% to approx. 41% for $E_p \leq 25$ μJ, and this dependency revealed a nonlinear behavior. In GaAs, the transmission drop was only about 2%, being nearly constant for pulse energies in the range $E_p \leq 24.5$ μJ. However, at $E_p \geq 25$ μJ an abrupt decrease in transmission was observed, which most probably indicates the onset of the modification processes, that is, surface damage.

The results of the Z-scan measurements showed that two-photon absorption is an important mechanism of energy transfer resulting in material modification of Si, whereas three-photon absorption in GaAs apparently does not contribute to the material modification significantly. Consequently, an alternative absorption mechanism must be assumed to generate sufficient seed electron densities.

10.2.2.3 Alternative material modification mechanisms

Iwata *et al.* [10] proposed an alternative reasonable physical model for surface-enhanced absorption. Absorption that is higher than expected could result from a very thin absorbing layer located at the surface of the wafer. Enhanced absorption in thin layers, which cannot be detected by conventional absorption measurement techniques, can result from near-surface structural modifications caused, for example, by chemo-mechanical surface processing during wafer preparation (cutting, grinding, and polishing) and consequently the reduced mechanical integrity of the material. Therefore, absorption process within such a layer can be initiated by micro- and nanoscale sites including dangling bonds, native oxide layers, and various lattice defects. These sites are able to absorb light even with the photon energies below the bandgap.

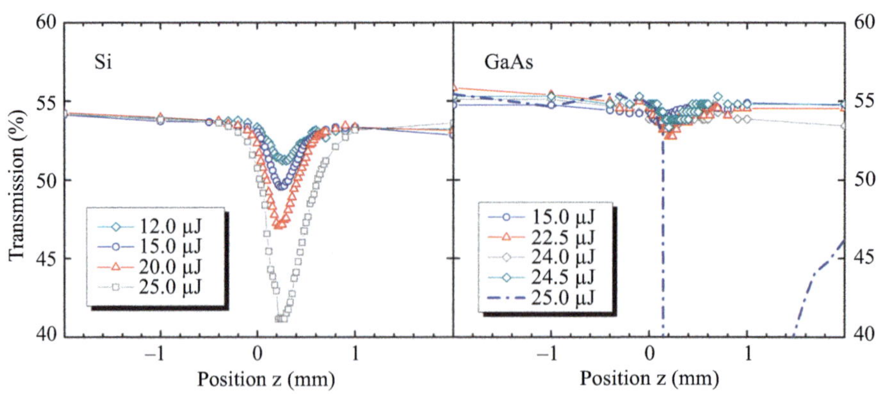

Figure 10.3 Results of the open-aperture Z-scan for Si and GaAs

These experiments confirmed Iwata's model in multiple experiments on surface modification using 2 μm light. Isolated random micro-damage zones and other defects were always present even in highest-quality materials. With sufficiently high spatial overlap of laser pulses these zones occupy larger surface areas, and the damage is gradually accumulated. This damage accumulation is assumed one of the driving mechanisms leading to surface modifications observed in the 2-μm laser processing experiments. As a consequence of enhanced light absorption at these defect sites, large densities of additional charge carriers could be created for the subsequent avalanche ionization and thermal modification processes.

10.2.2.4 Temperature dependence of absorption processes

The band gap of semiconductors can decrease significantly as a consequence of the rising material temperature. This effect could have significant implications on modification behavior. For a basic approximation, the common Varshni relation [11] can be used providing the following temperature dependency of the energy band gap $E_g(T)$:

$$E_g(T) = E_g(0) - \frac{\alpha T^2}{T + \beta} \tag{10.3}$$

where $\alpha = 4.73 \cdot 10^{-4}$ eV/K, $\beta = 636$ K for Si ($\alpha = 5.41 \cdot 10^{-4}$ eV/K, $\beta = 204$ K for GaAs), and $E_g(0)$ gives the material's bandgap energy at $T = 0$ K. This effect can have an impact on absorption behavior and create more favorable conditions for absorption at defect sites with increasing material temperature. Moreover, when the temperature of GaAs reaches approx. 420 °C, the reduced E_g will be small enough to enable two-photon absorption at the wavelength 2 μm. It should be noted, however, that the 2PA coefficient is fairly small at the corresponding energy threshold. Therefore, the contribution of the two-photon absorption in GaAs should be accurately evaluated based on specific experimental conditions.

In addition, elevated material temperature may cause changes in electronic and lattice heat conduction parameters of Si and GaAs. As a general trend, the total conductivity and the mobility of electrons and holes decrease with increasing temperature. Higher laser-induced charge carrier densities may lead to further mobility reductions due to electron-hole scattering [12]. On the one hand, these effects may contribute to a more localized energy deposition leading to surface modification through melting and solidification. On the other hand, high pressures can be created in the bulk material resulting in nonequilibrium phase transitions and defects generation, for example, voids and nano-/microcracks. Lastly, surface reflectivity of semiconductors increases with temperature (approx. $\Delta R \approx 0.2\%$ per $\Delta T = 100$ K), which can potentially impact energy deposition in the material and alter surface modification behavior.

10.2.3 Infrared optical materials

Laser light with the emission wavelength in infrared can be instrumental to fabrication of functional infrared Graded Index (GRIN) optics. Compared to conventional optics made of homogeneous materials, IR GRIN optics offer enhanced optical functionality in a much more compact package, and at a significantly reduced cost. Currently, the development of GRIN optics for modern commercial and defense platforms faces many challenges from material sciences, to optical design and processing technology. Both the selection of suitable host materials that can be used for the fabrication of GRIN optics, and the maturity of laser processes for space-selective refractive index modification are still fairly limited. Recent studies have shown that thulium fiber lasers can enable a unique processing regime capable of producing custom refractive index modifications in the volume of chalcogenide (ChG) glasses.

ChGs are a class of optical materials showing high transparency in a broad spectral range in the infrared, and the ability to adjust their refractive index through compositional variation [13,14]. The method of spatially varying the refractive index is based on partial crystallization of the glass, which leads to formation of optical nanocomposites. The amount of partial crystallization can be effectively tailored in 3D through alkali ion migration (poling). Alternatively, localized crystallization can be controlled through formation of nuclei and their growth within the ChG material by a laser process and subsequent heat treatment. In particular, vapor deposited metastable thin-film GAP-Se glasses [15] have shown the ability to modulate the refractive index through formation of high-index Pb-rich nanocrystals within the low-index glassy matrix. The material utilized in this study was $15GeSe_2$-$45As_2Se_3$-40PbSe (15-45-40 GAP-Se), a high-lead-content multicomponent glass. GAP-Se films with the thickness 20 μm were deposited onto Schott IRG 24 substrates featuring matched coefficient of thermal expansion (Figure 10.4).

In order to evaluate laser-matter interaction mechanisms, optical transmittance spectra were first collected from an uncoated substrate and a GAP-Se-coated substrate. The spectra indicate that the transmission cut-off of the substrate becomes red-shifted from approx. 0.8 μm toward ~1.1 μm once the 20 μm film is introduced (Figure 10.5). Therefore, laser light with the wavelength larger than 1.1 μm, such as that produced by a thulium fiber laser, is needed to produce in-volume modifications. Beyond the transmission cut-off, the base film remains transparent in the mid-wave infrared.

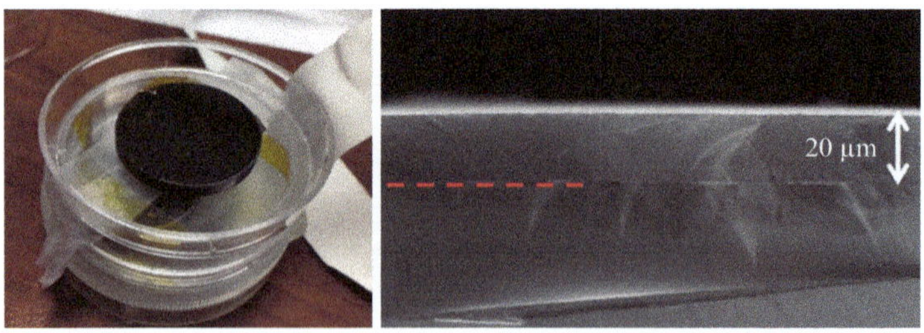

Figure 10.4 Image of a 30-mm substrate with a GAP-Se film (left), and a cross-section of the sample showing the 20-μm GAP-Se film (right)

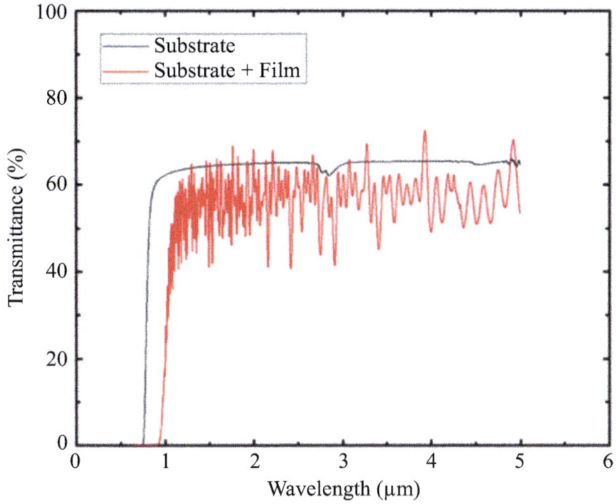

Figure 10.5 Optical transmittance spectra of GAP-Se film with the substrate, and the substrate only

Similar to semiconductors, absorption in GAP-Se films is governed by linear and nonlinear absorption. Due to the complex elemental composition of GAP-Se materials, the knowledge of the complete dispersion behavior is still very limited. Recent studies have shown that nonlinear absorption properties will differ for bulk materials and films. In addition, the optical performance of the GRIN optics produced from GAP-Se will strongly depend on heat treatment protocols.

10.3 Joining of polymers

Absorber-free transmission welding of polymers PE-LD, PE-HD, PMMA, PP, POM, and PETG was performed using thulium fiber laser radiation. Additionally, a feasibility study of transmission welding of PETG, PMMA, and combinations PMMA/PP and PP/PE-LD was performed. In all the experiments intrinsic absorption of polymers was utilized for welding without adding any IR-absorbers. The resulting weld quality was investigated by means of optical, phase-contrast, and polarization microscopy at the weld cross-sections. The tensile strength of butt-welded materials was measured and compared to that of a corresponding bulk material.

10.3.1 Experimental details

The 2-μm laser was operated as a CW source at power levels up to 27 W. The output laser radiation was nonpolarized. For the butt-welding and transmission welding experiments, the laser radiation was focused directly onto the interface between the parts, and the parts were pressed together (Figure 10.6).

After irradiation of the samples, the produced weld joints were sectioned across the weld interface using a diamond saw. Then the samples were lapped and

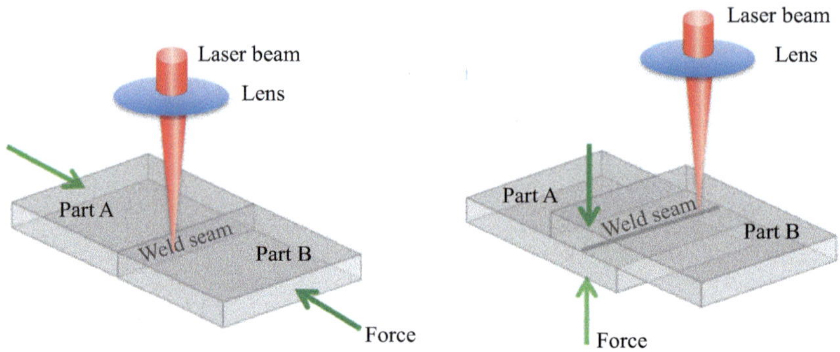

Figure 10.6 Experimental arrangements of parts for butt-welding (left) and transmission welding (right)

polished in a three-step process (Lapmaster 15ATP) to achieve optical surface quality for further characterization. Investigations of welded samples were performed using an optical microscope Olympus BX-51 with differential interference contrast (DIC) capability, and a polarizing microscope Leica DMR XE. Testing of mechanical strength was performed according to the ASTM D638 standard using a tensile tester (Zwick Z100) and the corresponding sample geometries.

10.3.2 Butt-welding experiments

For direct butt-welding experiments, two samples with dimensions 40 × 8 × 1.6 mm were used in each irradiation experiment, and welding was performed along the 8-mm edge. The roughness of the corresponding joining planes was measured before welding and ranged from $R_a = 0.26$ μm for low-density polyethylene to $R_a = 0.90$ μm for PP.

Differences in dilution behavior of different polymers (Figure 10.7) were observed. The cross-sections of the welds in PE-LD, PE-HD, and PP exhibit well-mixed interface areas without distinct boundaries between the joining partners. In contrary, the interface regions in PMMA, POM, and PETG still show clear edges at any laser power level adopted. Additionally, significant variations in the formation of the weld seam and bead can be observed compared to the melting experiments. Partially these variations can be explained by light deflection at the edge defects, and possibly insufficient positioning accuracy.

Furthermore, the evaluation of weld cross-sections enables a qualitative analysis of materials density changes in processed materials. DIC microscopy adopted for optical characterization allows a visualization of residual stresses in laser-processed materials, caused by the rapid heating and slow cooling of the interface region (Figure 10.7, e.g., PE-LD and PETG). These thermal gradients lead to local expansion, shrinkage, and material density changes.

Based on the results of the optical characterization of the weld cross-sections, the set of process parameters used for successful welding was further narrowed down to three laser power values and used for preparation of tensile-test samples.

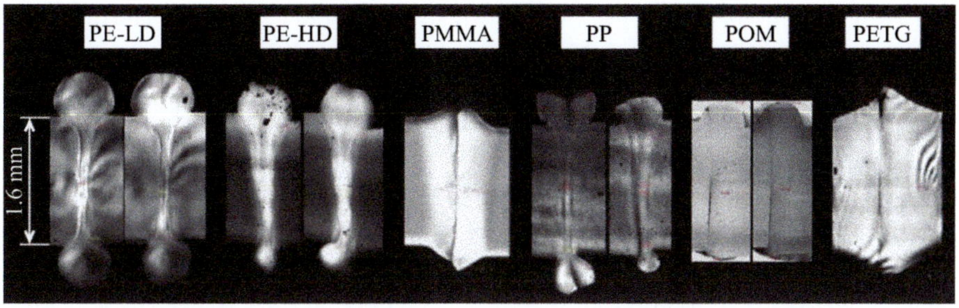

Figure 10.7 Examples for cross-sections of butt-welded polymers PE-LD (P = 8 W), PE-HD (P = 10 W), PMMA (P = 11 W), PP (P = 7 W), POM (P = 14 W), and PETG (P = 9 W) observed by phase-contrast microscopy

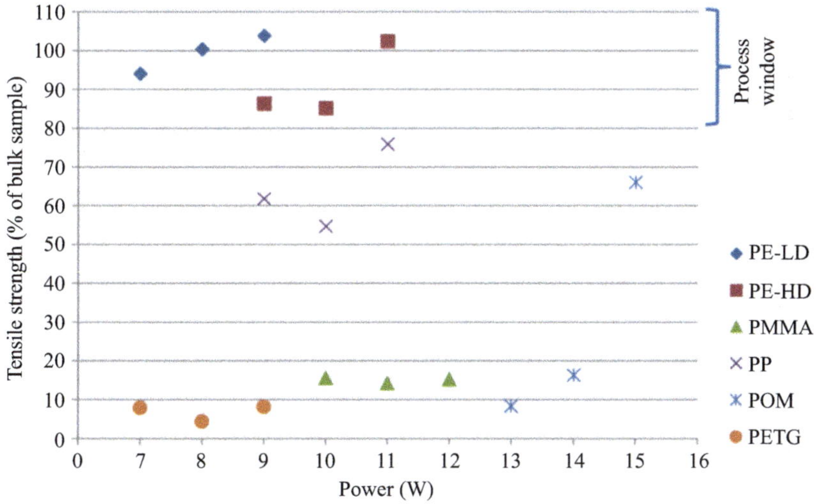

Figure 10.8 Tensile strength of laser-welded polymers

These samples were prepared in accordance with the ASTM D638 standard and subsequently cut into two equal parts using a diamond blade saw. After that each pair was laser-welded together using the process parameters determined earlier.

Tensile strength measurements were performed both on intact (bulk) and cut and welded samples to compare the mechanical strength of laser-processed and nonirradiated materials (Figure 10.8). The results show that for PE-LD and PE-HD the tensile strength of welded samples is more than 80% of that of the bulk samples. In some tests even a slightly higher tensile strength was measured in welded samples, which is possibly a result of laser-induced chemical changes in polymers, and physical changes, for example, increased joint area due to the constant pressure applied to joining partners. The welded samples of PETG and PMMA revealed less than 18% of the tensile strength of the corresponding bulk material. The analysis of the weld cross-sections for these materials indicates that the amount of molten material was insufficient to enable enough heat conduction, and therefore an effective gap bridging between two joining partners.

10.3.3 Transmission welding experiments

The transmission welding experiments were performed using the experimental arrangement shown in Figure 10.9, right, with four different polymers in the following combinations (top/bottom material): PETG/PETG, PMMA/PMMA, PMMA/PP, and PP/PE-LD. The same pressure $\sigma = 0.59$ N/mm^2 was applied to the joining partners as in the melting and butt-welding experiments, normal to the surface.

The transmission welding of similar polymers PETG/PETG and PMMA/PMMA resulted in a large-area material intermixing in the interface region. No interface boundaries can be detected in case of PETG/PETG joining. The joining of dissimilar materials PMMA/PP and PP/PE-LD exhibits no effective material intermixing, but the bead produced at the lower, stronger absorbing material penetrates the interface toward the weaker absorbing upper joining partner. Instead of a "material joint," a so-called "form joint" is produced in these material combinations. The maximum material penetration depth was measured at $h = 0.6$ mm for the PP/PE-LD combination. Due to the generation of pronounced form joints, high shear strength is expected as a result of transmission welding of dissimilar polymers.

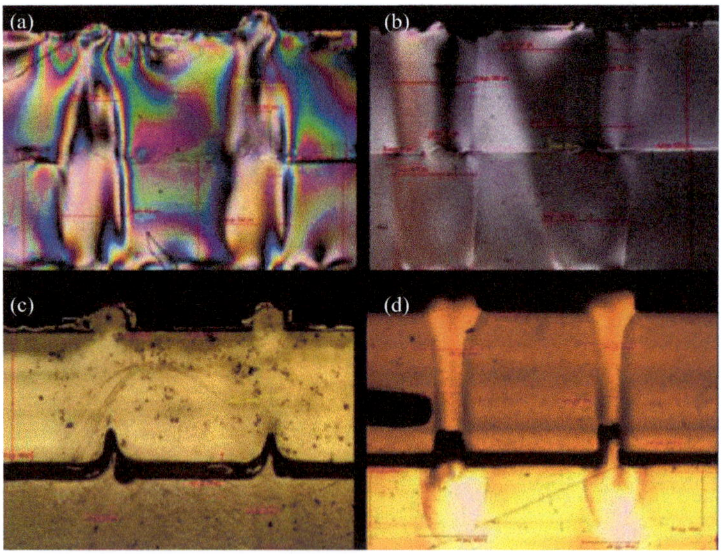

Figure 10.9 Examples for cross-sections of transmission-welded polymers: (a) PETG/PETG, (b) PMMA/PMMA, (c) PMMA/PP, and (d) PP/PE-LD, detected by optical and polarizing microscopy. Thickness for all samples is d = 1.6 mm

10.4 Processing of semiconductors

Experimental studies on laser processing of silicon and gallium arsenide with nanosecond laser pulses produced by a thulium fiber laser were performed. In this study it was shown that by utilizing semiconductors' transparency at $\lambda = 2$ μm modifications can be induced not only at the front (laser incident) surface, but also at the back surface of thin wafers. An experimental validation was provided that the back surface of Si and GaAs wafers could be modified independently, without inducing damage to the front surface. Therefore, this approach is referred to as the trans-wafer processing. The influence of process conditions on the resulting surface morphology change was studied by varying incident pulse energy and laser pulse duration. The laser-induced surface modifications were investigated by optical microscopy and scanning electron microscopy (SEM), and the modification thresholds for multiple-pulse irradiation were estimated for front and back surfaces.

10.4.1 Experimental details

The first amplifier stage of the self-developed MOPA system described in [16] was used for the semiconductor irradiation experiments in this study. The maximum peak power was >40 kW, with approx. 400 μJ maximum pulse energy at 6.8 ns and 1 kHz repetition rate. The output was linearly polarized, the initial beam diameter was approx. 4.5 mm, the beam quality was $M^2 < 1.2$, and the spectral width was <1 nm (full width at half maximum, FWHM) at the wavelength 1,985 nm. The pulse duration was adjustable in the range $t_p = 6.8$–100 ns.

For basic studies of front and back surface modifications, uncoated intrinsic double-side polished semiconductor wafers were used. The Si wafers (Cz-Si, Virginia Semiconductor) had a thickness of 500 μm with the normal axis along the {110} crystal direction, and the GaAs wafers (Wafer World) were 600 μm thick with the normal axis along the {100} crystal direction. For coating removal studies, four different material systems (ML 1–4) were used with either a single metallization layer, or a complex multilayer composition. The coatings were produced using commercial sputter coating and plasma-enhanced chemical vapor deposition (PECVD) equipment. The material compositions of the coatings are presented in Table 10.1.

Table 10.1 Material composition of multilayer-coated semiconductor samples

Notation	Substrate	Coating composition	Coating deposition method
ML 1	Intrinsic Si (1 mm)	Ni (100 nm)	Sputter coat
ML 2	Intrinsic Si (500 μm)	Ti (70 nm)	PECVD
ML 3	p-Si (180 μm)	SiO_2 (500 nm), TiN (100 nm), Ti (20 nm)	PECVD
ML 4	p-Si (180 μm)	SiO_2 (1,500 nm), Ti (100 nm), NiV (200 nm), Ti (20 nm)	PECVD

Material interfaces such as Ti/TiN and Ti/NiV deposited on Si are widely used as metallization layers in many current and prospective photovoltaic module designs. Substrates used in samples ML 1 and ML 2 were intrinsic double-side polished Si wafers, whereas samples ML 3 and ML 4 were based on 180 μm p-Si substrates that were wet-etched to reveal a random pyramid texture as commonly used in photovoltaic applications. The base resistivity of *p*-doped substrates was 4 Ωcm.

As shown in Figure 10.10, the output of the MOPA system was directed through a mechanical shutter (Uniblitz) in order to control the number of pulses incident on the target. The beam was then directed along a vertical rail to achieve normal incidence of the beam onto the horizontally mounted targets. The beam was focused to an approx. 10 μm spot size using an aspheric lens with a focal length $f = 7.5$ mm. Refraction properties of semiconductors were taken into account when determining the focus position at the back surfaces of the samples (refractive indices $n_{\mathrm{Si}} = 3.45$ and $n_{\mathrm{GaAs}} = 3.36$ at $\lambda = 2$ μm). Sequential ray-tracing modeling using Zemax was carried out to simulate the intensity distribution and the focus position at the back surface by taking into account linear refraction and aberration effects. The wafer targets were mounted on a 3-axis motion controller stage (Newport VP-25X), while the lens was stationary. A co-axial imaging arrangement comprising a CMOS camera and a 1.25x imaging objective was adopted to monitor back surface modifications.

The repetition rate was set to $f_{\mathrm{rep}} = 1$ kHz for all basic surface modification experiments. Irradiations were performed at pulse energies $E_{\mathrm{p}} < 350$ μJ, and at two pulse durations $t_{\mathrm{p}} = 7$ ns and $t_{\mathrm{p}} = 100$ ns. Unless stated otherwise, trenches were produced by translating the sample in the focal x–y plane at a constant speed $v = 1$ mm/s orthogonally to the incident laser beam. Alternatively, irradiation with a specified number of laser pulses carried out at fixed positions produced surface

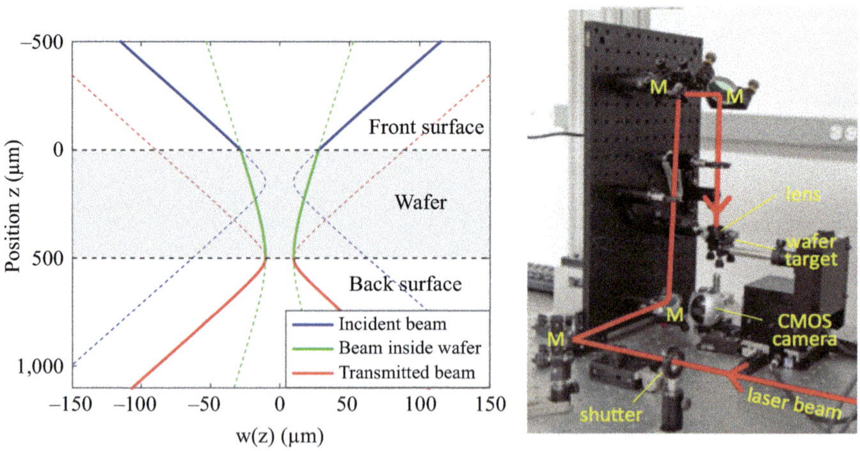

Figure 10.10 Simulation of laser beam propagation for different focus positions in wafer (left), and experimental irradiation setup (right)

craters. For the coating removal experiments the repetition rate was set to $f_{rep} = 20$ kHz, and the samples were translated at the speeds $v = 1$–2 mm/s.

10.4.2 Processing of uncoated semiconductor surfaces

Experimental conditions and the resulting surface morphology modifications were investigated both for the front and the back surface processing of the samples. In the first series of experiments, the laser radiation was focused on the front surface of the samples (z_0 position) to study the resulting morphology change by translating the sample and at fixed points, in dependence on the laser pulse energy and the pulse duration.

The morphology of craters produced in Si and GaAs with 1,000 pulses at 7 and 100 ns pulse duration revealed differences in material response to laser irradiation (Figure 10.11). Depending on the pulse duration, as the material undergoes melting and sublimation, its products are ejected and resolidified at the surface differently. In case of Si, the material removal process was accompanied with the dispersion of micrometer-sized, mostly spherically shaped debris particles to a wide surface area for $t_p = 7$ ns. For $t_p = 100$ ns pulses, the area surrounding the craters was covered with up to approx. 60 μm long, radially distributed re-solidified material "splatter" originating from the craters. Ring-like surface swellings were observed around the craters, probably resulting from the periodical melting and resolidification process, and indicating a larger contribution of thermal phenomena during irradiation with longer laser pulses. When the same experimental conditions were applied for

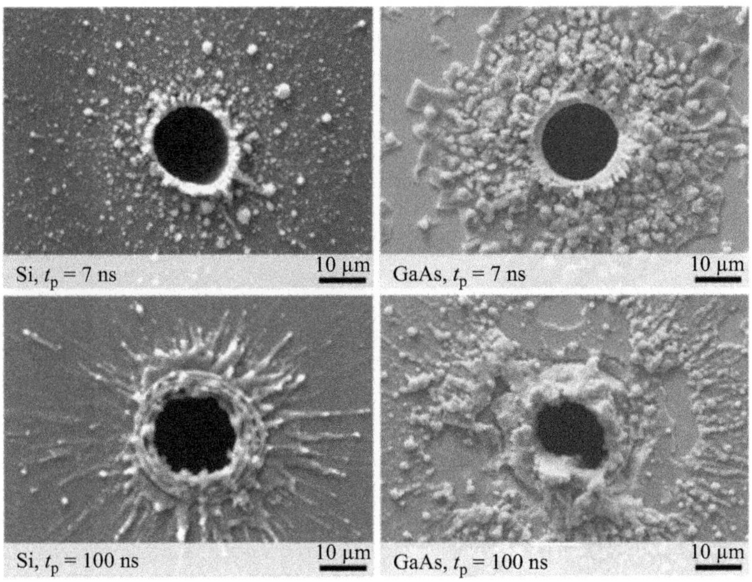

Figure 10.11 SEM images of front surface craters produced in Si and GaAs utilizing 1,000 pulses at $E_p = 120$ μJ

processing of GaAs samples, a larger amount of debris was generated compared to Si. A large round area around the craters was splash coated with ablation products. EDX measurements indicated up to 18 wt% of oxygen in close vicinity ($\approx$10 μm) of craters, with the oxygen fraction vanishing with increasing distance from the craters. The debris film appeared cracked, and also exposed sites where the debris stripped off the wafer surface. These sites did not seem to be significantly heat-affected, showing extremely smooth surface quality and approx. 0.5 wt% of oxygen according to the EDX analysis similar to the unprocessed bulk material. Irradiation of GaAs with $t_p = 100$ ns pulses produced an excessive amount of recast material.

Trenches were produced on the front surface of Si and GaAs samples by applying the same experimental conditions as for the irradiation at fixed positions, and by translating the samples in the x–y plane at a constant speed $v = 1$ mm/s resulting in a spatial pulse overlap of approx. 90%. The trenches obtained in Si (Figure 10.12 (left)) at $t_p = 7$ ns were surrounded by a large amount of melted and resolidified material and individual micrometer-sized melt droplets dispersed on the surface. Interestingly, Si trenches produced at $t_p = 100$ ns (Figure 10.12 (middle)) did not reveal a deep modification as compared to irradiation with shorter pulses. Instead, a more confined region with a smaller width and featuring apparently melted and resolidified material was detected. Similar to irradiation at fixed surface positions, processing of trenches in GaAs (Figure 10.12 (right)) produced a continuous debris film as well as sporadic splashes of resolidified material.

In the second series of experiments, transparency of semiconductors at the wavelength $\lambda = 2$ μm was utilized to induce surface modifications at the back surface. When a focused laser beam passes through a Si or a GaAs wafer, it is subject to considerable refraction and Fresnel reflections from both surfaces. In general, these effects can significantly affect focusing conditions at the back surface due to spherical aberrations and multiple reflections of the beam inside the wafer. In addition, semiconductors exposed to high-intensity electromagnetic fields are subject to nonlinear optical effects, such as absorption, beam focusing/defocusing, and scattering. Therefore irradiation conditions at the back surface cannot be compared to front surface irradiation directly.

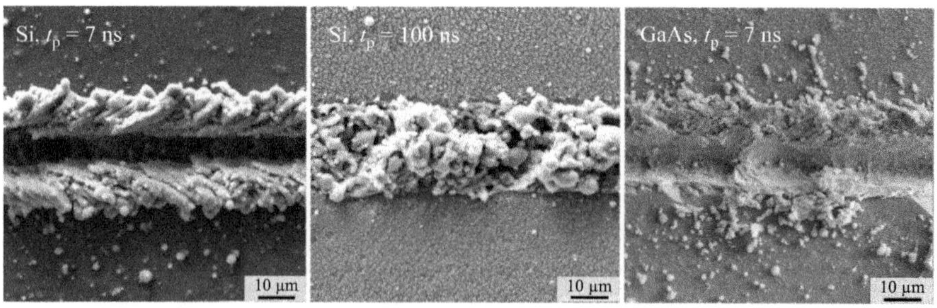

Figure 10.12 SEM images of front surface trenches induced at $v = 1$ mm/s. Si processed at $E_p = 145$ μJ, $t_p = 7$ ns (left) and $t_p = 100$ ns (middle). GaAs processed at $E_p = 120$ μJ and $t_p = 7$ ns (right)

To achieve back surface modifications the wafer was first translated approximately $\Delta z = 1$ mm along the *z*-axis toward the focusing lens. This procedure ensured the initial laser focus was well below the back surface of the wafer. Afterward the distance between the assumed focus position and the back surface was gradually reduced while concurrently translating the wafer in the *x–y* plane at 1 mm/s to produce trenches. Irradiations were performed at different pulse energies to determine the threshold value for the processing initiation. The onset of the back surface modification in GaAs (Figure 10.13 (right)) was observed at a position offset $\Delta z \approx 210 \pm 20$ μm measured from the initial z_0 position, which roughly corresponds to the focus being at the back surface according to the simulation data for the 600 μm thick wafer. The modification morphology revealed trenches apparently induced by a thermal melting process, predominantly. Compared to modifications achieved at the front surface, the areas surrounding the trenches were covered with a notably reduced amount of debris. Surface analysis by optical microscopy and SEM indicated no evidence of modification at the corresponding sites of the front surface in GaAs at pulse energies adopted. It should be noted that back surface modifications were only possible when producing trenches, as single-spot irradiations did not consistently result in reproducible ablation crater generation.

Back surface irradiation of Si (Figure 10.13 (left)) was carried out utilizing the same procedure as for GaAs. Following the same geometric considerations and taking into account possible spherical aberrations and the optical path length in a 500 μm thick Si wafer, the wafer position offset corresponding to the beam focus being at the back surface was estimated to $\Delta z \approx 180 \pm 20$ μm using the ray-tracing model produced by Zemax. However, in experiments, processing of Si at pulse energies comparable to those adopted for GaAs, did not result in any detectable back surface modification, and only the front surface could be modified with increasing pulse energy and/or decreasing Δz. By replicating the experiment while gradually increasing the pulse energy, the onset of the back surface modification was observed at a much larger offset of $\Delta z \approx 450$ μm than predicted by simulations. The resulting surface morphology represented trenches produced

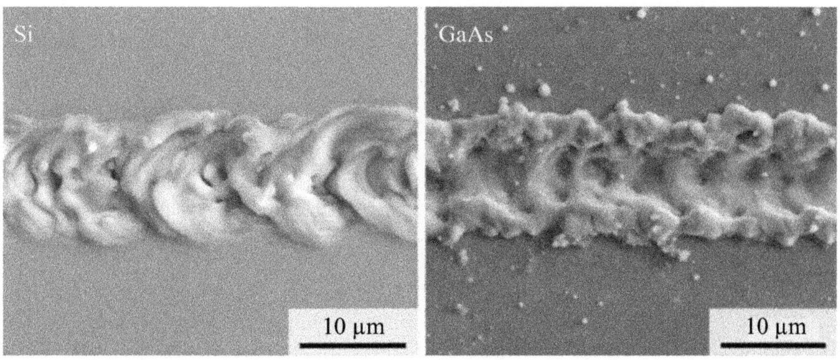

Figure 10.13 SEM images of back surface modifications induced at $t_p = 7$ ns, $v = 1$ mm/s. Left: Si processed at $E_p = 320$ μJ, and right: GaAs processed at $E_p = 120$ μJ

presumably by local surface remelting process; however, no evidence of debris generation was detected. Similar to GaAs, no visible damage was induced to the front surface of Si at these experimental conditions. While back surface modifications at a larger z-offset can be explained by higher intensities throughout the beam path, the lack of modifications at the geometrically estimated focus position around $\Delta z \approx 180$ is an indication of a complex spatial energy distribution in the laser-material interaction zone caused by nonlinear material response.

Similar to irradiation of front surfaces, different modification behavior was observed for the irradiation of back surfaces of Si wafers using longer laser pulses (Figure 10.14). While no deep modifications were detected both for 7 and 100 ns laser pulses, the morphology of the laser-induced surface change obtained with longer pulses revealed a larger contribution of thermal phenomena to the modification process. In case of irradiation with 100 ns laser pulses, periodic resolidified pattern was observed, with a period of approx. 1 μm, and apparently resulting from overlapping single pulse modifications.

To determine the energy density thresholds for the front surface modifications, surface craters were produced by applying 20 pulses at varying pulse energies $E_p = 10$–50 μJ. After irradiation the diameters d of the craters were determined by optical microscopy using the Liu method [17]. The front surface threshold fluences in Si were 2.86 J/cm^2 for 7-ns pulses, and 3.22 J/cm^2 for 100-ns pulses. The corresponding front surface threshold fluences in GaAs were 2.30 J/cm^2 for 7-ns pulses, and 3.38 J/cm^2 for 100-ns pulses.

Irradiations performed with a smaller number of pulses did not deliver consistent results, which was caused by considerable discrepancies in the crater diameters obtained. However, the experiment confirmed that the fluence threshold values were in general larger with the smaller numbers of pulses (<20) applied. Similar behavior and dependency of the threshold values on the number of pulses were reported for nanosecond laser processing of semiconductors at other wavelengths.

A reliable determination of the back surface modification thresholds was not possible by measuring the crater diameters due to mainly probabilistic nature of their

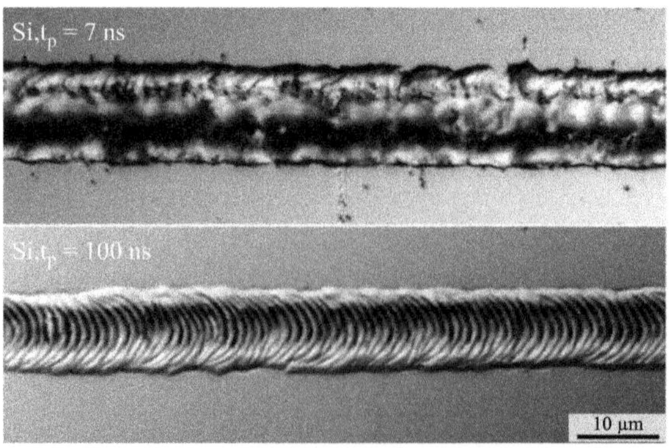

Figure 10.14 Optical microscopy images of back surface modifications obtained in Si at v = 1 mm/s, E_p = 294 μJ and different pulse durations

generation and large discrepancies in their diameters. The experiments showed that only wafer processing by translating the samples in the *x–y* plane resulted in consistent modifications induced at the back surface. In addition, while performing such experiments, occasional modifications were observed that apparently originated from laser beam interaction with microscopic irregularities of the wafer surface, for example, microcraters and scratches. This "initiation effect" implies damage accumulation phenomena could play an important role in the physical mechanism of back surface processing. Therefore, based on the experimental results obtained only an upper limit estimation of the modification thresholds could be provided.

The minimum pulse energy necessary to induce back surface trenches in GaAs was $E_p = 18$ μJ. Without an exact knowledge of light propagation and absorption conditions in the bulk it was impossible to determine what fraction of the incident optical energy was effectively contributing to the back surface modification process. Thus, the only measurable parameter was the transmittance $T_{GaAs} = 54.6\%$ through the entire wafer. Assuming the laser beam waist of $w_0 = 10$ μm as resulted from ray-tracing modeling, taking into account the wafer transmittance, and considering the accumulated energy applied to a single spot being equivalent to 10 pulses, the corresponding fluence threshold was estimated to $F_{th} < 3.2$ J/cm^2 for 10 pulses at $t_p = 7$ ns.

For Si, the minimum pulse energy required for consistent back surface modification was $E_p \approx 100$ μJ; however, sporadic modifications were also possible at lower pulse energies. With the total wafer transmittance $T_{Si} = 53.4\%$ and the offset Δz corresponding to the processing initiation being much larger than predicted by ray tracing modeling, only a rough estimation for the modification threshold can be given. Accordingly, the back surface modification threshold for Si is $F_{th} < 17.2$ J/cm^2 for ten pulses with the pulse duration $t_p = 7$ ns.

While the utilized threshold determination method itself is not sensitive to uncertainties resulting from the beam size characterization as long as the spatial beam profile is nearly Gaussian, its accuracy depends critically on the measurement uncertainty of the laser-induced crater diameters. Therefore, surface damage that is not detectable by optical microscopy or SEM could increase the measurement errors. Taking into account the calibration uncertainty of pulse energy measurements in the experimental series presented here, the errors were estimated to be within ±6% and ±15%, for all front and back surface fluence threshold measurements, respectively. A higher uncertainty in back surface threshold data resulted from larger measurement discrepancies of laser-induced back surface features, because more irregular surface structures were developed. Another potential source of measurement error is the possible effect of the destructive sample preparation method, for example, wafer grinding and polishing, which might induce additional damage and is difficult to quantify.

10.4.3 Processing of coated semiconductor surfaces

Utilizing the process knowledge obtained, trans-wafer processing method was applied to the removal of single- and multimaterials (metallization) coatings deposited on p-Si wafers featuring base resistivity levels of 4 Ωcm. Unlike intrinsic semiconductors, the *p*-doped substrates reveal equilibrium charge carriers possibly resulting in linear absorption of laser energy. However, previous research has shown that the aforementioned resistivity level is high enough, so that no substantial absorption occurs. For the experiments, the removal of metallization layers was first studied for the front (laser-incident) surfaces, in order to find out process conditions for selective coating ablation without damaging the substrate. Figure 10.15 shows examples of metallization removal from the front surface of the sample ML 3. Depending on the process conditions, the combined TiN/Ti layers were successfully removed from the surface without causing visible substrate damage. The minimum laser fluence needed to completely remove the layers was $F_{thr} < 1.6$ J/cm^3. The resistance of the processed sample measured for 10 mm long modifications (full separation of a metallization layer in two separate portions) was $R > 76$ kΩ.

In the following series of experiments, trans-wafer processing was utilized to remove various material layers form the back surface of Si wafers (Figure 10.16). The focus position was chosen to compensate for light refraction in the wafer, according to the thickness of the wafer adopted. Compared to front surface processing, the removal of the back surface metallization produced less consistent surface modifications. At low pulse energies the processed areas were still partially covered with coating residues. With increasing pulse energy, a gradual disintegration and removal of the coating from the surface was observed. Significant differences in modification behavior were also observed for different thicknesses of the coatings. The consistent removal of thick coatings (ML 4 samples) was not possible without partially damaging the substrate with the process conditions adopted (Figure 10.16(b) and (d)), whereas processing of thin layers (ML 1 samples) within well-defined modification boundaries was enabled.

In order to study the electrical properties of the separated metallization films, a series of experiments was performed using ML 3 samples (4×25 mm^2), and by processing the samples at different pulse energies in single-passes across the width and leaving approx. 1 mm distance between the individual lines. Then the electrical resistance between the separated portions was measured using the 4-terminal method (Figure 10.17). The measurements were performed on ML 4 samples

Figure 10.15 Examples of metallization removal from the front surface of ML 3 samples. Modifications were produced at the pulse energy $E_p = 3$ *μJ (left) and* $E_p = 8$ *μJ (right)*

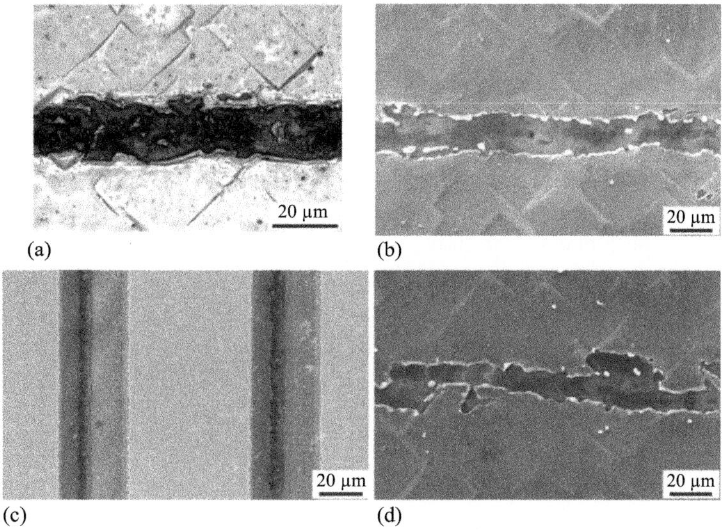

Figure 10.16 Examples of trans-wafer metallization removal. (a) ML 3 processed at $E_p = 35$ μJ, (b) ML 4 processed at $E_p = 32$ μJ, (c) ML 1 processed at $E_p = 35$ μJ, and (d) ML 4 processed at $E_p = 22$ μJ

prepared using the same process conditions, and each resistance value was measured at three different locations along the corresponding separation line induced on the metallization layer.

Metallization removal from back surfaces of Si wafers in multiple overlapping tracks was adopted to enable large-area processing (Figure 10.18). The distance between the adjacent tracks (track overlap) was varied to achieve processing with a minimum of debris left. Because the surfaces of the wet-etched samples (ML 3 and 4) exhibited a pyramid pattern, a complete removal of the metallization layers was not possible even at large track overlaps >90%. Processing of the Ti-coated polished sample (ML 2) however produced large areas free of metallization layer on the order of several mm^2.

10.5 Processing of chalcogenide glasses

Chalcogenide glasses (ChGs) are a class of optical materials featuring high transparency in a broad IR spectral range and the ability to tailor their refractive index through compositional variation [2]. Recent studies have demonstrated spatially varied refractive index through partial crystallization of the glass leading to formation of optical nanocomposites, or through a micropoling process. Experiments have shown that the amount of crystallization or surface compositional change can be effectively controlled in 3D through alkali ion migration (poling) or formation of nuclei and/or their growth within the ChG material by spatially selective laser irradiation and heat treatment. Among other high-index ChG compositions, vapor

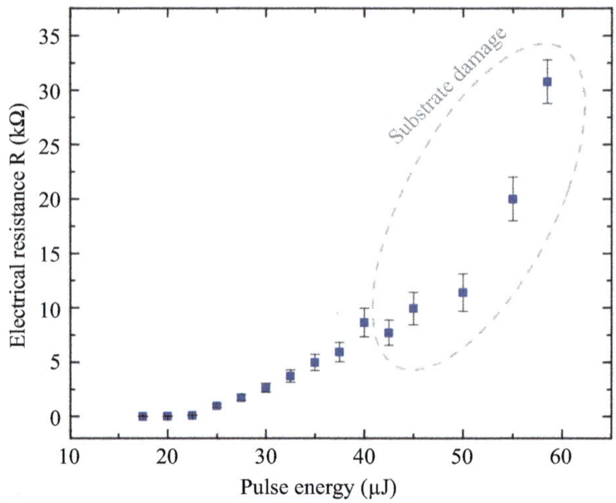

Figure 10.17 Electrical resistance of the separated metallization layer in dependence on the laser pulse energy

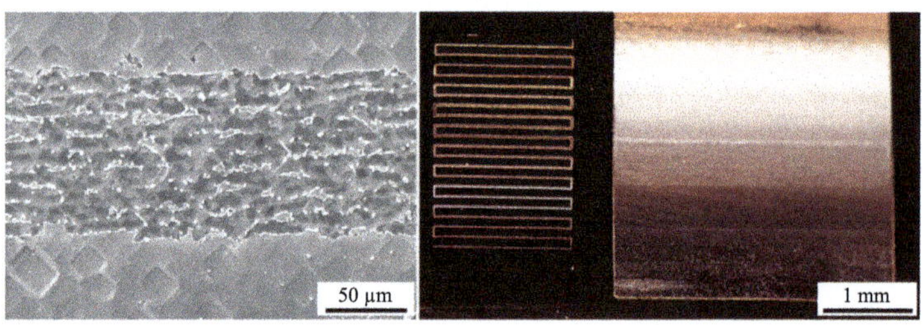

Figure 10.18 Examples of large-area metallization removal from back surfaces of Si wafers. Sample ML 4 processed at $E_p \approx 32\ \mu J$ (left). Sample ML 2 processed at $E_p \approx 150\ \mu J$ and using an $f = 25$ mm triplet lens (right)

phase deposited metastable thin-film GAP-Se glasses have shown the ability to modulate the refractive index through formation of high-index Pb-rich nanocrystals within the low-index glassy matrix. A detail review of the state of the art in GRIN materials can be found in [13] which illustrates the potential and limitations of various methods beyond those referred to here.

In this study, a two-step photo-thermal process comprising irradiation of GAP-Se films in scanning patterns at precisely controlled energy densities and a subsequent heat treatment (HT) was adopted to locally change the refractive index.

10.5.1 Experimental conditions

The pulse duration was set to $t_p = 100$ ns, and the repetition rate was set to $f_{rep} = 100$ kHz. GAP-Se films were irradiated in $5 \times 5\ mm^2$ square regions using line scan patterns at a constant translation speed of 1 mm/s, and with a line pitch of 50 μm, to avoid spatial overlap between adjacent lines. Using this approach both uniform large-area refractive index modifications and index step profiles can be implemented within a square by changing irradiation conditions used for individual lines. In addition, the number of scans can be varied in the range $N = 1.4$ in order to control the energy dose deposited in the film. For example, the total irradiation time for a $5 \times 5\ mm^2$ square is 100 s for $N = 1$, and the total radiation energy can be calculated using the average laser power.

Upon laser irradiation, heat treatment was carried out in a muffle furnace (ThermoScientific 48000) in air atmosphere calibrated with a type K thermocouple that was placed next to the films on the same refractory brick. In the present study, three types of single-step heat treatment were conducted for the films. All heat treatments were performed at the temperature 190 °C for initially 30 min, 90 min, and 180 min. For a limited set of measurements, an additional heat treatment for a total duration of 270 min at 190 °C was performed. The films were isothermally treated by placing them inside the furnace after the furnace was at the targeted temperature for nominally 30 min. The surface topography of the films was investigated over an area of $1.6 \times 1.6\ mm^2$ using white light interferometry (Zygo) before and after heat treatment to determine whether there was any photo- and thermal-expansion upon laser irradiation and post heat treatment, respectively. Transmission spectra of the films were measured at wavelengths from 1.5 to 25 μm, using a ThermoFisher Nicolet iS5 FTIR spectrometer. Room temperature linear refractive indices of the films were measured at a wavelength of 4.515 μm, using a Metricon prism coupler modified for use in the infrared. Chemical composition of the GAP-Se films was determined using energy-dispersive X-ray spectroscopy (EDX, Zeiss Ultra-55 SEM).

10.5.2 Characterization of the film composition and morphology

Compositional uniformity of thermally evaporated GAP-Se films with the thickness 25 μm was studied at several different locations on the film surface across a 30-mm length, as well as several different locations along the cross-section of the specimen. The results (Figure 10.19) show that the average measured atomic percentages of the main chemical constituents Ge, As, Pb, and Se are within the 1% tolerance range. Therefore, the film's chemical composition can be assumed uniform throughout the specimen geometry.

In the first series of laser irradiation experiments, modification thresholds of ChG films were determined. Films of various thicknesses were processed by inducing line modifications on film surfaces at varying pulse energies, while keeping all other process parameters constant, that is, the translation speed set to $v = 1$ mm/s and the pulse repetition rate set to $f_{rep} = 100$ kHz. Initially, the average power between lines was increased in increments of 10 mW until a visible damage occurred to the film. Subsequently, the experiment was repeated with an increment of 2 mW in a narrower interval of average power around the assumed modification

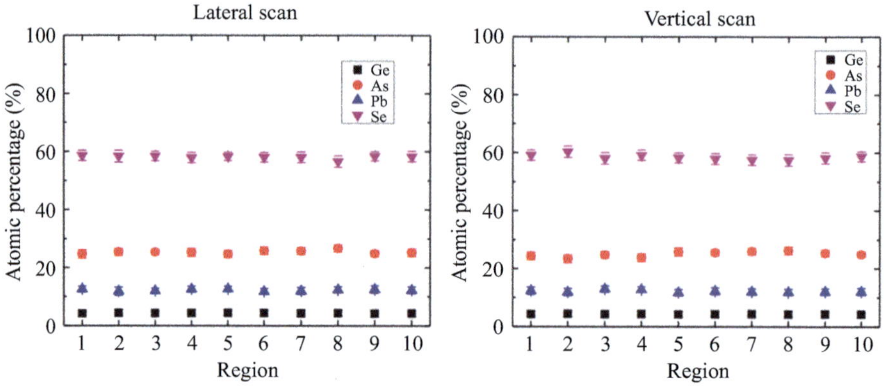

Figure 10.19 Evaluation of EDX data showing chemical composition of GAP-Se films obtained in a lateral scan across a 30-mm distance at the surface of the specimen (left), and a vertical scan across the film thickness (right)

threshold value. Surface modifications were then investigated using optical microscopy.

With increasing laser power, initial damage occurred at preexisting surface damage sites such as micro-scratches and other surface irregularities, which is often referred to as the damage accumulation. At higher power levels and in the absence of preexisting damage, the film morphology changed due to local heating exhibiting thermally expanded areas; however, no ablation or trench formation was observed. Lastly, at even higher laser powers, film ablation through evaporation and melt ejection occurred. No visible film delamination due to laser processing was observed in these experiments. In films of all thicknesses, initial film damage occurred at approx. $P_{thr} = 185$ mW (1.85 μJ in terms of pulse energy), which provided the power threshold value not to be exceeded in the subsequent experiments.

The goal of the second series of irradiation experiments was to determine if large-area laser irradiation would result in photo-induced material expansion or other morphology changes due to repetitive surface irradiation and possible damage accumulation. Surface roughness was measured in films irradiated at different powers, and exposed areas were compared to unexposed reference regions. The results (Figure 10.20) show that the RMS surface roughness of the most intensively irradiated region ($P = 150$ mW) is still very similar to that of the pristine reference region, and therefore the photo-expansion is negligible at process conditions selected.

10.5.3 Refractive index changes

Films of various thicknesses (1.1–25 μm) were irradiated in 5×5 mm^2 patterns using varying average power ($P = 30$–150 mW) and the number of scans ($N = 1$–4). For each set of the process parameters the refractive index was measured before heat treatment, and after 30 min, 90 min, and 180 min of heat

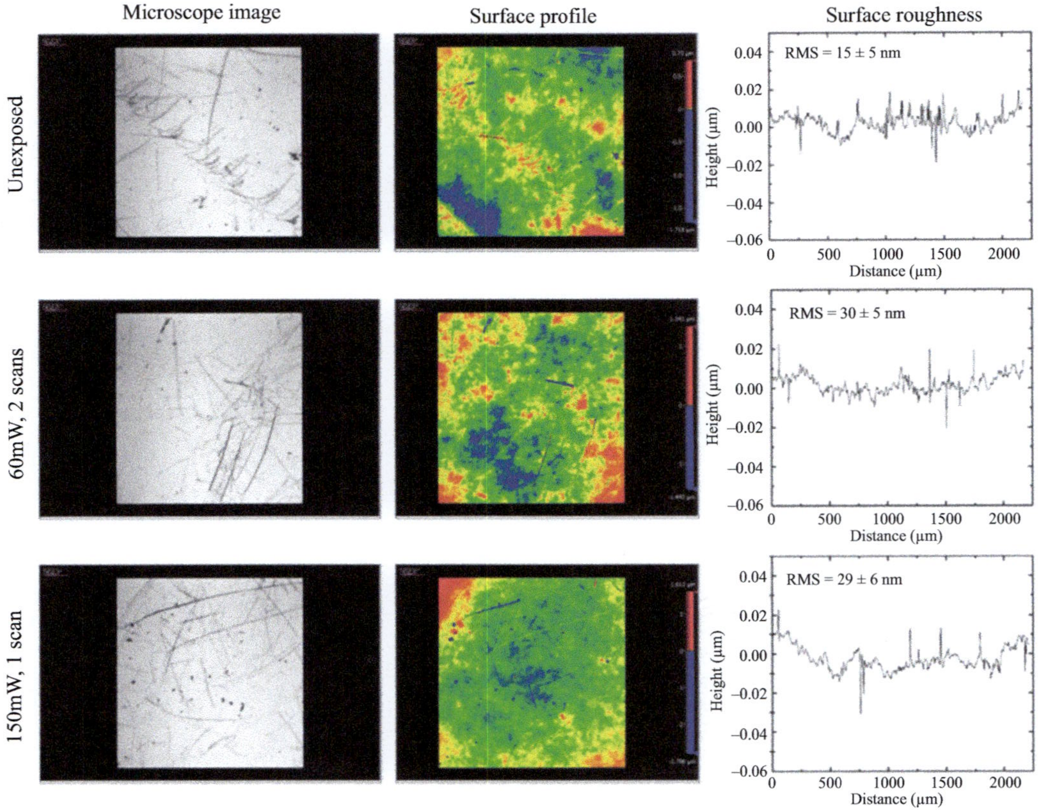

Figure 10.20 *Results of surface morphology study for unexposed and laser-irradiated film. The areas shown have dimensions of* $1.6 \times 1.6\ mm^2$*, and the linear height profiles were extracted from upper left corner to bottom right corner of each Zygo image*

treatment. Figure 10.21 shows the refractive index increase due to the combined laser irradiation and heat treatment processing. The maximum index increase of $\Delta n \approx 0.071$ was found for the pattern irradiated at 150 mW (1 scan) and heat-treated for 180 min. The index increased by $\Delta n \approx 0.051$ for the pattern irradiated at 60 mW (2 scans) after the same heat treatment time.

Experimental results show that film's refractive index only increases after a heat treatment of irradiated films is performed. Film exposure with high-intensity laser light leads to a partial phase separation of the initially homogeneous glass film into Pb-rich and Pb-deficient amorphous phases. Upon film exposure these two phases are still glassy (amorphous), and the effective index of the nanocomposites remains largely unchanged. Once heat treatment is performed at a temperature between approx. 150 °C and 200 °C, the Pb-rich phases will crystallize becoming high-refractive-index phases in the surrounding low-index glassy Pb-deficient matrix. This results in the increase of the effective index of the glass-ceramic nanocomposite due to a localized formation of high-index crystals. Heat treatment at higher temperatures above approx. 200 °C would eventually lead to a spontaneous mass crystallization of the entire film, leading to a loss of the spatial processing selectivity and making fabrication of custom GRIN profiles impossible.

After evaluating the trend, heat treatment process was applied for additional 90 min (270 min total duration), and the resulting refractive index was measured for two sets of laser irradiation parameters that resulted in highest index increase. However, the index increase was only slightly larger than that determined after 180 min of heat treatment. The index changes were $\Delta n \approx 0.052$ and $\Delta n \approx 0.072$ for the 60 mW (2 scans) pattern and the 150 mW pattern, respectively. This indicates that the crystallization process of the Pb-rich amorphous phase was largely

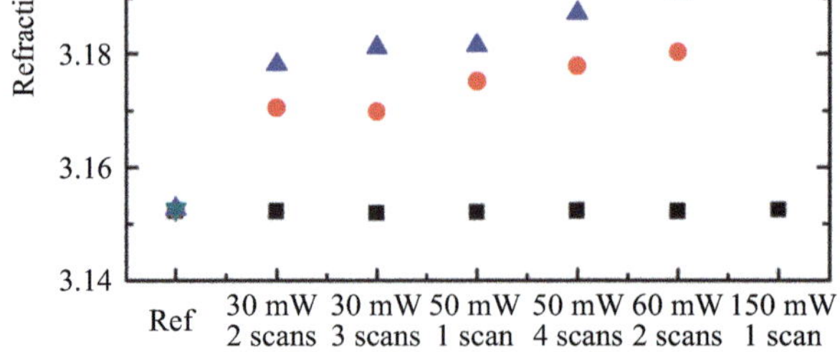

Figure 10.21 Refractive index (measured at the wavelength 4.5 μm) of the 20-μm film for different irradiation and heat-treatment regimes

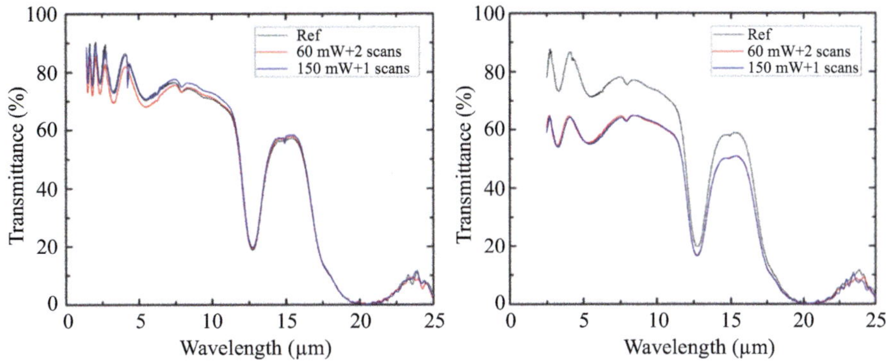

Figure 10.22 Optical transmittance of the irradiated 20-μm film heat-treated for 180 min (left) and heat-treated for 270 min (right). All heat treatments were performed at 190 °C

completed after the initial 180 min of heat treatment, and the high-index crystalline phase was quasi-saturated at laser irradiation conditions given.

Maintaining optical properties such as the sufficient transmittance is essential for the performance of the GRIN optics. Therefore, in addition to determining the refractive index change, raw optical transmittance was determined without taking into account Fresnel losses (Figure 10.22). The results showed that transmittance remains mostly unchanged compared to nonirradiated reference material after the initial 180 min of heat treatment. However, additional heat treatment for the total duration of 270 min leads to a moderate drop in raw optical transmittance, for example, a reduction of approx. 20% at the wavelength $\lambda \approx 2.5$ μm, and a reduction of approx. 12% at the wavelength $\lambda \approx 10$ μm.

References

[1] Mingareev I, Gehlich N, Bonhoff T, *et al.* Principles and applications of trans-wafer processing using a 2-μm thulium fiber laser. *The International Journal of Advanced Manufacturing Technology*. 2016; **84**(9–12):2567–78.

[2] Kang M, Sisken L, Cook J, *et al.* Refractive index patterning of infrared glass ceramics through laser-induced vitrification. *Optical Materials Express*. 2018;8(9):2722–33.

[3] Mingareev I, Weirauch F, Olowinsky A, Shah L, Kadwani P, and Richardson M. Welding of polymers using a 2 μm thulium fiber laser. *Optics and Laser Technology*. 2012;44(7):2095–9.

[4] Olowinsky A, Boglea A, Gedicke J. Innovative laser welding processes: TWIST and SHADOW in microtechnology. *Laser Technik Journal*. 2008;5(3): 48–51.

[5] Horn W. Strategies for polymer welding with high-power diode lasers. In *Key Engineering Materials* (Vol. 447). Switzerland: Trans Tech Publications; 2010; pp. 277–81. https://www.scimagojr.com/journalsearch.php?q=12378&tip=sid&clean=0
[6] Jaeschke P, Herzog D, Haferkamp H, Peters C, and Herrmann AS. Laser transmission welding of high-performance polymers and reinforced composites-a fundamental study. *Journal of Reinforced Plastics and Composites*. 2010; 29(20):3083–94.
[7] Meyer JR, Bartoli FJ, and Kruer MR. Optical heating in semiconductors. *Physical Review B*. 1980;21(4):1559.
[8] Hurlbut WC, Lee Y-S, Vodopyanov KL, Kuo PS, and Feyer MM. Multiphoton absorption and nonlinear refraction of GaAs in the mid-infrared. *Optics Letters*. 2006;32:668.
[9] Hurlbut WC, Lee YS, Vodopyanov KL, Kuo PS, and Fejer MM. Multiphoton absorption and nonlinear refraction of GaAs in the mid-infrared. *Optics Letters*. 2007;32(6):668–70.
[10] Iwata H, and Asakawa K. Accumulative damage of GaAs and InP surfaces induced by multiple-laser-pulse irradiation. *Japanese Journal of Applied Physics*. 2008;47(4R):2161.
[11] Varshni YP. Temperature dependence of the energy gap in semiconductors. *Physica*. 1967;34(1):149–54.
[12] Jellison Jr GE, and Burke HH. The temperature dependence of the refractive index of silicon at elevated temperatures at several laser wavelengths. *Journal of Applied Physics*. 1986;60(2):841–3.
[13] Richardson K, Kang M, Sisken L, *et al.* Advances in infrared GRIN: a review of novel materials towards components and devices. In *Advanced Optics for Defense Applications: UV through LWIR III* (Vol. 10627). Bellingham, WA: International Society for Optics and Photonics: 2018; p. 106270A.
[14] Richardson K, Buff A, Smith C, *et al.* Engineering novel infrared glass ceramics for advanced optical solutions. In *Advanced Optics for Defense Applications: UV through LWIR* (Vol. 9822). Bellingham, WA: International Society for Optics and Photonics: 2016; p. 982205.
[15] Kang M, Swisher AM, Pogrebnyakov AV, *et al.* Ultralow dispersion multicomponent thin-film chalcogenide glass for broadband gradient-index optics. *Advanced Materials*. 2018;30(39):1803628.
[16] Gaida C, Gebhardt M, Kadwani P, *et al.* Amplification of nanosecond pulses to megawatt peak power levels in Tm 3+-doped photonic crystal fiber rod. *Optics Letters*. 2013;38(5):691–3.
[17] Liu JM. Simple technique for measurements of pulsed Gaussian-beam spot sizes. *Optics Letters*. 1982;7(5):196–8.

Chapter 11
High-power GHz linewidth diode lasers and their applications

Ivan Divliansky[1] *and Oussama Mhibik*[1]

This chapter will discuss the development and applications of semiconductor laser systems with output reaching several hundred watts and with spectral widths of the emission down to 10 GHz.

11.1 GHz linewidth high-power diode laser sources

11.1.1 Introduction

During the last decade, high-power diode lasers have become an alternative to the long developed solid state and fiber lasers. One of the main characteristics of such high-power diode systems is their broad emission which can be a disadvantage in some particular application cases. In this chapter, we will describe an approach for narrowing the linewidth of this lasers down to GHz linewidths. The combination of high-power and such narrow emission have enormous potential in the areas of optically pumped solid state (thulium, erbium) lasers, optically pumped alkali-vapor (cesium, potassium, and rubidium) lasers, and in atom cooling and Raman spectroscopies [1–3]. Especially, in the cases where matching of the pump source emission characteristics to the absorption linewidth of the given lasing medium is of significant importance, the capability to make a pump source with given specific parameters is of great benefit and importance. The development and advances in the past 10 years of high-quality volume Bragg gratings (VBGs) recorded in photo-thermo-refractive (PTR) glass have offered an alternative direction for designing and building narrow line high-power laser systems using externally locked semiconductor bars [4–8]. The properties of such VBGs will be discussed in the chapter but it is worth mentioning here that they are capable of handling kW of laser power and upon subject to heating the Bragg wavelength change can be used for smooth adjustment of the laser emission wavelength. In the chapter, we will discuss the fundamental VBG properties [9], the design of VBG locked diode laser systems

[1]CREOL, The College of Optics and Photonics, University of Central Florida, Orlando, FL, USA

and finally, a group of applications that greatly benefit from the development of such laser systems.

11.1.2 Volume Bragg gratings in PTR glass [9]

11.1.2.1 Volume Bragg gratings—description and properties

Volume grating as its name suggests is a grating that occupies the volume of a medium. Typically, for such gratings, the term volume Bragg gratings (VBGs) is used in relation to Sir William Bragg who in 1915 used diffraction of light propagating through a crystal to determine the crystal's lattice structure [10]. What he found was that at certain conditions the light is strongly diffracted by the crystal. Such condition is called "resonant condition" or also "Bragg condition." Here is also the place to make the distinction between surface and VBG. If we start with a surface grating and start increasing its thickness at some point the different diffraction orders will reduce to a moment where there will be only one order. This defines the transition to a volume grating behavior [11].

For simplicity, in Figure 11.1 for both types of volume gratings, the angle of incidence is the same θ_i and the tilt of the volume grating inside the medium is θ_{tilt}. Resonant diffraction from each of these VBGs happen upon satisfaction of their Bragg conditions which are defined by (11.1) and (11.2):

$$\lambda_{TBG} = 2 \wedge \sin(\theta_i + \theta_{tilt}) \tag{11.1}$$

$$\lambda_{RBG} = 2 \wedge \cos(\theta_i + \theta_{tilt}) \tag{11.2}$$

where, Λ is the period of the VBG and λ_{VBG} is the wavelength of the incident light which for the particular θ_i and θ_{tilt} satisfies the Bragg condition. The gratings depicted in Figure 11.1 are uniform VBGs that can be recorded in a photosensitive material by simple interference of two collimated laser beams. The recording wavelength, angle of interference, and the refractive index of the material determine the grating's parameters. There are techniques capable of recording more

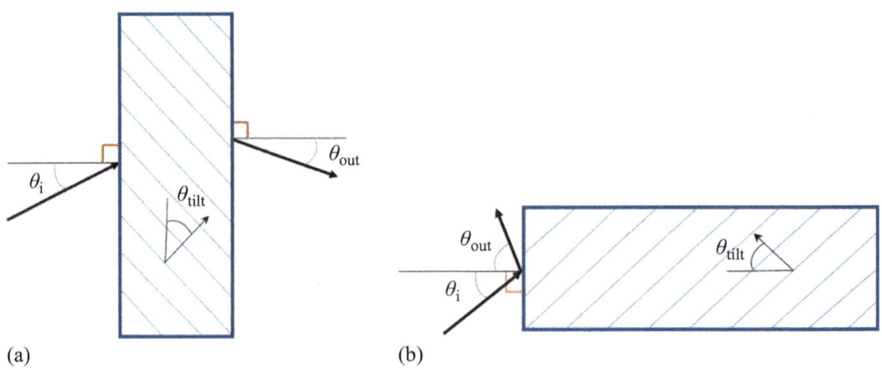

Figure 11.1 (a) Beam geometries for transmission Bragg grating and (b) for reflective Bragg grating

complex volume gratings which have nonuniform period and for which the Bragg condition will be different depending on the space coordinates [11]. Regardless, if the variation of the period is negligible when compared to the probe beam size used for characterization of the VBG, (11.1) and (11.2) can still provide the resonance wavelength.

Figure 11.2 demonstrates one example of beam geometry for recording uniform transmitting and reflecting volume gratings. In this recording approach, two plane waves (purple beams) illuminate the sample from one side at a half-angle of interference φ. Depending on the direction from which the grating is used it can either work as a TBG (green beams) or as an RBG (orange beams). Before recording, the grating parameters such as period and modulation needs to be calculated so the recording is done accordingly.

The main model describing VBG was introduced by H. Kogelnik in 1969 [12]. His model which describes diffraction from a VBG is based on coupled wave theory (CWT) and provides analytical solutions for RBGs and TBGs including tilted ones. Figures 11.3(a) and (b) and 11.4(a) and (b) present examples of the wavelength and angular responses of an RBG and a TBG correspondingly, calculated using Kogelnik's theoretical approach.

Figures 11.3 and 11.4 give a good overview of the main properties of TBGs and RBGs. In particular, RBGs are much more suited for implementation as narrow wavelength filters. For example, the full-width half-maximum (FWHM) wavelength selectivity of the reflective grating simulated in Figure 11.3(a) is around 225 pm but it can reach down to 15–20 pm if designed accordingly. TBGs, in contrast, have

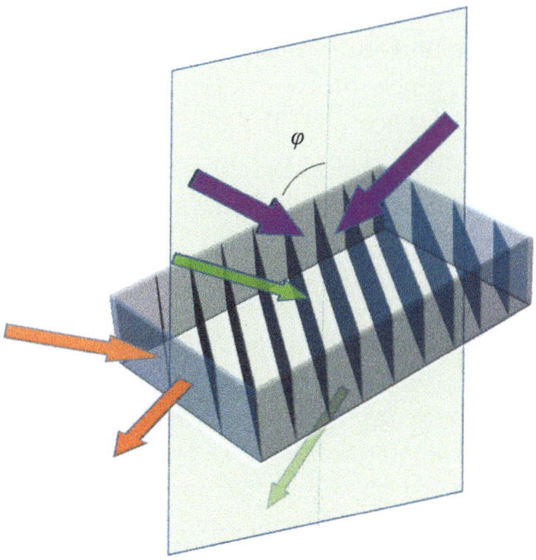

Figure 11.2 Recording geometry used for recording reflective and transmitting Bragg gratings. The recording beams are shown in purple and the half-angle of interference φ determines the properties of the grating. The VBG is utilized as an RBG if probed as the orange beams or as TBG if probed as the green beams [9]

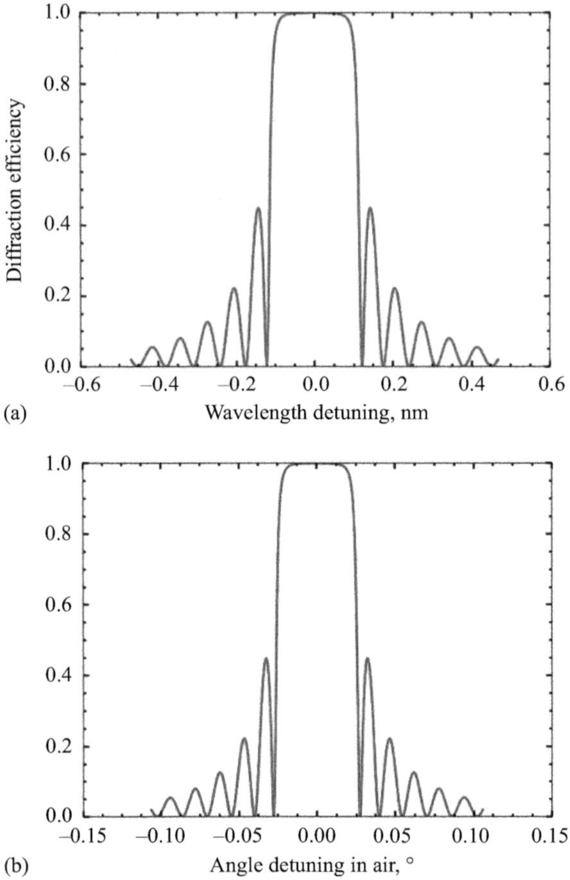

Figure 11.3 Wavelength (a) and angular (b) response for an RBG. The VBG is 5.5 mm thick, it is 20° tilted, and has a 240 ppm refractive index modulation [9]

much wider wavelength acceptance starting at few hundred picometers and reaching several nanometers. Figure 11.4(a) shows the spectral response of 1.5 mm thick TBG with modulation of 330 ppm. For these parameters its FWHM of 2.3 nm is close to an order magnitude larger if compared to the RBG one.

TBGs, alternatively, can be used as narrow angular filters with acceptance values as low as 0.1 mrad, while RBGs have typical angular acceptance of more than 10 mrad all the way to 100 mrad. These properties of the two types of VBGs define their use in different applications. For example, the narrow angular selectivity of the TBGs makes them great angular filter that can be used to suppress higher order modes generation in laser cavities while keeping them very compact [13]. RBGs with their narrow wavelength selectivity can be used for narrow wavelength beam combining where the diffracted by and the transmitted through an

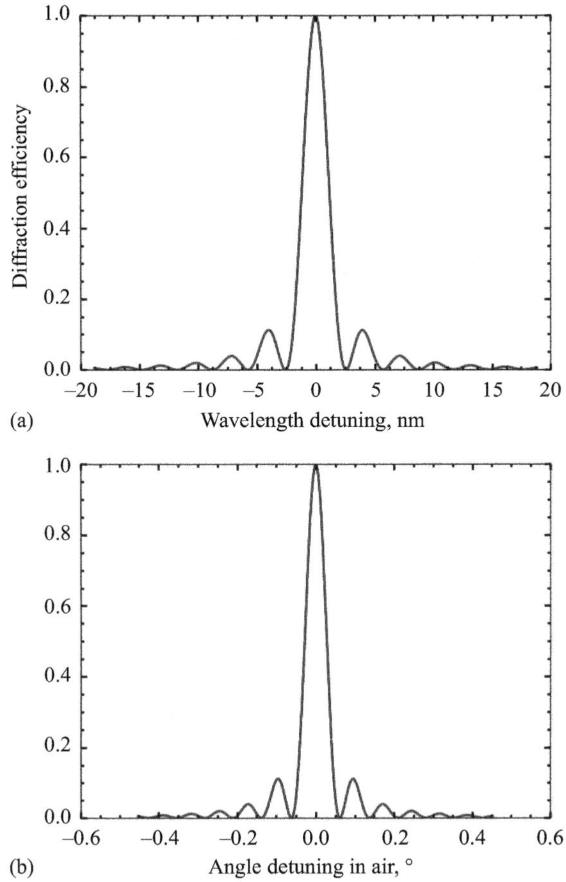

Figure 11.4 Wavelength (a) and angular (b) response for a TBG. The VBG is 1.5 mm thick, it is 20° tilted, and has a 333 ppm refractive index modulation [9]

RBG beams can be separated by only few hundred picometers [14]. Regardless of the close wavelength separation, the RBG does not diffract the transmitted beam even though both beams have a common propagating direction.

Table 11.1 summarizes the TBG's and RBG's characteristics and their typical range. Until now, we were discussing wavelength and angular selectivity of VBGs but the third very important parameter is the VBG's efficiency. There are two generally accepted ways to define a VBG's efficiency and the more widely used one is the so-called "relative diffraction efficiency." It is defined as normalization of the diffracted to the transmitted by a VBG power. Its advantage is that it removes any losses introduced by the medium. The second way is called "absolute diffraction efficiency," where the diffracted power is normalized to the incident power.

Table 11.1 TBG and RBG characteristics [9]

Transmitting VBG (TBG)	Reflecting VBG (RBG)
Angular selectivity: 0.1 to 10 mrad	Angular selectivity: 10 to 100 mrad
Spectral selectivity: 0.3 to 20 nm	Spectral selectivity: 0.02 to 2 nm
Diffraction efficiency: up to 100%	Diffraction efficiency: >99.9%

11.1.2.2 Holographic recording materials

Recoding of a volume grating requires the use of a material that is photosensitive. The modulation of the recording light intensity should create a corresponding refractive index change in the recording material which on a macroscopic level will be in fact the VBG. There is a wide variety of photosensitive materials that can be used for recording VBGs [15,16]. The main requirement that they need to fulfill is to have enough spatial resolution that will allow the recording of gratings with particular periods. The other two factors are the photosensitivity of the material and its dynamic range. The photosensitivity determines the exposure duration and given that VBGs are most commonly recorded by the interference of light, it is of great benefit to keep the exposure time as short as possible. The material's dynamic range provides the maximum refractive index change that can be achieved. This property affects the VBGs thickness and the maximum number of volume gratings that can be multiplexed together in the same volume. Other properties that depending on the particular application may be important are the optical damage threshold, the maximum physical dimensions of the material, its losses, its environmental sensitivity, and others.

The most common recording materials are dichromated gelatins, photopolymers, photorefractive crystals, photosensitive fibers, and photo-thermo refractive glasses. We will not discuss in detail the properties of each of these materials because they have been well investigated in the literature [17–25]. The applications and experimental results shown further in the chapter are based on using PTR glass due to its capabilities to handle high-power laser radiation because of its low losses, its environmental robustness, and extremely high resolution [23–25].

PTR glass is a relatively new photosensitive material well suited for phase hologram recording. It combines high sensitivity achieved by two-step hologram formation process and high optical quality resulting from its technological development. PTR glass is a Na_2O-ZnO-Al_2O_3-SiO_2 glass doped with silver (Ag), cerium (Ce), and fluorine (F). It is transparent from 350 to 2,500 nm. The chain of processes, which produce refractive index variation is as follows. The first step is the exposure of the glass to UV radiation, somewhere in the range from 280 to 350 nm. This exposure results in photo-reduction of silver ions Ag^+ to atomic state Ag^0. This stage is similar to the formation of a latent image in a conventional photo film because no significant changes in the optical properties of the glass occur. The final formation of the holographic recording is secured by subjecting the glass to thermal development. During this step, at elevated temperatures, a number of silver containing clusters are formed in the exposed regions of the glass due to the increased mobility of Ag^0 atoms. These silver-containing clusters serve as nucleation centers for NaF crystals growth. Interaction of those nanocrystals with the surrounding glass matrix causes the decrease of refractive index. Refractive index change Δn of about 1.5×10^{-3} (1,500 ppm) can be achieved and is enough to allow the recoding of high-efficiency hologram into glass wafers with thickness exceeding several hundred microns.

The second consequence of the crystalline phase precipitation in PTR glass is related to its physical properties and is extremely valuable. The NaF crystalline particles in the glass matrix are almost impossible to destroy by any type of radiation which makes PTR holograms stable under exposure to IR, visible, UV, X-ray, and gamma-ray irradiation. For example, laser damage threshold for 8 ns laser pulses at 1,064 nm is in the range of 40 J/cm^2. Also, the nonlinear refractive index of PTR glass is the same as that for fused silica which allows the use of PTR diffractive elements in all types of pulsed lasers. Another PTR advantage is its very low losses—on the order of 10^{-5} cm^{-1}. Testing of VBG recorded in PTR glass done under irradiation of 9 kW CW with a 6-mm-diameter spot showed heating that did not exceed 15 K [26]. Even though small, heating effects lead to thermal variations of the refractive index of the glass ($dn/dt = 5 \times 10^{-8}$ K^{-1}). In the case of Bragg grating written inside a PTR glass, this feature leads to thermal shift of the Bragg wavelength of around 10 pm/K. It is worth mentioning also, that due to the melting temperature of the NaF crystals being almost 1,000 °C, PTR holograms are stable at elevated temperatures and could tolerate thermal cycling up to 400 °C. This temperature is determined by the plasticity point of the glass matrix. Typically, Bragg gratings in PTR glass are recorded by an exposure to interference pattern of radiation from a He-Cd laser operating at 325 nm. The spatial frequency of the gratings can vary from 50 mm^{-1} up to about 10,000 mm^{-1}, their thickness from 0.5 to 25 mm, and diffraction efficiency up to 99.9%.

11.1.3 100 W 20 GHz spectral width laser diode system operating at 1,550 nm

The approach of resonant pumping of Er-doped lasers have attracted interest in the last several years. Simultaneously, high-power diode bars emitting in the region of 1,550 nm have reached power and price level which brings the possibility of developing a diode pump source close to reality.

The system described in this section produces 100 W of output power with emission spectral width of 20 GHz [27]. The laser diode bars chosen for the system were made by Princeton Lightwave and were operating in the range of 1,525 to 1,538 nm. Each bar had a maximum output of 35 W at 100 A pump current. In order to condition the emission of each laser diode bar so it is suitable for fiber coupling, the authors used a commercially available optical system (LIMO) consisting of several optical elements to collimated and re-arrange the emission of the laser bar. After this step, a reflective VBG (RBG) was aligned to lock the laser in an external resonator scheme. The RBG provided the desired narrowing of the emission spectrum and a precise central wavelength. Figure 11.5 provides a very good comparison between the spectra of the free running laser bar and the VBG locked bar at different pump currents.

Four different 1,550 nm laser bars were locked using VBGs that had spectral widths of 110–120 pm (~14 GHz) and spectral uniformity of 30–50 pm across the aperture. Upon locking of the bars in external resonator scheme the authors measured minimum spectral widths of 160 pm (20 GHz). The wider output spectra were attributed to two factors: to the nonoptimized for external resonator work antireflection coatings of the front facets of the bars and to the quite wide initial free-running spectra of the diode bars. Regardless, the RBGs provided a very good suppression of 25 dB or better of the side lobes.

The approach of resonant pumping requires that the pumping light frequency matches precisely the absorption band of the laser medium. In addition, to increase even further the efficiency, the pump source's linewidth should be close to the excited state transition bandwidth. The great advantage of using VBGs to externally lock diode bars is that the final spectral width and central wavelength are governed by the VBG's internal properties. Although the spectral width of a VBG is fixed after recording, the central Bragg wavelength can be adjusted with very high precision. For the purpose, the glass into which the VBG is recorded needs to be heated and due to thermal expansion of the material the grating period and therefore its resonant wavelength will change accordingly. Applying an active feedback control allows precise tuning and locking of all diode bars to a specific wavelength with high precision. The first experiments that the authors performed showed that upon heating of the VBGs the output spectra of the bars got wider. This

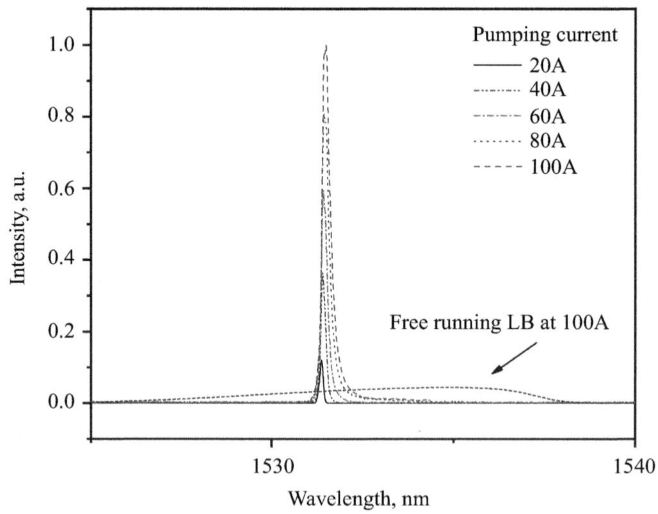

Figure 11.5 Comparison of the emission spectra of free running diode bar and a locked with a VBG on at different pumping currents [27]

was caused by the asymmetric heating of the gratings which was a result of using only one thermo-electric heater on one side of each VBG. The VBGs were recorded in 6 mm thick glass which was enough to develop temperature gradient along which the period of the Bragg grating was changing. Due to this change, different parts of the beam were locked at different wavelengths which at the end resulted in a wider total spectrum. To solve the issue, a second heater positioned symmetrically to the first one was added to each VBG and Figure 11.6 demonstrates the comparison between the spectral widths at different temperatures for two laser diode bars. The addition of the second heater brought the fluctuations of the spectral width in the range of 5–10 pm for a 60 °C temperature change.

Figure 11.7(a) shows the stability of the spectral widths, upon heating, of all four diode laser bars and Figure 11.7(b) demonstrates the tuning capabilities of the bars under the same heating conditions. All four bars demonstrated very good thermal stability while being tuned over more than 100 GHz. After achieving such stability and tunability, the system that the authors developed was calibrated so all four bars were capable of tuning and locking to a particular wavelength with precision better than 10 pm. The system's final spectral width was 204 pm (20 GHz) at 100 W output power.

The final system is presented in Figure 11.8 with all four laser bars coupled to 1 mm fibers that are bundled together into a single end in order to make one high power and high brightness pump source.

This part of the chapter presented the initial development of high power, narrow linewidth, tunable laser diode system that is uniquely suited for resonant

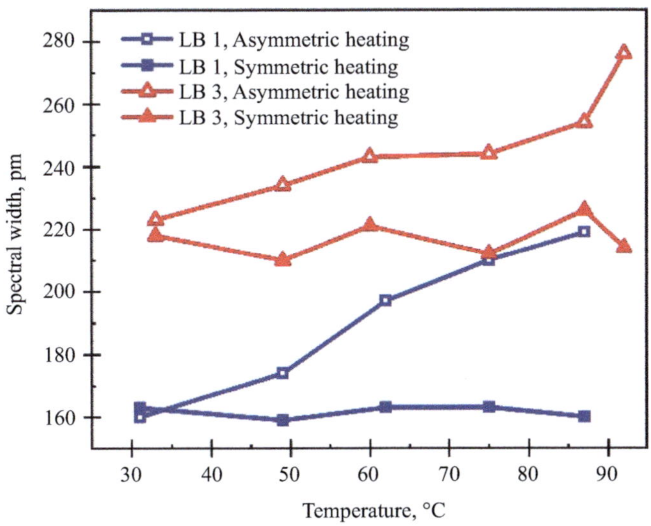

Figure 11.6 Behavior of two LD bars' spectral widths upon heating of the locking VBGs [27]

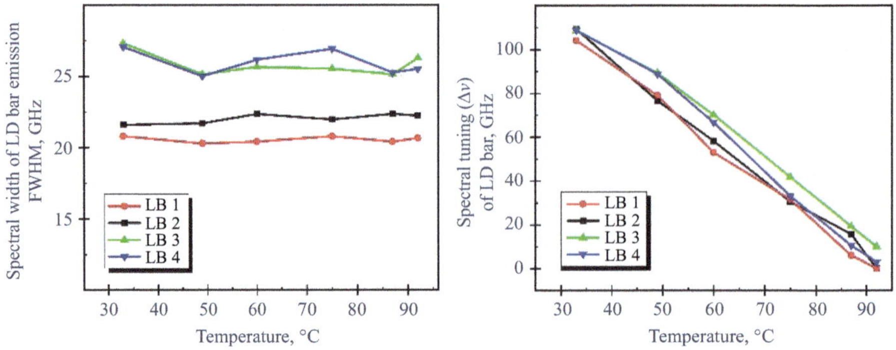

Figure 11.7 (a) Stability of the spectral widths upon heating of the VBGs, and (b) spectral tuning of the LD bars [27]

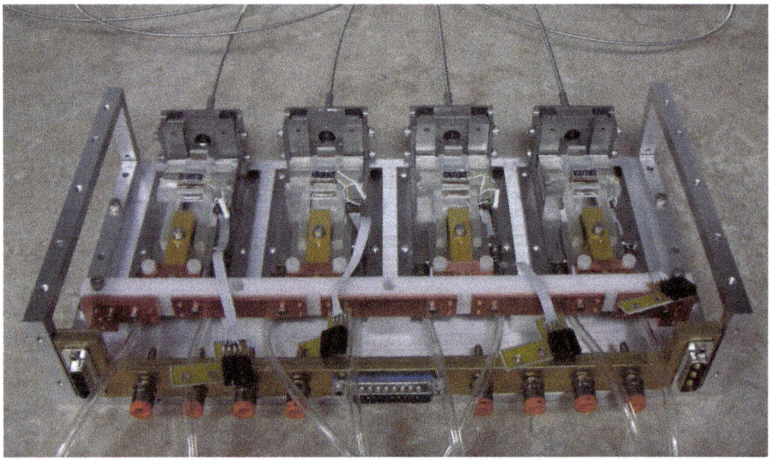

Figure 11.8 Photo of the module consisting of four LBs with their emission coupled into 1 mm optical fibers [27]

pumping at 1,550 nm. In section 11.1.4 of the chapter, we will show the next generation system and discuss its applications.

11.1.4 250 W 10 GHz laser diode system operating at 780 nm

Diode-pumped alkali vapor lasers (DPALs) appeared in the last few years as a new high-power laser source alternative [1]. Their main advantage is the possibility to achieve easily hundreds of watts or more of output power with minimum heat released because of their very small quantum defect parameter. It is imperative for DPAL systems to be pumped exactly in the resonant transition in order to use fully the benefit of the small quantum defect. For this purpose narrow line and high-power diode pump systems are of great interest. Here we describe the development of such system producing 250 W at 780 nm with spectral width of 10 GHz [28].

The system was developed by collaborators working at Optigrate Corp., Orlando, Florida, The College of Optics and Photonic—CREOL at the University of Central Florida and Air Force Research Laboratory, Kirtland AFB. New Mexico [28]. The system's design is identical to the one described in section 11.1.3 but seven diode bar modules were used instead. The achieved locked spectral width of each of the laser bars was less than 20 pm which corresponds to 10 GHz at 780 nm.

The central emission wavelengths of all the bars were within 5 pm of each other and secured absorption of more than 90% in a low-pressure Rb-gain medium. The VBG used to lock each of the diode bars had the capability to shift its resonant Bragg wavelength by being thermally tuned. This thermal wavelength adjustment of the central wavelength of each bar served two purposes—first to make sure that all the bars emit at the same wavelength and second to achieve general tuning of the central wavelength at which the system as a whole emitted. With a side lobe suppression of 30 dB the spectral brightness of the diodes was increased more than 33 times with slope efficiency of approximately 90% of that of a free running diode bar.

Figure 11.9 demonstrates the emission spectrum of the one of the laser diode bar modules tuned to different central wavelengths upon change of the temperature of the VBG. The resolution of the measurements was limited by the maximum resolution of the spectrum analyzer used.

At the end, the fibers into which the bars were coupled formed a 1 mm fiber bundle in order to create one unified source. The final system produced 250 W with spectral width of 19 pm, with tunability of 140 pm. The software developed for the system allowed simultaneous tuning of all modules to the desired central wavelength. The system was also tested with a low-pressure Rb vapor cell in order to

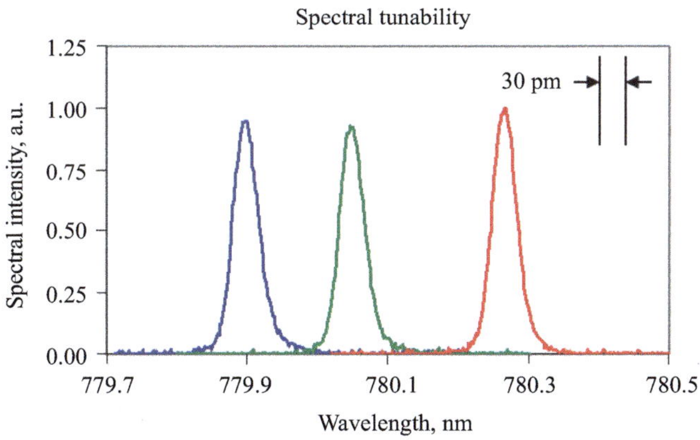

Figure 11.9 Emission spectra of a volume Bragg module for different temperatures of the VBG. The spectral width is determined by the spectral resolution of the optical spectrum analyzer [28]

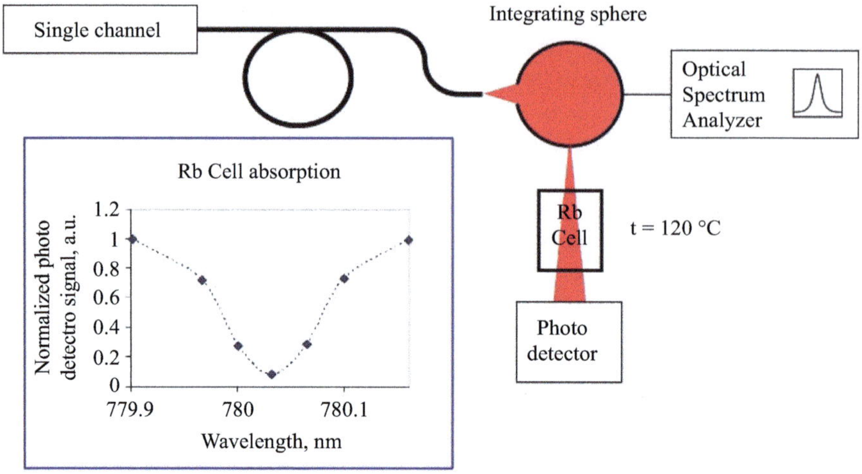

Figure 11.10 Emission spectra of a volume Bragg module for different temperatures of the VBG. The spectral width is determined by the spectral resolution of the optical spectrum analyzer [28]

measure the absorption efficiency. The cell was heated to 120 °C and the measured absorption was more than 90% (Figure 11.10).

In summary, the progress of high-power narrow line laser sources based on externally locked laser diode bars offers a more cost-efficient and less complex approach for making laser systems with such parameters. In section 11.2 of the chapter, several applications where such systems are used are discussed in more details.

11.2 Applications of narrow-line high-power diode laser systems

11.2.1 Spin-exchange optical pumping (SEOP)

In spin-exchange optical pumping, resonant circularly polarized light is absorbed by a saturated vapor of alkali-metal atoms in a glass cell. The cell contains also a much larger quantity of noble gas. Collisions between the alkali-metal atoms gas and rare gas atoms transfer some of the electron-spin polarization to the nuclei of the noble gas. In a well-designed system, almost half of the spin angular momentum of the absorbed photons is transferred to the alkali-vapor atoms.

Using alkali-metals is a convenient choice for optical pumping because they have a single unpaired valence electron, whose resonance wavelengths lie in the visible and near-infrared part of the spectrum, compatible with inexpensive and powerful diode laser sources. Additionally, their relatively high volatility enables dense vapors to be produced at moderate temperatures.

SEOP took a considerable advantage of the recent advances in high-power diode-laser arrays, which can provide tens to hundreds of watts of pumping light. The use of hyperpolarized noble gases such as ^{3}He and ^{129}Xe have gained an increased attention in a variety of scientific and medical applications such as magnetic resonance imaging (MRI) of lungs [29–32] and other human organs [33,34], spin-polarized targets in fundamental research [35,36], quantum computing, high precision metrology [37], and surface physics [38].

In order to achieve high-efficiency optical pumping in a SEOP system, the spectral bandwidth of the high-power pump lasers is required to match the pressure-broadened D1 absorption line of the alkali atom. In other words, the spectral bandwidth of the exciting laser should exactly match the pressure-broadened absorption line-width of the alkali vapor; a small drift of the center wavelength will greatly reduce the process efficiency. The pump wavelength long-term stability is also crucial to the SEOP efficiency.

Figure 11.11(a) shows the experimental setup used by G. Norquay *et al.* [32] to generate a hyperpolarized ^{129}Xe gas. The polarizer system consists of five main modules: a wavelength narrowed laser-diode array used as a pump source (150 W at 794.77 nm with $\Delta\lambda = 0.3$ nm), a long cylindrical optical cell contained inside an oven, an electromagnetic coil assembly, a gas-handling manifold, and a cryostat within a permanent magnetic field.

The gas dose is dispensed into a polyvinyl fluoride bag and is inhaled through a filtered mouthpiece. The duration of inhalation is 2 s. Figure 11.11(b) shows

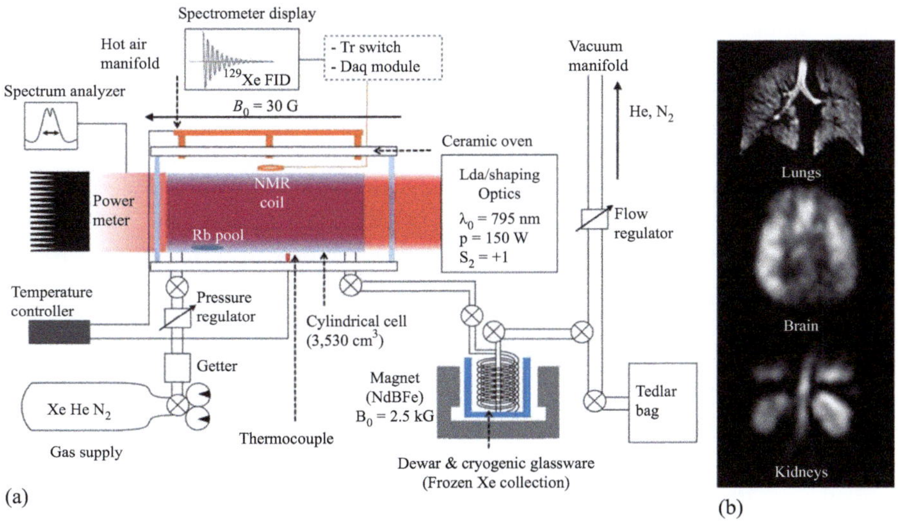

Figure 11.11 (a) Schematic of the key functional components of the Rb^{-129}Xe polarizer and (b) magnetic resonance images of hyperpolarized ^{129}Xe gas in human lungs and dissolved-phase ^{129}Xe in the human brain and kidneys [32]

examples of the high-definition human lung, brain, and kidney MR images using ^{129}Xe-hyperpolarized gas SEOP system.

The development of such medically applied SEOP systems has opened the possibility to perform MRI with naturally rich Xe (26% ^{129}Xe) at higher doses, which is a huge gain toward lowering the cost of such MRI procedures when compared to the case of using enriched Xe. This should also allow to further expand the use of hyperpolarized ^{129}Xe MRI for large clinical trials.

11.2.2 Diode-pumped alkali laser

The concept of DPAL was introduced in 2003 by Krupke *et al.* [39]. The DPAL active medium is a mixture of a vapor of neutral alkali atoms and one or more buffer gases. The Alkali laser is very promising for many reasons. Firstly, in contrast to its chemical counterpart; the DPAL does not need the storage or the replenishment and the manipulation of highly toxic chemicals. Instead, the DPAL gain medium is an alkali vapor enclosed inside a sealed glass cell. The alkali vapor is reusable as one single emitting atom is cycled through the process 100 million times every second. The DPAL is not a solid-state laser as well. Consequently, it does not have the thermal issues that could be faced in such a device. Actually, the heat generated in the entire process is negligible since its only source is the relaxing of the atom from the $n^2P_{3/2}$ state (where $n = 3, 4, 5$, and 6 corresponding to Na, K, Rb, and Cs respectively) down to the $n^2P_{1/2}$.

An alkali laser is a three-level system and there are two allowed electric-dipole transitions in this system. These transitions, the $^2S_{1/2}$–$^2P_{1/2}$ and the $^2S_{1/2}$–$^2P_{3/2}$, are more commonly called the D_1 and the D_2 lines, respectively. Table 11.2 shows these transitions for five alkali atoms. It also highlights their quantum defect values ($\Delta E/E_{pump}$) because they are significantly smaller than those of the most common solid-state lasers Nd (0.27) and Yb (0.1) doped YAGs.

Using the cesium atom as an example, if a pump laser at 852 nm is used, it drives the atoms from the ground state ($6^2S_{1/2}$) to $6^2P_{3/2}$. Collisional relaxations bring the atoms to the $6^2P_{1/2}$ state. Population inversion then takes place between this level and the ground state where lasing occurs at 895 nm (Figure 11.12).

One of the acknowledged challenges of the DPAL systems is the spectral width narrowness of the D_2 transition (≈10 GHz at 1 atmosphere). This becomes a problem when developing a high-power pump laser, because the spectral width from

Table 11.2 D_1 and D_2 transition wavelengths and spin-orbit splitting energies of the neutral alkali atoms [40]

Atom	D_2 λ_{pump} (nm)	D_1 λ_{laser} (nm)	ΔE (cm^{-1})	$\Delta E/E_{pump}$
Li	670.96	670.98	0.34	0.000023
Na	589.16	589.76	17.2	0.00101
K	766.7	770.11	57.7	0.0044
Rb	780.25	794.98	237.5	0.019
Cs	852.35	894.59	554.1	0.047

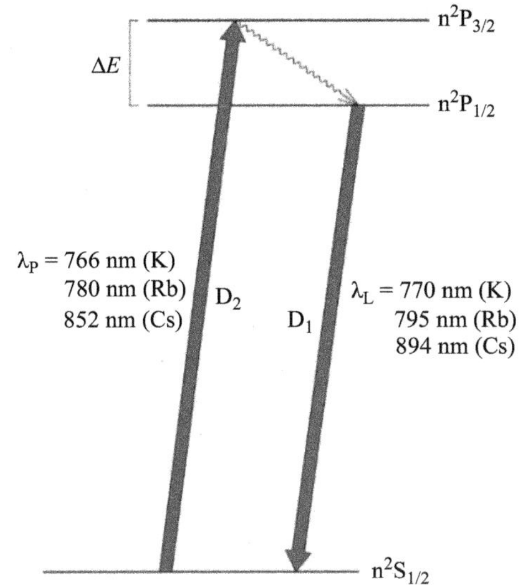

Figure 11.12 D-line transitions for alkali elements. D1 designates the $n^2P_{1/2} \leftrightarrow n^2S_{1/2}$ transition, and D_2 designates the $n^2P_{3/2} \leftrightarrow n^2S_{1/2}$ transition where $n = 4$, 5, and 6 and the energy defect, ΔE, is 58, 238, and 544 cm^{-1} for potassium, rubidium, and cesium, respectively [1]

diode bars and stacks at high powers is about 100 times larger than that. This is the reason for the development of the described earlier narrow-line high-power diode system. Regardless of the considerable advances that have been made to develop high-power DPAL systems, still a critical technical challenge on the way further is the chemically aggressive nature of alkali-metal vapor.

11.2.3 Rare gas lasers applications

Optically pumped rare gas lasers have the potential to provide high-power and exhibit excellent beam quality. In terms of emission wavelengths, they operate near the boundary between the infrared and the visible spectral ranges. Their emission falls in the atmospheric transparency window which is a great applications benefit as well. Their optical characteristics are very similar to the DPAL systems described in section 11.2.2. Rare gas atoms excited to $np^5(n + 1)s^3P_2$ configuration are metastable and have spectral properties that are very similar to those of the alkali metals. Thanks to the chemical inertness of the rare gases they do not react with the buffer gas in the system. A rare gas laser scheme that mimics the DPAL system is depicted in Figure 11.13. The rare gas atoms are driven from the $(n + 1)[3/2]_2$ sate to the $(n + 1)p[5/2]_3$ state. The lasing takes place on the $(n + 1)[1/2]_1 \rightarrow (n + 1)s[3/2]_2$ transition.

For such rare atom gas laser systems, diode pumping is a great candidate, due to the advantages of having high laser pump power and high optical conversion

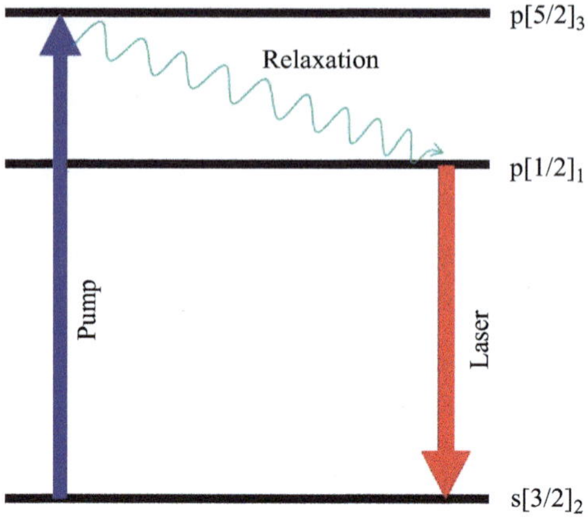

Figure 11.13 Energy diagram for an optically pumped rare gas laser

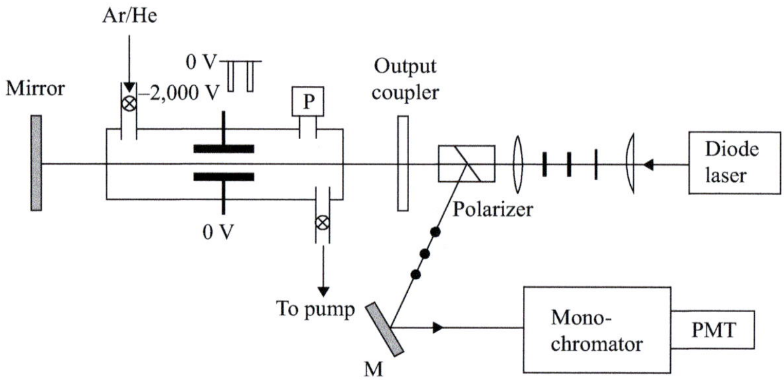

Figure 11.14 Experimental setup of an Ar-based diode-pumped CW rare gas laser [41]

efficiency if narrow line pumping is used. Figure 11.14 shows the experimental setup used by Han *et al.* [41] to create such laser system (DPRGL).

In this work, a pulsed discharge was used achieved by two 2.5 × 2.5 cm parallel stainless steel electrodes separated by 0.5 cm. The electrodes were contained in a cylindrical glass cell with a diameter of 5 cm and a length of 15 cm. The gas flowed through the cell and the plasma was stable at pressures up to 130 mbar for pure Ar and 460 mbar for Ar/He mixtures with freshly cleaned electrodes. The discharge pulses were 1μs long at a frequency of 1 kHz. Using −1,000 to −2,000 V pulses, the reduced electric field in the discharge at 200 mbar was 41 and 83 Td, respectively. (If E is the electric field and N is the gas density, E/N is in units of Townsend or [Td] where 1 Td = 10^{-21} V m^2.) At the highest operating pressure of 460 mbar the reduced electric field was 18 Td and 36 Td, respectively. Pulsed lasing was observed for pure Ar at 90 mbar and for Ar/He mixtures at total pressures of 395 and 460 mbar.

The high-power pump source was a diode bar locked with a VBG to emit at 811 nm with a spectral width of 10 GHz. Lasing of the pure Ar was achieved at threshold of 3 W and the output power determined to be linearly dependent on the pump power.

The same group recently achieved more than 4 W of lasing power from a similar diode pumped Ar metastable laser using VBG locked diode pump source with 40 W of power [42].

11.3 Conclusion remarks

The development of externally locked narrow line high-power diode laser systems and their use as pump sources unlocks a lot of new applications that needed such optical pumping but was hard to achieve using solid state or fiber lasers. The flexibility of making semiconductor lasers emit at variety of wavelengths and the progress of making diode bars capable of delivering tens of Watts boosted the development of such pump systems. The simultaneous advances in making high-quality ultra-selective VBGs made possible the locking of high-power laser diode bars and thus creating laser pump sources with unique wavelengths and very narrow spectral widths. This opens the field of having very efficient resonant pumping for various laser systems, medical imaging, and different spectroscopies. The fast pace of the semiconductor industry guarantees that even more powerful and with greater application possibilities systems similar to the ones described in this chapter will be conceivable in the near future.

References

[1] Pitz G. A. and Anderson M. D., "Recent advances in optically pumped alkali lasers" *Appl. Phys. Rev.* **4**, 041101, 2017.

[2] Koechner W., *Solid-State Laser Engineering*. Springer, Berlin, 2006.

[3] Claps R., Sabbaghzadeh J., and Fink M., "Raman spectroscopy with a single-frequency, high-power, broad-area laser diode," *Appl. Spectrosc.* **53**, 491–6, 1999.

[4] Zhdanov B. V., Ehrenreich T., and Knize R. J., "Narrowband external cavity laser diode array," *IEE Electron. Lett.* **43**(4), 221–2, 2007.

[5] Efimov O. M., Glebov L. B., and Smirnov V., "High efficiency volume diffractive elements in photo-thermo-refractive glass," United States Patent 6,673,497, January 6, 2004.

[6] Efimov O. M., Glebov L. B., Glebova L. N., and Smirnov V. I., "Process for production of high efficiency volume diffractive elements in photothermo-refractive glass," United States Patent 6,586,141, July 1, 2003.

[7] Venus G., Sevian A., Smirnov V., and Glebov L., "High-brightness narrow-line laser diode source with volume Bragg-grating feedback," *Proc SPIE* **5711**, 10.1117/12.590425, 2005.

[8] Glebov A. L., Mokhun O., Rapaport A., Vergnole S., Smirnov V., and Glebov L. B., "Volume Bragg gratings as ultra-narrow and multiband optical filters," *Proc. SPIE* **8428**, 84280C, 2012.

[9] Divliansky I., *Volume Bragg gratings: fundamentals and applications in laser beam combining and beam phase transformations, holographic materials and optical systems, IntechOpen*, 2017. DOI: 10.5772/66958. Available from: https://www.intechopen.com/books/holographic-materials-and-optical-systems/volume-bragg-gratings-fundamentals-and-applications-in-laser-beam-combining-and-beam-phase-transform

[10] Bragg W. H. and Bragg W. L., *X Rays and Crystal Structure*. London: G Bell and Sons, Ltd., 1915.

[11] Syms R. R., *Practical Volume Holography*. Oxford, UK: Clarendon Press, **48**, 64, 132–62, 1990.

[12] Kogelnik H., "Coupled wave theory for thick hologram grating," *Bell Syst. Tech. J.*, **48**(9), 2909–45, 1969.

[13] Anderson B. M., Venus G., Ott D., *et al.*, "Higher order mode selection for power scaling in laser resonators using transmitting Bragg gratings," *Proc. SPIE* **9466**, 94660C, 2015.

[14] Drachenberg D., Divliansky I., Venus G., Smirnov V., and Glebov L., "High-power spectral beam combining of fiber lasers with ultra high-spectral density by thermal tuning of volume Bragg gratings," *Proc. SPIE* **7914**, 79141F, 2011.

[15] Lin L. H., "Method of characterizing hologram-recording materials," *J. Opt. Soc. Am.*, **61**(2), 203–08, 1971.

[16] Collier R., Burchardt C., and Lin L., *Optical Holography*. New York: Academic, 1971.

[17] Chang B. J. and Leonard C. D., "Dichromated gelatin for the fabrication of holographic optical elements," *Appl. Opt.* **18**(14), 2407–17, 1979.

[18] Colburn W. S. and Haines K. A., "Volume hologram formation in photopolymer materials," *Appl. Opt.* **10**(7), 1636–41, 1971.

[19] Wu S.-D. and Glytsis E. N., "Holographic grating formation in photopolymers: analysis and experimental results based on a nonlocal diffusion model and rigorous coupled-wave analysis," *J. Opt. Soc. Am. B* **20**(6), 1177, 2003.

[20] Chen F. S., LaMacchia J. T., and Fraser D. B., "Holographic storage in lithium niobate," *Appl. Phys. Lett.* **13**(7), 223, 1968.

[21] Stepanov S. I., "Applications of photorefractive crystals," *Rep. Prog. Phys.* **57**, 39–116, 1994.

[22] Dewra S., Grover V., and Grover A., "Fabrication and applications of fiber Bragg grating – a review," *Adv. Eng. Tec. Appl.* **4**(2), 15–25, 2015.

[23] Glebov L. B., "Kinetics modeling in photosensitive glass," *Opt. Mater. (Amst)* **25**(4), 413–18, 2004.

[24] Efimov O. M., Glebov L. B., and Smirnov V. I., "High-frequency Bragg gratings in a photothermorefractive glass," *Opt. Lett.* **25**(23), 1693–5, 2000.

[25] Efimov O. M., Glebov L. B., and Smirnov V. I., "Diffractive optical elements in photosensitive inorganic glasses," *Inorg. Opt. Mater. III* **4452**, 39–47, 2001.

[26] Anderson B., Kaim S., Venus G., *et al.*, "Forced air cooling of volume Bragg gratings for spectral beam combination," *Proc. SPIE* **8601**, 86013D, 2013. doi:10.1117/12.2005951.

[27] Divliansky I., Smirnov V., Venus G., Gourevitch A., and Glebov L., "High-power semiconductor lasers for applications requiring GHz linewidth source," *Proc. SPIE* **7198**, 71981N, 2009.

[28] Podvyaznyy A., Venus G., Smirnov V., Hostutler D., and Glebov L., "250-W LD bar pump source with 10-GHz spectral width for rubidium vapor medium," *Proc. SPIE* **7686**, 76860P, 2010.

[29] Albert M. S., Tseng C. H., Williamson D., *et al.*, "Hyperpolarized ^{129}Xe MR imaging of the oral cavity," *J. Magn. Reson.* **111**, 204–07, 1996.

[30] Abdeen N., Cross A., Cron G., *et al.*, "Measurement of xenon diffusing capacity in the rat lung by hyperpolarized ^{129}Xe MRI and dynamic spectroscopy in a single breath-hold," *Magn. Reson. Med.* **56**, 255–64, 2006. doi:10.1002/mrm.20943

[31] Mugler J. P. and Altes T. A., "Hyperpolarized ^{129}Xe MRI of the human lung." *J. Magn. Reson. Imaging* **37**, 313–31, 2013. doi:10.1002/jmri.23844

[32] Norquay G., Collier G. J., Rao M., Stewart N. J., and Wild J. M., "^{129}Xe-Rb spin-exchange optical pumping with high photon efficiency," *Phys. Rev. Lett.* **121**, 153201, 2018.

[33] Rao M. R., Stewart N. J., Griffiths P. D., Norquay G., and Wild J. M., "Imaging human brain perfusion with inhaled hyperpolarized ^{129}Xe MR imaging," *Radiology* **286**, 659, 2018.

[34] Chacon-Caldera J., Maunder A. M., Rao M., *et al.*, *Proceedings of International Society of Magnetic Resonance in Medicine*, Paris, France, 2018, p. 4470.

[35] Anthony P. L., Arnold R.G., Band H.R., *et al.*, "Determination of the neutron spin structure function," *Phys. Rev. Lett.* **71**, 959–62, 1993.

[36] Xu W., Dutta D., Xiong F., *et al.*, "Transverse asymmetry $A_{T'}$ from the quasielastic ^{3}He(e,e′) process and the neutron magnetic form factor," *Phys. Rev. Lett.* **85**, 2900–04, 2000.

[37] Kornack T. W., Ghosh R. K., and Romalis M. V., "Nuclear spin gyroscope based on an atomic comagnetometer," *Phys. Rev. Lett.* **95**, 230801, 2005.

[38] Wu Z., Happer W., Kitano M., and Daniels J., "Experimental studies of wall interactions of absorbed spin-polarized ^{131}Xe nuclei," *Phys. Rev. A* **42**, 2774, 1990.

[39] Beach R. J., Krupke W. F., Keith Kanz V., Payne S. A., Dubinskii M. A., and Merkle L. D., "End-pumped continuous-wave alkali vapor lasers: experiment, model, and power scaling," *J. Opt. Soc. Am. B* **21**, 2151–63, 2004.

[40] Krupke W. F., "Diode pumped alkali lasers (DPALs)—A review," *Prog. Quant. Electron.* **36**, 4–28, 2012.

[41] Han J., Glebov L., Venus G., and Heaven M. C., "Demonstration of a diode-pumped metastable Ar laser," *Opt. Lett.* **38**, 5458–61, 2013.

[42] Han J., Heaven M. C., Moran P. J., *et al.*, "Demonstration of a CW diode-pumped Ar metastable laser operating at 4 W," *Opt. Lett.* **42**, 4627–30, 2017.

Index